ELECTROMAGNETIC SHIELDING

ELECTROMAGNETIC SHIELDING

Kenneth L. Kaiser
Kettering University
Flint, Michigan

CRC Taylor & Francis
Taylor & Francis Group
Boca Raton London New York

A CRC title, part of the Taylor & Francis imprint, a member of the Taylor & Francis Group, the academic division of T&F Informa plc.

This material was previously published in *The Electromagnetic Compatibility Handbook* © CRC Press, LLC 2004.

Published in 2006 by
CRC Press
Taylor & Francis Group
6000 Broken Sound Parkway NW, Suite 300
Boca Raton, FL 33487-2742

Printed in the United States of America on acid-free paper
10 9 8 7 6 5 4 3 2 1

International Standard Book Number-10: 0-8493-6372-1 (Hardcover)
International Standard Book Number-13: 978-0-8493-6372-6 (Hardcover)
Library of Congress Card Number 2005050484

Library of Congress Cataloging-in-Publication Data

Kaiser, Kenneth L.
Electromagnetic shielding / Kenneth L. Kaiser.
p. cm.
ISBN 0-8493-6372-1
1. Shielding (Electricity) I. Title.

TK7867.8.K35 2005
621.382'24--dc22 2005050484

T&F informa

Taylor & Francis Group
is the Academic Division of T&F Informa plc.

Visit the Taylor & Francis Web site at
http://www.taylorandfrancis.com

and the CRC Press Web site at
http://www.crcpress.com

Preface

My interest in the important topic of shielding began in the 1960s while watching science fiction shows such as *Star Trek* and *Lost in Space*. I was mystified and fascinated by the various "shields" introduced on these shows. Although many of the questions that I had concerning the how and why of real-life electromagnetic shielding have now been answered, the engineering and science of shielding still fascinate me several decades later.

The shielding-related topics contained in this book were selected from my reference and textbook previously published by CRC Press entitled *Electromagnetic Compatibility Handbook*. I hope this "spinoff" book, being only a small fraction of the *Handbook*, can be more conveniently used and more easily owned by individuals interested mainly in shielding. One of the main purposes of this particular book is to demystify shielding and to help explain the source and limitations of the approximations, guidelines, models, and rules of thumb seen in this field. For further reference and personal edification, I wrote many of the examples contained in this book to document the answers to questions I had about the subject matter. Although the chapters are fairly self contained, it is assumed that the reader has a rudimentary background in electromagnetics, including vector analysis.

Although many well-researched sources have been used in the writing of this book, I mention two that have been particularly valuable and inspirational: Dr. Clayton R. Paul's *Introduction to Electromagnetic Compatibility* and Drs. Hermann A. Haus and James R. Melcher's *Electromagnetic Fields and Energy*. I recommend both of these fine textbooks.

I have tried to be diligent in crediting all of the sources that were used in solving a problem, generating tabled results, or understanding an unfamiliar or confusing concept, and I apologize in advance for any oversights. (If by chance you locate some material in this book not appropriately referenced, which is possible considering the book's length and the number of editing iterations, please e-mail me. I will include the addition on my web site and in any future editions of this book.)

So as not to burden the reader with frequent citation interruptions, in most cases the references are grouped together at the end of each problem statement and located between brackets, [ ]. The references are listed by last name roughly from most used to least used. (The date of publication is also given when necessary to avoid confusion with other authors with the same surname or different publications by the same author.) In some cases, as with many of the tables, specific references are provided for particular equations or results. When possible, I have tried to use "original" sources.

The program Mathcad was used to generate most of the plots and solve many of the equations. The major reason this program was selected is that it is easily understood even with little or no prior experience with it. Although it is somewhat unorthodox, entire Mathcad programs have been provided in many cases. This allows all of the variable assignments, assumptions, equations, and possible mistakes to be clearly seen. I used the program VISIO, sometimes with an embedded Mathcad output, to generate *all* of the figures.

An enormous amount of time was devoted to crafting the tables. A brief comment concerning the accuracy of the many expressions contained in this book, including in the tables, is appropriate. What is not seen are the many derivations and checks. Although most of the equations in this book were personally derived, obviously not every equation has been personally obtained from basic principles. Extensive use has been made of the years of hard work of myriad scholars. It can be stated, however, that nearly every expression has been either derived or checked — sometimes through much effort. A typical analytical and numerical check might involve taking a limit on one or more variables in an expression and then comparing it with a reliable result. The number of approximate expressions contained in the book prohibited an indication of their relative error. Sets of approximations are provided, in part, to help show the relationship of the individual expressions to each other.

As in the preface of *Electromagnetic Compatibility Handbook*, I again thank my friend, mentor, and colleague Professor James C. McLaughlin (Kettering University) for his careful review of these chapters, his many suggestions, and his steadfast support.

Your comments, suggestions, and corrections are most welcomed and encouraged. They are invaluable to the current edition and any possible future editions of this book. At the site http://www.klkaiser.com, student problems, errata, and a few extras are provided for the *Electromagnetic Compatibility Handbook*. I sincerely hope this book on electromagnetic shielding is helpful and inspirational to you. Unless specified otherwise, Système International (SI) units are used throughout this book. Enjoy.

Kenneth L. Kaiser
Electrical and Computer
Engineering Department
Kettering University

The Author

Kenneth L. Kaiser's interest in electrical engineering began in high school with his involvement in amateur radio. While obtaining a solid theoretical background in a number of fields in electrical engineering, he obtained additional inspiration and practical experience working in several nonacademic positions. Since obtaining a Ph.D. in electrical engineering from Purdue University, he has focused his attention on the researching of topics of personal and industrial interest, on effective teaching methods, and on the writing of this book. He is currently at Kettering University (formerly GMI Engineering & Management Institute), continuing with his research and stimulating his students to excel.

Contents

1

Plane Wave Shielding

In the far field of a time-varying electromagnetic source, the radiation appears like a plane wave, and the electric and magnetic fields are perpendicular to the direction of propagation. When a plane wave is incident to a shield, reflection from, absorption into, and transmission through the shield occurs. The degree of reflection is determined by the impedance of the incident wave and intrinsic impedance of the shield material. The degree of absorption is a function of the thickness of the shield and frequency-dependent skin depth in the shield. The degree of transmission is determined by the reflections from both sides of the shield and absorption through the shield.

1.1 The "Magic" of Shielding Waves Revealed

Qualitatively explain in terms of electron movement, current density, and induced fields how a shield "blocks out" far-field plane waves.

For many individuals, the ability of a material, which is often a metal, to shield or screen plane waves seems "magical." Although many electrical engineers are exposed to Maxwell's equations and the use of boundary conditions to solve simple reflection problems, the physical reason behind how materials shield is often neglected. Although a high-level qualitative discussion rarely helps to provide the "answer," it does build a degree of confidence in the mathematical results and reduce the magical aspect of shielding.

When a shield is sufficiently far from an electromagnetic source, referred to as the far field, the field incident to the shield can be modeled as a plane wave. Conveniently, the electric and magnetic fields are perpendicular to each other in the far field, and these fields are perpendicular (or transverse) to the direction to propagation. In Figure 1.1, a plane wave is shown normally incident to a flat shield. When an incident wave hits the shield, the electrons and other charged particles in the shield respond to the electric and magnetic fields according to Lorentz's force law:

$$\vec{F} = q\vec{E} + q\vec{v} \times \mu_o \vec{H} \tag{1.1}$$

where q is the charge of each particle of velocity $\vec{v}$. In other words, the charges in the shield move in response to the electric and magnetic fields of the incident wave. These moving charges also generate a magnetic and an electric field, referred to as the scattered field or induced field. The induced fields, of course, change the total field inside the shield, which affects the charge motion. Eventually, assuming the incident field is in steady state, a steady-state condition is reached inside the shield. The fascinating conclusion for good shields is that the sum of the fields inside the shield (all of the induced and incident fields) is much less than the initial incident field. The shield has effectively reduced the field strength by generating a counter field.

The power of the shield to generate this counter field is a function of the electrical parameters of the material, which include the conductivity since the induced current density, $\vec{J}$, is a function of the shield's conductivity, σ:

$$\vec{J} = \sigma \vec{E} \tag{1.2}$$

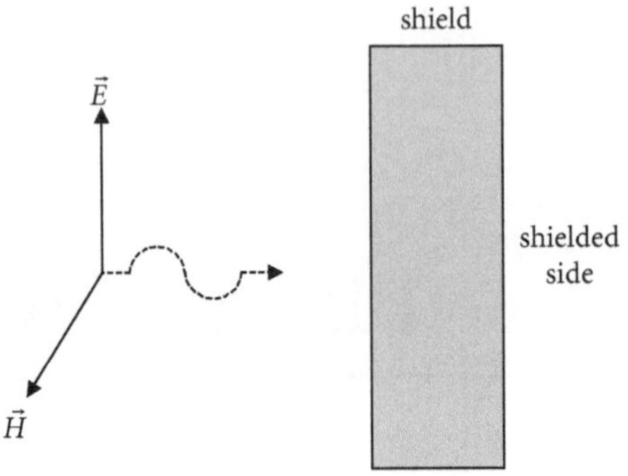

FIGURE 1.1 Plane wave normally incident to a flat shield.

For perfect conductors, the induced current is entirely along the shield's surface. For real materials, the current and field penetrate a nonzero distance into the material. This depth of penetration is described by a parameter known as the skin depth of the material, which is defined for good conductors as $\delta = 1/\sqrt{\pi f \mu \sigma}$. If the material is thin compared to this skin depth, then a significant fraction of the incident field could penetrate through the shield. This should seem reasonable for a fraction of the induced currents will appear on the surface of the shield not directly exposed to the incident wave. These induced currents on the shielded side are a source of electromagnetic fields. An effective shield would reduce these shielded-side surface currents to a negligible level.

1.2 The Impedance of a Wave

Discuss the relationship between the intrinsic impedance of a material and characteristic impedance of a transmission line.

When a shield is in the far field of an electromagnetic source, the electromagnetic wave at the shield appears like a plane wave: the electric and magnetic fields are perpendicular to each other *and* perpendicular (or transverse) to the direction of propagation. The intrinsic impedance is defined as the ratio of the electric field to the magnetic field for a wave traveling in one direction. This simple ratio is extremely helpful in determining the degree of reflection of an incoming wave to an object or a shield. The unit of the intrinsic impedance is E/H = (V/m)/(A/m) = V/A = Ω. The general expression for the intrinsic impedance of a material is

$$\eta = \sqrt{\frac{j\omega\mu}{\sigma + j\omega\varepsilon}} \tag{1.3}$$

where σ is the conductivity, μ is the permeability, and ε is the permittivity of the material. Notice that it is quite similar in form to the characteristic impedance of a lossy transmission line:

$$Z_o = \sqrt{\frac{R + j\omega L}{G + j\omega C}} \tag{1.4}$$

The concept of a wave in a medium possessing an impedance may seem strange, but it is merely a convenient parameter similar to reactance or circuit impedance.

The conductivity, σ, of a shield material corresponds to the conductance, G, of a lossy dielectric separating the conductors of a transmission line. The permeability, μ, of a shield material corresponds to the inductance, L, of the conductors of a transmission line. The permittivity, ε, of a shield material corresponds to the capacitance, C, between the conductors of a transmission line. The only parameter that does not have a correspondence is the resistance, R, of the two conductors of a transmission line.

In the far field, a plane wave does not have an electric field component in the direction of propagation. For a lossless transmission line, the electric and magnetic fields between the conductors are also entirely transverse to the direction of propagation. However, when the conductor losses of the line are included, through R, there is an electric field component in the direction of propagation along the conductors. Also, a voltage drop exists along each electrically-short segment of the lossy conductors. For far-field waves, there are no field components in the direction of propagation. Even when there are losses in the medium, specifically a nonzero conductivity, the electric field is still transverse to the direction of propagation for plane waves. The conductance, G, corresponds to these losses in the medium. The voltage drop between the conductors of a transmission line implies an electric field component between the conductors, transverse to the direction of propagation.

1.3 Impedance of Air, Real Metals, and Real Insulators

Determine the intrinsic impedance of air and approximate expressions for the intrinsic impedance of real metals and real insulators. Determine the impedance of a real conductor and insulator. [Ramo; Kraus, '92]

The intrinsic impedance, or wave impedance,[1] of free space or air is

$$\eta_o = \sqrt{\frac{j\omega\mu}{\sigma + j\omega\varepsilon}} = \sqrt{\frac{j\omega\mu_o}{0 + j\omega\varepsilon_o}} = \sqrt{\frac{\mu_o}{\varepsilon_o}} = \sqrt{\frac{4\pi \times 10^{-7}}{8.854 \times 10^{-12}}} \approx 377\ \Omega \tag{1.5}$$

since the conductivity, σ, of air is quite small. As the conductivity of the medium increases, the intrinsic impedance of the medium decreases. For perfect ideal conductors, the conductivity is infinite and intrinsic impedance is zero:

$$\eta = \sqrt{\frac{j\omega\mu}{\sigma + j\omega\varepsilon}} = \sqrt{\frac{j\omega\mu_o}{\infty + j\omega\varepsilon_o}} = 0\ \Omega \tag{1.6}$$

This may seem quite reasonable since the resistance of a perfect conductor is also zero, but this analogy should not be taken too far. For example, the resistance of a perfect insulator is infinite, but the intrinsic impedance of a perfect insulator is *not* infinite:

$$\eta = \sqrt{\frac{j\omega\mu}{\sigma + j\omega\varepsilon}} = \sqrt{\frac{j\omega\mu}{0 + j\omega\varepsilon}} = \sqrt{\frac{\mu_r\mu_o}{\varepsilon_r\varepsilon_o}} \approx 377\sqrt{\frac{\mu_r}{\varepsilon_r}}\ \Omega \tag{1.7}$$

where μ_r is the relative permeability and ε_r is the relative permittivity of the medium. The intrinsic impedance can actually be greater than 377 Ω for magnetic materials where the relative permeability is greater than one.

Approximate expressions for the intrinsic impedance that do not involve the square root of a complex number can be readily obtained by recalling that the binomial expression is approximately equal to

$$(1 + x)^n \approx 1 + nx$$

[1]The terms wave impedance and intrinsic impedance are occasionally used interchangeably. It is more descriptive, however, to use only the term intrinsic impedance to describe the impedance of a wave in the far field where it is a plane wave.

when x is much less than one. The intrinsic impedance expression can be easily rewritten so that this approximation can be utilized:

$$\eta = \sqrt{\frac{j\omega\mu}{\sigma + j\omega\varepsilon}} = (j\omega\mu)^{1/2}(\sigma + j\omega\varepsilon)^{-1/2} = (\omega\mu\angle 90°)^{1/2}\sigma^{-1/2}\left(1+\frac{j\omega\varepsilon}{\sigma}\right)^{-1/2}$$

For good conductors,

$$\frac{\omega\varepsilon}{\sigma} \ll 1 \tag{1.8}$$

where $j\omega\varepsilon/\sigma$ corresponds to x. Therefore, with $n = -1/2$

$$\begin{aligned}\eta &\approx [(\omega\mu)^{1/2}\angle 45°]\sigma^{-1/2}\left(1-\frac{1}{2}\frac{j\omega\varepsilon}{\sigma}\right) = \sqrt{\frac{\omega\mu}{\sigma}}\left(\frac{1}{\sqrt{2}}+j\frac{1}{\sqrt{2}}\right)\left(1-\frac{1}{2}\frac{j\omega\varepsilon}{\sigma}\right)\\ &\approx \sqrt{\frac{\omega\mu}{2\sigma}}\left[\left(1+\frac{\omega\varepsilon}{2\sigma}\right)+j\left(1-\frac{\omega\varepsilon}{2\sigma}\right)\right]\end{aligned} \tag{1.9}$$

This expression can be further approximated as

$$\eta \approx \sqrt{\frac{\omega\mu}{2\sigma}}(1+j) \tag{1.10}$$

which is valid for good conductors. An approximate expression for the intrinsic impedance can also be obtained for good insulators:[2]

$$\eta = (j\omega\mu)^{1/2}(\sigma + j\omega\varepsilon)^{-1/2} = (\omega\mu\angle 90°)^{1/2}(j\omega\varepsilon)^{-1/2}\left(\frac{\sigma}{j\omega\varepsilon}+1\right)^{-1/2}$$

Since for good insulators

$$\frac{\omega\varepsilon}{\sigma} \gg 1$$

$$\begin{aligned}\eta &= (\omega\mu\angle 90°)^{1/2}(\omega\varepsilon\angle 90°)^{-1/2}\left(\frac{\sigma}{j\omega\varepsilon}+1\right)^{-1/2} \approx \left(\frac{\omega\mu\angle 90°}{\omega\varepsilon\angle 90°}\right)^{1/2}\left(1-\frac{1}{2}\frac{\sigma}{j\omega\varepsilon}\right)\\ &\approx \sqrt{\frac{\mu}{\varepsilon}}\left(1+j\frac{\sigma}{2\omega\varepsilon}\right)\end{aligned} \tag{1.11}$$

This expression can be further approximated as

$$\eta \approx \sqrt{\frac{\mu}{\varepsilon}} = \sqrt{\frac{\mu_r\mu_o}{\varepsilon_r\varepsilon_o}} = 377\sqrt{\frac{\mu_r}{\varepsilon_r}} \tag{1.12}$$

[2]In practice, the terms insulator and dielectric are used interchangeably. The term insulator is probably more closely tied to the conductivity of the material while the term dielectric is probably more closely tied to the dielectric constant of the material.

The intrinsic impedance of copper, a good conductor, is

$$\eta \approx \sqrt{\frac{\omega(4\pi\times10^{-7})}{2(5.8\times10^{7})}}(1+j) \approx 10^{-7}\sqrt{\omega}(1+j) \tag{1.13}$$

As is seen by this expression, the impedance of copper is much less than one for virtually all frequencies of current interest to the EMC engineer. At 1 GHz, for example,

$$\eta \approx 1.04\times10^{-7}\sqrt{2\pi\times10^{9}}(1+j) \approx 0.0082(1+j)$$

At lower frequencies, the impedance is smaller than this value. This impedance can be viewed as a resistance in series with an inductance ($+j$ term). The intrinsic impedance is defined as the ratio of the electric field to the magnetic field in the far field; hence, the electric field is 90° out of phase with the magnetic field for a wave traveling in one direction inside a good conductor. Of course, the approximate expression can be used since copper is a good conductor at 1 GHz:

$$\frac{\omega\varepsilon}{\sigma} \approx 9.6\times10^{-10} \ll 1$$

For that matter, copper is a good conductor for frequencies much less than the optical range. The intrinsic impedance of a good insulator such as Bakelite is

$$\eta \approx 377\sqrt{\frac{\mu_r}{\varepsilon_r}} = 377\sqrt{\frac{1}{5.2}} \approx 170\ \Omega$$

This impedance is totally real; hence, the electric field is in phase with the magnetic field for good insulators. The approximate expression can be used since Bakelite is a good insulator at 1 kHz:

$$\frac{\omega\varepsilon}{\sigma} = \frac{2\pi\times10^{3}(5.2\varepsilon_o)}{1.3\times10^{-8}} \approx 20 \gg 1$$

Although the conductivity of copper is fairly constant over a wide frequency range, the permittivity, conductivity, and permeability of many materials often vary with frequency. It is important that the correct frequency-based value for these parameters be used in these expressions.

A summary of the various forms for the exact expressions for the intrinsic impedance is provided in Table 1.1. In Table 1.2, various approximations seen for the intrinsic impedance and its magnitude are provided as a function of the σ and $\omega\varepsilon$. The intrinsic impedance has units of ohms (both its real and

TABLE 1.1 Exact Expressions for the Intrinsic Impedance

$$\eta = |\eta|\angle\theta = \sqrt{\frac{j\omega\mu}{\sigma + j\omega\varepsilon}} = \sqrt{\frac{\omega^2\mu\varepsilon + j\omega\mu\sigma}{\sigma^2 + (\omega\varepsilon)^2}} = \frac{[(\omega^2\mu\varepsilon)^2 + (\omega\mu\sigma)^2]^{\frac{1}{4}}}{\sqrt{\sigma^2 + (\omega\varepsilon)^2}} \angle \frac{1}{2}\tan^{-1}\left(\frac{\sigma}{\omega\varepsilon}\right)$$

$$= \frac{[(\omega^2\mu\varepsilon)^2 + (\omega\mu\sigma)^2]^{\frac{1}{4}}}{\sqrt{\sigma^2 + (\omega\varepsilon)^2}}\left\{\cos\left[\frac{1}{2}\tan^{-1}\left(\frac{\sigma}{\omega\varepsilon}\right)\right] + j\sin\left[\frac{1}{2}\tan^{-1}\left(\frac{\sigma}{\omega\varepsilon}\right)\right]\right\}$$

TABLE 1.2 Approximate Expressions for the Intrinsic Impedance and its Magnitude

$\sigma \to 0$ Ideal insulator Perfect insulator Lossless dielectric Lossless media	η	$\sqrt{\frac{\mu}{\varepsilon}}$
	$\lvert\eta\rvert$	$\sqrt{\frac{\mu}{\varepsilon}}$
$\frac{\sigma}{\omega\varepsilon} \ll 1$ Good insulator Good dielectric Imperfect dielectric Lossy dielectric Weakly/slightly conducting Lossy media	η	$\sqrt{\frac{\mu}{\varepsilon}}\left[\left(1-\frac{3\sigma^2}{8\omega^2\varepsilon^2}\right)+j\frac{\sigma}{2\omega\varepsilon}\right] \approx \sqrt{\frac{\mu}{\varepsilon}}\left(1+j\frac{\sigma}{2\omega\varepsilon}\right) \approx \sqrt{\frac{\mu}{\varepsilon}}$
	$\lvert\eta\rvert$	$\sqrt{\frac{\mu}{\varepsilon}\left[\left(1-\frac{3\sigma^2}{8\omega^2\varepsilon^2}\right)^2+\left(\frac{\sigma}{2\omega\varepsilon}\right)^2\right]} \approx \sqrt{\frac{\mu}{\varepsilon}}$
$\sigma \approx \omega\varepsilon$ Poor conductor Poor insulator Quasiconductor Quasi-insulator Lossy media	η[a]	$\sqrt{\frac{\frac{\mu}{\varepsilon}}{\sqrt{1+\left(\frac{\sigma}{\omega\varepsilon}\right)^2}}}\left[\sqrt{\frac{\sqrt{1+\left(\frac{\sigma}{\omega\varepsilon}\right)^2}+1}{2}}+j\sqrt{\frac{\sqrt{1+\left(\frac{\sigma}{\omega\varepsilon}\right)^2}-1}{2}}\right].$
	$\lvert\eta\rvert$[b]	$\sqrt{\frac{\frac{\mu}{\varepsilon}}{\sqrt{1+\left(\frac{\sigma}{\omega\varepsilon}\right)^2}}}$
$\frac{\sigma}{\omega\varepsilon} \gg 1$ Good conductor Imperfect conductor Lossy conductor Highly conducting Lossy media	η	$\sqrt{\frac{\omega\mu}{2\sigma}}\left[\left(1+\frac{\omega\varepsilon}{2\sigma}\right)+j\left(1-\frac{\omega\varepsilon}{2\sigma}\right)\right] \approx \sqrt{\frac{\omega\mu}{2\sigma}}(1+j)$
	$\lvert\eta\rvert$	$\sqrt{\frac{\omega\mu}{2\sigma}\left[\left(1+\frac{\omega\varepsilon}{2\sigma}\right)^2+\left(1-\frac{\omega\varepsilon}{2\sigma}\right)^2\right]} \approx \sqrt{\frac{\omega\mu}{\sigma}}$
$\sigma \to \infty$ Ideal conductor Perfect conductor	η	0
	$\lvert\eta\rvert$	0

[a]This equation is equivalent to the exact expression for η.
[b]This equation is equivalent to the exact expression for the magnitude of η.

imaginary parts) and

$$\varepsilon = \text{permittivity} = \varepsilon_r\varepsilon_o = \varepsilon_r(8.854\times10^{-12}) \quad \text{where } \varepsilon_r = \text{relative permittivity}$$

$$\mu = \text{permeability} = \mu_r\mu_o = \mu_r(4\pi\times10^{-7}) \quad \text{where } \mu_r = \text{relative permeability}$$

$$\sigma = \text{conductivity} = \sigma_r\sigma_o = \sigma_r(5.8\times10^{7}) \quad \text{where } \sigma_r = \text{conductivity relative to copper}$$

The expressions in these tables were obtained by analogy with the exact and approximate expressions for the characteristic impedance of a transmission line ($L \Leftrightarrow \mu$, $C \Leftrightarrow \varepsilon$, $G \Leftrightarrow \sigma$, $R \Leftrightarrow 0$).

For completeness, the intrinsic impedance of a material (e.g., ferrites) is sometimes seen as

$$\eta = \sqrt{\frac{\mu^*}{\varepsilon^*}} \tag{1.14}$$

where the complex permittivity is defined as

$$\varepsilon^* = \varepsilon_o(\varepsilon' - j\varepsilon'') = \varepsilon_o\left(\varepsilon_r - \frac{j\sigma}{\omega\varepsilon_o}\right) \tag{1.15}$$

and the complex permeability is defined as

$$\mu^* = \mu_o(\mu' - j\mu'') \tag{1.16}$$

The imaginary portions of the complex permittivity and complex permeability are a function of the electric and magnetic losses, respectively, of the material. There are several different ways of defining the complex permittivity and permeability. For example, the intrinsic impedance can also be defined as

$$\eta = \sqrt{\frac{j\omega\mu^*}{\sigma + j\omega\varepsilon^*}} = \sqrt{\frac{j\omega(\mu' - j\mu'')}{\sigma + j\omega(\varepsilon' - j\varepsilon'')}} \tag{1.17}$$

In this case, the complex permittivity and complex permeability have incorporated their free-space values.

1.4 Reflection and Transmission Coefficients

What are the reflection and transmission coefficients for plane wave propagation normally incident to a shield? When is the magnitude of the reflection coefficient approximately one and approximately zero?

Assuming that the incoming wave is normally incident to the object or shield, the reflection coefficient has the same form as the voltage reflection coefficient for transmission lines. For a transmission line, the voltage reflection coefficient is

$$\rho_L = \frac{Z_L - Z_o}{Z_L + Z_o}$$

For normally incident plane wave propagation, the electric field reflection coefficient is

$$\rho_L = \frac{\eta_L - \eta_o}{\eta_L + \eta_o} \tag{1.18}$$

where η_L is the intrinsic impedance of the material/shield and η_o is the intrinsic impedance of the medium preceding the object or shield. For a transmission line, the voltage transmission coefficient is

$$\mathrm{T} = 1 + \rho_L = \frac{2Z_L}{Z_L + Z_o}$$

while for a normally incident plane wave, the electric field transmission coefficient is

$$\mathrm{T} = 1 + \rho_L = \frac{2\eta_L}{\eta_L + \eta_o} \tag{1.19}$$

The effectiveness of a shield in the far field is a strong function of these coefficients. For example, if an electric field of amplitude A is incident upon a material (many skin depths thick) having a reflection coefficient ρ_L and transmission coefficient T, then the reflected electric field is

$$A\rho_L \tag{1.20}$$

and the electric field transmitted into the material is

$$AT \tag{1.21}$$

Not surprisingly, the magnetic field reflection and transmission coefficients are similar in form to the current coefficients for transmission lines.

The magnitude of the reflection coefficient is approximately one when there is a large impedance "mismatch" between the mediums:

$$|\rho_L| = \left|\frac{\eta_L - \eta_o}{\eta_L + \eta_o}\right| \approx 1 \quad \text{if } \eta_L \ll \eta_o \; \left(\frac{\eta_L}{\eta_o} \ll 1\right) \quad \text{or} \quad \eta_L \gg \eta_o \; \left(\frac{\eta_L}{\eta_o} \gg 1\right) \tag{1.22}$$

Since the impedance of good metals is small compared to the impedance of air (377 Ω), the impedance "mismatch" between metal and air is large. This large mismatch is why metals are excellent reflectors of plane waves. A metal shield can be viewed as the short on a short-circuited transmission line. All incident energy is reflected from a short-circuited load since a short circuit cannot dissipate electrical energy.

The magnitude of the reflection coefficient is approximately zero when the mediums are "matched" at the interface:

$$|\rho_L| = \left|\frac{\eta_L - \eta_o}{\eta_L + \eta_o}\right| \approx 0 \quad \text{if } \eta_L \approx \eta_o \; \left(\frac{\eta_L}{\eta_o} \approx 1\right) \tag{1.23}$$

The electric field passes mostly into the medium when this reflection coefficient is small. In this case, if the wave is to be effectively shielded, the absorption loss of the medium must be high or additional reflective interfaces and materials should be present.

1.5 Plane Wave Power

At frequencies greater than about 10 GHz, the maximum safe power-density exposure for humans is given as 5 mW/cm^2 at a distance of 5 cm. Convert this power density to an equivalent electric field. State all assumptions. [Gandhi, '90]

The electromagnetic power expressions are similar to the power expressions for circuit variables and can be remembered by analogy. For an electrical circuit, the instantaneous power, which is the power as a function of time, absorbed by an element is given by

$$p(t) = v(t)i(t)\ \text{W} \tag{1.24}$$

When the electric and magnetic fields are known, the instantaneous power density is given by the Poynting vector, $\vec{P}(t)$,

$$\vec{P}(t) = \vec{E}(t) \times \vec{H}(t)\ \text{W/m}^2 \tag{1.25}$$

where both the electric field, $\vec{E}$, and magnetic field, $\vec{H}$, in this expression are functions of time. Note that Equation (1.25) is an instantaneous power *density* (power per unit area), and it is a vector with a direction. If the total power is desired, it must be integrated over the area of interest (or multiplied by the area if the power density is not varying much over the region). Since the electric field is analogous to the voltage and the magnetic field is analogous to the current, the circuit relationship is very similar to the field relationship. The units associated with the Poynting vector are V/m (from the electric field) multiplied by A/m (from the magnetic field) or W/m².

The average power for a periodic signal with a period of T is given by the integral of the instantaneous power, over one period, divided by the period:

$$p_{avg} = \frac{1}{T}\int_0^T v(t)i(t)\,dt \text{ W} \tag{1.26}$$

(If the average of N discrete values is desired, these values are first summed, corresponding to integration, and then divided by N, corresponding to the division by T.) Of course, the average power is no longer a function of time, t, since it has been integrated out of the expression. In terms of field variables,

$$\vec{P}_{avg} = \frac{1}{T}\int_0^T [\vec{E}(t)\times\vec{H}(t)]\,dt \text{ W/m}^2 \tag{1.27}$$

If the voltage and current are given in phasor form, in the frequency domain, the average power is not merely the voltage phasor, V_s, multiplied by the current phasor, I_s, but

$$P_{avg} = \frac{1}{2}\text{Re}\left(V_s I_s^*\right) \text{ W} \tag{1.28}$$

where I_s^* is the complex conjugate of the current phasor (and peak phasors are being used). In terms of phasor field variables, the average power is equal to

$$\vec{P}_{avg} = \frac{1}{2}\text{Re}\left(\vec{E}_s \times \vec{H}_s^*\right) \text{ W/m}^2 \tag{1.29}$$

where $\vec{H}_s^*$ is the complex conjugate of the magnetic field phasor. Notice that this average power is a vector. When working with phasors, the *phasor* Poynting vector is defined as

$$\vec{P}_s = \vec{E}_s \times \vec{H}_s^* \text{ W/m}^2 \tag{1.30}$$

This expression is analogous to complex power in circuits. The "conversion" of this phasor version of the Poynting vector to the time domain is a meaningless operation. If the time-domain version of the Poynting vector is desired, the phasor version of the electric field should be converted to the time domain and the phasor version of the magnetic field should be converted to the time domain. Then, the two time-domain expressions should be substituted into the expression

$$\vec{P}(t) = \vec{E}(t)\times\vec{H}(t) \text{ W/m}^2 \tag{1.31}$$

The previous expressions for the power in terms of the field variables are valid for any field distribution, whether in the near field or far field. In the far field, the average power expression can be simplified by using the definition for intrinsic impedance for a wave traveling in the positive z direction

$$\eta = \frac{E_{xs}}{H_{ys}}$$

in the definition for average power:

$$\vec{P}_{avg} = \frac{1}{2}\mathrm{Re}\left[E_{xs}\hat{a}_x \times \left(\frac{E_{xs}}{\eta}\right)^* \hat{a}_y\right] = \frac{1}{2}\mathrm{Re}\left(\frac{|E_{xs}|^2}{\eta}\hat{a}_z\right) = \frac{|E_{xs}|^2}{2\eta}\hat{a}_z \text{ W/m}^2 \tag{1.32}$$

(The final simplified result assumes that η is real.) The variables E_{xs} and H_{ys} are the electric and magnetic field components in the frequency domain, respectively. To reiterate, (1.32) is the average power expression for a single plane wave traveling in the z direction. If there are multiple waves involved, for example an incident and a reflected wave, this expression is generally not valid. Because of the simple relationship between the electric and magnetic fields for plane waves, the average power can also be written as

$$\vec{P}_{avg} = \frac{\eta|H_{ys}|^2}{2}\hat{a}_z \text{ W/m}^2 \tag{1.33}$$

When fields are measured, rms values are typically used:

$$\vec{P}_{avg} = \frac{E^2_{x,rms}}{\eta}\hat{a}_z = \eta H^2_{y,rms}\hat{a}_z \text{ W/m}^2 \tag{1.34}$$

To be in the far field of a 10 GHz source, the receiver should be several wavelengths from the source for a wire antenna. It is more likely at this frequency that a horn, dish, or other high-frequency antenna is used. For these antennas, the receiver is in the far field when the distance, d, between the source and receiver satisfies the inequality

$$d > \frac{2D^2}{\lambda_o}$$

where D is the largest dimension of the antenna. For a horn antenna, the aperture dimension is often about a wavelength, so the distance, d, is again several wavelengths from the source. The free-space wavelength corresponding to 10 GHz is 3 cm. The maximum safe power-density exposure for humans is 5 mW/cm^2 at a distance of 5 cm.[3] Five centimeters is in the transition region between the near field and far field. If it is assumed that the receiver is in the far field, however, the electric field amplitude corresponding to this power density is easily obtained:

$$50 \text{ W/m}^2 = \frac{E_x^2}{2(377)} \quad \Rightarrow E_x \approx 194 \text{ V/m} \tag{1.35}$$

This electric field magnitude is occasionally used in place of the power density limit. The relationships between the average power density, in various units, and the corresponding rms values of the electric and magnetic fields for free-space conditions are given in Table 1.3. Again, this assumes far-field, plane wave conditions where the electric and magnetic fields are perpendicular and transverse to the direction of wave propagation. For near-field conditions, the relationship between the power density and the electric and magnetic fields is usually much more complex. Table 1.4 lists typical values for the power density, which should help provide a "feeling" for the numbers. Some power densities are provided at very low frequencies where it is difficult to be in the far field during the measurement. These "far-field equivalents" are provided for comparison purposes and for calibration of some field measuring instruments. In the near field, it is usually more meaningful to use electric and magnetic field data.

[3]For important and critical applications, up-to-date power density and electric and magnetic field strength standards, such as the IEEE standard for safe field exposure levels, should be consulted.

TABLE 1.3 Relationship between Average Power Density and rms Value of the Electric and Magnetic Fields (in the far field)

P_{avg}, $\eta = \sqrt{\mu_o/\varepsilon_o} \approx 377\ \Omega$				
W/m^2	mW/cm^2	μW/cm^2	E_{rms} V/m	H_{rms} A/m
0.01	0.001	1	1.94	0.00515
0.1	0.01	10	6.14	0.0163
1	0.1	100	19.4	0.0515
10	1	1,000	61.4	0.163
100	10	10,000	194	0.515
1,000	100	100,000	614	1.63

TABLE 1.4 Typical Average Power Densities [IPCS, '92; Cleveland, '99; Gandhi, '86; Tell]

Source	Approximate Power Density
"Near" air traffic control radar (stationary)	10's MW/m^2 @ few m's from main beam
Cataracts in rabbits	1,000–2,000 W/m^2 (30–60 minute RF exposure)
FCC limit for controlled/occupational exposure	1,000 W/m^2 (0.3–3 MHz) (9,000/f^2)[a] W/m^2 (3–30 MHz) 10 W/m^2 (30–300 MHz) (f/30)[a] W/m^2 (300 MHz–1.5 GHz) 50 W/m^2 (1.5–100 GHz)
FCC limit for uncontrolled/general population exposure	1,000 W/m2 (0.3–1.34 MHz) (1,800/f^2)[a] W/m^2 (1.34–30 MHz) 2 W/m^2 (30–300 MHz) (f/150)[a] W/m^2 (300 MHz–1.5 GHz) 10 W/m^2 (1.5–100 GHz)
Heating of biological tissue	10–100 W/m^2 (RF exposure)
Human perception of very warm to hot	87 W/m^2 (infrared exposure)
Threshold of human perception for warming	6.7 W/m^2 (infrared exposure)
Microwave oven leakage	< 1.5 W/m^2 @ 0.3 m <0.10 W/m^2 @ 1 m
"Vicinity" of air traffic control radar	0.03–0.8 W/m^2
Microwave alarm (10 mW)	0.1 W/m^2 @ 0.5 m
Cellular tower	≤ 0.01 W/m^2 (near tower along the surface)
Median background exposure for large cities (U.S.)	50 μW/m^2

[a] f in MHz

1.6 Single-Layer Conducting Shield

Determine the exact expression for the ratio of the electric field transmitted through to the electric field incident to a flat shield of conductivity σ, permittivity ε, permeability μ, and thickness d. Assume the incident field is a plane wave and the shield is a good conductor. [Paul, '92(b)]

It is worthwhile to repeat the analysis given in the stated reference especially since it is not uncommon in EMC books to neglect the multiple reflection term to be defined later. The shielding effectiveness (SE) of a shield is defined in dB as

$$SE_{dB} = 20\log\left(\frac{E_i}{E_t}\right) \tag{1.36}$$

where E_i is the amplitude of the electric field component of the incident plane wave and E_t is the amplitude of the electric field component of the plane wave transmitted through the shield. When a plane wave is incident upon a shield, a percentage of the wave is reflected from the shield and a percentage is transmitted into the shield. Of the field that is transmitted into the shield, a fraction of it is absorbed by the shield through ohmic loss, which is a function of the conductivity of the shield medium. When the attenuated transmitted wave hits the backside of the shield, a percentage of the wave is transmitted through this second interface and a percentage is reflected off this second interface. This reflected signal from the second interface passes back toward the first interface and, of course, is again reduced by ohmic losses in the process. When it hits the front interface, a percentage of the field is transmitted through this interface and a percentage is rereflected. The entire process is repeated until a steady-state condition is reached. This process is similar to what occurs in a transmission line with a transient input signal.

Fortunately, sinusoidal steady state is assumed in this analysis so that not every single reflected and transmitted term, both inside and outside the shield, has to be tracked. The forward-traveling and backward-traveling field components in sinusoidal steady state include all these terms. For example, assume that the total electric field inside the shield (at the first interface) can be approximated as

$$e_t(t) \approx 5\cos(10^6 t) + 0.2\cos(10^6 t + 15°) - 0.006\cos(10^6 t + 65°)$$

The first term represents the first term transmitted into the shield, and the second and third terms represent the signals due to multiple reflections. Since these terms are sinusoids of the same frequency, they can be written in phasor form and the sum reduced to one term:

$$E_{ts} = 5\angle 0° + 0.2\angle 15° - 0.006\angle 65° \approx 5 + j0 + 0.193 + j0.052 - 0.00254 - j0.00544$$

$$\approx 5.19 + j0.0466 \approx 5.2\angle 0.51°$$

Therefore, the transmitted field in the time domain is

$$e_t(t) = 5.2\cos(10^6 t + 0.51°)$$

In general, any number of same frequency sinusoids of different amplitudes and phases can be reduced to a single sinusoidal term:

$$e_t(t) = A_m \cos(\omega t + \theta) \quad \text{or} \quad E_{ts} = A_m \angle \theta = A_m e^{j\theta}$$

The various sinusoidal steady-state terms for this single shield problem are shown in Figure 1.2. To obtain the desired expression for the transmitted term, the various field components will be described

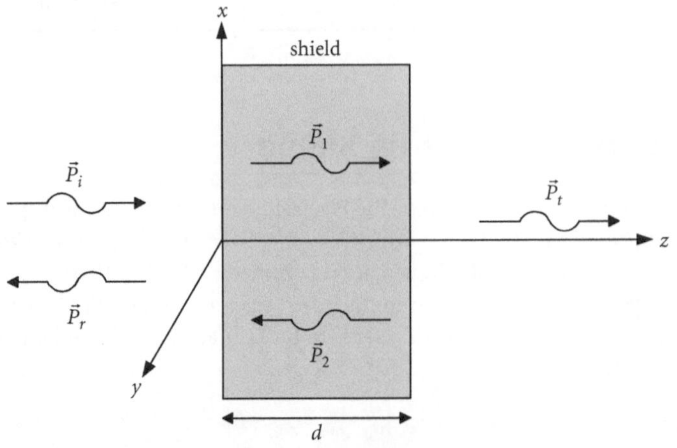

FIGURE 1.2 Poynting vectors inside and outside a shield of thickness d.

as phasors. Assume that the incident electric field is in the x direction:

$$\vec{E}_{is} = A_s e^{-\gamma_o z}\hat{a}_x \tag{1.37}$$

The negative sign in the argument of the exponential indicates that the incident field is traveling in the *positive z* direction. The variable A_s is a phasor that includes both the magnitude and phase of the incident waveform. If the medium surrounding the shield is assumed to be free space, then the propagation constant, γ_o, is equal to

$$\gamma_o = \alpha_o + j\beta_o = 0 + j\omega\sqrt{\mu_o\varepsilon_o} \tag{1.38}$$

and

$$\vec{E}_{is} = A_s e^{-j\beta_o z}\hat{a}_x \tag{1.39}$$

There is no attenuation of the field in free space, only a phase shift, since the propagation constant is entirely imaginary. The magnetic field can be easily obtained for a plane wave with a single component since the ratio of the electric field to the magnetic field is equal to the intrinsic impedance of the medium in which the wave is traveling:

$$\vec{H}_{is} = \frac{A_s}{\eta_o}e^{-j\beta_o z}\hat{a}_y \quad \text{where } \eta_o = \sqrt{\frac{\mu_o}{\varepsilon_o}} \tag{1.40}$$

Notice that the magnetic field, although it is still traveling in the positive z direction, is perpendicular to the electric field and transverse to the direction of propagation z. As a check, the direction of the time-average power is also in the $+z$ direction:

$$\vec{P}_{iavg} = \frac{1}{2}\text{Re}\left(\vec{E}_{is}\times\vec{H}_{is}^*\right) = \frac{1}{2}\text{Re}\left(A_s e^{-j\beta_o z}\hat{a}_x \times \frac{A_s^*}{\eta_o}e^{j\beta_o z}\hat{a}_y\right) = \frac{|A_s|^2}{2\eta_o}\hat{a}_z \tag{1.41}$$

The reflected electric field traveling in the $-z$ direction is also in the x direction:

$$\vec{E}_{rs} = B_s e^{j\beta_o z}\hat{a}_x \tag{1.42}$$

The amplitude and phase of this negative-going signal are not usually the same as the incident signal, so it is assigned the variable B_s. Again, this variable is not necessarily real: it is a phasor with both a magnitude and phase. The positive argument of the exponential indicates that the signal power is in the *negative z* direction. The corresponding magnetic field is

$$\vec{H}_{rs} = -\frac{B_s}{\eta_o}e^{j\beta_o z}\hat{a}_y \tag{1.43}$$

The magnetic field is still traveling in the negative z direction. The most confusing point for most students is the negative sign in front of the overall function. This sign is required if the power flow is in the negative z direction, which is only reasonable because this portion of the wave is traveling in this direction. To verify this, the time-average power for this negative-going signal is

$$\vec{P}_{ravg} = \frac{1}{2}\text{Re}\left(\vec{E}_{rs}\times\vec{H}_{rs}^*\right) = \frac{1}{2}\text{Re}\left(B_s e^{j\beta_o z}\hat{a}_x \times \frac{-B_s^*}{\eta_o}e^{-j\beta_o z}\hat{a}_y\right) = -\frac{|B_s|^2}{2\eta_o}\hat{a}_z \tag{1.44}$$

If an individual is unsure of the direction of a field, it can be determined through trial-and-error by checking the direction of the resultant power. If the power flow is not in the expected direction, the signs and directions of the fields should be appropriately modified.

The remaining field components are obtained in the same manner. The only real difference is in the shield's propagation constant and intrinsic impedance:

$$\gamma = \alpha + j\beta = \sqrt{j\omega\mu(\sigma + j\omega\varepsilon)},\ \eta = \sqrt{\frac{j\omega\mu}{\sigma + j\omega\varepsilon}} \tag{1.45}$$

Both of these variables can be complex. When the propagation constant contains a real term, attenuation in the shield is expected. The other electric and magnetic field terms, in the frequency domain, are determined in a similar manner:

$$\vec{E}_{1s} = C_s e^{-\gamma z}\hat{a}_x,\ \vec{H}_{1s} = \frac{C_s}{\eta}e^{-\gamma z}\hat{a}_y \tag{1.46}$$

$$\vec{E}_{2s} = D_s e^{\gamma z}\hat{a}_x,\ \vec{H}_{2s} = -\frac{D_s}{\eta}e^{\gamma z}\hat{a}_y \tag{1.47}$$

$$\vec{E}_{ts} = E_s e^{-j\beta_o z}\hat{a}_x,\ \vec{H}_{ts} = \frac{E_s}{\eta_o}e^{-j\beta_o z}\hat{a}_y \tag{1.48}$$

The total field for each of the three regions is the sum of the positive-going (or forward-traveling) and negative-going (or backward-traveling) terms in each of the regions:

$$\vec{E}_{is} + \vec{E}_{rs} = A_s e^{-j\beta_o z}\hat{a}_x + B_s e^{j\beta_o z}\hat{a}_x,\ \vec{H}_{is} + \vec{H}_{rs} = \frac{A_s}{\eta_o}e^{-j\beta_o z}\hat{a}_y - \frac{B_s}{\eta_o}e^{j\beta_o z}\hat{a}_y \quad \text{for } z < 0 \tag{1.49}$$

$$\vec{E}_{1s} + \vec{E}_{2s} = C_s e^{-\gamma z}\hat{a}_x + D_s e^{\gamma z}\hat{a}_x,\ \vec{H}_{1s} + \vec{H}_{2s} = \frac{C_s}{\eta}e^{-\gamma z}\hat{a}_y - \frac{D_s}{\eta}e^{\gamma z}\hat{a}_y \quad \text{for } 0 < z < d \tag{1.50}$$

$$\vec{E}_{ts} = E_s e^{-j\beta_o z}\hat{a}_x,\ \vec{H}_{ts} = \frac{E_s}{\eta_o}e^{-j\beta_o z}\hat{a}_y \quad \text{for } z > d \tag{1.51}$$

Since there is no reflected term for $z > d$, the total field past the shield consists of only one positive-going term. To obtain the ratio of the incident to transmitted electric field, the variable E_s is required as a function of the incident phasor A_s. Provided that the properties of the shield and A_s are given, there are four unknowns (B_s, C_s, D_s, and E_s). To obtain the four equations required to solve for these four unknowns, boundary conditions are required. These boundary conditions are conditions that relate the electric and magnetic fields on one side of a boundary to the fields on the other side of the same boundary. They are valid only at the boundary. For this single shield, there is a boundary at $z = 0$ and $z = d$.

One of the classical boundary conditions is that the *total* tangential electric field on one side of a boundary is equal to the *total* tangential electric field on the other side of the same boundary. Because the electric field is tangential to both boundaries,

$$\begin{aligned} E_{tang}(z = 0^-) &= E_{tang}(z = 0^+) \\ E_{tang}(z = d^-) &= E_{tang}(z = d^+) \end{aligned} \tag{1.52}$$

For the $z = 0$ boundary, the negative superscript on the zero indicates a position slightly before the first boundary (in the free space), and the positive superscript indicates a position slightly after the first boundary (in the shield). For the second boundary condition at $z = d$, the superscripts have the same basic meaning but are centered on $z = d$. The second classical boundary condition is that the difference between the tangential magnetic field on one side of a boundary and tangential magnetic field on the other side of the same boundary is equal to the magnitude of the surface current density at this boundary:

$$\begin{aligned} H_{tang}(z=0^-)-H_{tang}(z=0^+)&=K(z=0) \\ H_{tang}(z=d^-)-H_{tang}(z=d^+)&=K(z=d) \end{aligned} \tag{1.53}$$

If the medium on either side of a boundary is not a perfect conductor, then the surface current density, K, is zero and the magnetic field boundary condition is identical in form to the electric field condition. In this particular example, the magnetic field is also entirely tangential to the interfaces.

First, the electric and magnetic boundary conditions are applied at the first interface:

$$A_s e^{-j\beta_o 0}\hat{a}_x + B_s e^{j\beta_o 0}\hat{a}_x = C_s e^{-\gamma 0}\hat{a}_x + D_s e^{\gamma 0}\hat{a}_x \tag{1.54}$$

$$\frac{A_s}{\eta_o} e^{-j\beta_o 0}\hat{a}_y - \frac{B_s}{\eta_o} e^{j\beta_o 0}\hat{a}_y = \frac{C_s}{\eta} e^{-\gamma 0}\hat{a}_y - \frac{D_s}{\eta} e^{\gamma 0}\hat{a}_y \tag{1.55}$$

Second, the electric and magnetic boundary conditions are applied at the second interface:

$$C_s e^{-\gamma d}\hat{a}_x + D_s e^{\gamma d}\hat{a}_x = E_s e^{-j\beta_o d}\hat{a}_x \tag{1.56}$$

$$\frac{C_s}{\eta} e^{-\gamma d}\hat{a}_y - \frac{D_s}{\eta} e^{\gamma d}\hat{a}_y = \frac{E_s}{\eta_o} e^{-j\beta_o d}\hat{a}_y \tag{1.57}$$

A common student error is leaving the variable, in this case z, in the expressions rather than substituting the appropriate boundary locations in the boundary conditions.

The four unknowns can be determined from the previous four equations either by hand or through a symbolic manipulation package. These tedious steps will not be shown. When the shield is a good conductor, the propagation constant is a simple function of the skin depth,

$$\gamma = \alpha + j\beta \approx \frac{1}{\delta} + j\frac{1}{\delta} \tag{1.58}$$

and the exact expression for the ratio A_s/E_s for good conductors is (relabeled E_{is}/E_{ts})

$$\frac{E_{is}}{E_{ts}} = \frac{(\eta_o+\eta)^2}{4\eta_o\eta}\left[1-\left(\frac{\eta_o-\eta}{\eta_o+\eta}\right)^2 e^{-2\frac{d}{\delta}} e^{-j2\frac{d}{\delta}}\right] e^{\frac{d}{\delta}} e^{j\frac{d}{\delta}} e^{-j\beta_o d} \tag{1.59}$$

Since the impedance of the plane wave on both sides of (and exterior to) the slab is identical, this expression is also valid for magnetic fields since

$$\eta_o = \frac{E_{is}}{H_{is}} = \frac{E_{ts}}{H_{ts}} \Rightarrow \frac{E_{is}}{E_{ts}} = \frac{H_{is}}{H_{ts}} \tag{1.60}$$

TABLE 1.5 Exact Expressions for the Propagation Constant and Velocity

Propagation Constant

$$\gamma = \alpha + j\beta = \sqrt{j\omega\mu(\sigma + j\omega\varepsilon)} = \sqrt{-\omega^2\mu\varepsilon + j\omega\mu\sigma}$$

$$= [(\omega^2\mu\varepsilon)^2 + (\omega\mu\sigma)^2]^{\frac{1}{4}}\cos\left[\frac{1}{2}\tan^{-1}\left(\frac{\sigma}{-\omega\varepsilon}\right)\right] + j[(\omega^2\mu\varepsilon)^2 + (\omega\mu\sigma)^2]^{\frac{1}{4}}\sin\left[\frac{1}{2}\tan^{-1}\left(\frac{\sigma}{-\omega\varepsilon}\right)\right]$$

$$= \sqrt{\frac{\sqrt{(\omega^2\mu\varepsilon)^2 + (\omega\mu\sigma)^2} - \omega^2\mu\varepsilon}{2}} + j\sqrt{\frac{\sqrt{(\omega^2\mu\varepsilon)^2 + (\omega\mu\sigma)^2} + \omega^2\mu\varepsilon}{2}} \quad \text{(assuming } \alpha \text{ and } \beta \text{ are} > 0\text{)}$$

Velocity of Propagation

$$v = \frac{\omega}{\beta} = \frac{\omega}{\sqrt{\frac{\sqrt{(\omega^2\mu\varepsilon)^2 + (\omega\mu\sigma)^2} + \omega^2\mu\varepsilon}{2}}} \quad \text{(assuming } \beta > 0\text{)}$$

A number of books will assume (even if not explicitly stated) that the shield is sufficiently lossy or thick so that the reflection terms in the shield can be neglected, reducing the number of unknowns to three (B_s, C_s, and E_s). The simpler expression for the shielding effectiveness in this special case can be determined using the reflection and transmission coefficients at the boundaries. The reflected signals in the shield are ignored in this case.

For future reference, the exact and approximate expressions for the propagation constant and velocity are given in Table 1.5 and Table 1.6. The propagation constant, γ, has units of m^{-1}, attenuation coefficient, α, has units of nepers per meter or just m^{-1}, phase constant, β, has units of rad/m, and velocity, v, has units of m/s. Also,

$$\varepsilon = \text{permittivity} = \varepsilon_r\varepsilon_o = \varepsilon_r(8.854 \times 10^{-12}) \quad \text{where } \varepsilon_r = \text{relative permittivity}$$

$$\mu = \text{permeability} = \mu_r\mu_o = \mu_r(4\pi \times 10^{-7}) \quad \text{where } \mu_r = \text{relative permeability}$$

$$\sigma = \text{conductivity} = \sigma_r\sigma_o = \sigma_r(5.8 \times 10^{7}) \quad \text{where } \sigma_r = \text{conductivity relative to copper}$$

As with the intrinsic impedance expressions, these results were obtained by analogy with the exact and approximate expressions for the α, β, and v of a transmission line ($L \Leftrightarrow \mu$, $C \Leftrightarrow \varepsilon$, $G \Leftrightarrow \sigma$, $R \Leftrightarrow 0$).

It is interesting to note that the velocity of a plane wave over a range of frequencies in metals and other good conductors such as water and soil (at lower frequencies) can be quite low since the plane wave velocity is given by the expression

$$v = \frac{1}{\sqrt{\mu\varepsilon}\sqrt{\frac{\sqrt{1+\left(\frac{\sigma}{\omega\varepsilon}\right)^2}+1}{2}}} = \frac{c}{\sqrt{\mu_r\varepsilon_r}\sqrt{\frac{\sqrt{1+\left(\frac{\sigma}{\omega\varepsilon_r\varepsilon_o}\right)^2}+1}{2}}}$$

TABLE 1.6 Approximate Expressions for the Attenuation Constant, Phase Constant, and Velocity

$\sigma \to 0$ Ideal insulator Perfect insulator Lossless dielectric Lossless media	α	0
	β	$\omega\sqrt{\mu\varepsilon}$
	v	$\frac{1}{\sqrt{\mu\varepsilon}}$
$\frac{\sigma}{\omega\varepsilon} \ll 1$ Good insulator Good dielectric Imperfect dielectric Lossy dielectric Weakly/slightly conducting Lossy media	α	$\frac{\sigma}{2}\sqrt{\frac{\mu}{\varepsilon}}$
	β	$\omega\sqrt{\mu\varepsilon}\left(1+\frac{\sigma^2}{8\omega^2\varepsilon^2}\right) \approx \omega\sqrt{\mu\varepsilon}$
	v	$\frac{1}{\sqrt{\mu\varepsilon}\left(1+\frac{\sigma^2}{8\omega^2\varepsilon^2}\right)} \approx \frac{1}{\sqrt{\mu\varepsilon}}$
$\sigma \approx \omega\varepsilon$ Poor conductor Poor insulator Quasiconductor Quasi insulator Lossy media	α[a]	$\omega\sqrt{\mu\varepsilon}\sqrt{\frac{\sqrt{1+\left(\frac{\sigma}{\omega\varepsilon}\right)^2}-1}{2}}$
	β[b]	$\omega\sqrt{\mu\varepsilon}\sqrt{\frac{\sqrt{1+\left(\frac{\sigma}{\omega\varepsilon}\right)^2}+1}{2}}$
	v[c]	$\frac{1}{\sqrt{\mu\varepsilon}\sqrt{\frac{\sqrt{1+\left(\frac{\sigma}{\omega\varepsilon}\right)^2}+1}{2}}}$
$\frac{\sigma}{\omega\varepsilon} \gg 1$ Good conductor Imperfect conductor Lossy conductor Highly conducting Lossy media	α	$\sqrt{\frac{\omega\sigma\mu}{2}}\left(1-\frac{\omega\varepsilon}{2\sigma}\right) \approx \sqrt{\frac{\omega\sigma\mu}{2}}$
	β	$\sqrt{\frac{\omega\sigma\mu}{2}}\left(1+\frac{\omega\varepsilon}{2\sigma}\right) \approx \sqrt{\frac{\omega\sigma\mu}{2}}$
	v	$\frac{\sqrt{\frac{2\omega}{\sigma\mu}}}{1+\frac{\omega\varepsilon}{2\sigma}} \approx \sqrt{\frac{2\omega}{\sigma\mu}}$
$\sigma \to \infty$ Ideal conductor Perfect conductor	α	∞
	β	∞
	v	0

[a]This equation is equivalent to the exact expression for α.
[b]This equation is equivalent to the exact expression for β.
[c]This equation is equivalent to the exact expression for v.

where c is the speed of light ($c = 3\times10^8$ m/s). For good conductors this expression reduces to

$$v \approx \frac{c}{\sqrt{\mu_r\varepsilon_r}\sqrt{\frac{\left(\frac{\sigma}{\omega\varepsilon_r\varepsilon_o}\right)}{2}}} = \frac{c}{\sqrt{\frac{\mu_r\sigma}{2\omega\varepsilon_o}}} = \sqrt{\frac{2\omega}{\sigma\mu}}$$

As the conductivity and permeability increase, the velocity decreases. Good shielding metals thus reflect, absorb, and slow down plane waves.

1.7 Thin Shields and Reflection Loss

Using (1.59), show that the expression for the *magnitude* of the electric field transmitted through a thin (much less than the skin depth) good-conducting shield is

$$E_t \approx E_i \frac{2\sqrt{2}}{377\sigma d}$$

where E_i is the amplitude of the plane wave normally incident to the slab of thickness d and conductivity σ.

When a shield is thin relative to the skin depth,

$$\frac{d}{\delta} \ll 1 \tag{1.61}$$

Also, when a shield is a good conductor, its impedance is much less than that of air (377 Ω):

$$\eta \ll \eta_o \quad \text{and} \quad \eta \approx \sqrt{\frac{\omega\mu}{2\sigma}}(1+j) \tag{1.62}$$

These relationships dramatically reduce the complexity of the general shielding equation, (1.59):

$$\frac{E_{is}}{E_{ts}} \approx \frac{(\eta_o)^2}{4\eta_o\eta}\left[1-\left(\frac{\eta_o}{\eta_o}\right)^2 e^{-j2\frac{d}{\delta}}\right]e^{\frac{d}{\delta}}e^{j\frac{d}{\delta}}e^{-j\beta_o d} = \frac{\eta_o}{4\eta}\left(1-e^{-j2\frac{d}{\delta}}\right)e^{\frac{d}{\delta}}e^{j\frac{d}{\delta}}e^{-j\beta_o d}$$

The complex exponential can be reduced to a simpler form if Euler's formula is resurrected:

$$e^{-jx} = \cos x - j\sin x \approx 1 - jx \quad \text{if } x \text{ is small}$$

($\cos(x) \approx 1$ and $\sin(x) \approx x$ if x is small). Also, the magnitude of a complex exponential is one

$$|e^{-jx}| = \sqrt{\cos^2 x + \sin^2 x} = 1$$

and a real exponential raised to a small power (of d/δ) is about one. Therefore, the magnitude of the previous ratio reduces to

$$\left|\frac{E_{is}}{E_{ts}}\right| \approx \left|\frac{\eta_o}{4\eta}\left[1-\left(1-j2\frac{d}{\delta}\right)\right]\right| \approx \left|\frac{\eta_o}{2\eta}\right|\frac{d}{\delta} = \left|\frac{377}{2\sqrt{\frac{\omega\mu}{2\sigma}}(1+j)}\right| d\sqrt{\pi f\mu\sigma} = \frac{377\sigma d}{2\sqrt{2}} \tag{1.63}$$

$\mu_o := 4\cdot\pi\cdot 10^{-7} \quad \varepsilon_o := 8.854\cdot 10^{-12} \quad \sigma_o := 5.8\cdot 10^{7} \quad \mu_r := 1 \quad \sigma_r := 1 \quad \eta_o := \sqrt{\frac{\mu_o}{\varepsilon_o}} \quad j := \sqrt{-1}$

$x := 40, 40.1 .. 90$

$$f(x) := \left(x + 1 - 10\cdot \mathrm{floor}\left(\frac{x}{10}\right)\right)\cdot 10^{\mathrm{floor}\left(\frac{x}{10}\right)}$$

$$\delta(f) := \frac{1}{\sqrt{\pi\cdot f\cdot \mu_r\cdot \mu_o\cdot \sigma_r\cdot \sigma_o}} \quad \eta(f) := \sqrt{\frac{2\cdot\pi\cdot f\cdot\mu_r\cdot\mu_o}{2\cdot\sigma_r\cdot\sigma_o}}\cdot(1+j) \quad \beta_o(f) := \frac{2\cdot\pi}{\left(\frac{3\cdot 10^8}{f}\right)} \quad SE_{app}(d) := 20\cdot\log\left(\frac{377\cdot\sigma_r\cdot\sigma_o\cdot d}{2\sqrt{2}}\right)$$

$$SE_{dB}(f, d) := 20\cdot\log\left[\left|\frac{(\eta_o+\eta(f))^2}{4\cdot\eta_o\cdot\eta(f)}\cdot\left[1-\left(\frac{\eta_o-\eta(f)}{\eta_o+\eta(f)}\right)^2\cdot e^{-2\cdot\frac{d}{\delta(f)}}\cdot e^{-j\cdot 2\cdot\frac{d}{\delta(f)}}\right]\cdot e^{\frac{d}{\delta(f)}}\cdot e^{j\cdot\frac{d}{\delta(f)}}\cdot e^{-j\cdot\beta_o(f)\cdot d}\right|\right]$$

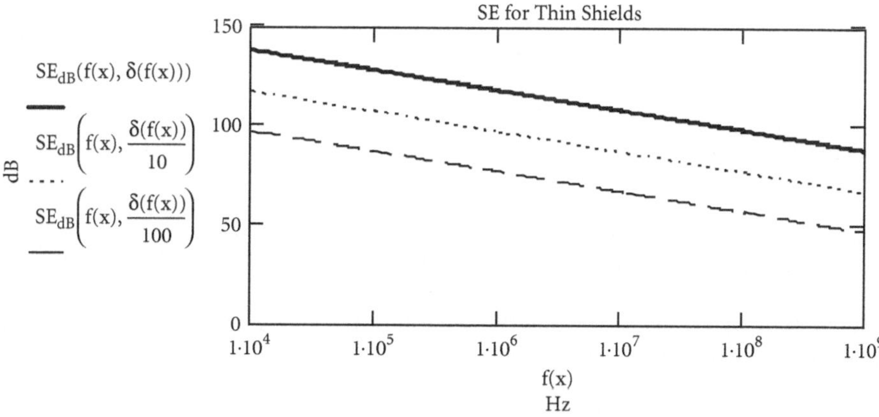

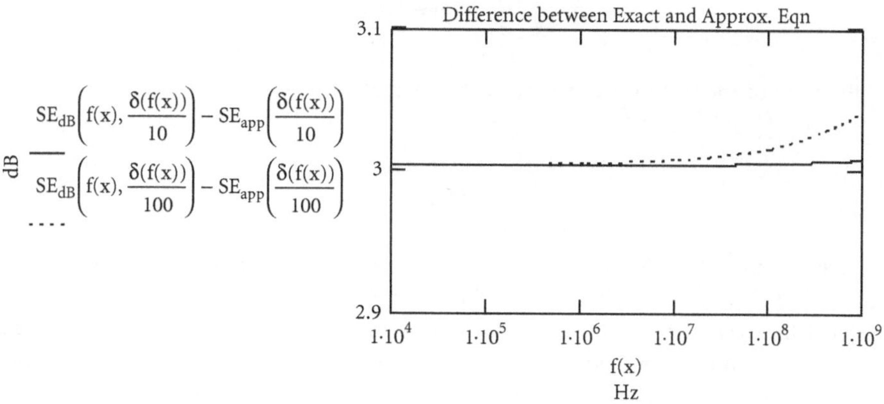

MATHCAD 1.1 SE vs. frequency as a function of shield thickness. Also shown is the difference between the exact SE and approximate thin-shield SE expressions.

For a thin shield, the transmitted electric field decreases as the conductivity or thickness of the shield increases. For very thin shields or metallic films, the conductivity is not equal to the bulk conductivity normally used throughout most of this and other books.

The shielding effectiveness of a thin shield is mainly due to reflection not absorption. The absorption of the plane wave is negligible since the shield is so thin. In Mathcad 1.1 it is seen that the shielding effectiveness for a flat copper shield is quite high even for very thin shields (relative to the skin depth, δ). However, the defects in the shield such as openings, cracks, and penetrating wires frequently limit the shielding effectiveness unless the shield is very thin (e.g., in the μm range). Note that the difference between the exact equation and the much simpler approximate equation for shielding effectiveness is around 3 dB or a factor of 1.4.

Using (1.63), determine the ratio of the electric field transmitted through the human body to the incident electric field for a 60 Hz electric field. Assume that the body is in the far field, and it is 0.2 m thick. Why is this an unlikely situation? (Let $\sigma = 0.2/\Omega\text{-m}$ and $\varepsilon_r = 81$.)

To use (1.63), the shield thickness must be much less than the skin depth and the shield's impedance must be less than that of air (377 Ω):

$$\frac{d}{\delta} \ll 1, \; \eta \approx \sqrt{\frac{\omega\mu}{2\sigma}}(1+j)$$

Obviously, the human body is an excellent conductor at 60 Hz,

$$\frac{\sigma}{\omega\varepsilon} \approx 7.4 \times 10^5 \gg 1$$

so its skin depth for plane waves is

$$\delta = \frac{1}{\sqrt{\pi f \mu \sigma}} = \frac{1}{\sqrt{\pi(60)(4\pi \times 10^{-7})0.2}} \approx 150\,\text{m}$$

This skin depth is much, much greater than the body thickness. The impedance of the body at 60 Hz is also much less than the impedance of free space:

$$\eta \approx \sqrt{\frac{\omega\mu}{2\sigma}}(1+j) \approx 0.034(1+j)$$

Therefore, the ratio of the transmitted to incident 60 Hz plane wave signal is

$$\frac{E_t}{E_i} \approx \frac{2\sqrt{2}}{377\sigma d} \approx 0.19$$

However, this result has little value since it is unlikely that the human body is in the presence of a 60 Hz *plane wave*. The free-space wavelength corresponding to 60 Hz is 5 Mm. Even if the body was in the far field of a 60 Hz signal, the magnitude of the signal is likely to be minuscule. The plane wave shielding expressions must be carefully used in the near field, which includes using the proper incident wave impedance.

1.8 Thick Shields and Absorption Loss

Show that the expression for the *magnitude* of the electric field directly inside a thick (much greater than the skin depth), good-conducting nonmagnetic shield is

$$E_{di} \approx 2E_i\sqrt{\frac{\omega\varepsilon_o}{\sigma}}$$

where E_i is the amplitude of the plane wave impinging normally upon the slab of thickness d and conductivity σ. Assume that the slab is so thick that the reflected wave from the second interface (shield-to-air) is negligible by the time it returns to the first interface (air-to-shield). In other words, the second interface is "decoupled" from the first interface.

When the shield is thick, the wave reflected off the second interface (the metal-to-air interface) that returns to the first interface (the air-to-metal interface) may be neglected. (Obviously, it follows that all other consequent reflections may also be neglected.) Thick shields are so absorbent that the amplitude of this reflected wave is negligible after traveling a round-trip distance of $2d$. The transmission coefficient may be used at the first interface to determine the amplitude of the electric field on the metal side of the first interface. As with thin good-conducting shields,

$$\eta \ll \eta_o \quad \text{and} \quad \eta \approx \sqrt{\frac{\omega\mu}{2\sigma}}(1+j)$$

where $\mu = \mu_o$ for nonmagnetic conductors. The electric field directly inside the thick shield (not past the shield) is

$$E_{dis} = E_{is}\mathrm{T}_{am} = E_{is}\frac{2\eta}{\eta+\eta_o} \approx E_{is}\frac{2\eta}{\eta_o} = E_{is}\frac{2\sqrt{\dfrac{\omega\mu}{2\sigma}}(1+j)}{\eta_o} \tag{1.64}$$

$$\left|\frac{E_{dis}}{E_{is}}\right| = \left|\frac{2\sqrt{\dfrac{\omega\mu_o}{2\sigma}}(1+j)}{\sqrt{\dfrac{\mu_o}{\varepsilon_o}}}\right| = 2\sqrt{\frac{\omega\varepsilon_o}{\sigma}} \tag{1.65}$$

As expected, as the conductivity increases, this electric field decreases due to increased reflection (not absorption).

Since the thickness, d, of the slab is several skin depths thick, it is fairly easy to obtain a very good approximation for the electric field transmitted past the entire shield. If the air-to-metal interface of the slab is located in the xy plane, the field while passing through the shield in the $+z$ direction attenuates exponentially as $e^{-\alpha z}$:

$$|E_s| = |C_s e^{-\gamma z}| = |C_s e^{-(\alpha+j\beta)z}| = |C_s e^{-\alpha z} e^{-j\beta z}| = |C_s| e^{-\alpha z}$$

where $\alpha = 1/\delta$ for good conductors, $|e^{-jx}| = 1$, and the variable C_s is a phasor. Once the field reaches the metal-to-air interface, the transmission through it is determined by

$$\mathrm{T}_{ma} = \frac{2\eta_o}{\eta_o + \eta}$$

(For good metals, this transmission coefficient is about 2.) Thus,

$$\begin{aligned} |E_{ts}| &= \left|E_{is}\mathrm{T}_{am}e^{-\frac{d}{\delta}}\mathrm{T}_{ma}\right| = \left|E_{is}\left(\frac{2\eta}{\eta+\eta_o}\right)e^{-\frac{d}{\delta}}\left(\frac{2\eta_o}{\eta_o+\eta}\right)\right| \\ &\approx \left|E_{is}\left(\frac{2\eta}{\eta_o}\right)e^{-\frac{d}{\delta}}\left(\frac{2\eta_o}{\eta_o}\right)\right| = \left|4E_{is}e^{-\frac{d}{\delta}}\frac{\eta}{\eta_o}\right| = \left|4E_{is}e^{-\frac{d}{\delta}}\sqrt{\frac{\omega\varepsilon_o}{2\sigma}}(1+j)\right| \end{aligned} \tag{1.66}$$

$$\left|\frac{E_{ts}}{E_{is}}\right| = 4e^{-\frac{d}{\delta}}\sqrt{\frac{\omega\varepsilon_o}{\sigma}} \tag{1.67}$$

This approach can also be used for multilayer shields. However, each layer must be several skin depths thick and the layers must be in direct contact, with no air or other low-loss medium between them. As a check on (1.67), this result will be obtained starting with the exact expression for a single-layer shield, (1.59):

$$\left|\frac{E_{is}}{E_{ts}}\right| = \left|\frac{(\eta_o+\eta)^2}{4\eta_o\eta}\left[1-\left(\frac{\eta_o-\eta}{\eta_o+\eta}\right)^2 e^{-2\frac{d}{\delta}}e^{-j2\frac{d}{\delta}}\right]e^{\frac{d}{\delta}}e^{j\frac{d}{\delta}}e^{-j\beta_o d}\right|$$

$$\approx \left|\frac{\eta_o^2}{4\eta_o\eta}\left[1-\left(\frac{\eta_o}{\eta_o}\right)^2 e^{-2\frac{d}{\delta}}e^{-j2\frac{d}{\delta}}\right]e^{\frac{d}{\delta}}\right| \approx \left|\frac{\eta_o}{4\eta}\left[e^{\frac{d}{\delta}} - e^{-\frac{d}{\delta}}e^{-j2\frac{d}{\delta}}\right]\right| \approx \left|\frac{\eta_o}{4\eta}e^{\frac{d}{\delta}}\right|$$

since $e^{d/\delta} \gg e^{-d/\delta}$ for a thick shield.

In Mathcad 1.2, the exact expression is used to determine the shielding effectiveness of two copper shields of different thicknesses. As expected, the shielding effectiveness is extremely large. Usually, openings in the shield prevent real shields from obtaining such large theoretical shielding values.

$\mu_o := 4\cdot\pi\cdot10^{-7}$ $\varepsilon_o := 8.854\cdot10^{-12}$ $\sigma_o := 5.8\cdot10^{7}$ $\mu_r := 1$ $\sigma_r := 1$ $\eta_o := \sqrt{\frac{\mu_o}{\varepsilon_o}}$ $j := \sqrt{-1}$

$x := 40, 40.1 .. 90$

$$f(x) := \left(x + 1 - 10\cdot\text{floor}\left(\frac{x}{10}\right)\right)\cdot 10^{\text{floor}\left(\frac{x}{10}\right)}$$

$$\delta(f) := \frac{1}{\sqrt{\pi\cdot f\cdot\mu_r\cdot\mu_o\cdot\sigma_r\cdot\sigma_o}} \qquad \eta(f) := \sqrt{\frac{2\cdot\pi\cdot f\cdot\mu_r\cdot\mu_o}{2\cdot\sigma_r\cdot\sigma_o}}\cdot(1+j) \qquad \beta_o(f) := \frac{2\cdot\pi}{\left(\frac{3\cdot10^{8}}{f}\right)}$$

$$SE_{dB}(f,d) := 20\cdot\log\left[\left|\frac{(\eta_o+\eta(f))^2}{4\cdot\eta_o\cdot\eta(f)}\cdot\left[1-\left(\frac{\eta_o-\eta(f)}{\eta_o+\eta(f)}\right)^2\cdot e^{-2\cdot\frac{d}{\delta(f)}}\cdot e^{-j\cdot2\cdot\frac{d}{\delta(f)}}\right]\cdot e^{\frac{d}{\delta(f)}}\cdot e^{j\cdot\frac{d}{\delta(f)}}\cdot e^{-j\cdot\beta_o(f)\cdot d}\right|\right]$$

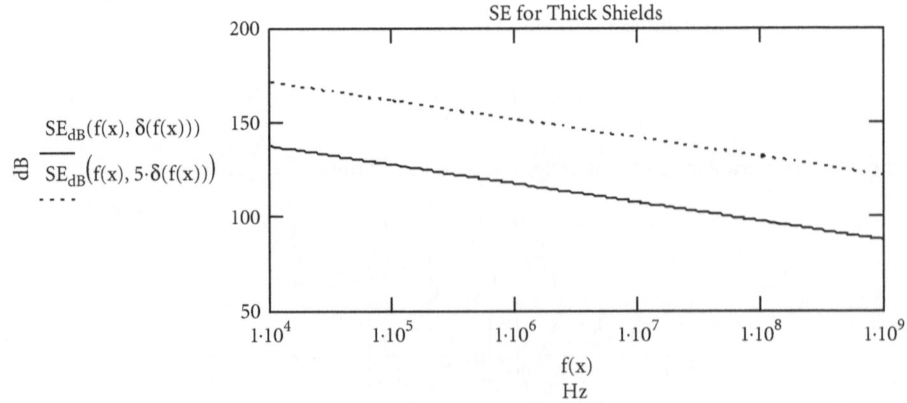

MATHCAD 1.2 SE as a function of the frequency for a shield one skin depth and five skin depths thick.

1.9 Skin Depth

When is the commonly used skin depth expression

$$\delta = \frac{1}{\sqrt{\pi f \mu \sigma}}$$

valid? Derive the general expression for the skin depth

$$\delta = \frac{1}{\omega \sqrt{\frac{\mu\varepsilon}{2}\left[\sqrt{1+\left(\frac{\sigma}{\omega\varepsilon}\right)^2}-1\right]}} \tag{1.68}$$

from the expression for the propagation constant. [Reilly]

When a plane wave passes through a material, its amplitude is attenuated exponentially as a function of the skin depth. The commonly used skin depth equation is valid for plane waves inside good conductors:

$$\frac{\sigma}{\omega\varepsilon} \gg 1$$

Under this condition, the general skin depth equation reduces to

$$\delta = \frac{1}{\omega \sqrt{\frac{\mu\varepsilon}{2}\left[\sqrt{1+\left(\frac{\sigma}{\omega\varepsilon}\right)^2}-1\right]}} \approx \frac{1}{\omega \sqrt{\frac{\mu\varepsilon}{2}\left[\left(\frac{\sigma}{\omega\varepsilon}\right)-1\right]}} \approx \frac{1}{\omega\sqrt{\frac{\mu\varepsilon}{2}\frac{\sigma}{\omega\varepsilon}}} = \frac{1}{\sqrt{\pi f \mu \sigma}} \tag{1.69}$$

The propagation constant for a plane wave is defined as

$$\gamma = \alpha + j\beta = \sqrt{j\omega\mu(\sigma + j\omega\varepsilon)}$$

The real part of the propagation constant, α, is referred to as the attenuation constant. A plane wave traveling in the positive z direction through a medium decays or attenuates in an exponential fashion:

$$E_o e^{-\alpha z} \tag{1.70}$$

where E_o is the initial amplitude of the wave. The reciprocal of α is defined as the skin depth:[4]

$$\delta = \frac{1}{\alpha} \tag{1.71}$$

It is somewhat analogous to the time constant for *RC* and *RL* circuits except that the skin depth is decay with respect to distance while the time constant is decay with respect to time.

[4]This expression was described by Oliver Heaviside in 1888!

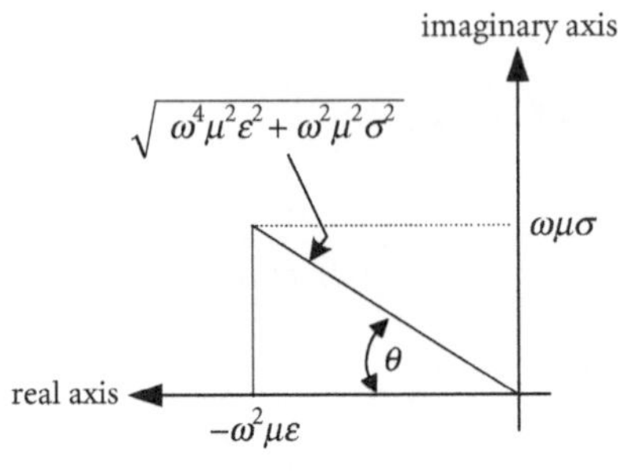

FIGURE 1.3 Relationship between θ and the electrical parameters defined in 1.72.

The real part of the propagation constant is determined by first placing the argument of the square root in polar form so that the square root of the complex function can be determined:

$$\gamma=\sqrt{-\omega^2\mu\varepsilon+j\omega\mu\sigma}=\sqrt{\sqrt{\omega^4\mu^2\varepsilon^2+\omega^2\mu^2\sigma^2}\angle\theta}\quad\text{where }\theta=\tan^{-1}\left(\frac{\omega\mu\sigma}{-\omega^2\mu\varepsilon}\right)$$
$$=\sqrt{\sqrt{\omega^4\mu^2\varepsilon^2+\omega^2\mu^2\sigma^2}}\angle\frac{\theta}{2}=\sqrt{\sqrt{\omega^4\mu^2\varepsilon^2+\omega^2\mu^2\sigma^2}}\left[\cos\left(\frac{\theta}{2}\right)+j\sin\left(\frac{\theta}{2}\right)\right]\tag{1.72}$$

The relationship between the cosine of θ and the electrical parameters follows directly from the definition of tangent. Referring to Figure 1.3,

$$\cos\theta=\frac{-\omega^2\mu\varepsilon}{\sqrt{\omega^4\mu^2\varepsilon^2+\omega^2\mu^2\sigma^2}}$$

The half-angle formula for cosine is

$$\cos\left(\frac{\theta}{2}\right)=\sqrt{\frac{1+\cos\theta}{2}}=\sqrt{\frac{1+\frac{-\omega^2\mu\varepsilon}{\sqrt{\omega^4\mu^2\varepsilon^2+\omega^2\mu^2\sigma^2}}}{2}}$$

The real part of the propagation constant is equal to the attenuation constant:

$$\alpha=\sqrt{\sqrt{\omega^4\mu^2\varepsilon^2+\omega^2\mu^2\sigma^2}}\sqrt{\frac{1+\frac{-\omega^2\mu\varepsilon}{\sqrt{\omega^4\mu^2\varepsilon^2+\omega^2\mu^2\sigma^2}}}{2}}=\omega\sqrt{\frac{\mu\varepsilon}{2}\left[\sqrt{1+\left(\frac{\sigma}{\omega\varepsilon}\right)^2}-1\right]}\tag{1.73}$$

The reciprocal of α is equal to the given general expression for the skin depth.

1.10 Skin Depth for Good Insulators

Determine an approximate expression for the skin depth for good insulators from (1.68).

For good insulators,

$$\frac{\sigma}{\omega\varepsilon}\ll 1$$

Since

$$\sqrt{1+x^2}-1\approx 1+\frac{1}{2}x^2-1=\frac{1}{2}x^2\quad\text{if } x \text{ is small}$$

(the binomial series was used), an approximate expression for the skin depth for good insulators is immediately obtained:

$$\delta = \frac{1}{\omega\sqrt{\frac{\mu\varepsilon}{2}\left[\sqrt{1+\left(\frac{\sigma}{\omega\varepsilon}\right)^2}-1\right]}} \approx \frac{1}{\omega\sqrt{\frac{\mu\varepsilon}{2}\frac{1}{2}\left(\frac{\sigma}{\omega\varepsilon}\right)^2}} = \frac{2}{\sigma}\sqrt{\frac{\varepsilon}{\mu}} \tag{1.74}$$

Similar to the good-conductor approximation, the skin depth for good insulators decreases with conductivity. However, (1.74) is not a function of the frequency but a function of the permittivity. The effects of the permittivity are masked by the conductivity for good conductors. The skin depth for good conductors is much smaller than the skin depth for good insulators.

1.11 Skin Depth for Several Good Metals

On the same set of axes, plot the skin depth of silver, copper, and aluminum from 60 Hz to 1 GHz. [Harper, '72; Browne; Weeks, '64; Baboian; Chatterton]

Before plotting the skin depth of these materials vs. frequency, a few important characteristics of several good conductors will be presented. The conductivity of these and other good metals is fairly independent of frequency over the stated frequency range, although the measurement of the conductivity at higher frequencies is difficult because of dielectric and magnetic effects.[5] For many other nonmetals such as human fat and soil, the conductivity (and permittivity) is *not* constant with frequency. Frequently, only one value is provided in text and reference books for the conductivity of metals *and* nonmetals, implying to the novice that it is constant with frequency. A constant conductivity should not be assumed unless the material is a good metal; however, the conductivity of good metals is affected by the temperature, chemical composition, and heat treatments and stresses during fabrication. For example, annealing copper, the heating of copper above a critical temperature and then slowly cooling it, will affect its conductivity by a few percent.

A brief listing of the properties of several common nonmagnetic ($\mu_r = 1$) metals follow. The relative conductivities provided, σ_r, are relative to annealed copper, which has a conductivity of 5.8×10^7 1/Ω-m at 20°C.

Aluminum: inexpensive, light weight (30% that of copper), soft but brittle at high temperatures, low melting point, should not be used if arcing is expected, good strength, easily formed and milled, tendency to corrode, oxide film thin but tough, galvanic action with metals such as copper and brass in salty environments, poor solderability, connection by bolting, crimping, or welding

applications: antennas, heat sinks, large power conductors (electrical buses)

pure: $\sigma_r = 0.65$

hard-drawn: $\sigma_r = 0.61$

soft-annealed: $\sigma_r = 0.45$–0.50

heat-treated: $\sigma_r = 0.30$–0.45

Brass: low conductivity, high-tensile strength but stress-corrosion cracking can occur, greater fatigue lifetime than copper

applications: lugs between steel and copper, bushings, chassis, microwave ground planes, flexible straps and gaskets

[5]For frequencies less than about 1,000 GHz, the conductivity of copper is relatively constant and equal to its low-frequency value. At optical frequencies, copper's conductivity is many orders of magnitude less than its low-frequency value. When the period of the wave approaches the mean time between collisions of the conduction electrons, the low-frequency conductivity value should not be used.

copper and 9.1% zinc: $\sigma_r = 0.45$
copper and 34.2% zinc: $\sigma_r = 0.26$

Bronze: excellent resistance to corrosion, strong
applications: high-quality antenna wire, trolley wire, plating of microwave filter cavities
copper and 12% tin: $\sigma_r = 0.26$

Copper: excellent conductor, machinable but harder than aluminum, low tensile strength, brittleness with repeated bending, excellent solderability, low contact resistance when oxide not present, oxide formation reduced when plated with silver
applications: circuit board traces, cable conductors, bulk current-carrying conductors
hard-drawn: $\sigma_r = 0.97$
soft-drawn: $\sigma_r = 0.98$
annealed: $\sigma_r = 1.00$

Gold: expensive, very soft, chemically inert (noncorrosive), self-healing (diffuses with time)
applications: plating over conductors such as edge connectors, high-frequency applications in corrosive environments, fine wires in IC's
$\sigma_r = 0.70$

Silver: highest conductivity of any metal, relatively soft and ductile, low melting point, expensive, readily migrates (whisker and dendrite formation that can cause unexpected leakage currents)
applications: lower current and voltage switch contacts, plating material
hard-drawn: $\sigma_r = 0.985$
annealed: $\sigma_r = 1.09$

$\mu_o := 4\cdot\pi\cdot 10^{-7} \qquad \sigma_o := 5.8\cdot 10^{7}$

$x := 10, 11 .. 90$

$$f(x) := \left(x + 1 - 10\cdot\text{floor}\left(\frac{x}{10}\right)\right)\cdot 10^{\text{floor}\left(\frac{x}{10}\right)}$$

$$\delta(f, \sigma_r) := \frac{1}{\sqrt{\pi\cdot f\cdot \mu_o\cdot \sigma_r\cdot \sigma_o}}$$

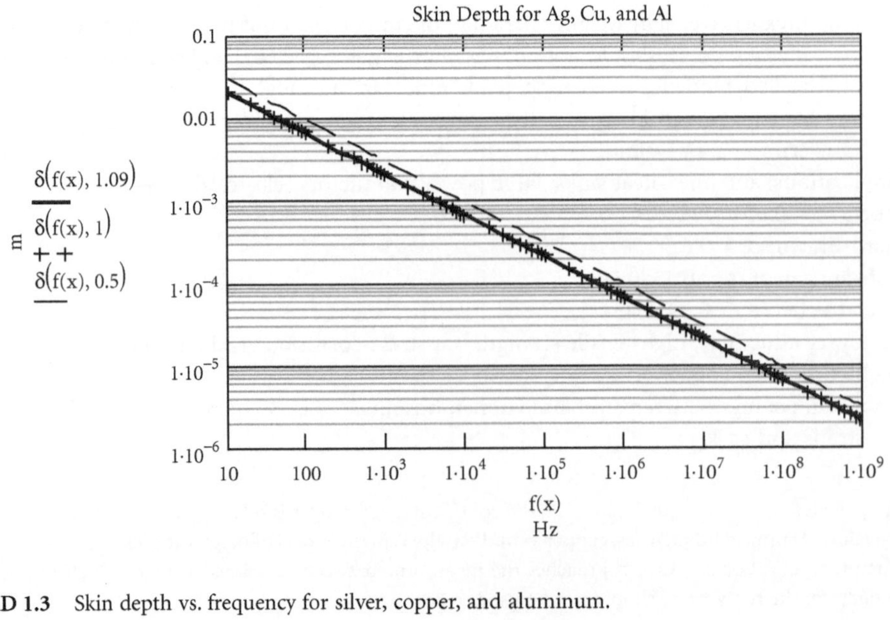

MATHCAD 1.3 Skin depth vs. frequency for silver, copper, and aluminum.

Notice that the conductivity of gold is less than copper and silver. Also, the conductivity of silver is often greater than copper. In some situations, a metal object is coated with silver to increase its conductivity. However, silver can produce a sulfide layer, which can counter the effects of its higher conductivity. Unless the silver coating is very thick compared to the skin depth or a thin gold coating or other protective layer is applied to the silver, the effective resistance of the silver-coated object at high frequencies may not be significantly different from a copper object.

The permittivity of metals is difficult to measure and is usually assumed to be equal to that of free space. Over this frequency range, the conduction current is much greater than the displacement current. For aluminum (hard-drawn) at 1 GHz

$$\frac{\sigma}{\omega\varepsilon} = \frac{0.61(5.8\times10^{7})}{2\pi(10^{9})(8.854\times10^{-12})} = 6.4\times10^{8} \gg 1$$

so that the standard equation for the skin depth may be used:

$$\delta = \frac{1}{\sqrt{\pi f \mu \sigma}}$$

The variation of the skin depth with frequency for annealed silver, annealed copper, and soft-annealed aluminum is given in Mathcad 1.3. At 60 Hz the skin depth of these metals is in the centimeter range, while at 1 GHz it is in the micrometer range.

1.12 Complex Permittivity and RF through Human Fat

Determine the necessary thickness of body fat to absorb a 27 MHz plane wave. Determine the reflection loss. The electrical properties of the fat at this frequency are given in (1.75). [Gandhi, '90; IPCS, '92]

$$\varepsilon^{*} = \varepsilon_{o}(\varepsilon' - j\varepsilon''),\ \ \varepsilon' = 20,\ \ \varepsilon'' = 7.23 \text{ to } 28.65 \tag{1.75}$$

The electrical properties of materials are often given in terms of a complex permittivity. The definition for the complex permittivity can be readily obtained from one of Maxwell's equations:

$$\nabla\times\vec{H} = \vec{J} + \frac{\partial\vec{D}}{\partial t} \tag{1.76}$$

In the frequency domain, valid for sinusoidal steady-state conditions, this equation is equal to

$$\nabla\times\vec{H}_s = \vec{J}_s + j\omega\vec{D}_s = \sigma\vec{E}_s + j\omega\varepsilon\vec{E}_s \tag{1.77}$$

where $\vec{J}_s$ represents the conduction current and $j\omega\vec{D}_s$ represents the displacement current. For good insulators, the conduction current should be small compared to the displacement current:

$$\sigma\vec{E}_s \ll \omega\varepsilon\vec{E}_s \quad \text{or} \quad \frac{\sigma}{\omega\varepsilon} \ll 1 \tag{1.78}$$

For good conductors, the conduction current should be large compared to the displacement current:

$$\sigma\vec{E}_s \gg \omega\varepsilon\vec{E}_s \quad \text{or} \quad \frac{\sigma}{\omega\varepsilon} \gg 1 \tag{1.79}$$

Many materials, such as human flesh, are good conductors at low frequencies (e.g., 60 Hz) but poor conductors at high frequencies (e.g., 2 GHz). The concept of complex permittivity stems directly from (1.77):

$$\nabla \times \vec{H}_s = j\omega\left(\varepsilon + \frac{\sigma}{j\omega}\right)\vec{E}_s = j\omega\left(\varepsilon - \frac{j\sigma}{\omega}\right)\vec{E}_s = j\omega\varepsilon_o\left(\varepsilon_r - \frac{j\sigma}{\omega\varepsilon_o}\right)\vec{E}_s$$

As previously stated, for insulators the displacement current dominates. The right side of the equation was rearranged into a form such that it "appears" like displacement current. The complex permittivity is defined as

$$\varepsilon^* = \varepsilon_o\left(\varepsilon_r - \frac{j\sigma}{\omega\varepsilon_o}\right) = \varepsilon_o(\varepsilon' - j\varepsilon'') \quad \Rightarrow \varepsilon' = \varepsilon_r = k, \ \varepsilon'' = \frac{\sigma}{\omega\varepsilon_o} \tag{1.80}$$

where k is the dielectric constant of the medium. The conductivity in this expression is sometimes referred to as the effective conductivity, since it may include several different loss mechanisms.

The complex permittivity can be defined in other ways. For example, in some books the complex permittivity includes the free-space term, ε_o, directly

$$\varepsilon^* = \varepsilon' - j\varepsilon'' \tag{1.81}$$

while other books will assume that the imaginary portion of the complex permittivity does not include the standard conductivity, σ. The relative size of the terms might provide a clue to which definition is being used.

For good insulators, the complex permittivity is mostly real since the conductivity is small compared to $\omega\varepsilon_o$. The losses or nonideal properties of the insulator are a function of the imaginary portion of the complex permittivity. Other losses associated with a real insulator, besides those due to conduction (i.e., σ), can also be incorporated into the imaginary portion of the complex permittivity. Additional loss mechanisms include phase delays and drag associated with the molecular dipole orientation in the insulation (ideally, the electric dipoles in the material respond immediately to and track the applied electric field with zero molecular drag loss). These losses are difficult to separate in a dielectric and are usually just combined into one dielectric loss. The losses are frequency dependent for most materials. The electrical properties of a material are frequently given in terms of the effective loss tangent, the ratio of the imaginary portion to the real portion of the complex permittivity:

$$\tan\delta_{eff} = \frac{\varepsilon''}{\varepsilon'} = \frac{\sigma}{\omega\varepsilon_r\varepsilon_o} \tag{1.82}$$

where δ_{eff} is the effective loss angle. As the losses increase, the loss tangent increases. This ratio is also called the dissipation factor.[6]

Given the complex permittivity of human fat in (1.75), the relative permittivity and conductivity of human fat at 27 MHz are

$$\varepsilon' = \varepsilon_r = 20, \ \sigma = \omega\varepsilon_o\varepsilon'' \approx 0.011 \text{ to } 0.043 \ 1/\Omega\text{-m}$$

[6]The dissipation factor is also defined as the ratio of the energy dissipated to the energy stored in an element and as the inverse of the Q of an element. Therefore, a dissipation factor can also be defined for real inductors. The loss factor is defined as the ratio of the average power loss to the peak-load loss. It is also sometimes defined as the product of the dielectric constant and the tangent of the loss angle, δ_{eff}.

Notice that

$$\frac{\sigma}{\omega\varepsilon} \approx 0.37 \text{ to } 1.4$$

so human fat at this frequency is neither a good conductor nor a good insulator. The general definition for the skin depth should be used to determine the necessary thickness to absorb the electric and magnetic fields:

$$\delta = \frac{1}{\omega\sqrt{\frac{\mu\varepsilon}{2}\left[\sqrt{1+\left(\frac{\sigma}{\omega\varepsilon}\right)^2}-1\right]}} \approx 0.66 \text{ to } 2.2 \text{ m}$$

For human fat to absorb effectively a 27 MHz signal, it should be several skin depths thick, which is obviously much too large at this frequency. However, if the reflection from the fat is large, then the electric field transmitted inside the body will be small, and the absorption losses will be less important.

The reflection coefficient for electric fields is

$$\rho_L = \frac{\eta_L - \eta_o}{\eta_L + \eta_o}$$

where η_L is the intrinsic impedance of the body fat. This reflection coefficient will only provide an estimate of the strength of the total reflection since it does not include reflected waves originating from multiple reflections inside the body or reflections from the fat-to-air interface. It is also only the reflection for the electric fields not the magnetic fields. The calculation of the intrinsic impedance of human fat and the magnitude of the reflection coefficient is shown in Mathcad 1.4. These results indicate that a large fraction of the incident signal (between 0.65 and 0.74) will be reflected. However, the transmitted signal is not negligible since the transmission coefficient is between 0.29 and 0.36. Thus, it appears that human fat is not a very good shield against 27 MHz signals.

If the concept of complex permittivity of a material is still confusing or difficult to swallow, it may be helpful first to model a material as a parallel *RC* circuit. If a sinusoidal voltage source $v(t) = A\cos(\omega t)$ is applied across the *RC* circuit model for the material, the current out of this source is

$$i(t) = \frac{v(t)}{R} + C\frac{dv(t)}{dt}$$

$$\mu_o := 4\cdot\pi\cdot 10^{-7} \qquad \varepsilon_o := 8.854\cdot 10^{-12} \qquad \omega := 2\cdot\pi\cdot 27\cdot 10^6 \qquad \eta_o := \sqrt{\frac{\mu_o}{\varepsilon_o}} \qquad j := \sqrt{-1}$$

$$\eta_1 := \sqrt{\frac{j\cdot\omega\cdot\mu_o}{0.011 + j\cdot\omega\cdot 20\cdot\varepsilon_o}} \qquad \eta_2 := \sqrt{\frac{j\cdot\omega\cdot\mu_o}{0.043 + j\cdot\omega\cdot 20\cdot\varepsilon_o}}$$

$$\eta_1 = 80.378 + 14.253j \qquad \eta_2 = 56.532 + 29.467j$$

$$\left|\frac{\eta_1 - \eta_o}{\eta_1 + \eta_o}\right| = 0.649 \qquad \left|\frac{\eta_2 - \eta_o}{\eta_2 + \eta_o}\right| = 0.74$$

$$\left|\frac{2\cdot\eta_1}{\eta_1 + \eta_o}\right| = 0.357 \qquad \left|\frac{2\cdot\eta_2}{\eta_2 + \eta_o}\right| = 0.294$$

MATHCAD 1.4 Reflection and transmission coefficients for an air-fat interface at 27 MHz.

In the frequency domain, this equation is equal to

$$I_s = \frac{V_s}{R} + Cj\omega V_s = j\omega C\left(1 - \frac{j}{\omega RC}\right)V_s \tag{1.83}$$

Second, imagine that the material is placed between parallel plates so that the ubiquitous definitions for the capacitance and resistance for two parallel plates of area *A*, separated by a distance of *d*,

$$C = \frac{\varepsilon_r \varepsilon_o A}{d}, \quad R = \frac{d}{\sigma A}$$

can be used. Substituting these definitions into (1.83),

$$I_s = j\omega \frac{\varepsilon_r \varepsilon_o A}{d}\left(1 - \frac{j}{\omega \frac{d}{\sigma A}\frac{\varepsilon_r \varepsilon_o A}{d}}\right)V_s = j\omega \frac{\varepsilon_o A}{d}\left(\varepsilon_r - \frac{j\sigma}{\omega \varepsilon_o}\right)V_s$$

Finally, the expressions for the uniform current density, $J_s = I_s/A$, and uniform electric field, $E_s = V_s/d$, between the parallel plates are used:

$$J_s = \frac{I_s}{A} = j\omega\varepsilon_o\left(\varepsilon_r - \frac{j\sigma}{\omega\varepsilon_o}\right)\frac{V_s}{d} = j\omega\varepsilon_o\left(\varepsilon_r - \frac{j\sigma}{\omega\varepsilon_o}\right)E_s = j\omega\varepsilon^* E_s$$

By retracing a few of the steps, it is obvious that the real part of the complex permittivity essentially corresponds to *C* while the imaginary part corresponds to *R*.

In adult humans, RF energy is absorbed most efficiently around 35 MHz if the person is grounded and around 70 MHz if the person is insulated from ground (or floating).[7] Various parts of the body can also resonate based on the body part size (similar to resonant cavities), sometimes referred to as partial body resonances. For example, the resonant frequency of an adult human head is around 400 MHz while for a baby head it is around 700 MHz. At these higher resonant frequencies, the energy is absorbed most efficiently, resulting in a high degree of heating or "hot spots."

1.13 Microwaves through Human Fat

Determine the necessary thickness of body fat to absorb a 2.45 GHz plane wave. Determine the reflection loss. The electrical properties of the fat at this frequency are given in (1.84). [Gandhi, 1990]

$$\varepsilon = \varepsilon_o(\varepsilon' - j\varepsilon''), \ \varepsilon' = 5.5, \ \varepsilon'' = 0.71 \text{ to } 1.56 \tag{1.84}$$

The dielectric constant and conductivity of human tissues, including muscle and fat, are strongly a function of the frequency. From the complex permittivity, the relative permittivity and conductivity of human fat at 2.45 GHz can be obtained:

$$\varepsilon' = \varepsilon_r = 5.5, \ \sigma = \omega\varepsilon_o\varepsilon'' \approx 0.097 \text{ to } 0.21 \ 1/\Omega\text{-m}$$

[7] A mouse absorbs energy very efficiently near 2.45 GHz, the frequency of a microwave oven.

Notice that

$$\frac{\sigma}{\omega\varepsilon} \approx 0.13 \text{ to } 0.28$$

so human fat at this frequency is not necessarily a good conductor or good insulator. The general definition for the skin depth should be used to determine the necessary thickness to absorb the electric and magnetic fields:

$$\delta = \frac{1}{\omega\sqrt{\frac{\mu\varepsilon}{2}\left[\sqrt{1+\left(\frac{\sigma}{\omega\varepsilon}\right)^2}-1\right]}} = 0.06 \text{ to } 0.13 \text{ m}$$

Compared to 27 MHz, the skin depth at 2.45 GHz is about an order of magnitude smaller. To absorb the electric and magnetic fields, the fat should be several skin depths thick. Obviously, the thickness of body fat is much less than several skin depths. As at 27 MHz, human fat is not a very good absorber of electromagnetic fields. However, if the reflection from the fat is large, then the electric field transmitted inside the body will be small, and the absorption losses will be less important.

The calculation of the intrinsic impedance of human fat and the magnitude of the reflection and transmission coefficients is given in Mathcad 1.5. These results indicate that a large fraction of the incident signal (about 0.4) will be reflected. However, the transmitted signal is not negligible since the magnitude of the transmission coefficient is about 0.6. Thus, human fat is not a very good shield against 2.45 GHz signals.

The reason the electric field was emphasized in this problem rather than the magnetic field is that only the internal electric field inside a body produces heat inside the body. The magnetic field does not transfer energy to the body, but it does "produce" an electric field if it is time varying. The time-average power transferred to a body is given by

$$P_{avg} = \frac{1}{2}\iiint_{vol} \sigma \vec{E}_s \cdot \vec{E}_s^* dv = \frac{1}{2}\iiint_{vol} \sigma \left|\vec{E}_s\right|^2 dv \tag{1.85}$$

This integral expression corresponds to ohmic (resistive) power dissipation, and it is the power used in moving charges in the body ("friction" and collisions). This equation is valid for sinusoidal signals in

$\mu_o := 4\cdot\pi\cdot10^{-7}$ $\quad \varepsilon_o := 8.854\cdot10^{-12}$ $\quad \omega := 2\cdot\pi\cdot2.45\cdot10^9$ $\quad \eta_o := \sqrt{\frac{\mu_o}{\varepsilon_o}}$ $\quad j := \sqrt{-1}$

$\eta_1 := \sqrt{\frac{j\cdot\omega\cdot\mu_o}{0.097 + j\cdot\omega\cdot5.5\cdot\varepsilon_o}}$ $\qquad \eta_2 := \sqrt{\frac{j\cdot\omega\cdot\mu_o}{0.21 + j\cdot\omega\cdot5.5\cdot\varepsilon_o}}$

$\eta_1 = 1.596\times10^2 + 10.286j$ $\qquad \eta_2 = 1.562\times10^2 + 21.461j$

$\left|\frac{\eta_1 - \eta_o}{\eta_1 + \eta_o}\right| = 0.405$ $\qquad \left|\frac{\eta_2 - \eta_o}{\eta_2 + \eta_o}\right| = 0.416$

$\left|\frac{2\cdot\eta_1}{\eta_1 + \eta_o}\right| = 0.596$ $\qquad \left|\frac{2\cdot\eta_2}{\eta_2 + \eta_o}\right| = 0.591$

MATHCAD 1.5 Reflection and transmission coefficients for an air-fat interface at 2.45 GHz.

steady state and involves the integration of the square of the magnitude of the electric field (in phasor form) over the given volume of the body. Notice that the power absorbed is directly proportional to the conductivity of the body. Because generally the skin depth is smaller and conductivity is larger at 2.45 GHz compared to 27 MHz for human fat, more energy of the plane wave will be absorbed near the surface at higher frequencies. Deep heating of tissue is thus more difficult at 2.45 GHz because of the smaller skin depth.

It is interesting to note that at 37°C the dielectric constant of tissues such as liver, kidney, and brain decrease from values of over 1,000 at 100 kHz to values near 40 at 10 GHz! (The dielectric constant of distilled water is around 80.) The conductivity increases from values around 0.2 at 100 kHz to 9 at 10 GHz.

1.14 Table of Dielectric Constants and Loss Tangents

Provide a comprehensive table of the dielectric constants and loss tangents vs. frequency for a number of materials of electrical engineering interest. [Hippel; Fink; Howard W. Sams; Johnson, '61; Lo; Metaxas; Plastics; Gandhi, '90; Ceramic; Uitert; Snelling; Stuchly; Reilly; product data sheets]

The dielectric constant (or relative permittivity) for mediums is relative to air or free space, which has a dielectric constant of one. Virtually all mediums have a dielectric constant greater than one. The loss tangent (or dissipation factor) is a measure of the loss in a medium, and it is a function of the frequency, conductivity, and relative permittivity, ε_r, or dielectric constant:

$$\tan\delta_{eff} = \frac{\sigma}{\omega\varepsilon_r\varepsilon_o} \quad \text{where } \varepsilon_o = 8.854\times10^{-12}\ \text{F/m} \tag{1.86}$$

Actually, the conductivity term in this expression is an effective conductivity representing all the various loss terms in the medium. Sometimes references use the complex permittivity defined as

$$\varepsilon^* = \varepsilon_o(\varepsilon' - j\varepsilon'') \tag{1.87}$$

when defining the loss tangent:

$$\tan\delta_{eff} = \frac{\varepsilon''}{\varepsilon'} \tag{1.88}$$

This definition could possibly be considered a more fundamental definition for the loss tangent. Notice that when

$$\tan\delta_{eff} = \frac{\sigma}{\omega\varepsilon_r\varepsilon_o} \gg 1 \tag{1.89}$$

the medium is considered a good conductor (or resistive). The variable δ_{eff} is the phase angle between the conduction and displacement currents. Some references state that when $\tan\delta_{eff}$ is greater than 100, the medium is a good conductor. When

$$\tan\delta_{eff} = \frac{\sigma}{\omega\varepsilon_r\varepsilon_o} \ll 1 \tag{1.90}$$

the medium is considered a good insulator (or capacitive). Some references state that when $\tan\delta_{eff}$ is less than 0.01, the medium is a good insulator. For values of the loss tangent between 100 and 1, the medium might be considered a poor conductor while for values between 1 and 0.01 the medium might be considered a poor insulator. Because the loss tangent is a function of the frequency and the losses can vary with frequency, a conductor's classification can change with frequency. For example, a material that is considered a good conductor at one frequency may become a poor conductor or even a poor insulator at another frequency.

Because in some references the loss or power factor is given instead of the loss tangent or dissipation factor, the definition for the loss factor will be provided. The loss or power factor is defined as $\sin\delta_{eff}$. The following trigonometric identities relate the loss factor to the dissipation factor:

$$\text{loss factor} = \sin\delta_{eff} = \frac{\tan\delta_{eff}}{\sqrt{1+\tan^2\delta_{eff}}} \tag{1.91}$$

$$\text{dissipation factor} = \tan\delta_{eff} = \frac{\sin\delta_{eff}}{\sqrt{1-\sin^2\delta_{eff}}} \tag{1.92}$$

When the material is a good insulator, δ_{eff} is small and, hence, $\sin\delta_{eff}$ is approximately equal to δ_{eff}. The dissipation factor, $\tan\delta_{eff} = \sin\delta_{eff}/\cos\delta_{eff}$, is also approximately equal to δ_{eff} when δ_{eff} is small ($\cos\delta_{eff}$ is approximately one for small δ_{eff}). Hence, for good insulators, the loss and dissipation factors are about the same:

$$\sin\delta_{eff} \approx \delta_{eff} \quad \text{and} \quad \tan\delta_{eff} \approx \delta_{eff} \tag{1.93}$$

Many different sources were used to compile the information contained in Table 1.7. The frequency corresponding to the relative permittivity, ε_r, (or dielectric constant) and loss tangent given in the table is obtained by referring to the corresponding heading frequency of the same type style (e.g., 1 kHz, *10 kHz*, **100 kHz**). Although most sources explicitly provided the relative permittivity and/or the loss tangent, in those cases where the conductivity and relative permittivity were provided, the loss tangent was determined using (1.86). Only a small fraction of the many different materials of engineering interest are tabulated. The materials are sometimes listed under their trade name (e.g., Plexiglass) and sometimes under their generic name (e.g., acrylic). Also, note that there are often many different variations of any type of material. For example, polyethylene can have a low, medium, or high density and the filler can be carbon black, carbon fiber, or a polymer. It is recommended that current commercial data sheets be consulted. Although not specifically given in this table, the dielectric constant of gases is a function of several factors including the density of the gas. As long as a large number of ions are not present, which is *not* true when the gas is breaking down, the dielectric constant is nearly that of free space (one). Although almost always not important, the dielectric constant of metals is usually assumed to be about one. The dielectric constant and loss tangent are sometimes a strong function of factors such as frequency, temperature, moisture content, field strength, field orientation, homogeneity, thickness, and impurities. (The conductivity of insulators is also often a function of the applied voltage.) Finally, although the conductivities of some highly insulating materials are listed, it is important to state that Ohm's law might not apply to these materials. In these cases, the conductivity values listed should only be used as a rough estimate. Unless otherwise noted, T = 24–26°C.

1.15 Loss in dB per Skin Depth

Provide the absorption loss in dB per skin depth for a plane wave normally incident to a flat shield. [Paul, '92(b)]

The effectiveness of a good-conducting shield is often given in dB in terms of the reflection loss, R_{dB}, absorption loss, A_{dB}, and multiple-reflection loss, MR_{dB}. The total loss in dB (or shielding effectiveness, SE_{dB}) of the shield is the sum of these losses in dB:

$$TOT_{dB} = R_{dB} + A_{dB} + MR_{dB} \tag{1.94}$$

TABLE 1.7 Relative Permittivity and Loss Tangent for Materials at Various Frequencies

Material		60 Hz *100 Hz*	1 kHz *10 kHz* **100 kHz**	1 MHz *10 MHz* **27 MHz**	40 MHz *70 MHz* **100 MHz**	1 GHz *2.5 GHz* **3 GHz**	9.4 GHz *10 GHz* **25 GHz**
Ceramics							
Alsimag 243	ε_r	*6.30*	*6.28*	6.22	**6.10**	**5.78**	*5.76*
	$\tan\delta_{eff}$	*0.00125*	*0.0004*	0.00037	**0.0003**	**0.0006**	*0.00085*
Alsimag 393	ε_r	*4.95*	*4.95*	4.95	**4.95**		*4.95*
	$\tan\delta_{eff}$	*0.0038*	*0.0016*	0.0010	**0.0010**		*0.00097*
Aluminum oxide/alumina	ε_r		8.2	8.2	**8.2**	8.2	*8.2*
(85% Al_2O_3)	$\tan\delta_{eff}$		0.0014	0.0009	**0.0009**	0.0014	*0.0019*
Aluminum oxide/alumina	ε_r		9.0	9.0	**9.0**	8.9	*8.9*
(96% Al_2O_3)	$\tan\delta_{eff}$		0.0011	0.0001	**0.0002**	0.0001	*0.0006*
Aluminum oxide/alumina	ε_r		10.1	10.1	**10.1**		*10.1*
(99.9% Al_2O_3)	$\tan\delta_{eff}$		0.00050	0.00004	**0.00006**		*0.00009*
Barium titanate	ε_r	*1240*	1200	1445		**600**	**100**
	$\tan\delta_{eff}$	*0.0360*	0.0130	0.015		**0.30**	**0.60**
Beryllium oxide/beryllia	ε_r		6.7	6.7	**6.6**	6.6	*6.6*
(99.5% BeO)	$\tan\delta_{eff}$		0.0010	0.0002	**0.0002**	0.0002	*0.0004*
Magnesium manganese	ε_r						12
ferrite	$\tan\delta_{eff}$						0.0008
Nickel zinc ferrite	ε_r		*50*				12.5
	$\tan\delta_{eff}$		*700*				0.001
Porcelein (dry process)	ε_r	*5.5*	5.36	5.08	**5.04**	**5.02**	*4.74*
	$\tan\delta_{eff}$	*0.022*	0.0140	0.0075	**0.0078**	**0.0098**	*0.0156*
Silicon nitride	ε_r						*5.50*
	$\tan\delta_{eff}$						*0.0030*
Steatite 410	ε_r	*5.77*	5.77	5.77	**5.77**	**5.7**	*5.7*
	$\tan\delta_{eff}$	*0.0055*	0.0030	0.0007	**0.0006**	**0.00089**	*0.0022*
Titania	ε_r	*100*	*100*	100	**100**		*90*
	$\tan\delta_{eff}$	*0.0023*	*0.00062*	0.0003	**0.00025**		*0.0020*
Food Items							
Bacon fat	ε_r					2.6, **2.5**	
	$\tan\delta_{eff}$					0.062, **0.052**	
Beef steak	ε_r		*24,400*	*50*		**40**	**15**
(bottom round)	$\tan\delta_{eff}$		*40.5*	*26*		**0.3**	**0.4**
Beef steak (0°C)	ε_r					4.4, **3.95**	
(frozen lean)	$\tan\delta_{eff}$					0.17, **0.076**	
Butter (35°C)	ε_r					**4.15**	
	$\tan\delta_{eff}$					**0.11**	
Potato raw	ε_r			*80*		65.1, **53.7**	
	$\tan\delta_{eff}$			*0.60*		0.30, **0.29**	
Turkey cooked	ε_r					**40**	
	$\tan\delta_{eff}$					**0.35**	
Suet	ε_r		750	14	**2.6**	**2.5**	**2.4**
	$\tan\delta_{eff}$		3	1.7	**0.15**	**0.07**	**0.05**

TABLE 1.7 Relative Permittivity and Loss Tangent for Materials at Various Frequencies (Continued)

Material		60 Hz *100 Hz*	1 kHz *10 kHz* **100 kHz**	1 MHz *10 MHz* **27 MHz**	40 MHz *70 MHz* **100 MHz**	1 GHz *2.5 GHz* **3 GHz**	9.4 GHz *10 GHz* **25 GHz**
Forest Products							
Balsawood	ε_r	*1.4*	1.4	1.37	**1.30**	**1.22**	*1.20*
(field ⊥ to grain)	$\tan\delta_{eff}$	*0.0050*	0.0040	0.012	**0.0135**	**0.0100**	*0.0083*
Board (22°C)	ε_r			*2.8*		**2.7**	
(230 g/m2 field ‖ to grain)	$\tan\delta_{eff}$			*0.11*		**0.11**	
Douglas fir plywood	ε_r	*2.1*	2.1	1.90		**1.7**	**1.5**
(field ⊥ to grain)	$\tan\delta_{eff}$	*0.012*	0.0105	0.0230		**0.0360**	**0.0220**
Mahogany	ε_r	*2.42*	2.40	2.25	**2.07**	**1.88**	**1.6**
(field ⊥ to grain)	$\tan\delta_{eff}$	*0.0086*	0.0120	0.025	**0.032**	**0.025**	**0.020**
Paper (royalgrey)	ε_r	*3.30*	3.29	2.99	**2.77**	**2.70**	*2.62*
	$\tan\delta_{eff}$	*0.0058*	0.0077	0.038	**0.066**	**0.056**	*0.0403*
Paper (white wove 22°C)	ε_r			*3.1*		**3.2**	
(78 g/m^2 field ‖ to grain)	$\tan\delta_{eff}$			*0.080*		**0.16**	
Yellow poplar	ε_r	*1.84*	1.79	1.75		**1.50**	**1.4**
(field ⊥ to grain)	$\tan\delta_{eff}$	*0.0042*	0.0054	0.019		**0.015**	**0.017**
Yellow birch	ε_r	*2.91*	2.88	2.70	**2.47**	**2.13**	**1.87**
(field ⊥ to grain)	$\tan\delta_{eff}$	*0.0072*	0.0090	0.0290	**0.0400**	**0.0330**	**0.0260**
Leather sole (dry)	ε_r	*4.1*	3.9	3.20	**3.1**		
	$\tan\delta_{eff}$	*0.0450*	0.0350	0.0280	**0.0380**		
Glasses							
D-glass	ε_r			3.95			
	$\tan\delta_{eff}$						
E-glass	ε_r	*6.43*	*6.39*	6.32	**6.22**		*6.11*
	$\tan\delta_{eff}$	*0.0042*	*0.0027*	0.0015	**0.0023**		*0.0060*
Fused quartz	ε_r	*3.78*	3.78	3.78	**3.78**	**3.78**	**3.78**
	$\tan\delta_{eff}$	*0.00085*	0.00075	0.0002	**0.0001**	**0.00006**	**0.00025**
Fused silica	ε_r	*3.78*	*3.78*	3.78	**3.78**	**3.78**	*3.78*
	$\tan\delta_{eff}$	*0.00066*	*0.00011*	0.00001	**0.00003**	**0.00005**	*0.00017*
Iron-sealing glass	ε_r	*8.40*	8.38	8.30	**8.20**	**7.99**	**7.84**
	$\tan\delta_{eff}$	*0.0004*	0.0004	0.0005	**0.0009**	**0.00199**	**0.0112**
S-glass	ε_r			5.2			
	$\tan\delta_{eff}$	0.0129		0.0034			
Miscellaneous							
Amber (fossil resin)	ε_r	*2.7*	2.7	2.65		**2.6**	
	$\tan\delta_{eff}$	*0.0013*	0.0018	0.0056		**0.0090**	
Cotton (210 kg/m^3)	ε_r			**1.5**			
	$\tan\delta_{eff}$			**0.02**			
Marble (dry)	ε_r			*9.0*		**9.0**	
	$\tan\delta_{eff}$			*0.037*		**0.024**	
Mica (glass-bonded)	ε_r		7.45	7.39			
	$\tan\delta_{eff}$		0.0019	0.0013			
Ruby mica	ε_r	*5.4*	5.4	5.4	**5.4**	**5.4**	
	$\tan\delta_{eff}$	*0.0025*	0.0006	0.0003	**0.0002**	**0.0003**	

(*Continued*)

TABLE 1.7 Relative Permittivity and Loss Tangent for Materials at Various Frequencies (Continued)

Material		60 Hz *100 Hz*	1 kHz *10 kHz* **100 kHz**	1 MHz *10 MHz* **27 MHz**	40 MHz *70 MHz* **100 MHz**	1 GHz *2.5 GHz* **3 GHz**	9.4 GHz *10 GHz* **25 GHz**
Sodium chloride	ε_r	*5.90*	5.90	5.90			**5.90**
	$\tan\delta_{eff}$	*<0.0001*	<0.0001	<0.0002			**<0.0005**
Shellac (28°C)	ε_r	*3.86*	3.81	3.47	**3.10**	**2.86**	
	$\tan\delta_{eff}$	*0.0065*	0.0074	0.031	**0.030**	**0.0254**	
Vaseline	ε_r	*2.16*	2.16	2.16	**2.16**	**2.16**	*2.16*
	$\tan\delta_{eff}$	*0.0003*	0.0002	<0.0001	**<0.0004**	**0.00066**	*0.001*
Wool (68 kg/m^3)	ε_r			*1.2*			
	$\tan\delta_{eff}$			*0.01*			
Wool fiber	ε_r					**3.9**	
	$\tan\delta_{eff}$					**0.074**	
Oils and Waxes							
Beeswax (yellow)	ε_r	*2.73*	2.66	2.49	**2.42**	**2.39**	
	$\tan\delta_{eff}$	*0.0240*	0.0140	0.0092	**0.0090**	**0.0075**	
Paraffin wax	ε_r	*2.25*	2.25	2.25	**2.25**	**2.25**	**2.2**
	$\tan\delta_{eff}$	*<0.0002*	<0.0002	<0.0002	**<0.0002**	**0.0002**	**<0.0003**
Vistawax (Polybutene)	ε_r	*2.34*	2.34	2.34	**2.30**	**2.27**	*2.26*
	$\tan\delta_{eff}$	*0.0002*	0.0003	0.00133	**0.00133**	**0.0009**	*0.0005*
Wax (Mitchell Rand)	ε_r	*2.36*	2.36	2.36	**2.36**	**2.31**	*2.31*
	$\tan\delta_{eff}$	*0.0022*	0.00088	0.00043	**0.00026**	**0.00049**	*0.0014*
Aviation gasoline (100 octane)	ε_r		*1.94*	1.94	**1.94**	**1.92**	
	$\tan\delta_{eff}$		*0.0001*		**0.00008**	**0.0014**	
Vegetable & mineral	ε_r	*2.3*	2.3	2.3	**2.3**	**2.25**	
	$\tan\delta_{eff}$	*0.0008*	0.0006	0.0004	**0.0004**	**0.00046**	
Plastics							
Acrylic	ε_r	3.2	3.2	2.8			
	$\tan\delta_{eff}$	0.028	0.03	0.023			
Alkyd (glass)	ε_r	6.3	6.0	5.8			
	$\tan\delta_{eff}$	0.02	0.01	0.01			
Bakelite	ε_r	*8.2*	*6.5*	5.4	**4.4**		*3.52*
	$\tan\delta_{eff}$	*0.1350*	*0.0630*	0.0600	**0.0770**		*0.0366*
Diallylphthalate (long glass)	ε_r	4.10	4.00				
	$\tan\delta_{eff}$	0.008	0.016				
Duroid 5650	ε_r						*2.65*
	$\tan\delta_{eff}$						*0.0030*
Eccofoam S	ε_r	1.05		1.18			*1.47*
	$\tan\delta_{eff}$	0.002		0.0040			*0.0070*
Epoxy resin RN-48	ε_r	*3.64*	*3.61*	3.52	**3.32**	**3.04**	*2.91*
	$\tan\delta_{eff}$	*0.0031*	*0.0068*	0.0142	**0.0264**	**0.0210**	*0.0184*
Ethylcellulose	ε_r	*3.11*	3.09	3.01	**2.90**	**2.74**	*2.70*
	$\tan\delta_{eff}$	*0.0075*	0.0065	0.0113	**0.0160**	**0.0210**	*0.0250*
Fiberglass (laminated) BK-174	ε_r	*14.2*	*7.2*	5.3	**4.8**	**4.4**	*4.37*
	$\tan\delta_{eff}$	*0.2500*	*0.1600*	0.0460	**0.0260**	**0.0290**	*0.0360*
FR-4/glass epoxy (50% resin)	ε_r	*4.8*	*4.7*	*4.6*	**4.5**	4.5	*4.4*
	$\tan\delta_{eff}$	*0.009*	*0.015*	*0.022*	**0.024**	0.025	*0.025*

TABLE 1.7 Relative Permittivity and Loss Tangent for Materials at Various Frequencies (Continued)

			1 kHz	1 MHz	40 MHz	1 GHz	9.4 GHz
		60 Hz	*10 kHz*	*10 MHz*	*70 MHz*	*2.5 GHz*	*10 GHz*
Material		*100 Hz*	**100 kHz**	**27 MHz**	**100 MHz**	**3 GHz**	**25 GHz**
Hexcel HRP	ε_r						*1.09*
	$\tan\delta_{eff}$						*0.0030*
Hexcel HRH-10	ε_r						*1.12*
	$\tan\delta_{eff}$						*0.0030*
Lexan	ε_r						*2.86*
	$\tan\delta_{eff}$						*0.0060*
Nylon	ε_r	5.3	4.8	3.8			
	$\tan\delta_{eff}$	0.06	0.06	0.03			
Nylon 6	ε_r	3.6	3.5	3.3			
	$\tan\delta_{eff}$	0.01	0.01	0.02			
Nylon 11	ε_r	7.1	5.9	3.3			
	$\tan\delta_{eff}$	0.178	0.192	0.103			
Nylon 66	ε_r	*3.88*	*3.75*	3.33	**3.16**	**3.03**	
	$\tan\delta_{eff}$	*0.0144*	*0.0193*	0.0257	**0.0210**	**0.0128**	
Phenolic	ε_r	17.0	9.5	5.5			
	$\tan\delta_{eff}$	0.65	0.35	0.06			
Phenolic (mineral filler)	ε_r	9.4	7.4	5.6			
	$\tan\delta_{eff}$	0.07	0.04	0.03			
Plexiglas	ε_r	*3.40*	*2.95*	2.76		**2.60**	*2.59*
	$\tan\delta_{eff}$	*0.0605*	*0.0300*	0.0140		**0.0057**	*0.0067*
Polycarbonates	ε_r	3.17		3.02		2.93	
	$\tan\delta_{eff}$	0.009		0.0021		0.010	
Polyclad (PCL-LD-621)	ε_r			3.7	**3.6**	3.5	*3.5*
	$\tan\delta_{eff}$			0.004	**0.004**	0.005	*0.008*
Polyester	ε_r	3.3		3.4		3.3	
	$\tan\delta_{eff}$			0.005		0.03	
Polyester (glass/mineral)	ε_r	5.3	5.3	5.2			
	$\tan\delta_{eff}$	0.004	0.004	0.006			
Polyester sulfone	ε_r	3.5		3.5		3.5	
	$\tan\delta_{eff}$	0.001		0.0035		0.006	
Polyethylene	ε_r	2.35	2.3	2.35		2.35	
	$\tan\delta_{eff}$	0.0005	0.0005	0.0005		0.0005	
Polyimide (50% resin)	ε_r		3.62	3.55			
	$\tan\delta_{eff}$		0.0018	0.0034			
Polypropylene	ε_r	2.6		2.6		2.6	
	$\tan\delta_{eff}$	0.0005		0.0018		0.0018	
Polysulfone	ε_r	3.14		3.13		3.10	
	$\tan\delta_{eff}$	0.0008		0.0010		0.0034	
Polystyrene	ε_r	3.1		2.65		2.7	
	$\tan\delta_{eff}$	0.0006		0.0003		0.0004	
Polyurethane (TP) (polyether filler)	ε_r	6.0	5.6	4.2			
	$\tan\delta_{eff}$	0.048	0.043	0.075			
Polyvinyl chloride/PVC (flexible)	ε_r	9		8.0		4.5	
	$\tan\delta_{eff}$	0.15		0.16		0.14	
Polyvinyl chloride/PVC (rigid)	ε_r	4		3.8		3.1	
	$\tan\delta_{eff}$	0.020		0.017		0.019	

(*Continued*)

TABLE 1.7 Relative Permittivity and Loss Tangent for Materials at Various Frequencies (Continued)

Material		60 Hz *100 Hz*	1 kHz *10 kHz* **100 kHz**	1 MHz *10 MHz* **27 MHz**	40 MHz *70 MHz* **100 MHz**	1 GHz *2.5 GHz* **3 GHz**	9.4 GHz *10 GHz* **25 GHz**
Polyvinylidene fluoride	ε_r	8.4	7.72	6.43		2.98	
	$\tan\delta_{eff}$	0.049	0.019	0.16		0.11	
PTFE	ε_r	2.1		2.1		2.1	
	$\tan\delta_{eff}$	0.0002		0.0002		0.0002	
Rexolite 1422	ε_r	*2.55*	*2.55*	2.55	**2.55**	**2.54**	*2.54*
	$\tan\delta_{eff}$	*0.00021*	*0.0001*	0.00013	**0.00038**	**0.00048**	*0.00047*
Silicone	ε_r	4.0	4.0	3.9			
	$\tan\delta_{eff}$	0.0015	0.0018	0.0034			
Styrofoam	ε_r	*1.03*	*1.03*	1.03		**1.03**	*1.03*
	$\tan\delta_{eff}$	*<0.0002*	*<0.0001*	<0.0002		**0.0001**	*0.00015*
Teflon	ε_r	*2.1*	2.1	2.1	**2.1**	**2.1**	**2.08**
	$\tan\delta_{eff}$	*<0.0005*	<0.0003	<0.0002	**<0.0002**	**0.00015**	**0.0006**
Urea Formaldehyde	ε_r	7.9	7.0	6.6			
	$\tan\delta_{eff}$	0.04	0.03	0.03			
Rubbers							
Butyl	ε_r	*2.39*	2.38	2.35	**2.35**	**2.35**	*2.35*
	$\tan\delta_{eff}$	*0.0034*	0.0035	0.0010	**0.0010**	**0.0009**	*0.0008*
Neoprene	ε_r	*6.70*	*6.60*	6.26	**4.5**	**4.00**	*4.0*
	$\tan\delta_{eff}$	*0.016*	*0.011*	0.038	**0.090**	**0.034**	*0.025*
Silicone-rubber compound	ε_r	*3.40*	3.35	3.20	**3.16**	**3.13**	*3.08*
	$\tan\delta_{eff}$	*0.0064*	0.0067	0.0030	**0.0032**	**0.0097**	*0.0146*
Soils and Waters							
Fresh water (20°C)	ε_r		*80*	80	**80**	80	**40**
	$\tan\delta_{eff}$		*70*	0.7	**0.01**	0.2	**0.9**
Pure water (20°C)	ε_r		*80*	80	**80**	80	**33**
	$\tan\delta_{eff}$				**0.004**	0.03	**0.9**
Sea water (20°C)	ε_r		*70*	70	**70**	70	**23**
	$\tan\delta_{eff}$		*1×10^5*	1,000	**10**	1	**1**
Ice (−12°C) (from distilled water)	ε_r		**4.8**	4.15	**3.45**	**3.20**	*3.17*
	$\tan\delta_{eff}$		**0.8**	0.12	**0.035**	**0.0009**	*0.0007*
Snow (−20°C) (freshly fallen)	ε_r		3.33	1.20		**1.20**	
	$\tan\delta_{eff}$		0.492	0.0215		**0.00029**	
Snow (−6°C) (hard-packed followed by light rain)	ε_r		**1.9**	1.55		**1.5**	
	$\tan\delta_{eff}$		**1.53**	0.29		**0.0009**	
Very dry ground	ε_r		*3*	3	**3**	3	**3**
	$\tan\delta_{eff}$		*60*	0.6	**0.006**	0.0008	**0.07**
Medium dry ground	ε_r		*13*	13	**13**	13	**6**
	$\tan\delta_{eff}$		*14*	1	**0.02**	0.06	**0.8**
Wet ground	ε_r		*30*	30	**30**	30	**6**
	$\tan\delta_{eff}$		*600*	6	**0.09**	0.09	**1**
Sandy soil (dry)	ε_r	*3.42*	2.91	2.59		**2.55**	*2.53*
	$\tan\delta_{eff}$	*0.196*	0.08	0.017		**0.0062**	*0.0036*

TABLE 1.7 Relative Permittivity and Loss Tangent for Materials at Various Frequencies (Continued)

Material		60 Hz *100 Hz*	1 kHz *10 kHz* **100 kHz**	1 MHz *10 MHz* **27 MHz**	40 MHz *70 MHz* **100 MHz**	1 GHz *2.5 GHz* **3 GHz**	9.4 GHz *10 GHz* **25 GHz**
Clay soil (dry)	ε_r	*4.73*	3.94	2.57		**2.27**	*2.16*
	$\tan\delta_{eff}$	*0.12*	0.12	0.065		**0.015**	*0.013*
Loamy soil (dry)	ε_r	*3.06*	2.83	2.53		**2.44**	*2.44*
	$\tan\delta_{eff}$	*0.07*	0.05	0.018		**0.0011**	*0.0014*
Biological Materials							
Blood (whole)	ε_r		2,900	2,040	**71**	63	*45*
	$\tan\delta_{eff}$		4,200	6.3	**2.1**	0.41	*0.39*
Brain (37°C)	ε_r		**(1.96–3.8) × 10³**	(0.54–1.25) × 10³	**57–90**	37–55	*38–44*
	$\tan\delta_{eff}$		**5.7–16**	2.0–7.0	**0.96-3.0**	0.27–0.58	*0.33–0.51*
Bone (37°C)	ε_r	*3,800*	**280**	87	23	8	
	$\tan\delta_{eff}$	*615*	**9.0**	3.5	0.45	0.11	
Fat	ε_r	*1.5×10⁵*	*5 × 10⁴*		11–13	*3.9–7.2*	
	$\tan\delta_{eff}$		*14*		1.1–1.9	*0.093–0.35*	
Kidney (37°C)	ε_r		**(10.9–12.5) × 10³**	(2.39–2.69) × 10³	**66–95**	42–50	*40*
	$\tan\delta_{eff}$		**3.5–4.1**	2.4–2.8	**1.2–2.9**	0.34–0.43	*0.44*
Liver (37°C)	ε_r	*8.5×10⁵*	**(9.8–13.7) × 10³**	1,970	**65–82**	47–49	*35*
	$\tan\delta_{eff}$	*27*	**2.0–2.9**	2.5-2.7	**1.3–2.0**	0.35–0.42	*0.46*
Skeletal muscle (37°C)	ε_r		**(14.4–24.8) × 10³**	(1.9–2.5) × 10³	**64–90**	57–59	*43–45*
	$\tan\delta_{eff}$		**2.8–7.4**	4.2–8.0	**1.5–3.0**	0.42–0.46	*0.46–0.48*
Spleen (37°C)	ε_r		**3,300**	1,450	**69–101**	50–55	*41*
	$\tan\delta_{eff}$		**34**	7.8	**1.3–2.7**	0.4–4.1	*0.44*

These losses are obtained directly from the general shielding expression given in (1.59):

$$
\begin{aligned}
20\log\left|\frac{E_{is}}{E_{ts}}\right| &= 20\log\left|\frac{(\eta_o+\eta)^2}{4\eta_o\eta}\left[1-\left(\frac{\eta_o-\eta}{\eta_o+\eta}\right)^2 e^{-2\frac{d}{\delta}}e^{-j2\frac{d}{\delta}}\right]e^{\frac{d}{\delta}}e^{j\frac{d}{\delta}}e^{-j\beta_o d}\right| \\
&= 20\log\left|\frac{(\eta_o+\eta)^2}{4\eta_o\eta}\right| + 20\log\left(e^{\frac{d}{\delta}}\right) \\
&\quad + 20\log\left|\left[1-\left(\frac{\eta_o-\eta}{\eta_o+\eta}\right)^2 e^{-2\frac{d}{\delta}}e^{-j2\frac{d}{\delta}}\right]e^{j\frac{d}{\delta}}e^{-j\beta_o d}\right|
\end{aligned} \tag{1.95}
$$

As the sum of these terms becomes greater, the shielding effectiveness increases. The first term in (1.95) is the reflection loss in dB:

$$
R_{dB} = 20\log\left|\frac{(\eta_o+\eta)^2}{4\eta_o\eta}\right| \approx 20\log\left|\frac{\eta_o}{4\eta}\right| \tag{1.96}
$$

(The intrinsic impedance of a good-conducting shield, η, is much less than $\eta_o = 377\ \Omega$.) This represents the loss due to the reflection of the wave at both the air-to-shield and shield-to-air interfaces. For good conductors, as the conductivity increases or frequency decreases, the reflection loss increases, resulting in a greater impedance mismatch.

The second term in (1.95) is the absorption loss in dB:

$$A_{dB} = 20\log\left(e^{\frac{d}{\delta}}\right) = 20\frac{d}{\delta}\log e \approx 8.69\frac{d}{\delta} \tag{1.97}$$

This loss is due to the nonzero conductivity of the medium. As the thickness of the shield increases, this loss term increases. A relationship seen in many EMC short courses is the loss per skin depth for a shielding material:

$$A_{dB/\delta} = 8.69\frac{\delta}{\delta} \approx 8.7 \text{ dB}/\delta \tag{1.98}$$

Often a shield thickness is given in terms of skin depths. For example, a shield that is five skin-depths thick has an absorption loss of 5(8.7) ≈ 44 dB.

The last term in (1.95) is the multiple-reflection loss in dB:

$$\begin{aligned} MR_{dB} &= 20\log\left|\left[1-\left(\frac{\eta_o-\eta}{\eta_o+\eta}\right)^2 e^{-2\frac{d}{\delta}}e^{-j2\frac{d}{\delta}}\right]e^{j\frac{d}{\delta}}e^{-j\beta_o d}\right| \\ &= 20\log\left|1-\left(\frac{\eta_o-\eta}{\eta_o+\eta}\right)^2 e^{-2\frac{d}{\delta}}e^{-j2\frac{d}{\delta}}\right|\left|e^{j\frac{d}{\delta}}\right|\left|e^{-j\beta_o d}\right| \\ &= 20\log\left|1-\left(\frac{\eta_o-\eta}{\eta_o+\eta}\right)^2 e^{-2\frac{d}{\delta}}e^{-j2\frac{d}{\delta}}\right| \\ &\approx 20\log\left|1-e^{-2\frac{d}{\delta}}e^{-j2\frac{d}{\delta}}\right| \end{aligned} \tag{1.99}$$

(Again, the intrinsic impedance of a good-conducting shield, η, is much less than $\eta_o = 377\ \Omega$.) This loss is due to the multiple reflections inside the shield. In some EMC books, this term is neglected. If the shield is electrically thick, then this term is close to zero. For example, if the shield is five skin-depths thick, then

$$MR_{dB} \approx 20\log\left|1-e^{-2(5)}e^{-j2(5)}\right| \approx 0 \text{ dB}$$

If the shield is not thick, this term is negative. For example, if the shield is one-fifth of a skin depth thick, then

$$MR_{dB} \approx 20\log\left|1-e^{-2\left(\frac{1}{5}\right)}e^{-j2\left(\frac{1}{5}\right)}\right| \approx -6.7 \text{ dB}$$

The reflection loss (for good conductors) and absorption loss are both always positive. Thus, the multiple-reflection term reduces the effectiveness of the shield.

1.16 Reflection, Absorption, and Multiple-Reflection Losses

Determine the reflection, absorption, multiple-reflection, and total loss of a 15 mil[8] copper shield vs. frequency. Vary the frequency so that the shield is electrically thin and thick. Determine approximate expressions for these losses in terms of the electrical parameters of the shield.

For a good-conducting flat shield of thickness d, the approximate expressions for the reflection, absorption, and multiple-reflection loss in dB are

$$R_{dB} \approx 20\log\left|\frac{\eta_o}{4\eta}\right| \approx 20\log\left|\frac{\eta_o}{4\left[\sqrt{\frac{\omega\mu_r\mu_o}{2\sigma_r\sigma_o}}(1+j)\right]}\right| \approx 20\log\left|\frac{2.6\times10^8}{\sqrt{\frac{\mu_r f}{\sigma_r}}}\right| \approx 168-10\log\left(\frac{\mu_r f}{\sigma_r}\right) \tag{1.100}$$

$$A_{dB} \approx 8.69\frac{d}{\delta} = 8.69d\sqrt{\pi f\mu_r\mu_o\sigma_r\sigma_o} \approx 131d\sqrt{f\mu_r\sigma_r} \tag{1.101}$$

$$\begin{aligned} MR_{dB} &\approx 20\log\left|1-e^{-2\frac{d}{\delta}}e^{-j2\frac{d}{\delta}}\right| \approx 20\log\left|1-e^{-30.3d\sqrt{\mu_r\sigma_r f}}e^{-j30.3d\sqrt{\mu_r\sigma_r f}}\right| \\ &= 20\log\left|1-e^{30.3d\sqrt{\mu_r\sigma_r f}}\left[\cos(30.3d\sqrt{\mu_r\sigma_r f})-j\sin\left(30.3d\sqrt{\mu_r\sigma_r f}\right)\right]\right| \\ &= 20\log\sqrt{\left[1-e^{-30.3d\sqrt{\mu_r\sigma_r f}}\cos\left(30.3d\sqrt{\mu_r\sigma_r f}\right)\right]^2+\left[e^{-30.3d\sqrt{\mu_r\sigma_r f}}\sin\left(30.3d\sqrt{\mu_r\sigma_r f}\right)\right]^2} \\ &= 10\log\left|1-2e^{-30.3d\sqrt{\mu_r\sigma_r f}}\cos\left(30.3d\sqrt{\mu_r\sigma_r f}\right)+e^{-60.6d\sqrt{\mu_r\sigma_r f}}\right| \end{aligned} \tag{1.102}$$

The variables μ_r and σ_r are the permeability and conductivity, respectively, of the shield relative to copper. To convert from mils to meters, multiply by 2.54×10^{-5} (15 mils ≈ 380 μm). For copper, the skin depth is equal to the thickness of the shield at a frequency of about 30 kHz. As seen in Mathcad 1.6, a 15 mils thick shield of copper is about 0.18δ thick at 1 kHz (electrically thin) and about 18δ thick at 10 MHz (electrically thick). At the lower frequencies, the shield should be in the far field of the source if the plane wave assumption is to be satisfied. Although not shown, the approximations (1.100)–(1.102) are in excellent agreement (less than 1 dB over the given frequency range) with the more exact equations used in the Mathcad analysis. As seen, the reflection loss is the strongest component over most of the frequency range followed by the absorption loss. The multiple-reflection loss is negligible over the entire range since the wave's energy (in the shield) is mostly dissipated.

The absorption loss is more important for magnetic fields than electric fields since the electric field is mostly reflected at the first interface (air-to-metal). Notice that the absorption loss for the electromagnetic wave is proportional to the square root of the product of the relative permeability

[8]The thickness of a dollar bill is about 4 mils, a dime is about 53 mils, and a penny is about 59 mils.

$\mu_o := 4 \cdot \pi \cdot 10^{-7}$ $\varepsilon_o := 8.854 \cdot 10^{-12}$ $\sigma_o := 5.8 \cdot 10^{7}$ $\mu_r := 1$ $\sigma_r := 1$ $d := 381 \cdot 10^{-6}$ $\eta_o := \sqrt{\frac{\mu_o}{\varepsilon_o}}$ $j := \sqrt{-1}$

$x := 30, 30.1..70$

$$f(x) := \left(x + 1 - 10 \cdot \text{floor}\left(\frac{x}{10}\right)\right) \cdot 10^{\text{floor}\left(\frac{x}{10}\right)}$$

$$\delta(f) := \frac{1}{\sqrt{\pi \cdot f \cdot \mu_r \cdot \mu_o \cdot \sigma_r \cdot \sigma_o}} \qquad \eta(f) := \sqrt{\frac{j \cdot 2 \cdot \pi \cdot f \cdot \mu_r \cdot \mu_o}{\sigma_r \cdot \sigma_o + j \cdot 2 \cdot \pi \cdot f \cdot \varepsilon_o}} \qquad R_{dB}(f) := 20 \cdot \log\left[\left|\frac{(\eta_o + \eta(f))^2}{4 \cdot \eta_o \cdot \eta(f)}\right|\right]$$

$$A_{dB}(f) := 20 \cdot \log\left(e^{\frac{d}{\delta(f)}}\right) \qquad MR_{dB}(f) := 20 \cdot \log\left[\left|1 - \left(\frac{\eta_o - \eta(f)}{\eta_o + \eta(f)}\right)^2 \cdot e^{-\frac{2 \cdot d}{\delta(f)}} \cdot e^{-\frac{2 \cdot d}{\delta(f)} \cdot j}\right|\right]$$

$$TOT_{dB}(f) := R_{dB}(f) + A_{dB}(f) + MR_{dB}(f) \qquad \frac{d}{\frac{1}{\sqrt{\pi \cdot 10^3 \cdot \mu_r \cdot \mu_o \cdot \sigma_r \cdot \sigma_o}}} = 0.182 \qquad \frac{d}{\frac{1}{\sqrt{\pi \cdot 10^7 \cdot \mu_r \cdot \mu_o \cdot \sigma_r \cdot \sigma_o}}} = 18.231$$

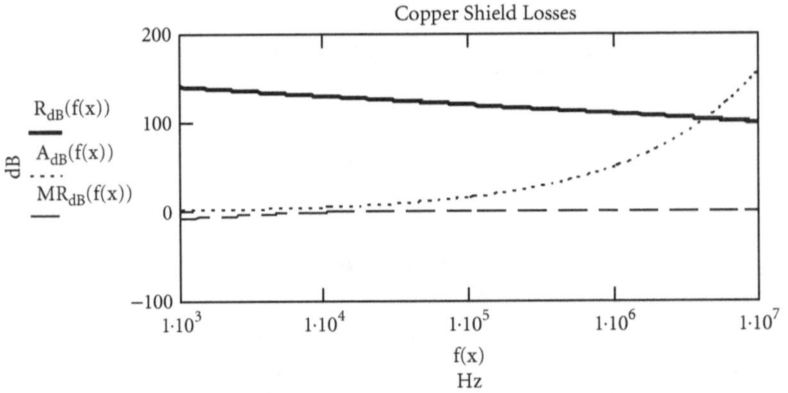

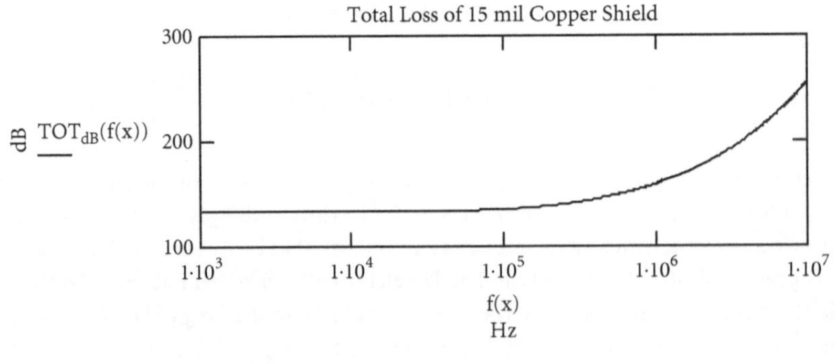

MATHCAD 1.6 Reflection, absorption, multiple-reflection, and total loss of a 15 mil thick copper shield.

and conductivity:

$$A_{dB} = 131 d \sqrt{f \mu_r \sigma_r}$$

Of course, for copper $\mu_r \sigma_r = 1$ for even very high frequencies. For other good nonmagnetic metals (with the exception of silver) this product is less than one. For magnetic materials this product can reach several thousand. For steel, this product is about 100. (The permeability is a function of several factors including the frequency, and it generally decreases with frequency.)

1.17 Effect of Dielectric Constant on Shielding

Why is the permittivity of the slab not a parameter in the equations previously derived for the reflection loss, R_{dB}, absorption loss, A_{dB}, and multiple-reflection loss, MR_{dB}? [Weeks, '64]

The effects of the permittivity are small since for good conductors the ratio of the displacement current to the conduction current is extremely small. If a known voltage is applied across a good conductor and the resultant current measured, the voltage and current should be in phase. The displacement current, if measurable and significant, would produce a phase difference between the applied voltage and current. Recall that the phase difference between the voltage across and current through a capacitor is 90°. Also, the conduction current density (representing the movement of free charges) is in phase with the electric field,

$$\vec{J}_s = \sigma \vec{E}_s$$

while the displacement current density (representing the movement of bound charges)

$$j\omega \vec{D}_s = j\omega \varepsilon \vec{E}_s$$

is 90° out of phase with the electric field. For metals, the phase difference between the voltage and current is extremely small up to the optical range. For this reason, the permittivity is difficult to measure for good conductors and is assumed to be equal to that of free space (a dielectric constant of one). There is no experimental evidence to assume otherwise.

1.18 Near Field or Far Field?

Analyze the various relationships used to determine whether a location of interest is in the near field or far field of an electromagnetic field source. [Lo; Rudge; Johnson, '61; Weeks, '68; Rulf]

When solving many electromagnetic field-based problems, one of the first questions that is asked by an experienced engineer is, "Is the location of interest in the near or far field?" The reason for this is that the method of attack for near-field situations is often different and more complicated than for far-field situations. In the far field, the electromagnetic field is considered a plane wave: the electric and magnetic fields are perpendicular and simply related. In the near field, however, the fields are often much more complex. Furthermore, in the near field, electric and magnetic coupling to nearby sources or objects can be important. Although the transition between the near and far fields is not abrupt or clearly defined, helpful rules-of-thumb are available to the engineer. This concept of near and far field applies not only to antennas but also to any field generated by oscillating charges (e.g., sinusoidal current on a printed circuit board trace).

Referring to Figure 1.4, the near-field region is that region closest to the antenna. It is often broken into two regions: the reactive and radiating near field. The region closest to the source is considered the reactive near field since the power in the fields is mostly reactive not real. Any field coupling to nearby objects is best modeled as capacitive and inductive coupling. The radiating fields are not necessarily zero, but they are small compared to the reactive fields. The reactive fields decay as the square or cube of the distance from the source. Once the source is turned off, the reactive fields disappear. The near fields are also referred to as the static or induction fields.

The radiating regions follow the reactive near-field region. The radiating fields dominate over the reactive fields in this region. (A radiating field has real power associated with it. Furthermore, after the source has been turned off, the radiating field will still be present or propagating in some direction.) Obviously, all antennas have a radiating far-field region. Some antennas, such as electrically-small antennas, do not have a radiating near-field region. In the radiating near field, the radiated *power* may be in both the radial and angular directions. In the radiating far field, the radiated *power* is only in the radial direction. (Far from the source, the distance from any point on an antenna to a specific far-field position is about the same.) A consequence of this radial power is that the radiation pattern is the same at any

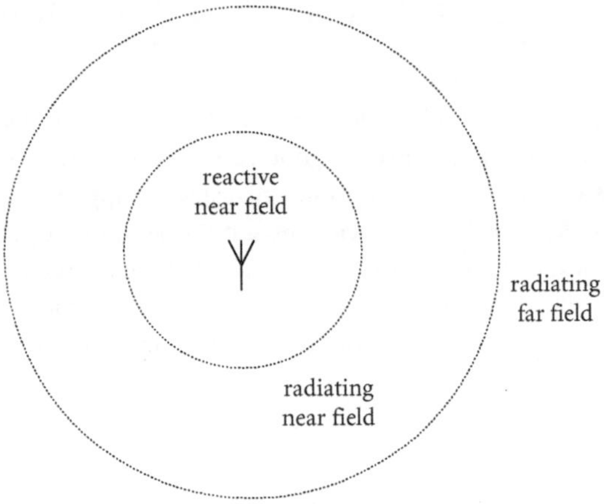

FIGURE 1.4 Reactive and radiating field regions.

fixed distance from the antenna when measured in the far field. It is nearly impossible to determine the radiation pattern (e.g., main beam direction or shape of side lobes) when in the near field because of the complexity of the field and Poynting vector. In the radiating far field, the field decays with the inverse of the distance from the source. The radiating far field is also sometimes referred to as just the radiation field.

To help clarify the distinction between the near and far field, the Poynting vector for an electrically-small current-carrying piece of wire (Hertzian dipole) will be determined. The phasor or complex power in a field is defined as

$$\vec{P}_s = \vec{E}_s \times \vec{H}_s^* \text{ W/m}^2$$

The real part of this vector corresponds to the radiated power, and the imaginary part of this vector corresponds to the reactive power (analogous to resistive and capacitive/inductive power, respectively, in circuits). The fields for a Hertzian dipole located along the z axis centered at the origin are

$$H_{\phi s} = \frac{I_m dl}{4\pi} \beta_o^2 \sin\theta \left(\frac{j}{\beta_o r} + \frac{1}{\beta_o^2 r^2} \right) e^{-j\beta_o r}$$

$$E_{rs} = \frac{2I_m dl}{4\pi} \eta_o \beta_o^2 \cos\theta \left(\frac{1}{\beta_o^2 r^2} - \frac{j}{\beta_o^3 r^3} \right) e^{-j\beta_o r}$$

$$E_{\theta s} = \frac{I_m dl}{4\pi} \eta_o \beta_o^2 \sin\theta \left(\frac{j}{\beta_o r} + \frac{1}{\beta_o^2 r^2} - \frac{j}{\beta_o^3 r^3} \right) e^{-j\beta_o r}$$

$$H_{rs} = 0,\ H_{\theta s} = 0,\ E_{\phi s} = 0$$

where

$$\eta_o = \sqrt{\frac{\mu_o}{\varepsilon_o}},\ \beta_o = \frac{2\pi}{\lambda_o},\ I_m = \text{current magnitude},\ dl = \text{length of dipole}$$

The complex Poynting vector for this Hertzian dipole is then

$$\vec{P}_s = (E_{rs}\hat{a}_r + E_{\theta s}\hat{a}_\theta)\times H^*_{\phi s}\hat{a}_\phi = -E_{rs}H^*_{\phi s}\hat{a}_\theta + E_{\theta s}H^*_{\phi s}\hat{a}_r$$

$$= -\left[\frac{2I_m dl}{4\pi}\eta_o\beta_o^2\cos\theta\left(\frac{1}{\beta_o^2 r^2}-\frac{j}{\beta_o^3 r^3}\right)e^{-j\beta_o r}\right]\left[\frac{I_m dl}{4\pi}\beta_o^2\sin\theta\left(\frac{j}{\beta_o r}+\frac{1}{\beta_o^2 r^2}\right)e^{-j\beta_o r}\right]^*\hat{a}_\theta$$

$$+\left[\frac{I_m dl}{4\pi}\eta_o\beta_o^2\sin\theta\left(\frac{j}{\beta_o r}+\frac{1}{\beta_o^2 r^2}-\frac{j}{\beta_o^3 r^3}\right)e^{-j\beta_o r}\right]\left[\frac{I_m dl}{4\pi}\beta_o^2\sin\theta\left(\frac{j}{\beta_o r}+\frac{1}{\beta_o^2 r^2}\right)e^{-j\beta_o r}\right]^*\hat{a}_r$$

$$\vec{P}_s = \frac{I_m^2(dl)^2\eta_o\beta_o^4}{8\pi^2}\cos\theta\sin\theta\left[j\left(\frac{1}{\beta_o^3 r^3}+\frac{1}{\beta_o^5 r^5}\right)\right]\hat{a}_\theta + \frac{I_m^2(dl)^2\eta_o\beta_o^4}{16\pi^2}\sin^2\theta\left[\left(\frac{1}{\beta_o^2 r^2}\right)-j\left(\frac{1}{\beta_o^5 r^5}\right)\right]\hat{a}_r \tag{1.103}$$

Far from this short antenna, the complex Poynting vector is entirely in the radial direction since the $1/r^2$ term dominates:

$$\vec{P}_s \approx \frac{I_m^2(dl)^2\eta_o\beta_o^4}{16\pi^2}\sin^2\theta\left(\frac{1}{\beta_o^2 r^2}\right)\hat{a}_r = \frac{I_m^2(dl)^2\eta_o\beta_o^2}{16\pi^2 r^2}\sin^2\theta\,\hat{a}_r$$
$$\approx 94 I_m^2\left(\frac{dl}{\lambda_o}\right)^2\frac{\sin^2\theta}{r^2}\hat{a}_r \tag{1.104}$$

This power is entirely real and is equal to the radiated power density in the far field. (The real average power density is one-half of this value.) This result is in agreement with the previous statements for far-field radiation. This power is due to the $1/r$ electric field and $1/r$ magnetic field terms. Close to the short antenna, however, the complex Poynting vector is entirely imaginary since the $1/r^5$ terms dominate:

$$\vec{P}_s \approx \frac{I_m^2(dl)^2\eta_o\beta_o^4}{8\pi^2}\cos\theta\sin\theta\left(j\frac{1}{\beta_o^5 r^5}\right)\hat{a}_\theta - \frac{I_m^2(dl)^2\eta_o\beta_o^4}{16\pi^2}\sin^2\theta\left(j\frac{1}{\beta_o^5 r^5}\right)\hat{a}_r \tag{1.105}$$

This reactive near-field power is in both the θ and r directions. Furthermore, for this electrically-small antenna, there are no terms corresponding to near-field radiation; that is, there are no real terms with a $1/r^5$ (or $1/r^3$) dependence. Again, this result is in agreement with the previous statements. Note that reactive power does exist in the far field, but it is negligible compared to the real power (the radiation). Also, real power does exist in the near field, but it is negligible compared to the reactive power.

For wire antennas with dimensions less than about one wavelength, the generally accepted criteria for the near-far field transition is $r = \lambda/2\pi \approx 0.16\lambda$. Notice that at this location for the Hertzian dipole, the $1/r$ electric field term (corresponding to the far-field radiation) for the E_θ component is equal in magnitude to the $1/r^2$ electric field term for the E_θ component. The corresponding magnetic field terms are also equal in magnitude. For the Hertzian dipole, the r component of the electric field does not contain a far-field term. Setting $r = \lambda/2\pi = 1/\beta_o$, these relationships are verified:

$$H_{\phi s}\left(r=\frac{\lambda}{2\pi}\right) = \frac{I_m dl}{4\pi}\beta_o^2\sin\theta\left(\frac{j}{\beta_o r}+\frac{1}{\beta_o^2 r^2}\right)e^{-j\beta_o r}\bigg|_{r=\frac{\lambda}{2\pi}} = \frac{I_m dl}{4\pi}\beta_o^2\sin\theta\,(j+1)\,e^{-j}$$

$$E_{rs}\left(r=\frac{\lambda}{2\pi}\right)=\frac{2I_m dl}{4\pi}\eta_o\beta_o^2\cos\theta\left(\frac{1}{\beta_o^2r^2}-\frac{j}{\beta_o^3r^3}\right)e^{-j\beta_o r}\Bigg|_{r=\frac{\lambda}{2\pi}}=\frac{2I_m dl}{4\pi}\eta_o\beta_o^2\cos\theta\,(1-j)\,e^{-j}$$

$$E_{\theta s}\left(r=\frac{\lambda}{2\pi}\right)=\frac{I_m dl}{4\pi}\eta_o\beta_o^2\sin\theta\left(\frac{j}{\beta_o r}+\frac{1}{\beta_o^2r^2}-\frac{j}{\beta_o^3r^3}\right)e^{-j\beta_o r}\Bigg|_{r=\frac{\lambda}{2\pi}}=\frac{I_m dl}{4\pi}\eta_o\beta_o^2\sin\theta\,(j+1-j)\,e^{-j}$$

Generally, a source is in the near field when $r \ll \lambda/2\pi$ and is in the far field when $r \gg \lambda/2\pi$. For other antennas, such as longer dipoles, the antenna is divided into many Hertzian dipoles, and superposition applied. This same transition guideline is generally used for longer antennas less than about one-half wavelength in dimension. When measuring the radiation pattern of an antenna, the far field is considered at 10λ. With this conservative estimate for the far field, the reactive fields are considered negligible.

Another guideline used to determine if a position is in the far field is whether the wave impedance of the field is near 377 Ω. For many wire-type antennas, the field's wave impedance is near this value for distances 3λ or greater. This concept is explored in a later discussion on the topic of wave impedance.

For electrically-large aperture antennas, a different guideline is used. Aperture antennas include horns and parabolic dishes. When the distance from an antenna is such that the spherical wave can be represented by a plane wave, this location is considered in the far field. (The electric and magnetic fields vary as $1/r$ in the far field.) This far-field guideline for aperture antennas can be quickly derived. However, rather than determining (by a small measuring device) this far-field distance for fields produced by a *transmitting* aperture antenna, this distance for fields produced by a transmitting small point source as *received* by an aperture antenna will be determined. These distances are the same since the transmitting and receiving radiation patterns of an antenna are identical (assuming linearity).

The flatness of the spherical wave received by the aperture antenna is determined by the size of the variable Δ as defined in Figure 1.5. This distance can be obtained by using Pythagorean's relation:

$$(r+\Delta)^2=r^2+\left(\frac{D}{2}\right)^2 \quad \Rightarrow r^2\left(1+\frac{\Delta}{r}\right)^2=r^2+\left(\frac{D}{2}\right)^2$$

where D is the largest dimension of the aperture. If $\Delta/r \ll 1$, the left side of this last expression can be approximated using the binomial expansion:

$$r^2\left(1+2\frac{\Delta}{r}\right)\approx r^2+\left(\frac{D}{2}\right)^2 \quad \Rightarrow \Delta\approx\frac{D^2}{8r}$$

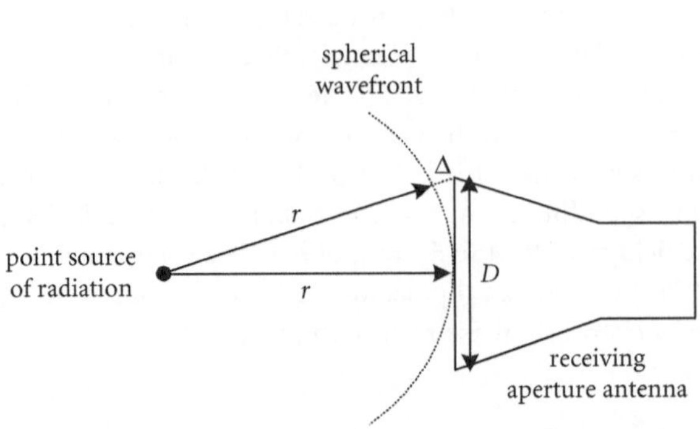

FIGURE 1.5 Spherical wave incident upon an aperture receiving antenna.

When the wavefront at the aperture antenna is an ideal plane wave, $\Delta = 0$, and there is no curvature in the wave. The wave is approximately a plane wave when

$$\Delta \le \frac{\lambda}{16} \tag{1.106}$$

which corresponds to a phase difference of 22.5° or less (360° corresponds to one wavelength). Therefore,

$$\frac{D^2}{8r} \le \frac{\lambda}{16}$$

(Sometimes a $\lambda/32$ or about 11.3° approximation is used.) When the distance to the antenna, r, satisfies the following inequality, this position is considered in the far field of the aperture antenna:

$$r \ge \frac{2D^2}{\lambda} \tag{1.107}$$

The conventional far-field criterion is given by (1.107).

Simple relationships for the far field for nonaperture electrically large antennas such as yagis, log periodics, long wires, and rhombics do not exist. However, it is common to use the previous aperture relationship because it tends to be a conservative estimate.

An FM radio station is interfering with a computer. Is this computer most likely in the near or far-field? [Johnson, '61]

FM radio broadcast stations in North America range from 88–108 MHz with a bandwidth of 0.2 MHz, which corresponds to a bandwidth of about 0.2%. These frequencies correspond to free-space wavelengths ranging from about 3.4 to 2.8 m. The radiation pattern for a commercial FM station is frequently omnidirectional. A vertical array of dipoles, loops, slots, or normal-mode helical antennas is used. For one half-wave dipole, application of the far-field guideline for a wire antenna results in

$$r \gg \frac{\lambda}{2\pi} = \frac{3.4}{2\pi} \approx 0.54\,\text{m}$$

If the antenna consists of four half-wave dipoles vertically stacked, the total length is about (accounting for the separation distance between the dipoles) 10 m. Using the conservative estimate obtained for aperture antennas, the distance is 71 m:

$$r \ge \frac{2D^2}{\lambda} = \frac{2(10)^2}{2.8} = 71\,\text{m} \approx 230\,\text{ft}$$

Given that the antennas are mounted on towers, buildings, or other high structures, the FM signal received by virtually all devices is in the far field.

A microwave antenna operating at 2 GHz has a diameter of 3 m. At what distance from the antenna would the field appear like a plane wave?

The free-space wavelength for a 2 GHz signal is

$$\lambda_o = \frac{3\times10^8}{2\times10^9} = 15\,\text{cm}$$

Using the aperture guideline,

$$r \geq \frac{2D^2}{\lambda} = \frac{2(3)^2}{0.15} = 120\,\text{m} \approx 390\,\text{ft}$$

For distances greater than 120 m, this microwave signal would be approximately a plane wave.

1.19 Wave Impedance

The electric and magnetic field components for a Hertzian dipole are

$$H_{\phi s} = \frac{I_m dl}{4\pi}\beta_o^2 \sin\theta\left(\frac{j}{\beta_o r} + \frac{1}{\beta_o^2 r^2}\right)e^{-j\beta_o r} \tag{1.108}$$

$$E_{rs} = \frac{2I_m dl}{4\pi}\eta_o\beta_o^2 \cos\theta\left(\frac{1}{\beta_o^2 r^2} - \frac{j}{\beta_o^3 r^3}\right)e^{-j\beta_o r} \tag{1.109}$$

$$E_{\theta s} = \frac{I_m dl}{4\pi}\eta_o\beta_o^2 \sin\theta\left(\frac{j}{\beta_o r} + \frac{1}{\beta_o^2 r^2} - \frac{j}{\beta_o^3 r^3}\right)e^{-j\beta_o r} \tag{1.110}$$

$$H_{rs} = 0,\; H_{\theta s} = 0,\; E_{\phi s} = 0$$

while the field components for a magnetic dipole are

$$H_{rs} = \frac{j2I_m\pi a^2\mu_o\omega}{4\pi}\frac{\beta_o^2}{\eta_o}\cos\theta\left(\frac{1}{\beta_o^2 r^2} - \frac{j}{\beta_o^3 r^3}\right)e^{-j\beta_o r} \tag{1.111}$$

$$H_{\theta s} = \frac{jI_m\pi a^2\mu_o\omega}{4\pi}\frac{\beta_o^2}{\eta_o}\sin\theta\left(\frac{j}{\beta_o r} + \frac{1}{\beta_o^2 r^2} - \frac{j}{\beta_o^3 r^3}\right)e^{-j\beta_o r} \tag{1.112}$$

$$E_{\phi s} = \frac{-jI_m\pi a^2\mu_o\omega}{4\pi}\beta_o^2 \sin\theta\left(\frac{j}{\beta_o r} + \frac{1}{\beta_o^2 r^2}\right)e^{-j\beta_o r} \tag{1.113}$$

$$H_{\phi s} = 0,\; E_{rs} = 0,\; E_{\theta s} = 0$$

where

$$\eta_o = \sqrt{\frac{\mu_o}{\varepsilon_o}},\; \beta_o = \frac{2\pi}{\lambda_o},\; I_m = \text{current magnitude},\; dl = \text{length of dipole},\; a = \text{radius of loop}$$

Using a numerical package, plot the "wave" impedance of these two sources of electromagnetic fields as the electrical distance from the source varies from one-tenth of a wavelength to two wavelengths.

The Hertzian dipole is an electrically-small antenna of one linear element. The given equations for the electric and magnetic field assume that this small antenna is located along the z axis centered at the origin. Because the antenna is small, the variation of the current along its length is assumed constant. The magnetic dipole is an electrically-small loop antenna. The equations provided describe the electric and magnetic fields from the loop assuming that the antenna is located in the xy plane centered about the origin. Because this antenna is also small, the current variation along its entire circumference is also assumed constant.

The wave impedance is usually defined as the ratio of the electric field component transverse to the direction of propagation to the magnetic field component transverse to the direction of propagation. Thus, the wave impedance of the Hertzian dipole is

$$\eta_w = \frac{E_{\theta s}}{H_{\phi s}} = \frac{\eta_o\left(\dfrac{j}{\beta_o r} + \dfrac{1}{\beta_o^2 r^2} - \dfrac{j}{\beta_o^3 r^3}\right)}{\dfrac{j}{\beta_o r} + \dfrac{1}{\beta_o^2 r^2}}\ \Omega \tag{1.114}$$

This impedance is independent of the angle, θ. Notice that very close to the antenna the magnitude of the wave impedance is equal to

$$|\eta_w| = \left|\frac{\eta_o\left(\dfrac{j}{\beta_o r} + \dfrac{1}{\beta_o^2 r^2} - \dfrac{j}{\beta_o^3 r^3}\right)}{\dfrac{j}{\beta_o r} + \dfrac{1}{\beta_o^2 r^2}}\right| \approx \left|\frac{\eta_o\left(\dfrac{j}{\beta_o^3 r^3}\right)}{\dfrac{1}{\beta_o^2 r^2}}\right| = \frac{\eta_o}{\beta_o r}\ \Omega \quad \text{if } \beta_o r \ll 1 \tag{1.115}$$

The wave impedance is large near the antenna. This near field

$$\beta_o r = \frac{2\pi}{\lambda_o} r \ll 1 \tag{1.116}$$

corresponds to the standard definition, electrically close relative to the wavelength. In the far field, the magnitude of the wave impedance, η_o, is about 377 Ω

$$|\eta_w| = \left|\frac{\eta_o\left(\dfrac{j}{\beta_o r} + \dfrac{1}{\beta_o^2 r^2} - \dfrac{j}{\beta_o^3 r^3}\right)}{\dfrac{j}{\beta_o r} + \dfrac{1}{\beta_o^2 r^2}}\right| \approx \left|\frac{\eta_o\left(\dfrac{j}{\beta_o r}\right)}{\dfrac{j}{\beta_o r}}\right| = \eta_o\ \Omega \quad \text{if } \beta_o r \gg 1 \tag{1.117}$$

and is referred to as the intrinsic impedance. The receiver is in the far field when

$$\beta_o r = \frac{2\pi}{\lambda_o} r \gg 1 \tag{1.118}$$

and the distance from the antenna is much greater than a wavelength. Notice that the wave impedance of the Hertzian dipole approaches the intrinsic impedance of a plane wave when far from the antenna. Antennas have this characteristic. As seen in Mathcad 1.7, when the ratio of the magnitude of the *total*

$\mu_o := 4\cdot\pi\cdot10^{-7}$ $\varepsilon_o := 8.854\cdot10^{-12}$ $v_o := 3\cdot10^8$ $I_m := 1\cdot10^{-6}$ $f := 100\cdot10^6$ $j := \sqrt{-1}$

$\eta_o := \sqrt{\frac{\mu_o}{\varepsilon_o}}$ $\lambda_o := \frac{v_o}{f}$ $\beta_o := \frac{2\cdot\pi}{\lambda_o}$ $dl := \frac{\lambda_o}{100}$

$r := \frac{\lambda_o}{10}, \frac{\lambda_o}{9.80} .. 2\cdot\lambda_o$

$$H_\phi(r,\theta) := \frac{I_m\cdot dl}{4\cdot\pi}\cdot\beta_o^{\,2}\cdot\sin(\theta)\cdot\left(\frac{j}{\beta_o\cdot r}+\frac{1}{\beta_o^{\,2}\cdot r^2}\right)\cdot e^{-j\cdot\beta_o\cdot r}$$

$$E_r(r,\theta) := \frac{2\cdot I_m\cdot dl}{4\cdot\pi}\cdot\eta_o\cdot\beta_o^{\,2}\cdot\cos(\theta)\cdot\left(\frac{1}{\beta_o^{\,2}\cdot r^2}-\frac{j}{\beta_o^{\,3}\cdot r^3}\right)\cdot e^{-j\cdot\beta_o\cdot r}$$

$$E_\theta(r,\theta) := \frac{I_m\cdot dl}{4\cdot\pi}\cdot\eta_o\cdot\beta_o^{\,2}\cdot\sin(\theta)\cdot\left(\frac{j}{\beta_o\cdot r}+\frac{1}{\beta_o^{\,2}\cdot r^2}-\frac{j}{\beta_o^{\,3}\cdot r^3}\right)\cdot e^{-j\cdot\beta_o\cdot r}$$

$\eta(r,\theta) := \left|\frac{E_\theta(r,\theta)}{H_\phi(r,\theta)}\right|$ $\eta_o = 376.734$ $\eta_{tot}(r,\theta) := \left|\frac{\sqrt{E_\theta(r,\theta)\cdot\overline{E_\theta(r,\theta)}+E_r(r,\theta)\cdot\overline{E_r(r,\theta)}}}{H_\phi(r,\theta)}\right|$

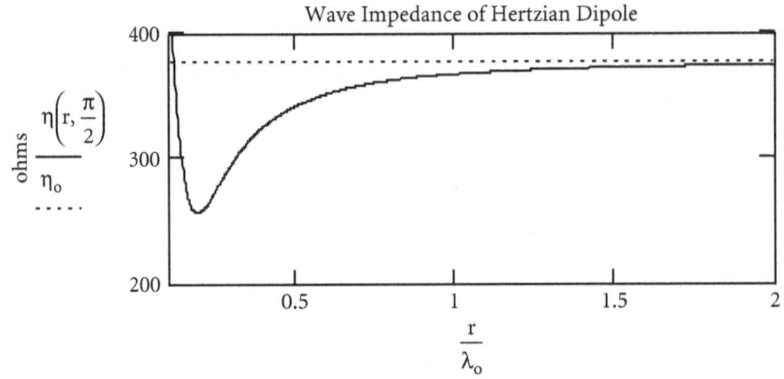

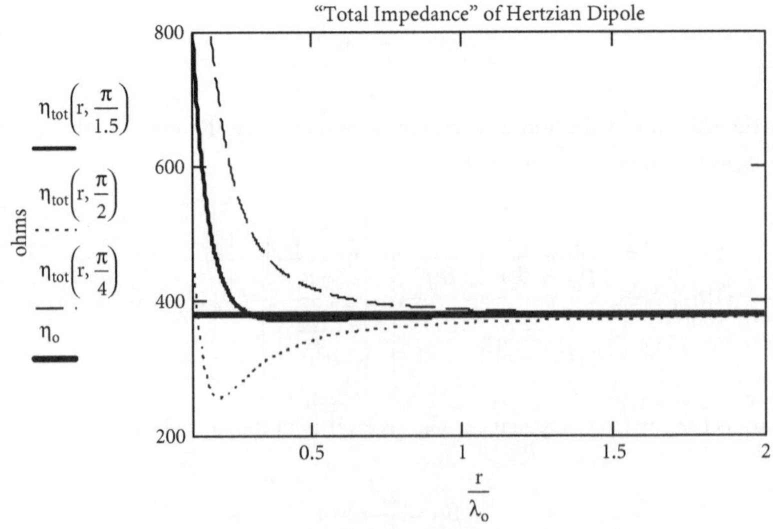

MATHCAD 1.7 Wave and "total impedance" magnitude vs. distance (relative to the wavelength) for a Hertzian dipole.

$\mu_o := 4\cdot\pi\cdot10^{-7}$ $\varepsilon_o := 8.854\cdot10^{-12}$ $v_o := 3\cdot10^{8}$ $I_m := 1\cdot10^{-6}$ $f := 100\cdot10^{6}$ $j := \sqrt{-1}$

$\eta_o := \sqrt{\dfrac{\mu_o}{\varepsilon_o}}$ $\lambda_o := \dfrac{v_o}{f}$ $\beta_o := \dfrac{2\cdot\pi}{\lambda_o}$ $a := \dfrac{\lambda_o}{100}$

$r := \dfrac{\lambda_o}{10}, \dfrac{\lambda_o}{9.80} .. 2\cdot\lambda_o$

$$H_r(r,\theta) := \frac{j\cdot2\cdot I_m\cdot\pi\cdot a^2\cdot\mu_o\cdot2\cdot\pi\cdot f}{4\cdot\pi}\cdot\frac{\beta_o^{\,2}}{\eta_o}\cdot\cos(\theta)\cdot\left(\frac{1}{\beta_o^{\,2}\cdot r^2}-\frac{j}{\beta_o^{\,3}\cdot r^3}\right)\cdot e^{-j\cdot\beta_o\cdot r}$$

$$H_\theta(r,\theta) := \frac{j\cdot I_m\cdot\pi\cdot a^2\cdot\mu_o\cdot2\cdot\pi\cdot f}{4\cdot\pi}\cdot\frac{\beta_o^{\,2}}{\eta_o}\cdot\sin(\theta)\cdot\left(\frac{j}{\beta_o\cdot r}+\frac{1}{\beta_o^{\,2}\cdot r^2}-\frac{j}{\beta_o^{\,3}\cdot r^3}\right)\cdot e^{-j\cdot\beta_o\cdot r}$$

$$E_\phi(r,\theta) := \frac{-j\cdot I_m\cdot\pi\cdot a^2\cdot\mu_o\cdot2\cdot\pi\cdot f}{4\cdot\pi}\cdot\beta_o^{\,2}\cdot\sin(\theta)\cdot\left(\frac{j}{\beta_o\cdot r}+\frac{1}{\beta_o^{\,2}\cdot r^2}\right)\cdot e^{-j\cdot\beta_o\cdot r}$$

$\eta(r,\theta) := \left|\dfrac{E_\phi(r,\theta)}{H_\theta(r,\theta)}\right|$ $\eta_o = 376.734$ $\eta_{tot}(r,\theta) := \left|\dfrac{E_\phi(r,\theta)}{\sqrt{H_\theta(r,\theta)\cdot\overline{H_\theta(r,\theta)}+H_r(r,\theta)\cdot\overline{H_r(r,\theta)}}}\right|$

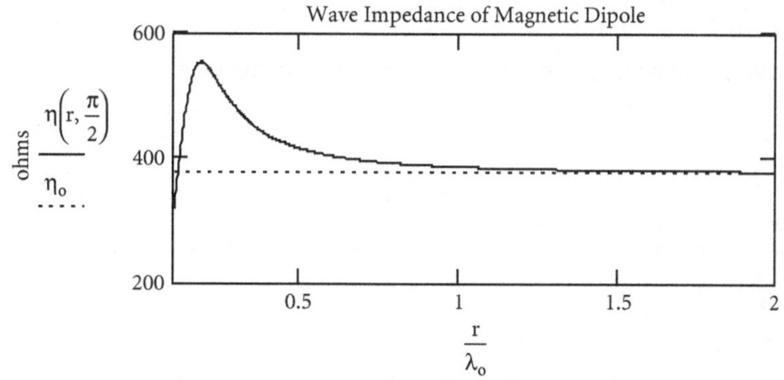

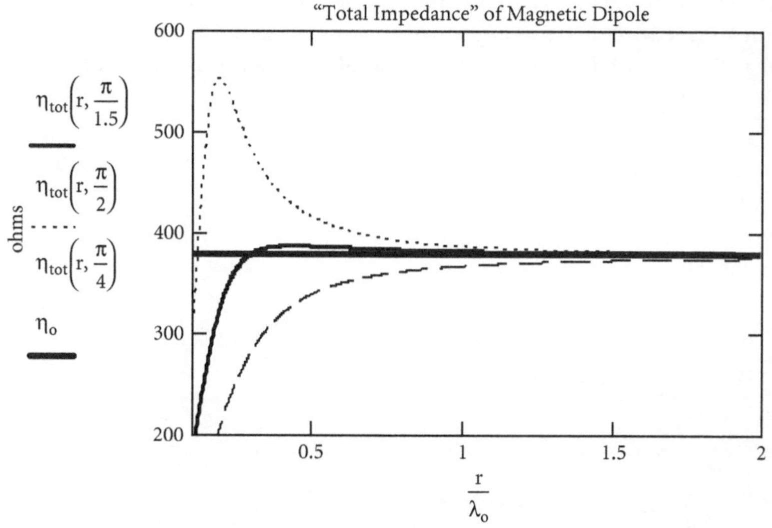

MATHCAD 1.8 Wave and "total impedance" magnitude vs. distance (relative to the wavelength) for a magnetic dipole.

electric field, including the radial component, to the magnetic field is plotted as a function of distance from the antenna, this "total impedance" is θ dependent. This total field definition is rarely used.

The wave impedance of the magnetic dipole is

$$\eta_w = \frac{E_{\phi s}}{H_{\theta s}} = \frac{-\eta_o\left(\dfrac{j}{\beta_o r} + \dfrac{1}{\beta_o^2 r^2}\right)}{\dfrac{j}{\beta_o r} + \dfrac{1}{\beta_o^2 r^2} - \dfrac{j}{\beta_o^3 r^3}}\ \Omega \tag{1.119}$$

This impedance is independent of the angle, θ. Notice that very close to the antenna the magnitude of the wave impedance is equal to

$$|\eta_w| = \left|\frac{-\eta_o\left(\dfrac{j}{\beta_o r} + \dfrac{1}{\beta_o^2 r^2}\right)}{\dfrac{j}{\beta_o r} + \dfrac{1}{\beta_o^2 r^2} - \dfrac{j}{\beta_o^3 r^3}}\right| \approx \left|\frac{-\eta_o\left(\dfrac{1}{\beta_o^2 r^2}\right)}{-\dfrac{j}{\beta_o^3 r^3}}\right| = \eta_o \beta_o r\ \Omega \quad \text{if } \beta_o r \ll 1 \tag{1.120}$$

The wave impedance is small near the antenna. In the far field, the magnitude of the wave impedance is about 377 Ω:

$$|\eta_w| = \left|\frac{-\eta_o\left(\dfrac{j}{\beta_o r} + \dfrac{1}{\beta_o^2 r^2}\right)}{\dfrac{j}{\beta_o r} + \dfrac{1}{\beta_o^2 r^2} - \dfrac{j}{\beta_o^3 r^3}}\right| \approx \left|\frac{-\eta_o\left(\dfrac{j}{\beta_o r}\right)}{\dfrac{j}{\beta_o r}}\right| = \eta_o\ \Omega \quad \text{if } \beta_o r \gg 1 \tag{1.121}$$

In the far field, the wave impedance is called the intrinsic impedance. As seen in Mathcad 1.8, the wave impedance of the magnetic dipole also approaches the intrinsic impedance of a plane wave when far from the antenna. As with the Hertzian dipole, the "total impedance" is θ dependent.

2

Electric Field Shielding

Electric field shielding refers to the shielding of near-field electric fields. Near-field electric fields are frequently described as having a high impedance and, therefore, easily reflect off low-impedance metals. It is important to realize that near-field electric fields are not plane waves, and their field distribution is typically complex and source dependent. Fortunately, it is relatively easy to shield against near-field electric fields.

2.1 The "Magic" of Electric Field Shielding Revealed

Qualitatively explain in terms of induced charge how a shield "blocks out" near-field electric fields.

In response to any electric field, applied, external, or interfering, charge on an electric field shield redistributes itself. This charge on the shield is also a source of an electric field. The interaction of all of the fields is such that the electric field in the region to be shielded is less than without the shield. A charge redistribution argument can be at times too fundamental to use to explain near-field electric field shielding. However, a one-time study of a simple scenario reduces or even eliminates the magic associated with this type of shielding. The ground reference in this specific discussion is modeled as a physical object of limited dimensions and not an infinitely large idealistic ground.

Referring to Figure 2.1, imagine that a voltage source charges an initially charge-neutral conducting spherical electrode to some positive potential. The ground plane is also initially charge neutral. The energy from the voltage source is sufficient to transfer electrons from the spherical electrode to the ground plane reference, leaving the spherical electrode positively charged and ground plane negatively charged. The positive and negative symbols in this figure are only crudely representing the location of the charge on the bodies and obviously do not represent the actual quantity of charge in the metallic objects.[1] Also, these charge distributions assume that there are no other nearby objects to influence the charge distribution. The conductivities of the two conductors are large so that the relaxation time, a measure of the time required for the charges to distribute themselves on the objects, is very small. If the observer waits several relaxation time constants, the steady-state charge distribution will appear on the surfaces of the two conductors. An electric field exists between the two objects due to these surface charges. In this particular example, this near electric field is to be "blocked" or reduced.

A charge-neutral trough-shaped metallic conductor is next placed around the spherical electrode, but it is left floating. This can be accomplished by supporting this floating conductor on a good insulator base material of very high relaxation time constant such as Teflon. As seen in Figure 2.2, an electric field does appear outside of the floating conductor. The free electrons in this floating conductor are attracted to the positively charged spherical electrode and appear along the inner surface of the floating conductor. Since the total charge on the charge-neutral conductor is zero, positive charge must exist somewhere on

[1]The charge and electric field distributions are only crudely sketched in the figures in this book. Although the electric fields may not appear normal to the perfectly conducting surfaces and electrodes in the sketches, the electric fields should be entirely normal.

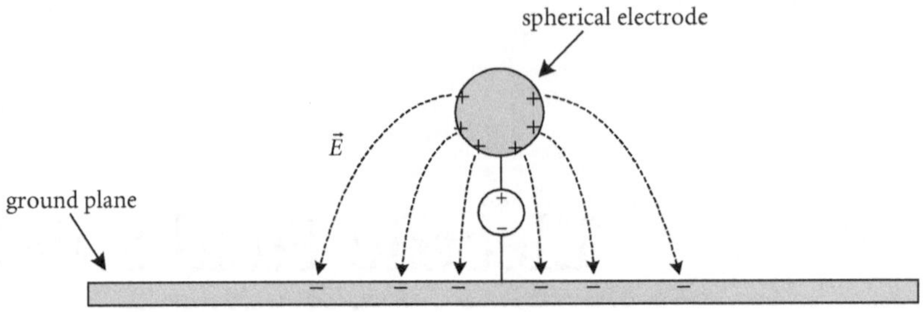

FIGURE 2.1 Charging a spherical electrode to a positive potential relative to a ground plane.

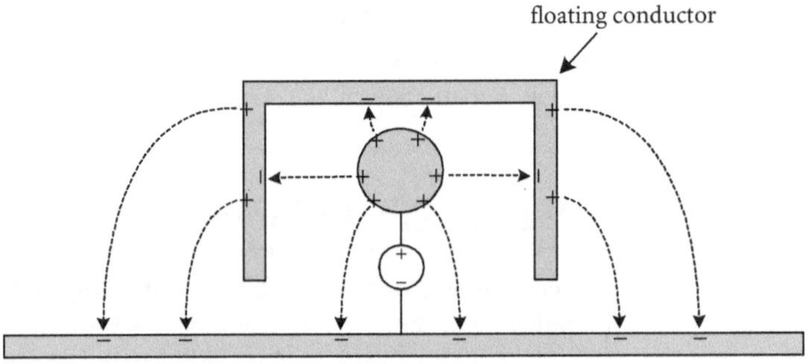

FIGURE 2.2 Static electric fields passing "through" the floating conducting "shield."

the surface of this conductor. In this case, the positive charge[2] is on the outside of the conductor. Since charge is a source of electric field, the electric field appears outside the trough-shaped conductor; thus, this floating conductor is not effective in shielding the near-field electric field originating from the electrode. (For the floating conductor to be considered in the near field, the distance between the spherical electrode and floating conductor must be electrically small.) Note that increasing the thickness of this floating conductor does not reduce the field. Of course, the field does decrease with increasing distance from the source of the field. The actual field distribution outside of the floating conductor is a function of the geometry of its outer surface.

Frequently, for mechanical convenience, safety, electrostatic discharge, current return, or electric field control, a conductor is "grounded." In this example, the flat conducting plane is the ground. Figure 2.3

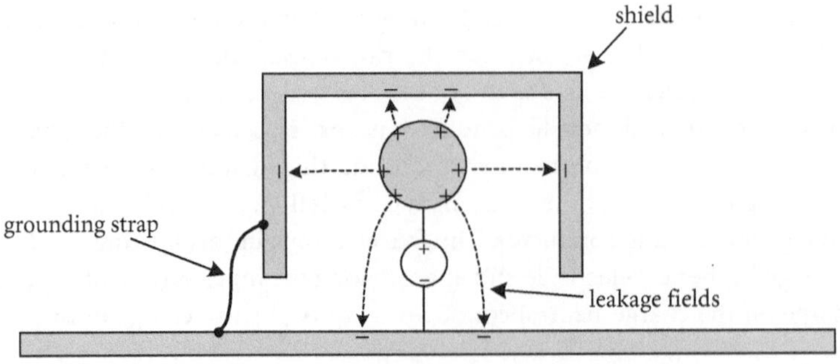

FIGURE 2.3 Grounding the shield conductor to help contain the electric fields.

[2]Positive charges in a metallic conductor are actually a lack or deficiency of negatively charged electrons.

illustrates the consequence of connecting the floating conductor to this ground plane, which is the voltage reference, via a grounding strap. The floating conductor is no longer floating but at a fixed potential of zero volts. This conductor, which partially surrounds the charged spherical conductor, is referred to as an electric field shield. This shield, in this particular case, is shielding or protecting the region exterior to itself from fields originating in the interior of the shield.[3] The negative charge on the flat reference conductor is attracted to the positive charge on the shield. After a few time constants, which is a very short period of time for metallic objects, a fraction of the total negative charge passes through the strap connected to the shield thereby neutralizing the positive charge on the shield. Thus, this grounded shield has "blocked" the near-field electric field: the net charge contained within an imaginary surface around the shield is zero, and no net surface charge exists on the outside of the shield. To repeat, it is necessary to ground a shield if it is to be effective in preventing near-field electric fields from passing beyond it. As shown in Figure 2.3, because the shield does not completely enclose the charged spherical electrode, some leakage fields exist between the sphere and ground plane. If the shield completely enclosed the charged spherical electrode (and the ground plane were very large or ideal), there would be no *leakage* fields.

Previously, it was assumed that the charging voltage source was dc. For most situations, the time of interest is usually much greater than the relaxation time of the metallic objects, and the period of a dc source is infinitely greater than the relaxation time. If the frequency of the source is not zero, its period is finite. An ac source will cyclically positively and negatively charge the spherical electrode. The grounded shield will be effective in "blocking" the near-field electric field if the charge redistribution can occur much faster than the frequency of the source. If the charge cannot be neutralized at a much faster rate than the source is changing, the electric field will not be "blocked" by the shield. Of course, if the frequency of the source is too great, the shield will no longer be in the near-field, and far-field techniques must be applied.

It is helpful to use the concepts of capacitance, inductance, and resistance to determine whether the charge can quickly distribute itself for ac sources. For example, the ability of the charge to disperse from the ground plane to the grounded shield is a function of the time constant ($\tau = RC$) of the system. For this example, at lower frequencies, the resistance of the strap and the capacitance between the spherical electrode and shield are often the dominant circuit elements. Multiple grounding straps are used to reduce the impedance for the electron transfer, allowing the charge to transfer more easily between the ground plane and shield. If the dimensions of the system are not electrically small, then a lumped model of the system is not appropriate.

2.2 Size is Important!

Positive charge is inside the cavity of a spherical conducting shell. Discuss why this charge produces an electric field outside of the cavity if the shell is floating but not if the shell is connected to an idealistic ground plane. If the shell is connected to a finite size "ground" about the size of the shell, is the field outside the shield still zero? If the charge is moved to a different location inside the shell, the electric field inside the cavity will change but the electric field outside will not. Explain why. [Rogers]

If the shell completely enclosing the positive charge is floating, then a negative charge is induced along the inner surface of the shield and a positive charge is induced on the outer surface of the shield. The total charge on the outer surface of the shell is equal but opposite to the charge on its inner surface if the shell was initially charge neutral. The electric field emanates from the positive charge (or appearance of positive charge) on the outer surface. Hence, this floating shell does not function as a shield to near-field electric fields. To shield the electric field, the shell must be connected to a good ground.

The size of the good ground is important. Imagine that the ground is about the same size as the shell as shown in Figure 2.4; that is, the outer surface area of the sphere is about equal to the surface area of the ground plane. Initially, both the shell and ground are charge neutral. After connecting the shell to this ground, negative charge (electrons) from this ground plane pass via the grounding strap to the shield.

[3]Of course, a shield can also protect its interior from exterior fields.

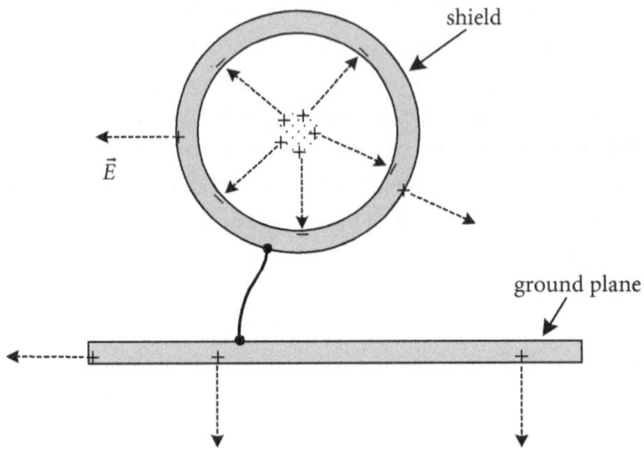

FIGURE 2.4 Finite ground plane and shield connected via a grounding strap.

The positive charge inside the cavity of the shell induces these negative charges on the inner surface of the shell. The charges are induced since there is no ohmic connection between the shell and positive charge inside the cavity. The net charge on the ground plane is now positive since it transferred some negative charge to the shell. Note that the net charge on the shield and ground plane is the same before and after connecting the ground strap. Grounding the shell results in a *redistribution* of charge. One of the first questions asked by students is, "When does the charge transfer between the shield and ground plane stop?" The charge transfer ceases when the potential of the shield is equal to the potential of the ground plane. A potential difference is required for charge flow (i.e., current) between two objects. A potential difference implies an electric field. (Zero surface charge is *not* required for two objects at the same potential. Similarly, an equipotential conductor does *not* require either zero charge along its surface or a uniform surface charge distribution.)

Electric fields exist outside and near the shield since the ground plane is not ideal. Although the potential difference between the shield and ground plane is zero, the overall system has a net charge. The net charge for the situation shown in Figure 2.3 was zero. In many academic problems, the ground plane is assumed infinitely large and an infinite reservoir of charge. Any charge on the ground plane is distributed over its surface, and these charges are a source of an electric field.[4] In Figure 2.4, positive charge exits on *both* the outer surface of the shield and along the surface of the ground plane. Charge exists along the outer surface of the shield since the ground plane is not large in comparison to the shield. When two conducting objects are connected together via a conducting strap, the quantity of charge on each object after the connection is a function of the relative size of the objects. The larger the object, the greater the percentage of the overall charge it will carry. The electric field distribution outside the shield is also a function of the relative size and shape of the shield and ground plane. If the ground plane is initially very large compared to the shield, increasing the size of the ground plane will have little effect on the charge distribution on the ground and the field emanating from this charge. If the ground plane is comparable or smaller than the shield, increasing the size of the ground plane can noticeably affect the charge and field distribution.

The position and strength of the charge inside the conducting spherical shell, which completely encloses the charge, will affect the electric field distribution inside the cavity. For example, if the charge is moved very close to the cavity's inner surface, the electric field between the charge and this surface will intensify while the electric fields elsewhere will diminish. Surprisingly, the electric field outside the cavity will remain the same for a given total enclosed charge no matter where the charges are located inside the shield cavity. The electric field distribution *outside* the shell is determined solely by the size, shape, and position of the

[4]The net charge is still conserved in the system. Unless the positive and negative charges "overlap" and neutralize each other, any unneutralized charge must produce an electric field.

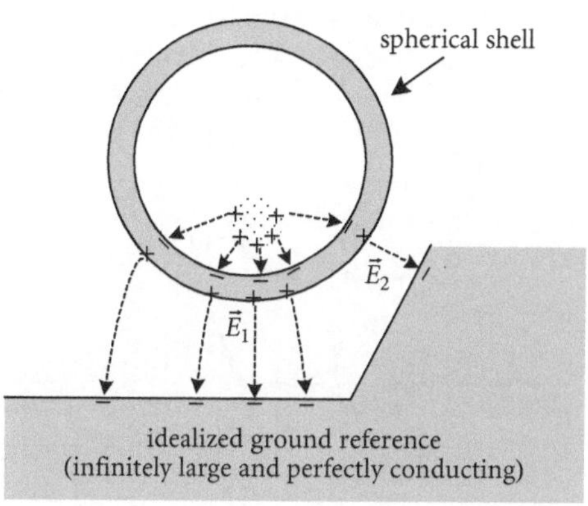

FIGURE 2.5 Reasonable electric field distribution within the cavity, but unreasonable field distribution outside the cavity.

outer surface of the shell relative to the size and shape of the ground plane and other objects outside the shield. The validity of this statement is easily checked. Referring to Figure 2.5, positive charges inside a cavity are shown close to a portion of the cavity's inner surface; hence, the electric field in the cavity is correctly shown as strongest between this charge distribution and the nearby inner surface. However, the electric field outside of the shell, as indicated by the density of the field lines, is *falsely* shown as strongest near $\vec{E}_1$. Again, the electric field $\vec{E}_1$ is *falsely* assumed greater than the electric field $\vec{E}_2$. It is not difficult to show that the electric field distribution between the shell and ground reference is incorrectly sketched. First, recall that the potential between any two locations a and b for an electrostatic situation is

$$V = -\int_a^b \vec{E} \cdot d\vec{L} \tag{2.1}$$

Second, since the electric field tangential to, or along, a perfectly conducting surface is zero, the potential *difference* between any two locations along a perfectly conducting object is also zero (i.e., the voltage along a perfectly conducting object is the same everywhere). Third, according to Kirchoff's voltage law, the net voltage drop around a closed path is zero. Therefore, referring to Figure 2.6,

$$\oint \vec{E} \cdot d\vec{L} = \int_a^b \vec{E}_1 \cdot d\vec{L} + \int_b^c 0 \cdot d\vec{L} + \int_c^d \vec{E}_2 \cdot d\vec{L} + \int_d^a 0 \cdot d\vec{L} = 0$$

$$\Rightarrow \int_a^b \vec{E}_1 \cdot d\vec{L} = -\int_c^d \vec{E}_2 \cdot d\vec{L} = \int_d^c \vec{E}_2 \cdot d\vec{L}$$

The second integral is zero since the voltage drop from b to c along the conducting ground reference is zero. The fourth integral is zero since the voltage drop from d to a along the conducting shield is also zero. As shown previously, the electric fields $\vec{E}_1$ and $\vec{E}_2$ are both directed from the outside of the shell to the ground reference. If both of these fields are assumed constant along these paths, then

$$\int_a^b \vec{E}_1 \cdot d\vec{L} = |\vec{E}_1| d_{ab} = \int_d^c \vec{E}_2 \cdot d\vec{L} = |\vec{E}_2| d_{cd} \text{ [5]} \tag{2.2}$$

[5]This equation is stating that the potential of the perfectly conducting sphere relative to the ground plane is the same everywhere.

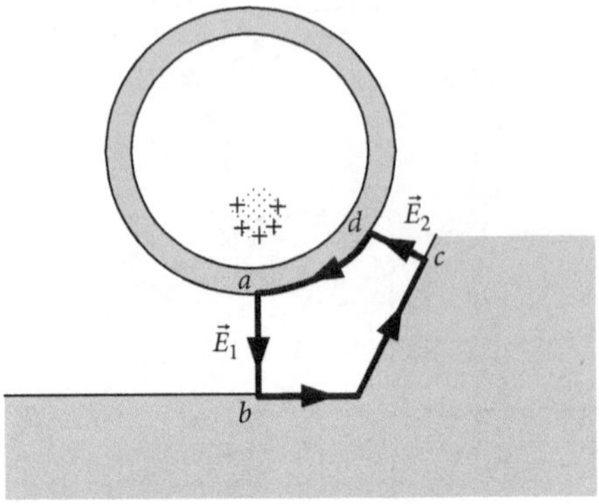

FIGURE 2.6 Net voltage drop along the given closed path must be zero.

where d_{ab} is the distance from a to b and d_{cd} is the distance from c to d. Since $\vec{E}_1$ was assumed larger than $\vec{E}_2$ and, by inspection, d_{ab} is greater than d_{cd}, it follows that $|\vec{E}_1|d_{ab} > |\vec{E}_2|d_{cd}$. This contradicts (2.2). To satisfy the equality $|\vec{E}_1|d_{ab} = |\vec{E}_2|d_{cd}$, $|\vec{E}_1|$ must be less than $|\vec{E}_2|$. This implies that the charge on the outer surface of the shell and the ground plane should be more concentrated near $\vec{E}_2$. This charge concentration is reasonable since the shell and ground are closest at this location. The actual position of the charges inside the cavity do not affect the field distribution outside the cavity.

2.3 Shielding Reciprocity?

A positive charge outside of an empty, spherical conducting shell cannot produce an electric field inside the shell (whether it is floating or grounded). By falsely assuming that a potential distribution does exist inside the shell due to this external charge, show that this is a true statement. [Rogers]

As was previously discussed, a floating conducting "shield" enclosing charge is not generally effective for reducing the strength of the fields exterior to itself,[6] although the floating shield can distort the electric field from the charge. It is probably surprising, then, that a static electric field source outside the shield cannot pass through the shield to the inner cavity, whether the shield is floating or grounded. When a conducting enclosure is used to "block" electric fields, it is sometimes referred to as a Faraday shield. A sketch of the charge distribution on a spherical shield in the presence of an exterior static charge distribution is shown in Figure 2.7. Notice the lack of surface charge along the inner surface of the shield. If both negative and positive surface charge did exist along the inner surface of the shield, then an electric field would have to exist inside the cavity as shown in Figure 2.8. A nonzero field inside the cavity is not

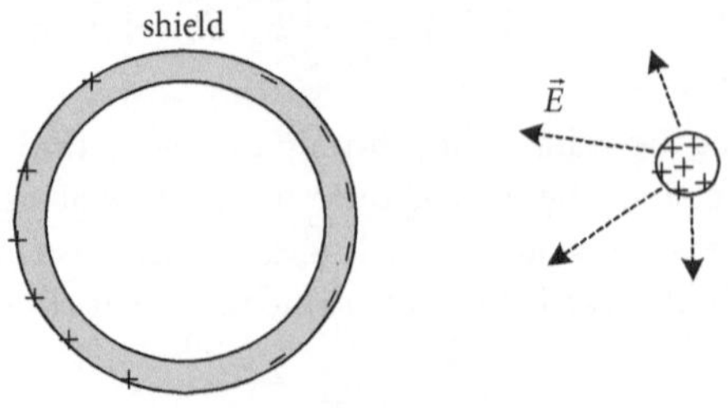

FIGURE 2.7 Surface charge distribution induced along a neutral floating shield when exposed to an external, static electric field.

[6] See the later discussion on time-varying charge inside a conducting shell.

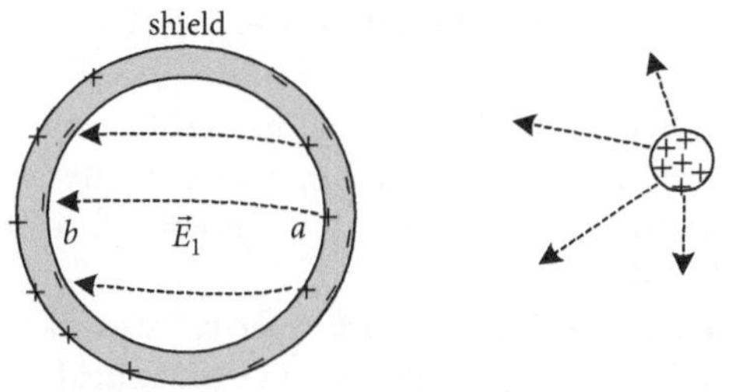

FIGURE 2.8 Electric field distribution inside the cavity of a shield *if* positive and negative surface charge existed along its inner surface.

possible since the voltage around a closed loop (for an electrostatic situation) must be zero:

$$\oint \vec{E}\cdot d\vec{L} = \int_a^b \vec{E}_1 \cdot d\vec{L} + \int_b^a 0\cdot d\vec{L} = 0 \;\Rightarrow\; \int_a^b \vec{E}_1 \cdot d\vec{L} = 0$$

The electric field inside the cavity must be zero. (Electrostatic fields originate on positive charge and terminate on negative charge.) The return path from b to a is along the inner surface of the shield. This second integrand is zero since the electric field tangential to a perfect conductor is zero.

If the external electric field did produce an electric field inside the cavity, then a potential distribution must also exist inside the cavity. Since the potential along the inner surface of the conducting shield must be the same everywhere, another potential contour can be drawn near this surface inside the cavity. Potential contours do not overlap or suddenly terminate. (Overlapping contours would imply multiple different potentials at one location). Inside this contour, another contour can be drawn. These and other hypothetical contours are shown in Figure 2.9. The electric field lines are perpendicular to the potential contours (a property of the gradient of a scalar function) since

$$\vec{E} = -\nabla\Phi \tag{2.3}$$

where Φ is the potential distribution function. In this figure, the field lines imply a source of charge exists inside the cavity, but there is no charge inside the cavity. For any field distribution inside the cavity, this contradiction exists. Thus, external static charges cannot produce an electric field inside a floating (or grounded) cavity.

The potential everywhere inside and on the conducting shield is the same. This implies zero electric field everywhere inside the shield and inside the cavity. When the shield is grounded, the potential inside and on the shield is zero by definition.

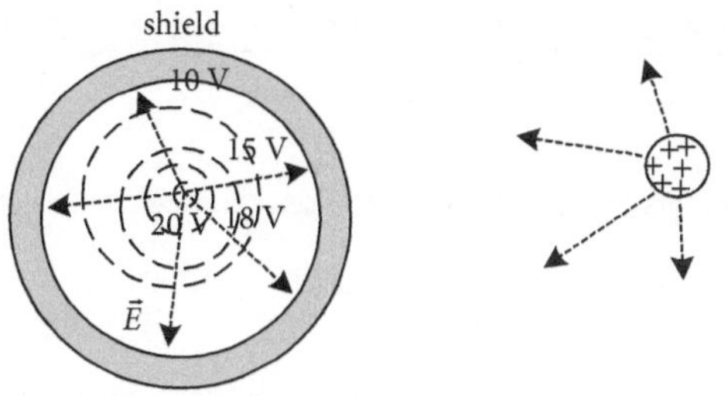

FIGURE 2.9 Equipotential distribution inside the cavity of a shield *if* surface charge existed along its inner surface.

2.4 Using Capacitance to Model Shielding

How can the circuits concept of capacitance be used to model shielding and shield leakage? [Morrison, '98]

Competency in electrostatic (or electroquasistatic[7]) shielding requires an understanding of the distribution of the electric fields between objects. At times, the concept of circuit capacitance can also be used to help understand electric field shielding.

With two isolated or separated conducting objects, a capacitance exists between the two objects. When there is a voltage difference between the two conductors, an electric field exists between them. With three or more isolated conductors, a capacitance can exist between each of the conductors. Sometimes, the (mutual) capacitance between two conductors is zero when one of the three (or more) conductors functions as a shield. Imagine that there are three conductors in space, and conductor 3 completely encloses conductor 1 as shown in Figure 2.10. A mutual capacitance exists between conductors 1 and 3 and conductors 2 and 3. However, the capacitance between conductors 1 and 2, labeled C_{12}, is zero.

The concept of capacitance is also used when the voltage across two objects is time varying. Although, most of the discussion hitherto assumes electro*static* conditions, as long as the dimensions involved (e.g., distance between and size of the conductors) are electrically short, the system can be considered nearly electrostatic or electroquasistatic. Imagine that the charge on object 2 is varied by connecting it to a time-varying voltage source referenced to a ground conductor as shown in Figure 2.11. Con-

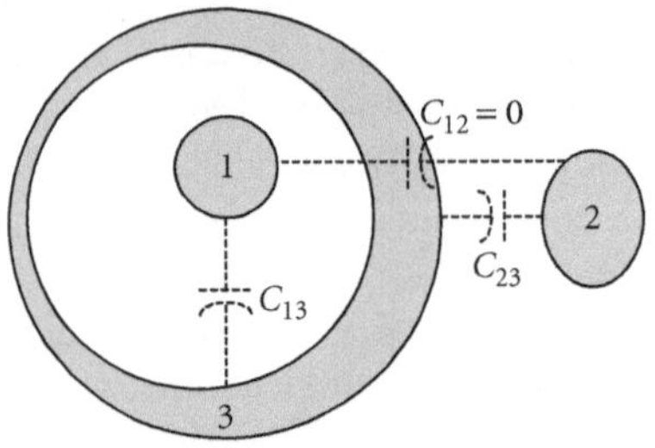

FIGURE 2.10 Conductor 3 completely enclosing conductor 1.

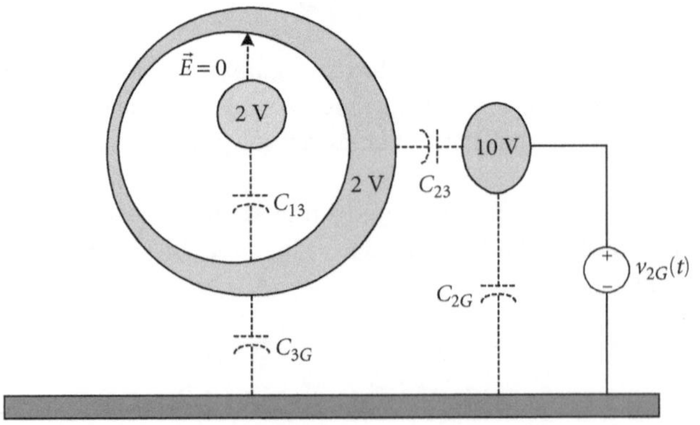

FIGURE 2.11 Potential of conductors 1 and 3 are identical when exposed to an electric field external to conductor 3.

[7]Quasi means seemingly or nearly. Those readers familiar with Victor Hugo's beautiful novel *The Hunchback of Notre-Dame* might recall the fascinating character Quasimodo, which might be roughly translated as "seemingly the devil."

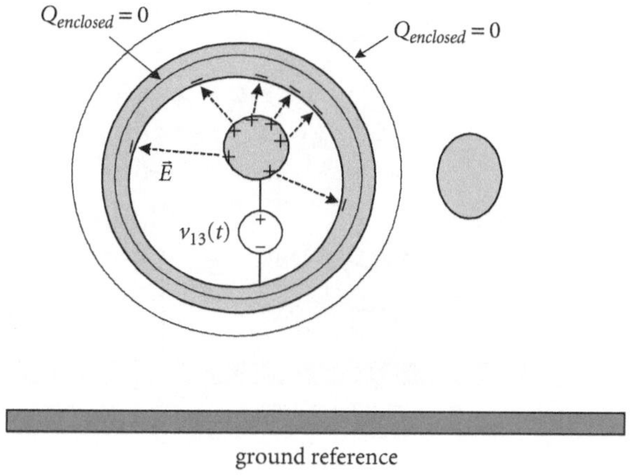

FIGURE 2.12 Net charge enclosed by an imaginary closed surface external to conductor 3 is zero.

ductor 3 is floating or not connected directly to any voltage source or the ground reference, and both conductors 1 and 3 are charge neutral. The hypothetical voltages shown are at one instant in time. The remaining capacitances between the conducting objects and ground are also shown. The potential of conductor 2 will affect the potential of conductors 1 and 3. A floating conductor will assume a constant potential determined by its electrical environment. However, as was shown previously, when conductor 1 is charge neutral, the potentials of conductors 1 and 3 are identical and the electric field between them is zero. Since the charge on conductor 2 cannot induce charge on conductor 1, the capacitance between them must be zero.

To place a time-varying charge or voltage on conductor 1, a floating or battery operated voltage source can be connected between conductors 1 and 3 both assumed to be initially charge neutral. Although not shown in Figure 2.12, the capacitances are still present. The floating voltage source transfers negatively charged electrons between conductor 3 and conductor 1. A voltage source does not generate or produce charge: it does provide the necessary energy to distribute charge. The net charge enclosed, $Q_{enclosed}$, by any closed imaginary surface surrounding the inner surface of conductor 3 is zero. Two imaginary surfaces are shown. For the imaginary surface contained within conductor 3, the electric field everywhere along the surface must obviously be zero: the electric field within a perfect conductor is zero. Therefore, the positive charges shown (for a particular instant of time) on conductor 1 and negative charges shown on conductor 3 generate no fields outside this imaginary surface. Furthermore, since there are no surface charges present on the outer surface of conductor 3, this floating voltage source generates no fields external to conductor 3. Although this is not a rigorous "proof," this should convince most readers that no external fields are generated by this floating voltage source.

The time-varying charge on the inner conductor can also be obtained by connecting a voltage source between the ground reference and inner conductor through a small opening in conductor 3. If conductor 3 is still floating, then $Q_{enclosed}$ is only zero when the voltage source is zero. The charge on the ground reference is equal but opposite to the charge on conductor 1. Hence, the electric fields pass through the floating shield. If the shield conductor 3 is connected to a good ground reference, the field passing through the shield is very small. Depending on the size of the small opening, there will be some field leakage through conductor 3.

An opening in the shield conductor 3 allows electric field coupling between conductors 1 and 2. This leakage is represented by the capacitance C_{12} in Figure 2.13. There are only a meager number of analytical expressions for the capacitances between three or more conducting objects. (However, if the capacitances are not too small, they can be measured.) Applying a positive voltage and, hence, charge to conductor 2 will affect the charge distribution on conductor 1 (and conductor 3) as shown in Figure 2.14. Conductors 1

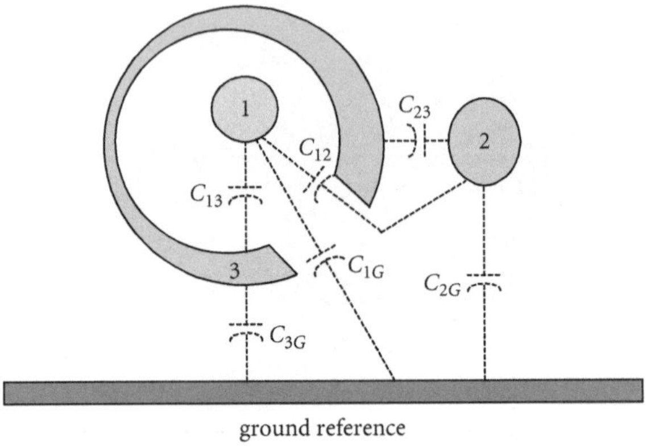

FIGURE 2.13 Capacitive coupling between conductors 1 and 2 through an opening in conductor 3.

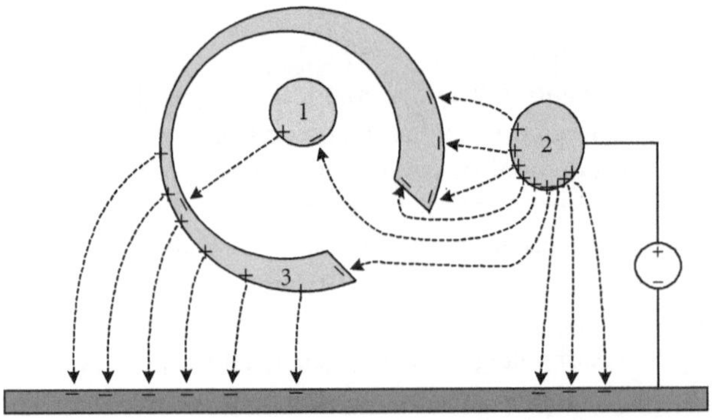

FIGURE 2.14 Induced charge on conductor 1 through an opening in conductor 3.

and 3 are charge neutral before and after the application of the voltage to conductor 2. (The number of "+" charges and "–" charges on conductor 1 and on conductor 3 are equal.) The position and size of the opening in conductor 3 will affect the coupling capacitance C_{12}. For example, if conductor 3 is rotated clockwise so that the opening is on the far left, farthest from conductor 2, the capacitance C_{12} will decrease. If conductor 3 is connected to the ground plane via a grounding strap, then more electric field lines from conductor 2 will terminate on conductor 3 (and possibly conductor 1). A grounded conductor 3 is physically closer to conductor 2 than the ground plane. A simple example of a partial shield is a grounded (not floating) metal roof of an instrument shed located beneath a radio relay tower.

2.5 Capacitor Shielding

Normally, nearby objects have negligible effect on the capacitance of a capacitor. However, if highly accurate capacitors are required, with a reproducibility of 0.1 pF, the capacitor should be shielded. If the capacitor is completely enclosed in a copper case, except for the two terminals to the capacitor leads, where should the case be connected? Should it just "float?" [Kibble]

A capacitance exists between isolated or separated conducting objects. Unless these objects and the medium between them were specifically created to generate a known capacitance, the capacitance between them is usually referred to as a parasitic capacitance. For example, if one of the objects is a highly accurate

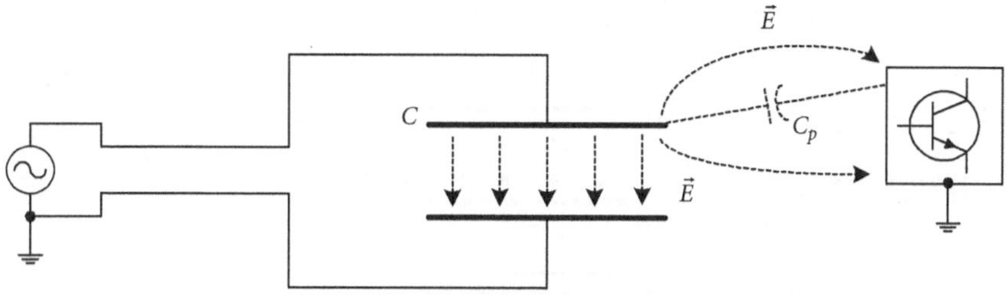

FIGURE 2.15 Parasitic capacitance between the ungrounded plate of the capacitor and grounded heat sink.

capacitor and the other object is a transistor with a grounded heat sink, as shown in Figure 2.15, then a parasitic capacitance exists between the capacitor and heat sink. When a time-varying voltage v exists between these two objects, a displacement current

$$i = C_p \frac{dv}{dt} \tag{2.4}$$

can "flow"[8] between the two objects. The parasitic capacitance is due to the stray electric fields between the capacitor and heat sink.

If this same capacitor were placed in a different circuit with a different environment, then the parasitic capacitance to nearby objects would likely be different. A shield placed around the capacitor is designed to fix the parasitic capacitance from the capacitor to its environment. In Figure 2.16, the parasitic capacitance between the capacitor and shield is given as C_{ip} and the parasitic capacitance between the shield and heat sink is given as C_{ep}. (The capacitance, C, is expected to be greater than either of these parasitic capacitances.) In this figure, the shield is floating. It is not connected to any particular voltage reference, and its potential is determined by nearby charged objects. An electric field exists between objects of different potentials. In this example, an electric field exists between the capacitor, C, and shield, and between the shield and heat sink.

To produce a capacitor with a capacitance of high reproducibility, it is necessary to control or minimize any stray electric fields to the capacitor. Any parasitic capacitance can result in stray current to other objects. From the standpoint of the voltage source in Figure 2.16, any additional current from the source is due to a net capacitance different from C.

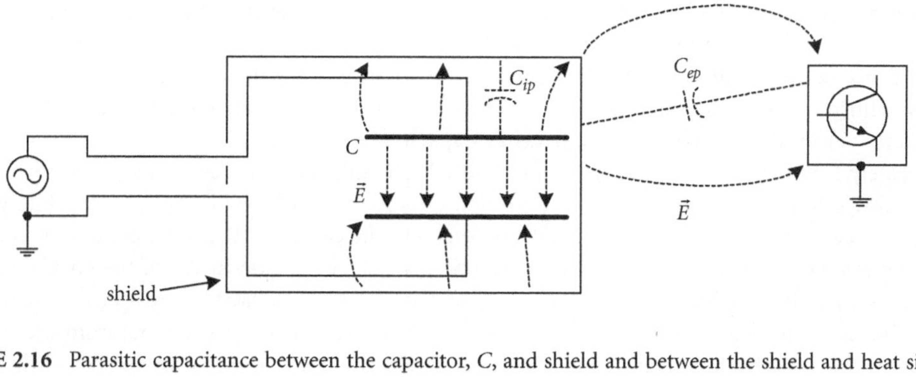

FIGURE 2.16 Parasitic capacitance between the capacitor, C, and shield and between the shield and heat sink.

[8]Current does not "flow." Current is the rate of charge flow since $i = dq/dt$.

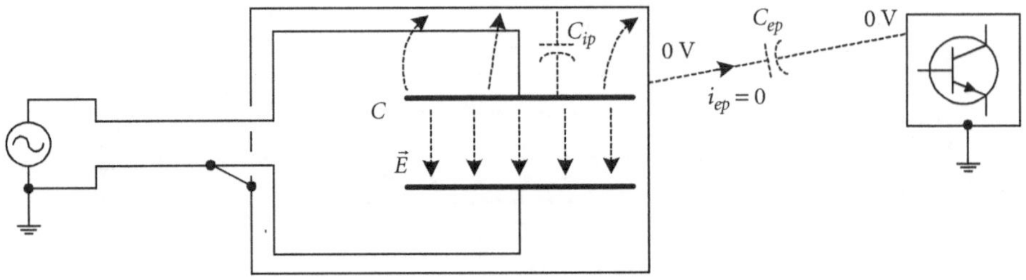

FIGURE 2.17 Grounding the shield around the capacitor, C, to obtain a more stable and predictable capacitance.

The shield for the capacitor should not float, but it should instead be connected to the low-potential side (or grounded side) of the source as shown in Figure 2.17. This will eliminate the electric fields between the shield and heat sink and nullify the effect of the parasitic capacitance, C_{ep} (assuming the heat sink is also at approximately zero volts). Although the parasitic capacitance between the capacitor and shield has not been reduced, the parasitic capacitance to C is at least fixed (assuming the physical relationship between the shield and capacitor is constant) and, if not too small, measurable.

Many capacitors are unbalanced. For example, for tubular capacitors, the outermost circular "plate" is closer to some external ground than the other inner "plate." On the other hand, a simple disc capacitor, composed of two circular parallel discs, is more likely to be a balanced device. Some unbalanced capacitors have a preferential "plate," which is referred to as the low side. A band on tubular capacitors indicates the lead that is connected to the outside foil and, thus, the lead that should be connected to the low potential or ground reference for shielding purposes.

2.6 Three-Terminal Capacitor

As an improvement over the shield shown in Figure 2.17, a "three-terminal" capacitor is used. The shield encloses the capacitor and has two coaxial ports. One of the coaxial terminals is connected to the shield and the other is not connected to the shield. Explain why this "three-terminal" capacitor is an improvement over the shielded capacitor previously presented. [Kibble]

In the scenario given in Figure 2.17, the nearby object was at ground potential and the low side of the capacitor was at ground potential. By placing the shield at the same potential a the nearby object, the electric fields and "capacitor current" between the shield and heat sink were eliminated. In many situations, nearby objects and the low side of the capacitor are not necessarily at the same potential. Merely connecting the shield to the low side of the capacitor in this case will not eliminate the effects of the parasitic capacitance: a potential difference can still exist between the shield and nearby objects. If the shield is connected to the object to eliminate this potential difference, an excellent path will exist for charge to flow from the capacitor via the parasitic capacitance C_{ip}.

Fortunately, some of these difficulties can be resolved. The shield can be kept at ground potential even when the low side of the capacitor is not connected directly to ground by bringing a separate ground line to the shield. By using a coaxial line from the source to the capacitor, the outer conductor can carry the ground reference to the shield as shown in Figure 2.18. The outer conductor of the coaxial line acts as an extension of the shield. The parasitic capacitance between the capacitor and shield now includes the capacitance of the coaxial line. Of course, external objects can also affect the line from the shielded capacitor to the load. For this reason, an additional shield is placed around this line to minimize the effects of the parasitic capacitance from the line to nearby objects. The parasitic capacitance between the capacitor and shield is now well controlled. If an object of nonzero potential is in close proximity to the shield or coaxial lines, the parasitic current between the capacitor and shield will tend to return along the good-conducting outer shield rather than through another path with a higher impedance.

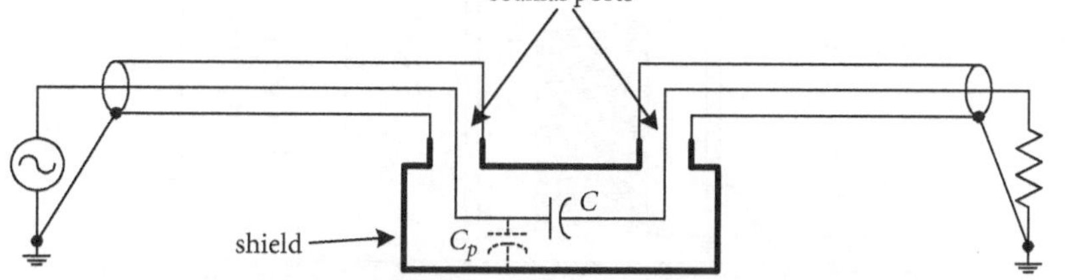

FIGURE 2.18 Ground reference for the shield is through the outer conductor of the coaxial cable.

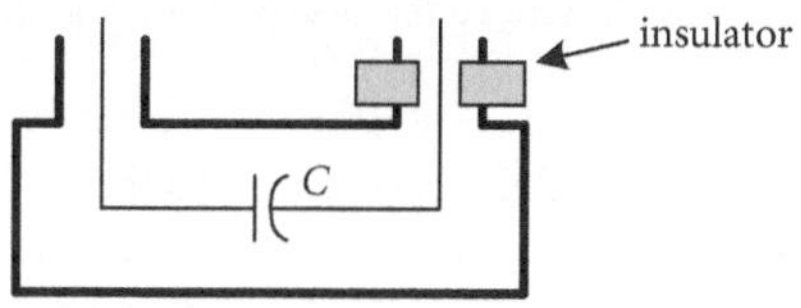

FIGURE 2.19 Current through the shield is reduced by eliminating the conductive path to one of the outer conductors of the connecting cables.

There is a potential difficulty associated with the given arrangement if the potential difference between the grounds at the source and at the load are significant. The probability of a substantial difference in potential increases as the distance between the source and load increases. A fraction of the signal return current will return through the shield and outer conductor of the coax, while the remaining current will return via the ground plane. Noise currents on the shield, of nonzero impedance, will generate a noise voltage along the shield. This noise voltage can couple into the inner conductor. For this reason, as shown in Figure 2.19, the outer conductor of the outgoing coaxial line is not connected to the shield, breaking the path for this current.

2.7 Shielding Cans

Inside radio receivers, inductors are frequently perpendicular to the printed circuit board and sometimes enclosed in shielding cans. Why? Should the metal can be grounded? What are the electrical costs of using this can? What other components might benefit from being in a shielding can? [Bogle; Moser, '68(b); Welsby; Howard W. Sams; Wheeler, '42; Macalpine; Watt; Simpson]

Larger solenoidal inductors are frequently placed perpendicular to the board (i.e., coil axis perpendicular to the board) to save board space and, possibly, to reduce their parasitic coupling to other components. Inductors tend to "pollute" electromagnetically much more than capacitors. In an attempt to reduce the parasitic coupling to nearby components, inductors (and some tuning transformers) are seen inside grounded metal cans. The metal can be magnetic or nonmagnetic.

By placing an inductor inside a metallic can, the parasitic electromagnetic coupling between the inductor and *can* is generally greater than the parasitic coupling between the inductor and its environment without the can! Probably the most obvious consequence of surrounding all or part of the inductor with a nearby metal can is the increase in parasitic capacitance between the inductor and its new environment. However, the inductor's new environment in the can is now well controlled and predictable. The capacitance of the coil to the can will vary little as the environment outside of the can changes. When designing tuning circuits containing inductance, this invariability of the parasitic capacitance is highly desirable. Furthermore, by grounding the can, the electric field interaction between the coil and components outside the can is greatly reduced. In Figure 2.20, the parasitic capacitances between the coil and shield and between the shield and external environment are shown. The parasitic capacitance between the coil and external environment, C_{ce}, is about zero. The capacitance between the various turns of the coil is not shown in Figure 2.20.

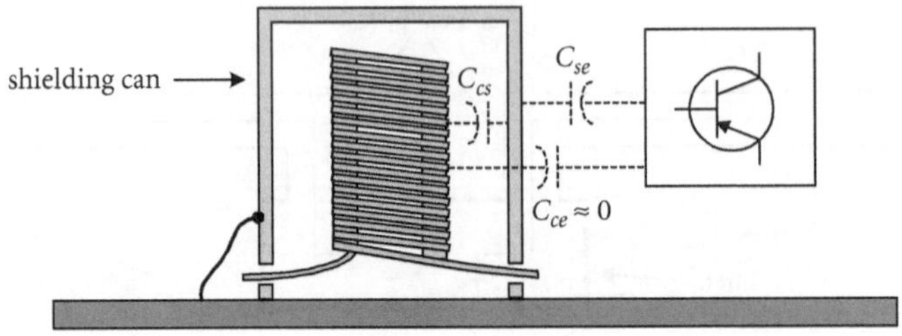

FIGURE 2.20 Grounded shielded enclosure to reduce the electric field coupling to objects external to the shield.

For multilayer coils without a shielding can, the connection to the outer layer is frequently made to the ground reference or lower voltage side to reduce some of the electric field coupling to nearby lower voltage objects. Of course, if the nearby objects are at a high potential, this would not necessarily be effective.

Another consequence of enclosing part or all of the inductor in a metallic can or shield is the Eddy currents generated in the can (i.e., the currents induced in the can by the time-varying magnetic field emanating from the coil). This current in the shield generates a counter magnetic field that reduces the effective inductance of the coil. A few steps of highly simplified analysis will illustrate this reduction in inductance. The magnetic field inside a N_c-turn tightly wound long solenoid carrying a current of I_c is essentially uniform, away from the ends. This uniform magnetic field inside the coil is given as

$$\vec{H} = \frac{N_c I_c}{l_c} \hat{a}_z \quad \rho < a$$

where a is the radius of the solenoid or coil and l_c is the coil's length. The unit vector $\hat{a}_z$ is along the coil's axis. This expression assumes that no other metallic objects, including the metallic can, are nearby. The inductance of this long isolated coil or solenoid is the ratio of the sum of the magnetic flux through each turn of the coil to the current in the coil generating this flux:

$$L_c = \frac{\Phi}{I} = \frac{\iint_S \vec{B} \cdot d\vec{s}}{I_c} = \frac{\mu_o \frac{N_c I_c}{l_c}(N_c \pi a^2)}{I_c} = \frac{\mu_o \pi a^2 N_c^2}{l_c}$$

The magnetic fields when leaving one end of the finite-length coil return along the outside of the coil and re-enter through the other end of the coil. If a concentric metallic shield is placed outside the coil, the magnetic fields external to the coil induce current in the shield. These fields from the coil can be considered approximately parallel to the shield's surface. Thus, the induced current in the shield is approximately circumferential but in the opposite direction to the current in the coil as shown in Figure 2.21. Rather than modeling this outer shield as a solid circular conductor with a given surface current, it will be temporarily modeled as a long tightly wound coil with N_s turns and length equal to l_s. The magnetic field inside this shield without the inner coil present is

$$\vec{H} = -\frac{N_s I_s}{l_s} \hat{a}_z \quad \text{if } \rho < b$$

where b is the radius of the shield modeled as a coil. The inductance of the shield is obtained as before:

$$L_s = \frac{\mu_o \pi b^2 N_s^2}{l_s}$$

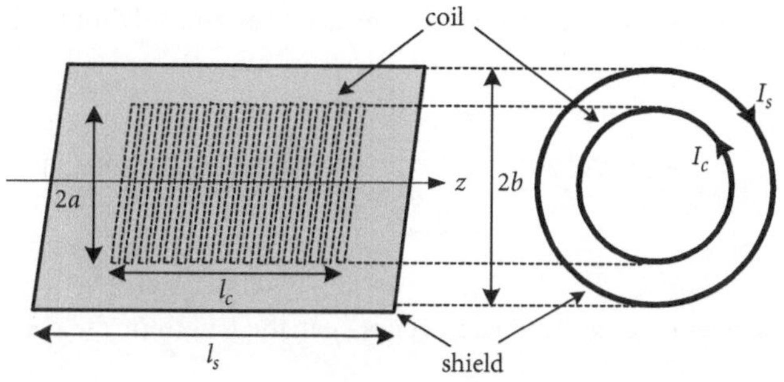

FIGURE 2.21 Current-carrying coil enclosed in a metallic shield. A top view of the current in the coil and induced counter current in the shield is also shown.

The magnetic interaction between the coil and shield is modeled as a mutual inductance, M. It is equal to the total flux passing through the inner coil due to the current in the shield:

$$M = \frac{\Phi}{I} = \frac{\iint_S \vec{B} \cdot d\vec{s}}{I_s} = \frac{\mu_o \frac{N_s I_s}{l_s}(N_c \pi a^2)}{I_s} = \frac{\mu_o \pi a^2 N_c N_s}{l_s}$$

The shield and inner coil can now be modeled as a transformer as shown in Figure 2.22. The primary corresponds to the inner coil and the secondary, which is short circuited, corresponds to the shield. The resistances of the coil and shield are also shown in Figure 2.22. The input impedance into this transformer, or the effective impedance of the coil, is

$$\begin{aligned} Z_{eff} &= R_c + j\omega L_c + \frac{\omega^2 M^2}{R_s + j\omega L_s} = R_c + j\omega L_c + \frac{\omega^2 M^2}{R_s + j\omega L_s}\frac{R_s - j\omega L_s}{R_s - j\omega L_s} \\ &= \left[R_c + \frac{\omega^2 M^2 R_s}{R_s^2 + (\omega L_s)^2}\right] + j\omega\left[L_c - \frac{\omega^2 M^2 L_s}{R_s^2 + (\omega L_s)^2}\right] \end{aligned} \tag{2.5}$$

By assuming that the ohmic losses are negligible, (2.5) reduces to

$$Z_{eff} = j\omega\left(L_c - \frac{M^2}{L_s}\right) = j\omega L_{eff} \quad \text{if } R_c = R_s = 0 \tag{2.6}$$

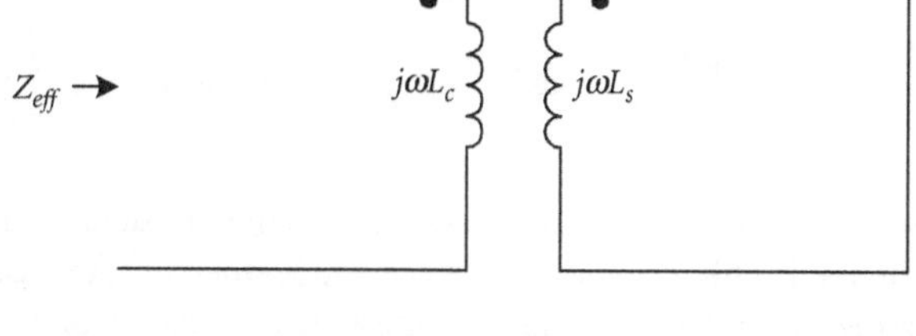

FIGURE 2.22 Transformer model of the shielded can.

Substituting the expressions for the self and mutual inductances and solving for L_{eff},

$$L_{eff} = \frac{\mu_o \pi a^2 N_c^2}{l_c} - \frac{\left(\dfrac{\mu_o \pi a^2 N_c N_s}{l_s}\right)^2}{\dfrac{\mu_o \pi b^2 N_s^2}{l_s}} = \frac{\mu_o \pi a^2 N_c^2}{l_c}\left(1 - \frac{a^2}{b^2}\frac{l_c}{l_s}\right) \tag{2.7}$$

If most of the induced current along the shield occurs near the length of the coil and $l_c \approx l_s$,

$$L_{eff} \approx \frac{\mu_o \pi a^2 N_c^2}{l_c}\left(1 - \frac{a^2}{b^2}\right) = L_c\left(1 - \frac{a^2}{b^2}\right) \tag{2.8}$$

Equation (2.8) is one expression seen for the effective inductance for a long shielded inductor. The shield reduces the effective inductance of the coil. The net flux through the coil is reduced by the counter magnetic field generated by the induced current in the shield. Although derived for a long lossless coil surrounded by a long, lossless concentric shield, this effective inductance expression is approximately correct for real shielded coils. Sometimes, the fractional change in inductance due to the addition of the shield is desired:

$$\frac{\Delta L}{L_c} = \frac{L_c - L_{eff}}{L_c} = \frac{a^2}{b^2} \tag{2.9}$$

This formula can be further improved by including the resistance of the shield:

$$L_{eff} = L_c - \frac{\omega^2 M^2 L_s}{R_s^2 + (\omega L_s)^2} \tag{2.10}$$

The effective inductance of the coil can change with frequency. The fractional change in inductance, again letting $l_c \approx l_s$, is

$$\frac{\Delta L}{L_c} = \frac{L_c - L_c + \dfrac{\omega^2 M^2 L_s}{R_s^2 + (\omega L_s)^2}}{L_c} = \frac{l_c}{\mu_o \pi a^2 N_c^2}\,\frac{\omega^2\left(\dfrac{\mu_o \pi a^2 N_c N_s}{l_s}\right)^2 \dfrac{\mu_o \pi b^2 N_s^2}{l_s}}{R_s^2 + \left(\omega \dfrac{\mu_o \pi b^2 N_s^2}{l_s}\right)^2}$$

$$= \frac{a^2}{b^2}\,\frac{1}{1 + \left(\dfrac{R_s l_s}{\omega \mu_o \pi b^2 N_s^2}\right)^2} \tag{2.11}$$

The effect of the resistance on the inductance is negligible when

$$\left(\frac{R_s l_s}{\omega \mu_o \pi b^2 N_s^2}\right)^2 \ll 1 \text{ or } R_s l_s \ll \omega \mu_o \pi b^2 N_s^2 \tag{2.12}$$

For many coils, which are used at higher frequencies, the effective inductance is not a function of the frequency, and (2.12) is satisfied. This dependency of the inductance on the frequency (and resistance) is somewhat analogous to the dependency of the characteristic impedance of a transmission line on the frequency (and line resistance) at lower frequencies.

Another consequence of enclosing the inductor with a shield is the increase in the ohmic losses. The real part of the effective impedance of the shielded coil is a measure of these losses:

$$R_{eff} = \mathrm{Re}(Z_{eff}) = R_c + \frac{\omega^2 M^2 R_s}{R_s^2 + (\omega L_s)^2} \tag{2.13}$$

The fractional change in resistance, again substituting the previous expressions for the inductances and letting $l_c \approx l_s$ and $N_s = N_c$, is

$$\Delta R = R_c - R_{eff} = -\frac{\omega^2 M^2 R_s}{R_s^2 + (\omega L_s)^2} = -\frac{\omega^2 \left(\dfrac{\mu_o \pi a^2 N_c N_s}{l_s}\right)^2 R_s}{R_s^2 + \left(\omega \dfrac{\mu_o \pi b^2 N_s^2}{l_s}\right)^2}$$

$$= -R_s \frac{a^4}{b^4} \frac{1}{1 + \left(\dfrac{R_s l_s}{\omega \mu_o \pi b^2 N_s^2}\right)^2} = -R_s \frac{a^2}{b^2} \frac{\Delta L}{L_c} \tag{2.14}$$

The shield, as expected, increases the losses due to the nonzero conductivity of the shield. This resistive loading is directly proportional to the resistance of the shield. Of course, for a real coil, the actual magnetic field distribution inside and outside the coil will vary along the length of the coil. Therefore, the magnetic field and ohmic losses along the length of the shield will also vary. Also, because of the skin effect, the resistances of the coil and shield are a function of the frequency. For sufficiently high frequencies, the shield current is mainly along its inner surface closest to the coil.

Usually some "rattle space" is necessary between the coil and can to prevent excessive loading of the coil. One guideline states that the distance between the coil and shield should be at least equal to the diameter of the coil. For sufficiently high frequencies, the effective inductance is then

$$L_{eff} = L_c\left(1 - \frac{a^2}{b^2}\right) > L_c\left[1 - \frac{a^2}{(3a)^2}\right] \approx 0.9 L_c \quad \text{if } b - a > 2a \ (\text{or } b > 3a) \tag{2.15}$$

The increase in the effective resistance is also reasonably small under this guideline for metallic shields. The Q or quality factor of the inductor away from resonance is defined as the ratio of the absolute value of the reactance to the resistance of Z_{eff}:

$$Q = \frac{|X_{eff}|}{R_{eff}} = \frac{\omega\left[L_c - \dfrac{\omega^2 M^2 L_s}{R_s^2 + (\omega L_s)^2}\right]}{R_c + \dfrac{\omega^2 M^2 R_s}{R_s^2 + (\omega L_s)^2}} = \frac{\omega L_{eff}}{R_{eff}} \tag{2.16}$$

Since the shield generally reduces the effective inductance and increases the effective resistance, the shield decreases the Q of the inductor.

To obtain a fast solution for the effective inductance and resistance of the shielded coil, it was conveniently assumed that $l_c \approx l_s$. A semi-empirical relationship is available relating the fractional change in the inductance and the length of the coil:

$$\frac{\Delta L}{L_c} = \frac{\dfrac{l_c}{b-a}}{\dfrac{l_c}{b-a} + 1.55} \frac{a^2}{b^2} \tag{2.17}$$

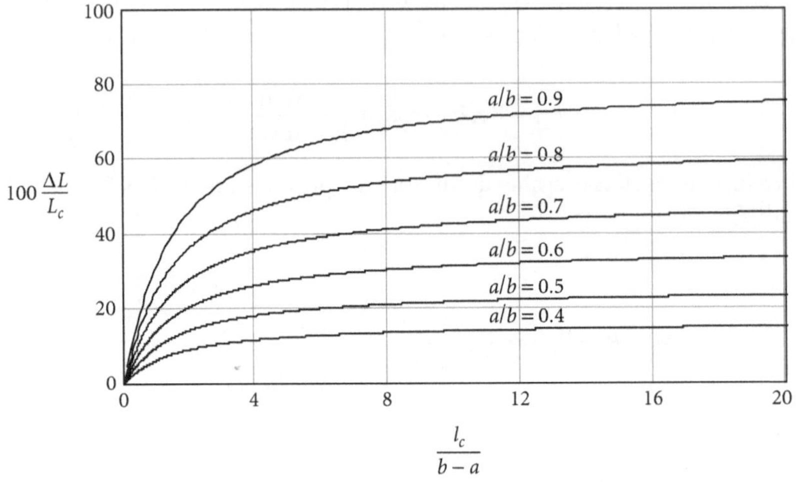

FIGURE 2.23 Percent change in the inductance of a shielded coil vs. the ratio of the coil length to the gap spacing (using Equation (2.17)).

This expression can be used when the shield length, l_s, equals or exceeds the value

$$l_s \geq l_c \frac{\frac{b}{a} - 0.84}{\frac{l_c}{a} - 0.38} + l_c = \frac{l_c}{b}\left(\frac{b^2 - 1.22ab + bl_c}{l_c - 0.38a}\right) \tag{2.18}$$

For shield lengths of this value or larger, the shield can be considered infinitely long. These relationships were based on coils with radii in the 2 to 4 cm range, coil lengths of about 2 to 27 cm, and various lengths of copper shields. When caps or ends were added to the shield, then for shield lengths where $l_s > l_c + 2(b-a)$, the expression for the fractional change in the inductance is approximately correct. The percent change in the inductance is shown in Figure 2.23 vs. the ratio of the coil length to gap spacing (i.e., $b - a$). As expected, the loading increases as the spacing between the shield and coil decreases.

The induced current generated in a conducting nonmagnetic shield will generate *some* counter or "bucking" magnetic field. This counter field does provide some magnetic shielding. However, it is generally difficult to "bottle up" magnetic fields. The shielding effectiveness is a function of the dimensions of the shield and its composition. Tests have been performed on various 1/32 and 1/16 inch thick, 3 inch diameter, 6 inch long cans of various magnetic and nonmagnetic materials surrounding a 2 inch diameter current-carrying coil. The voltage induced in a *magnetic* field sensor from the fields generated by the current in the coil with and without a shielding can was measured at various positions. The shielding effectiveness was taken as the dB value of the ratio of the voltage of the sensor without to with the shield present. The tests were performed from 50 Hz to 30 kHz. A summary of some of the results is given in Table 2.1. The values can vary considerably based on the position of the sensor. Louvers, notches, and holes in the can, including an opening on the top of the can to adjust the inductance, can dramatically reduce the effectiveness of the shield. The magnetic materials were the most effective in shielding.

Shielded inductors, including shielded surface mount inductors, can be purchased, but the actual effectiveness and composition of the shield are not clearly stated in many product specifications. A few manufacturers specifically indicate that the shield is constructed of ferrite. Sometimes the inductors are described as "self shielding," possibly indicating that the winding is on a toroidal form. In one instance,

TABLE 2.1 Maximum SE for Various Metallic Shielding Cans [Moser, '68(b)]

	$\approx$ max SE_{dB}		
Material (1/32" thick)	50–300 Hz	300 Hz–3 kHz	3–30 kHz
Stainless steel[a]	2 dB	2 dB	10 dB
Aluminum	4 dB	20 dB	38 dB
Copper	5 dB	24 dB	44 dB
Steel unannealed	18 dB	32 dB	85 dB
Mumetal	55 dB	120 dB	—

[a]According to the reference, this stainless steel was "probably" a 304 alloy, a commonly used nonmagnetic material.

the shielding is stated as "3% coupling between inductors," without specifying the distance, position, size, and type of these nearby inductors.

Crystals can also be enclosed in metallic shields or cans. However, to reduce the cost of crystals, nonmetallic enclosures are commonly used. For this nonmetallic enclosure, a ground plane under the crystal will produce a more well-defined and controlled parasitic capacitance. With the ground plane, a finger or nearby component will have less influence on the crystal's performance. Helical resonators also use a metallic shield with one side of an interior inductor connected to the shield.

2.8 Finite-Conductivity Spherical Bodies

The steady-state potential inside a solid spherical object of conductivity σ, permittivity ε, and radius R when placed in a uniform electric field

$$\vec{E}(t) = E_e \cos(\omega t)\,\hat{a}_z$$

is given by the expression

$$\Phi_i = -\mathrm{Re}\left\{3rE_e \cos\theta e^{j\omega t}\left[\frac{j\omega\varepsilon_o}{\sigma + j\omega(2\varepsilon_o + \varepsilon)}\right]\right\}$$

(The spherical object is surrounded by free space.) Determine the electric field inside this spherical body in the frequency domain. Determine approximate expressions for the electric field if the object is a good conductor and good insulator. [Haus; Bridges, '88; Chatterton; Weber]

Without the spherical object, the magnitude of the electric field between the two parallel plates is

$$\vec{E}(t) = \frac{V_o \cos(\omega t)}{d}\hat{a}_z = E_e \cos(\omega t)\,\hat{a}_z \tag{2.19}$$

where d is the distance between the plates as shown in Figure 2.24. This electric field is uniform and directed between the plates. Unless the object has zero conductivity and a permittivity equal to that of free space, the object will distort the field. If the object is small compared to the distance between and size of the plates, then this distortion of the field mainly occurs near the object. Far from the object, the field is approximately uniform. The expression for the potential inside the object

$$\Phi_i = -\mathrm{Re}\left\{3rE_e \cos\theta e^{j\omega t}\left[\frac{j\omega\varepsilon_o}{\sigma + j\omega\,(2\varepsilon_o + \varepsilon)}\right]\right\} \tag{2.20}$$

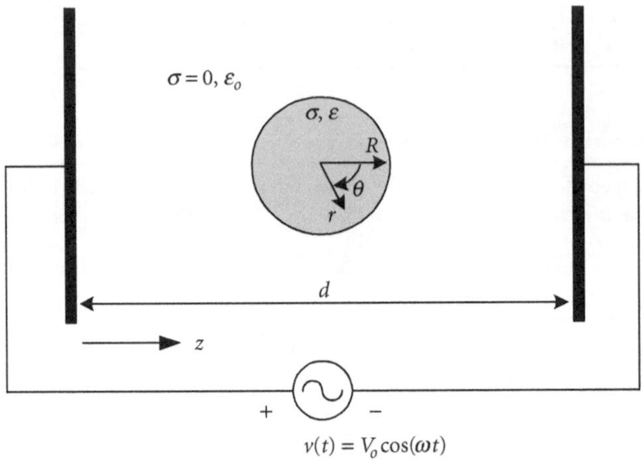

FIGURE 2.24 Spherical object between two charged parallel plates.

assumes that the electric field far from the object is uniform. The potential inside the argument of Re() is given in the frequency domain. The relationship between the frequency-domain expression and corresponding time-domain expression is

$$\Phi_i = \mathrm{Re}(A_s e^{j\omega t}) \quad \text{where } A_s = -3rE_e\cos\theta\left[\frac{j\omega\varepsilon_o}{\sigma + j\omega(2\varepsilon_o + \varepsilon)}\right]$$

The variable A_s includes both magnitude and phase information.

The electric field (a vector field) is related to the potential (a scalar field) via the gradient. In the spherical coordinate system, the relationship between the electric field and potential is

$$\vec{E} = -\nabla\Phi = -\left(\frac{\partial\Phi}{\partial r}\hat{a}_r + \frac{1}{r}\frac{\partial\Phi}{\partial\theta}\hat{a}_\theta + \frac{1}{r\sin\theta}\frac{\partial\Phi}{\partial\phi}\hat{a}_\phi\right)$$

Actually, this definition for the electric field is only valid for electrostatic or electroquasistatic situations. As a rough guideline, this equation should only be applied when the object is electrically small:

$$R < \frac{\lambda}{10}$$

In the frequency domain, the electric field is then

$$\begin{aligned}\vec{E}_{is} &= -\nabla\Phi_{is} = 3E_e\cos\theta\left[\frac{j\omega\varepsilon_o}{\sigma + j\omega(2\varepsilon_o + \varepsilon)}\right]\hat{a}_r - 3E_e\sin\theta\left[\frac{j\omega\varepsilon_o}{\sigma + j\omega(2\varepsilon_o + \varepsilon)}\right]\hat{a}_\theta + 0\hat{a}_\phi \\ &= 3E_e\left[\frac{j\omega\varepsilon_o}{\sigma + j\omega(2\varepsilon_o + \varepsilon)}\right](\cos\theta\hat{a}_r - \sin\theta\hat{a}_\theta)\end{aligned} \tag{2.21}$$

The spatial magnitude of the electric field is

$$|\vec{E}_{is}| = 3E_e\left[\frac{j\omega\varepsilon_o}{\sigma + j\omega(2\varepsilon_o + \varepsilon)}\right]\sqrt{\cos^2\theta + \sin^2\theta} = 3E_e\left[\frac{j\omega\varepsilon_o}{\sigma + j\omega(2\varepsilon_o + \varepsilon)}\right] \tag{2.22}$$

The magnitude of the electric field is not a function of position inside the material. Furthermore, the electric field is actually uniform inside the spherical body. A simple transformation to Cartesian coordinates results in

$$\vec{E}_{is} = 3E_e\left[\frac{j\omega\varepsilon_o}{\sigma + j\omega\,(2\varepsilon_o + \varepsilon)}\right]\hat{a}_z \tag{2.23}$$

The electric field inside the body is in the same direction as the applied field.

If the material is a good conductor, then the conduction current is much greater than the displacement current and

$$\frac{\sigma}{\omega\varepsilon} \gg 1$$

For a good conductor, an approximation for the electric field inside the material is

$$\vec{E}_{is} = 3E_e\left[\frac{\dfrac{j\omega\varepsilon_o}{j\omega(2\varepsilon_o + \varepsilon)}}{\dfrac{\sigma}{j\omega(2\varepsilon_o + \varepsilon)} + 1}\right]\hat{a}_z \approx 3E_e\left(\frac{j\omega\varepsilon_o}{\sigma}\right)\hat{a}_z \tag{2.24}$$

As the conductivity of the material increases, the electric field decreases. (The variable $j = \sqrt{-1}$ implies that the phase of the electric field inside the sphere is 90° out of phase with the applied field.) For good conductors, the internal electric field strength increases in direct proportion to the frequency of the applied field. The electric field should be zero inside a perfect conductor, and this equation confirms that the field is indeed zero when σ is equal to infinity. For a perfect conductor, the free charges in the medium redistribute themselves so that the sum of the applied field and field due to the induced surface charges on the sphere is zero inside the sphere. The charge distribution on the sphere (and on the two plates) at one particular time is sketched in Figure 2.25. Notice that the applied and induced fields are

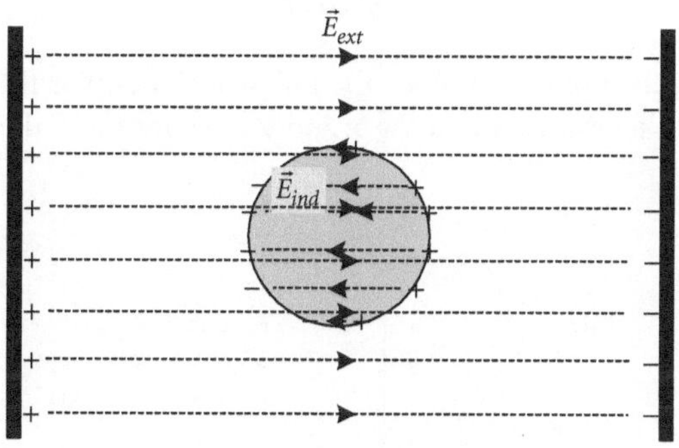

FIGURE 2.25 Fields from the charged plates and induced surface charge on the spherical object.

in opposite directions. The total field due to both the induced charges on the object and external charges on the two plates will not be uniform *outside* and near the spherical body. The applied field is distorted by the object.

If the material is a good insulator, then the conduction current is much less than the displacement current and

$$\frac{\sigma}{\omega\varepsilon} \ll 1$$

The electric field inside the good insulator is

$$\vec{E}_{is} = 3E_e \left[\frac{\dfrac{j\omega\varepsilon_o}{j\omega(2\varepsilon_o + \varepsilon)}}{\dfrac{\sigma}{j\omega(2\varepsilon_o + \varepsilon)} + 1} \right] \hat{a}_z \approx 3E_e \left[\frac{j\omega\varepsilon_o}{j\omega(2\varepsilon_o + \varepsilon)} \right] \hat{a}_z = E_e \frac{3}{2+\varepsilon_r} \hat{a}_z \quad (2.25)$$

where ε_r is the relative permittivity of the sphere. Since the magnitude of the applied field is E_e, the electric field inside the body is reduced by a factor of

$$\frac{3}{2+\varepsilon_r}$$

If the permittivity of the spherical body is large, then the electric field inside the body is small. The conductivity and number of free charges inside a good insulator is small. The number of bound charges, charges tied or bound to a nucleus, per unit volume, is a function of the permittivity. As the permittivity or relative permittivity increases, the number of these bound charges increase. These bound charges shield the inside of the spherical body in a way similar to free charges. These bound charges exist along the surface and generate a counter electric field to oppose the applied field. Unlike free charges, bound charges cannot be discharged. If an electromagnetic field simulator or "solver" is not capable of modeling a good conductor, the conductor can sometimes be approximated as a high-permittivity insulator. However, this simulated conductor will not have any free charges.

Expressions are also available for the electric field inside solid oblate and prolate spheroids of relative permittivity ε_{r2} surrounded by a medium of relative permittivity of ε_{r1}. The conductivity both inside and outside these spheroids is assumed zero (or the mediums are considered a good insulator at the frequency of interest). As before, the applied field far from the spheroid is uniform and in the z direction:

$$\vec{E}(t) = E_e \cos(\omega t)\hat{a}_z$$

The field inside the spheroid is uniform in all of the following cases. For the oblate spheroid shown in Figure 2.26 where the longer semi-axes are in the x and y directions and the axis of revolution (z axis)

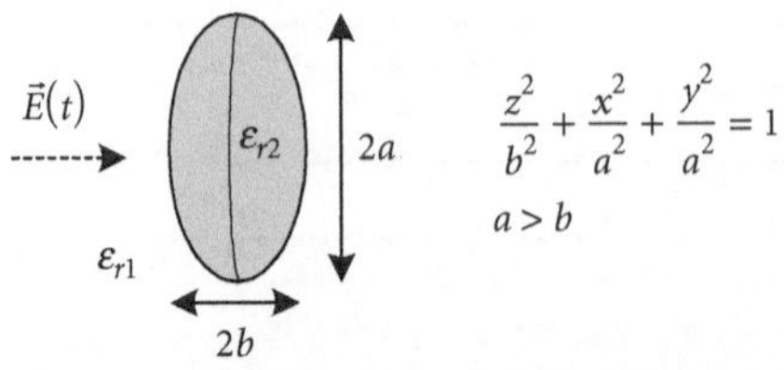

FIGURE 2.26 Oblate spheroid with the axis of revolution parallel to the applied field.

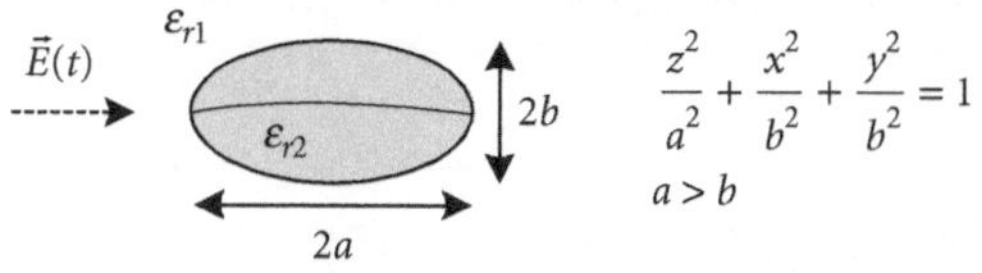

FIGURE 2.27 Prolate spheroid with the longer axis of revolution parallel to the applied field.

is parallel to the applied field, the electric field inside the spheroid is given by

$$\vec{E}_{is} = \frac{E_e}{1+\left(\frac{\varepsilon_{r2}}{\varepsilon_{r1}}-1\right)(\beta^2+1)[1-\beta\cot^{-1}(\beta)]}\hat{a}_z \quad \text{where } \beta=\sinh\left[\tanh^{-1}\left(\frac{b}{a}\right)\right] \tag{2.26}$$

$$\vec{E}_{is} \approx \frac{E_e}{1+0.53\left(\frac{\varepsilon_{r2}}{\varepsilon_{r1}}-1\right)}\hat{a}_z \quad \text{if } a=2b \tag{2.27}$$

$$\vec{E}_{is} \approx \frac{\varepsilon_{r1}E_e}{\varepsilon_{r2}}\hat{a}_z \quad \text{if } a \gg b \text{ (very flat disk)} \tag{2.28}$$

where $\cot^{-1}(x)=\tan^{-1}(1/x)$. For the prolate spheroid shown in Figure 2.27 where the shorter semi-axes are in the x and y directions and the longer axis of revolution (z axis) is parallel to the applied field, the electric field inside the spheroid is given by

$$\vec{E}_{is} = \frac{E_e}{1+\left(\frac{\varepsilon_{r2}}{\varepsilon_{r1}}-1\right)(\beta^2-1)[\beta\coth^{-1}(\beta)-1]}\hat{a}_z \quad \text{where } \beta=\frac{a}{\sqrt{a^2-b^2}} \tag{2.29}$$

$$\vec{E}_{is} \approx \frac{E_e}{1+0.17\left(\frac{\varepsilon_{r2}}{\varepsilon_{r1}}-1\right)}\hat{a}_z \quad \text{if } a=2b \tag{2.30}$$

$$\vec{E}_{is} \approx E_e\hat{a}_z \quad \text{if } a \gg b \text{ (very thin rod)} \tag{2.31}$$

where $\coth^{-1}(x)=\tanh^{-1}(1/x)$. For the prolate spheroid shown in Figure 2.28 where the shorter semi-axes are in the x and y directions and the longer axis of revolution (z axis) is perpendicular or normal

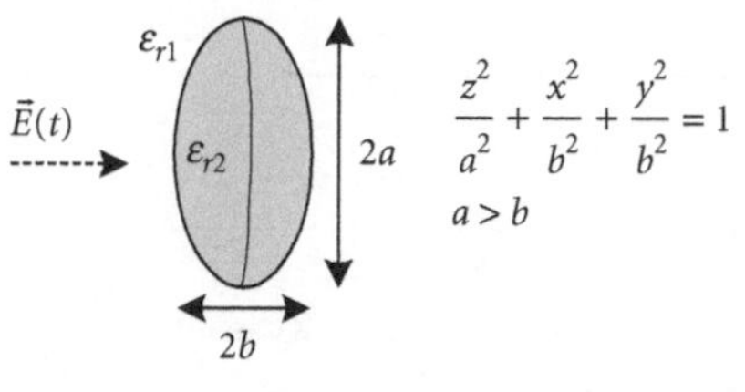

FIGURE 2.28 Prolate spheroid with the longer axis of revolution perpendicular to the applied field.

to the applied field, the electric field inside the spheroid is given by

$$\vec{E}_{is} = \frac{E_e}{1+\left(\frac{\varepsilon_{r2}}{\varepsilon_{r1}}-1\right)\frac{\beta}{2}[\beta-(\beta^2-1)\coth^{-1}(\beta)]}\hat{a}_z \quad \text{where } \beta = \frac{a}{\sqrt{a^2-b^2}} \tag{2.32}$$

$$\vec{E}_{is} \approx \frac{E_e}{1+0.41\left(\frac{\varepsilon_{r2}}{\varepsilon_{r1}}-1\right)}\hat{a}_z \quad \text{if } a = 2b \tag{2.33}$$

$$\vec{E}_{is} \approx \frac{E_e}{1+\frac{1}{2}\left(\frac{\varepsilon_{r2}}{\varepsilon_{r1}}-1\right)}\hat{a}_z \quad \text{if } a \gg b \text{ (very thin rod)} \tag{2.34}$$

where $\coth^{-1}(x) = \tanh^{-1}(1/x)$.

The expression for the electric field at the *center* of the thin, spherical conducting shell (not a solid sphere) of thickness Δ and conductivity σ shown in Figure 2.29 is also of interest. If the thickness of the shell is much less than the skin depth

$$\Delta \ll \delta = \frac{1}{\sqrt{\pi f \mu \sigma}} \tag{2.35}$$

and Δ is much less than the radius of the sphere (R), then the electric field at the center of the spherical shield is

$$\vec{E}_{cnts} = 3E_e\left(\frac{j\omega\varepsilon_o}{\sigma}\right)\left(\frac{R}{2\Delta}\right)\hat{a}_z \tag{2.36}$$

This equation is a low-frequency approximation for the electric field. As expected, the shielding effectiveness decreases as the shell thickness decreases. This electric field is greater than the uniform field inside a solid sphere. Even thin metallic shields, though, are *extremely* effective in shielding near-field electric fields. For example, for a copper 1 mil ($\approx$ 25.4 μm) thick shield of radius 0.5 m at a frequency

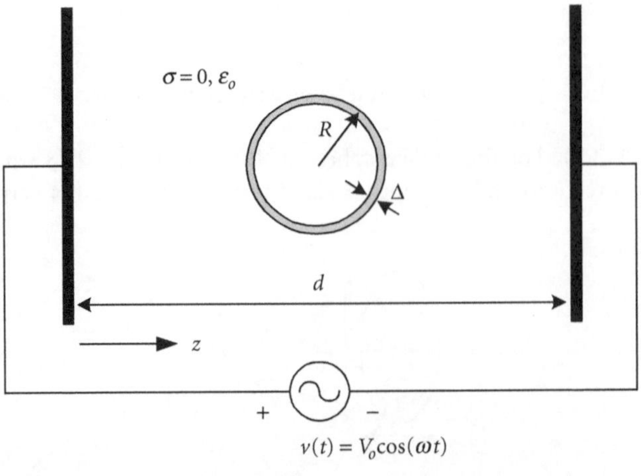

FIGURE 2.29 Thin, spherical conducting shell between two charged parallel plates.

of 1 kHz, the shielding effectiveness is

$$SE_{dB} = 20\log\left(\frac{E_e}{E_{cnts}}\right) = 20\log\left(\frac{\sigma}{3\omega\varepsilon_o}\frac{2\Delta}{R}\right)$$

$$= 20\log\left[\frac{5.8\times10^7}{3(2\pi\times10^3)(8.854\times10^{-12})}\frac{2(25.4\times10^{-6})}{0.5}\right] \approx 210\ \text{dB}$$

(The shell diameter is electrically small at 1 kHz.) When $\Delta \ll R$ and the frequency is sufficiently high so that

$$\Delta \gg \delta = \frac{1}{\sqrt{\pi f \mu \sigma}} \tag{2.37}$$

(but R is still electrically small), the electric field decays exponentially through the shield. In this case, the electric field at the center of the spherical shield is about

$$\vec{E}_{cnts} = 3E_e\left(\frac{j\omega\varepsilon_o}{\sigma}\right)\left(\frac{e^{-\frac{\Delta}{\delta}}R\sqrt{2}}{\delta}\right)\hat{a}_z \tag{2.38}$$

This equation is a high-frequency approximation for the electric field. At high frequencies, the current is confined to the outer surface due to the skin effect. As an example, at a frequency of 7 MHz ($\lambda_o \approx 43$ m), copper has skin depth of

$$\delta = \frac{1}{\sqrt{\pi f \mu \sigma}} = \frac{1}{\sqrt{\pi(7\times10^6)(4\pi\times10^{-7})(5.8\times10^7)}} \approx 25\ \mu\text{m}$$

Therefore, at the center of a copper 10 mil ($\approx 254\ \mu$m $\gg 25\ \mu$m) thick spherical shield of radius 0.5 m, which is electrically small, the shielding effectiveness is

$$SE_{dB} = 20\log\left(\frac{E_e}{E_{cnts}}\right) = 20\log\left(\frac{\sigma}{3\omega\varepsilon_o}\frac{\delta}{e^{-\frac{\Delta}{\delta}}R\sqrt{2}}\right)$$

$$= 20\log\left[\frac{5.8\times10^7}{3(2\pi\times7\times10^6)(8.854\times10^{-12})}\frac{25\times10^{-6}}{e^{-\frac{254}{25}}(0.5)\sqrt{2}}\right] \approx 210\ \text{dB}$$

Again, this metallic spherical shell is superb for shielding electric fields.

Although not shown in Figure 2.29, a flat equipotential contour exists parallel to the external electrodes, through the center of the sphere. The zero volt or ground reference can be selected along this specific

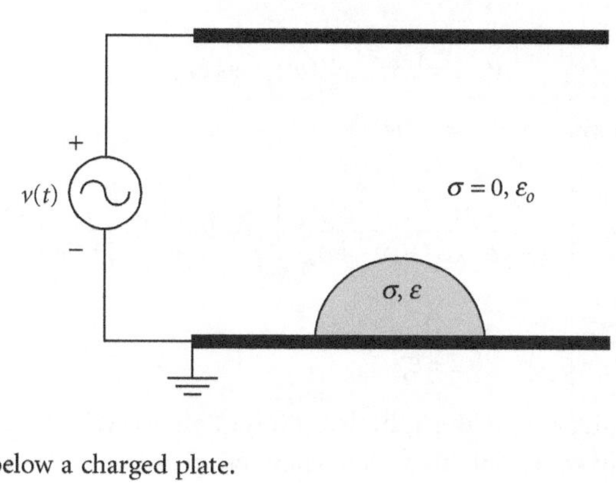

FIGURE 2.30 Hemisphere below a charged plate.

contour. Thus, because of the symmetry of the potential distribution, the given electric field expressions for a sphere can also be used for a hemispherical conductor on a large, flat ground plane (refer to Figure 2.30). Similarly, the electric field expressions given later for a cylinder can also be used for a half-cylinder on a flat ground plane. When modeling objects or shields along a ground plane, such as a hemispherical model of a human underneath an electric power line, these models can be useful.

2.9 Step Response of Spherical Bodies

A small spherical object of conductivity σ, permittivity ε, and radius R is in a field-free environment. Suddenly at $t = 0$, a uniform electric field is applied in a step-function manner. Using (2.23), determine the electric field in the object just after $t = 0$ and after a very long time. [Moore, '73]

The output of this system in this case is the electric field expression

$$\vec{Y}_s = 3E_e\left[\frac{j\omega\varepsilon_o}{\sigma + j\omega(2\varepsilon_o + \varepsilon)}\right]\hat{a}_z$$

while the input is the sinusoidal applied field:

$$\vec{X}_s = E_e\hat{a}_z$$

The sinusoidal steady-state transfer function, H_s, is

$$H_s = \frac{Y_s}{X_s} = 3\left[\frac{j\omega\varepsilon_o}{\sigma + j\omega(2\varepsilon_o + \varepsilon)}\right]$$

For a nonsinusoidal input signal, such as a step function, the output time response is obtained by multiplying the transfer function, H_s, by the Fourier transform of the input signal and then taking the inverse Fourier transform of this product. For a step function input signal of magnitude E_e, the output response is equal to

$$F^{-1}\left\{3\left[\frac{j\omega\varepsilon_o}{\sigma + j\omega(2\varepsilon_o + \varepsilon)}\right]\frac{E_e}{j\omega}\hat{a}_z\right\}$$

The resultant time-domain result involves exponentials.

In this example, the complete time-domain response of the electric field inside the object is not required. The responses just slightly after $t = 0$ and after a long time are required. Fortunately, the initial-value and final-value theorems can be used to determine the initial and final response without taking the inverse transform. The initial-value theorem states that if the frequency-domain response is given as $Y(\omega)$ then (lim = limit)

$$y(0^+) = \lim[j\omega Y(\omega)] \quad \text{as } \omega \to \infty$$

Therefore, the output of the system at $t = 0^+$ is

$$\vec{y}(0^+) = \lim_{\omega\to\infty}\left\{j\omega 3\left[\frac{j\omega\varepsilon_o}{\sigma + j\omega(2\varepsilon_o + \varepsilon)}\right]\frac{E_e}{j\omega}\hat{a}_z\right\} = \lim_{\omega\to\infty}\left\{3E_e\left[\frac{j\omega\varepsilon_o}{\sigma + j\omega(2\varepsilon_o + \varepsilon)}\right]\hat{a}_z\right\}$$
$$= 3E_e\frac{\varepsilon_o}{2\varepsilon_o + \varepsilon}\hat{a}_z = E_e\frac{3}{2 + \varepsilon_r}\hat{a}_z \tag{2.39}$$

where ε_r is the relative permittivity of the spherical body. This electric field is the same as for a good insulator in sinusoidal steady state. The high-frequency terms are important for a step function at the time of the discontinuity. At high frequencies, the displacement current is dominant, and the free charges in the spherical body do not have adequate time to respond (or relax) to the applied field to help shield the interior. Of course, if the frequency is too large, the system is not electroquasistatic, and this result should be used with caution.

The final-value theorem allows the determination of the response after a long time without actually determining the time-domain response.[9] This theorem states that if the frequency-domain response is given as $Y(\omega)$ then

$$y(\infty) = \lim[j\omega Y(\omega)] \quad \text{as } \omega \to 0$$

Therefore, the output of the system at $t = \infty$ is

$$\begin{aligned} \vec{y}(\infty) &= \lim_{\omega \to 0} \left\{ j\omega 3 \left[\frac{j\omega \varepsilon_o}{\sigma + j\omega(2\varepsilon_o + \varepsilon)} \right] \frac{E_e}{j\omega} \hat{a}_z \right\} \\ &= \lim_{\omega \to 0} \left\{ 3E_e \left[\frac{j\omega \varepsilon_o}{\sigma + j\omega(2\varepsilon_o + \varepsilon)} \right] \hat{a}_z \right\} = 0 \end{aligned} \tag{2.40}$$

The low-frequency terms are important for a step function after a long time. At low frequencies, the conduction current is dominant. If the applied field after a long time is at a very low frequency, in this case dc, then the free charges in the spherical body have adequate time to respond to the applied field and shield the interior. For a step function applied field, the electric field inside any spherical-shaped medium with a nonzero conductivity will *eventually* vanish if the observer is patient. (This statement assumes that charges are not fixed inside the medium and that the medium is homogeneous.) Whether a medium appears like a good conductor or good insulator is a function of the observation time relative to the time constant of the medium. For example, a material with a time constant of ten minutes will *appear* like a good conductor if the material is observed after one day! However, if the material is observed after only one second, it will appear like a good insulator!

2.10 Finite-Conductivity Cylindrical Body

Referring to Figure 2.31, the steady-state potential inside a solid cylindrical object of conductivity σ, permittivity ε, and radius R when placed in a uniform electric field (transverse to a plane passing through the axis of the cylinder)

$$\vec{E}(t) = E_e \cos(\omega t)\hat{a}_x$$

is given by the expression

$$\Phi_i = -\text{Re}\left\{ 2\rho E_e \cos\phi e^{j\omega t} \left[\frac{j\omega \varepsilon_o}{\sigma + j\omega(\varepsilon_o + \varepsilon)} \right] \right\}$$

(The cylindrical object is surrounded by free space.) Determine the electric field inside this cylindrical body in the frequency domain. Determine approximate expressions for the electric field if the object is a good conductor and good insulator. Then, determine the initial and final step response using the steady-state field expression. [Haus]

[9]Generally, with Fourier transforms, this theorem can be applied to one-sided (causal) time functions that have a single value after a long time.

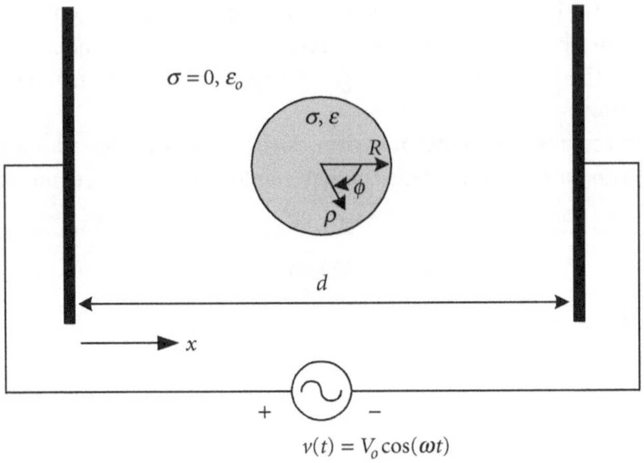

FIGURE 2.31 Cylindrical object between two charged parallel plates.

This analysis is very similar to that presented for the spherical body. The gradient in the cylindrical coordinate system is

$$\vec{E} = -\nabla\Phi = -\left(\frac{\partial\Phi}{\partial\rho}\hat{a}_\rho + \frac{1}{\rho}\frac{\partial\Phi}{\partial\phi}\hat{a}_\phi + \frac{\partial\Phi}{\partial z}\hat{a}_z\right)$$

The following results are valid if the frequency of the applied field is sufficiently low so that the system is electroquasistatic. Therefore,

$$\begin{aligned}\vec{E}_{ins} &= 2E_e\cos\phi\left[\frac{j\omega\varepsilon_o}{\sigma + j\omega(\varepsilon_o+\varepsilon)}\right]\hat{a}_\rho - 2E_e\sin\phi\left[\frac{j\omega\varepsilon_o}{\sigma + j\omega(\varepsilon_o+\varepsilon)}\right]\hat{a}_\phi + 0\hat{a}_z \\ &= 2E_e\left[\frac{j\omega\varepsilon_o}{\sigma + j\omega(\varepsilon_o+\varepsilon)}\right](\cos\phi\hat{a}_\rho - \sin\phi\hat{a}_\phi) = 2E_e\left[\frac{j\omega\varepsilon_o}{\sigma + j\omega(\varepsilon_o+\varepsilon)}\right]\hat{a}_x\end{aligned} \quad (2.41)$$

The electric field was converted from the cylindrical coordinate system to the Cartesian coordinate system to show clearly the uniformity of the field inside the body. If the cylindrical body is a good conductor, the electric field reduces to

$$\vec{E}_{is} = 2E_e\left[\frac{\dfrac{j\omega\varepsilon_o}{j\omega(\varepsilon_o+\varepsilon)}}{\dfrac{\sigma}{j\omega(\varepsilon_o+\varepsilon)}+1}\right]\hat{a}_x \approx 2E_e\left(\frac{j\omega\varepsilon_o}{\sigma}\right)\hat{a}_x \quad \text{if } \frac{\sigma}{\omega\varepsilon} \gg 1 \quad (2.42)$$

and for a good insulator

$$\vec{E}_{is} = 2E_e\left[\frac{\dfrac{j\omega\varepsilon_o}{\sigma}}{1+\dfrac{j\omega(\varepsilon_o+\varepsilon)}{\sigma}}\right]\hat{a}_x \approx 2E_e\left[\frac{j\omega\varepsilon_o}{j\omega(\varepsilon_o+\varepsilon)}\right]\hat{a}_x = E_e\frac{2}{1+\varepsilon_r}\hat{a}_x \quad \text{if } \frac{\sigma}{\omega\varepsilon} \ll 1 \quad (2.43)$$

where ε_r is the relative permittivity of the cylindrical object. These results are similar to the spherical-body field equations.

The initial-value theorem is used to determine the electric field inside the object just after $t = 0$:

$$\vec{y}(0^+) = \lim \left\{ j\omega 2\left[\frac{j\omega\varepsilon_o}{\sigma + j\omega(\varepsilon_o + \varepsilon)}\right]\frac{E_e}{j\omega}\hat{a}_x \right\} \quad \text{as } \omega \to \infty$$
$$= \lim_{\omega\to\infty} \left\{ 2E_e\left[\frac{j\omega\varepsilon_o}{\sigma + j\omega(\varepsilon_o + \varepsilon)}\right]\hat{a}_x \right\} = 2E_e\frac{\varepsilon_o}{\varepsilon_o + \varepsilon}\hat{a}_x = E_e\frac{2}{1+\varepsilon_r}\hat{a}_x \tag{2.44}$$

At high frequencies, the free charges associated with the conductivity of the medium cannot adequately respond to the external electric field. As a result, just after $t = 0$ for a step function input signal, the field lines penetrate through a conducting body as if the body was a perfect insulator.

The final-value theorem is used to determine the electric field inside the object after a long time:

$$\vec{y}(\infty) = \lim_{\omega\to 0} \left\{ j\omega 2\left[\frac{j\omega\varepsilon_o}{\sigma + j\omega(\varepsilon_o + \varepsilon)}\right]\frac{E_e}{j\omega}\hat{a}_x \right\}$$
$$= \lim_{\omega\to 0} \left\{ 2E_e\left[\frac{j\omega\varepsilon_o}{\sigma + j\omega(\varepsilon_o + \varepsilon)}\right]\hat{a}_x \right\} = 0 \tag{2.45}$$

At low frequencies, corresponding to the constant value of the step response after a long time, the free charges associated with the conductivity of the medium have sufficient time to relax toward the surface. These free surface charges shield the interior of the cylindrical body, and the body acts like a perfect conductor.

2.11 Electric Blankets and Infants

At a distance of 30 cm from an electric blanket, the electric field is 250 V/m rms. Determine a rough estimate of the electric field inside an infant from this electric blanket at this distance. [Gandhi, '90]

It will be assumed that the frequency of the current in the blanket is 60 Hz. Obviously, at this frequency, a distance of 30 cm is electrically short. Although the conductivity and permittivity of the various materials in a human body vary, values of $\sigma = 0.2/\Omega\text{-m}$ and $\varepsilon_r = 81$ will be used in this analysis. As expected, the human body is a good conductor at these low frequencies:

$$\frac{\sigma}{\omega\varepsilon} = \frac{0.2}{2\pi(60)(81\times 8.854\times 10^{-12})} \approx 7\times 10^5 \gg 1$$

The infant will be modeled as a solid conducting sphere. Therefore, the spatial magnitude of the sinusoidal steady-state electric field inside the baby is, using (2.23),

$$|\vec{E}_{is}| = 3E_e\left[\frac{j\omega\varepsilon_o}{\sigma + j\omega(2\varepsilon_o + \varepsilon)}\right]$$
$$= 3(250)\left[\frac{j2\pi(60)8.854\times 10^{-12}}{0.2 + j2\pi(60)(2\times 8.854\times 10^{-12} + 81\times 8.854\times 10^{-12})}\right]$$
$$\approx j0.013 \text{ mV/m rms}$$

The magnitude of the electric field, 0.013 mV/m rms, is much less than the external field. The human body is a very good conductor at these low frequencies. The free charges in the human body can respond very quickly, relative to 60 Hz, providing excellent electric field shielding to the interior of the body. The electric field, however, can generate a current along the surface of the body.

Although the concept of skin depth is usually used for plane wave or far-field shielding problems, the skin depth is also a parameter in some near-field electric and magnetic shielding expressions. Since the human body is a very good conductor at 60 Hz, the standard expression for the skin depth may be used to determine δ:

$$\delta = \frac{1}{\sqrt{\pi f \mu \sigma}} = \frac{1}{\sqrt{\pi(60)(4\pi \times 10^{-7})(0.2)}} \approx 150\,\text{m}$$

Regrettably, some individuals[10] mistakenly believe that 60 Hz electric fields easily penetrate through the body since the skin depth is large compared to the thickness of the human body. This is a misapplication of the skin depth. In shielding problems, skin depth is often considered a far-field effect. To be in the far field of a 60 Hz signal in free space, however, the individual should be about $(3\times10^8/60)/(2\pi) \approx 800\,\text{km}$ from the source. At these large distances from a 60 Hz source, the skin depth can be used to help determine the level of penetration. Of course, at these distances, the field strength from a 60 Hz source would be minuscule.

2.12 Typical Electric Field Strengths

Tabulate typical electric field magnitudes. [Taylor, '94; Gandhi, '90; Golde; Cleveland, '99; Caola; Hansson]

The data in Table 2.2 should help provide a "feeling" for typical electric field strengths. Although in critical applications, measurements should be performed, these numbers can frequently be used to determine whether a quantity is reasonable. Numerous other factors, including the shape of the waveform and the orientation and position of the source, will affect these values. Electric field exposure at 60 Hz is typically less than 100 V/m. It is interesting to note that the electric field from power lines can be partially shielded by houses, trees, bushes, and other (grounded) conductive objects. For free-space conditions, the relationship between the magnitudes of the electric flux density and electric field is $|\vec{D}| = \varepsilon_o |\vec{E}|$.

2.13 Current Through and Voltage Across a Field-Immersed Person

A hemispherical object, representing a person, of conductivity and permittivity σ_i and ε_i, respectively, sitting on a ground plane, is placed in a time-varying uniform electric field. Estimate the total current through and voltage across this object, assuming free space surrounds the hemisphere. [Haus; Zahn; Gandhi, '90; Deno, '77; IPCS, '92; Moon, '61]

Referring to Figure 2.32, the sinusoidal steady-state potential distributions outside and inside the solid sphere are

$$\Phi = \begin{cases} -\text{Re}\left[E_e R\cos\theta e^{j\omega t}\left\{\frac{r}{R} + \left[\frac{(\sigma_e - \sigma_i) + j\omega(\varepsilon_e - \varepsilon_i)}{(2\sigma_e + \sigma_i) + j\omega(2\varepsilon_e + \varepsilon_i)}\right]\left(\frac{R}{r}\right)^2\right\}\right] & \text{if } r > R \\ -\text{Re}\left[E_e R\cos\theta e^{j\omega t}\left\{3\frac{r}{R}\left[\frac{\sigma_e + j\omega\varepsilon_e}{(2\sigma_e + \sigma_i) + j\omega(2\varepsilon_e + \varepsilon_i)}\right]\right\}\right] & \text{if } r < R \end{cases} \quad (2.46)$$

[10]This includes some engineering graduates with just enough knowledge to be dangerous.

TABLE 2.2 Typical Electric Field Strengths

Source	Approximate $\|\vec{E}\|$ Field
Field emission of electrons from a room temperature metal surface	1,000 MV/m
Sparkover in air (uniform field)	3 MV/m
Hair "stands on end"	1 MV/m
Prickling sensation on hairs on back of hand	200 kV/m
Dust attraction to a charged surface	20 kV/m
Steady-precipitation in disturbed weather (along earth's surface)	0.5–3 kV/m
Thunderstorm environment (along earth's surface)	2–10 kV/m
Nonprecipitating clouds in disturbed weather (along earth's surface)	0.3–1 kV/m
Fair-weather, polluted atmosphere (along earth's surface)	100's V/m
Fair-weather along earth's surface	100 V/m
FCC limit for controlled/occupational exposure	614 V/m rms (0.3–3 MHz) $(1{,}842/f)^a$ V/m rms (3–30 MHz) 61.4 V/m rms (30–300 MHz)
FCC limit for uncontrolled/general population exposure	614 V/m rms (0.3–1.34 MHz) $(824/f)^a$ V/m rms (1.34–30 MHz) 27.5 V/m rms (30–300 MHz)
735 kV rms distribution lines	7 kV/m rms (60 Hz) (directly underneath the lines near the surface)
500 kV rms distribution lines	4 kV/m rms (60 Hz) (directly underneath the lines near the surface) 600 V/m rms (60 Hz) (20 m along the surface from the base of the lines) 80 V/m rms (60 Hz) (90 m along the surface from the base of the lines) 1–20 V/m rms (60 Hz) (in a house 90 m along the surface from the base of the lines)
130 kV rms distribution lines	800 V/m rms (60 Hz) (directly underneath the lines near the surface)
Residential building	54 V/m rms (50 Hz) Swedish houses 77 V/m rms (50 Hz) Norwegian houses
Electric blanket	250 V/m rms (60 Hz) @ 30 cm
Broiler	130 V/m rms (60 Hz) @ 30 cm
Stereo	90 V/m rms (60 Hz) @ 30 cm
Iron	60 V/m rms (60 Hz) @ 30 cm
Refrigerator	60 V/m rms (60 Hz) @ 30 cm
Hand mixer	50 V/m rms (60 Hz) @ 30 cm
Toaster	40 V/m rms (60 Hz) @ 30 cm
Television	30 V/m rms (60 Hz) @ 30 cm
Coffee pot	30 V/m rms (60 Hz) @ 30 cm
Vacuum cleaner	16 V/m rms (60 Hz) @ 30 cm
Fluorescent light	10 V/m rms (60 Hz) @ 30 cm
Electric range	4 V/m rms (60 Hz) @ 30 cm
Incandescent light bulb	2 V/m rms (60 Hz) @ 30 cm

[a] f in MHz

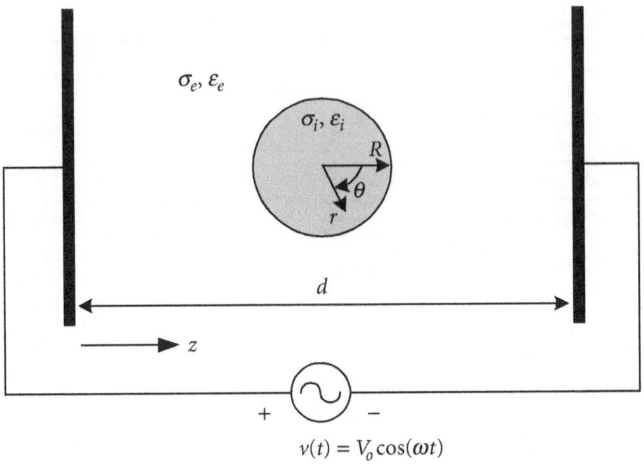

FIGURE 2.32 Spherical object between two charged parallel plates.

For $\theta = 90°$, the potential is zero for all values of *r*. The corresponding plane is the equipotential, zero volt, ground reference. Thus, these expressions also apply for a hemisphere sitting on a ground plane as shown in Figure 2.33. The electric field is obtained from the negative gradient (in the spherical coordinate system) of the potential:

$$\vec{E} = -\nabla\Phi = -\frac{\partial\Phi}{\partial r}\hat{a}_r - \frac{1}{r}\frac{\partial\Phi}{\partial\theta}\hat{a}_\theta - \frac{1}{r\sin\theta}\frac{\partial\Phi}{\partial\phi}\hat{a}_\phi$$

Fortunately, only the radial component of the electric field is required to determine the total current into the hemispherical surface:

$$E_r = -\frac{d\Phi}{dr} = \begin{cases} \mathrm{Re}\left[E_e R\cos\theta e^{j\omega t}\left\{\frac{1}{R} - 2\left[\frac{(\sigma_e - \sigma_i) + j\omega(\varepsilon_e - \varepsilon_i)}{(2\sigma_e + \sigma_i) + j\omega(2\varepsilon_e + \varepsilon_i)}\right]\frac{R^2}{r^3}\right\}\right] & \text{if } r > R \\ \mathrm{Re}\left[E_e R\cos\theta e^{j\omega t}\left\{3\frac{1}{R}\left[\frac{\sigma_e + j\omega\varepsilon_e}{(2\sigma_e + \sigma_i) + j\omega(2\varepsilon_e + \varepsilon_i)}\right]\right\}\right] & \text{if } r < R \end{cases} \quad (2.47)$$

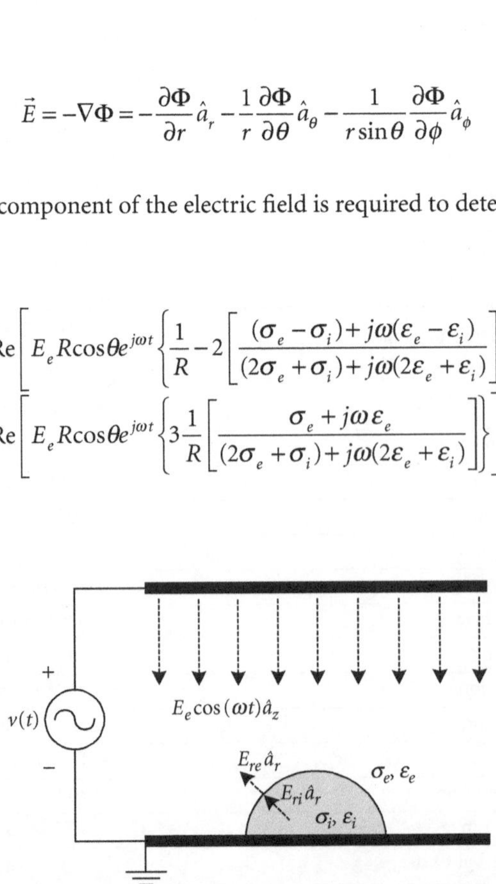

FIGURE 2.33 Far from the hemisphere, the electric field is uniform and perpendicular to the upper plate.

Along the hemispherical surface, $r = R$, with free-space conditions outside the hemisphere ($\sigma_e = 0, \varepsilon_e = \varepsilon_o$),

$$E_r = \begin{cases} \mathrm{Re}\left[E_e \cos\theta e^{j\omega t}\left\{1 - 2\left[\dfrac{-\sigma_i + j\omega(\varepsilon_o - \varepsilon_i)}{\sigma_i + j\omega(2\varepsilon_o + \varepsilon_i)}\right]\right\}\right] & \text{if } r = R^+ \\ \mathrm{Re}\left[E_e \cos\theta e^{j\omega t}\left\{3\left[\dfrac{j\omega\varepsilon_o}{\sigma_i + j\omega(2\varepsilon_o + \varepsilon_i)}\right]\right\}\right] & \text{if } r = R^- \end{cases} \tag{2.48}$$

where the ± superscripts refer to values slightly greater or less than R, respectively. The electric field expression for inside the sphere can be simplified by assuming that it is a good conductor:

$$E_{ri} \approx \mathrm{Re}\left(E_e 3\frac{j\omega\varepsilon_o}{\sigma_i}\cos\theta e^{j\omega t}\right) \quad \text{if } r = R, \frac{\sigma_i}{\omega\varepsilon_i} \gg 1 \tag{2.49}$$

At the 60 Hz power frequency, much of the human body is considered a good conductor. Converting to the time domain by using Euler's formula,

$$\begin{aligned} E_{ri} &= \mathrm{Re}\left\{E_e 3\frac{j\omega\varepsilon_o}{\sigma_i}\cos\theta[\cos(\omega t) + j\sin(\omega t)]\right\} \\ &= -\frac{3E_e\omega\varepsilon_o}{\sigma_i}\cos\theta\sin(\omega t) \quad \text{if } r = R, \frac{\sigma_i}{\omega\varepsilon_i} \gg 1 \end{aligned} \tag{2.50}$$

To determine the total current "into" the hemisphere toward the ground, the current density must be integrated over its entire surface:

$$\begin{aligned} I &= -\iint_S \vec{J}\cdot d\vec{s} = -\int_0^{2\pi}\int_0^{\frac{\pi}{2}} \sigma_i E_{ri}\hat{a}_r \cdot R^2\sin\theta\, d\theta d\phi \hat{a}_r = -\int_0^{2\pi}\int_0^{\frac{\pi}{2}} \sigma_i E_{ri} R^2 \sin\theta\, d\theta d\phi \\ &= \int_0^{2\pi}\int_0^{\frac{\pi}{2}} \sigma_i\left[\frac{3E_e\omega\varepsilon_o}{\sigma_i}\cos\theta\sin(\omega t)\right]R^2\sin\theta\, d\theta d\phi = 3\pi E_e\omega\varepsilon_o R^2\sin(\omega t) \\ &\approx 5.2\times10^{-10} E_e f R^2 \sin(\omega t) \end{aligned} \tag{2.51}$$

Although most objects, including humans, are not spherical in shape, this expression does provide a basis for comparison.

The electric field at the head of a person would probably be more intense or concentrated than the maximum field at the top surface of a hemisphere. Surprisingly, though, the hemisphere-based current expression is similar to an expression used for the current magnitude through a grounded human of length (or height) l_{th} standing in a uniform electric field E_e:

$$I = 1.08\times10^{-10} E_e f l_{th}^2 \tag{2.52}$$

For a human of height 1.75 m (≈ 5.7 ft), this expression is valid for frequencies up to about 40 MHz. (The free-space wavelength corresponding to 40 MHz is 7.5 m.) A 1.75 m human is then nearly electrically small. By equating (2.51) and (2.52), the relationship between R and l_{th} is determined to be $R \approx 0.5 l_{th}$. Although the electric field distortion produced by a hemisphere is not quite the same as a vertical cylinder in a uniform electric field, this relationship between R and l_{th} is reasonable. For a typical 50 V/m rms, 60 Hz signal, a 1.75 m tall barefooted (i.e., grounded) individual would experience a current of

$$I = 1.08\times10^{-10}(50)(60)(1.75)^2 \approx 1\,\mu\text{A rms}$$

Even standing under a high-voltage line, with a surface field of 1 kV/m rms, the current is small:

$$I = 1.08\times10^{-10}(10^3)(60)(1.75)^2 \approx 20\,\mu\text{A rms}$$

For a maintenance worker, the field may be perhaps 10–100 kV/m rms, corresponding to a current of 0.2–2 mA rms. For higher RF frequencies, a much lower field can generate a much higher current. For example, at 30 MHz, a field level of 50 V/m rms generates a current of about 500 mA. (This assumes the body is a good conductor at 30 MHz.) For frequencies above 100 kHz, one source recommends that the current should not exceed 50 mA rms to avoid the perception of pain. The threshold for burns is about 200 mA rms for RF.

It would seem more reasonable to model the human body as a long cylinder or prolate spheroid rather than a sphere. Similar to the steps given in the reference, the expression $I = 1.08\times10^{-10}E_e f l_{th}^2$ will now be derived. The capacitance between a long cylindrical rod of length l_{th} and radius R at a height of h above a large ground plane is

$$C = \frac{2\pi\varepsilon_o l_{th}}{\ln\left(\dfrac{l_{th}}{R}\sqrt{\dfrac{2h+\dfrac{l_{th}}{2}}{2h+\dfrac{3l_{th}}{2}}}\right)} \quad \text{if } l_{th} \gg R$$

If the time-varying voltage between the cylinder and flat plane is known, then the current in the frequency domain through this capacitance is

$$I_s = C(j\omega V_s)$$

When the floating conducting cylinder is placed in the uniform electric field, it acquires a voltage. The electric field lines will be perpendicular to the conducting cylinder. As shown by the sketch of the electric field given in Figure 2.34, the conducting cylinder definitely distorts the field near it. The cylinder's axis is parallel to the applied field. (A long conducting cylinder with its axis perpendicular to the applied

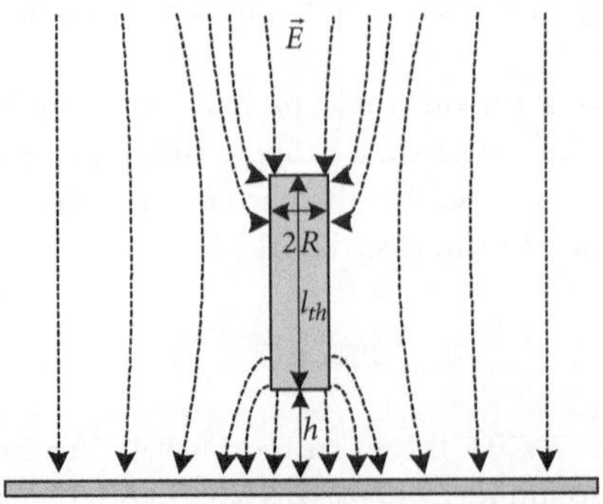

FIGURE 2.34 Distortion of a uniform electric field by a long conducting cylinder, close to a ground plane.

electric field would have less effect on the field since it would be along an equipotential contour.) Although far from being intellectually satisfying, the voltage of the cylinder is taken as the product of the field far from the cylinder, E_e, and $l_{th}/2$ (assuming $l_{th} \gg h$).[11] This approximation is essentially stating that the cylinder is assuming the average of the potential at a distance of l_{th} from the ground plane reference. A much more satisfying approach, if the field distribution were known, would be to determine the open-circuit voltage of the cylinder by integrating the electric field from the base of the cylinder to the ground plane beneath it. Nevertheless, the final expression, which appears to correlate well with measurements, is

$$|I_s| = \omega \frac{2\pi\varepsilon_o l_{th}}{\ln\left(\frac{l_{th}}{R}\sqrt{\frac{2h+\frac{l_{th}}{2}}{2h+\frac{3l_{th}}{2}}}\right)}\left(E_e\frac{l_{th}}{2}\right) = \frac{2\pi^2\varepsilon_o f l_{th}^2 E_e}{\ln\left(\frac{l_{th}}{R}\sqrt{\frac{4h+l_{th}}{4h+3l_{th}}}\right)} \approx \frac{1.75\times10^{-10} f l_{th}^2 E_e}{\ln\left(\frac{l_{th}}{R}\sqrt{\frac{4h+l_{th}}{4h+3l_{th}}}\right)} \quad (2.53)$$

If $l_{th} \gg h$ and $l_{th}/R = 9$, corresponding to a body diameter of about 0.39 m for a 1.75 m tall person, this expression is nearly identical to the previous result:

$$|I_s| \approx \frac{1.75\times10^{-10} f l_{th}^2 E_e}{\ln\left(\frac{9}{\sqrt{3}}\right)} \approx 1.06\times10^{-10} E_e f l_{th}^2 \quad (2.54)$$

If the cylinder is grounded or connected to the ground plane via a wire, then the short-circuit current through the cylinder is also taken as I_s. In this case, the capacitance is a measure of the area of an "equivalent charge-collecting" plate. The larger the capacitance of the object, the greater the induced charge and current.

The potential *across* the finite-conductivity hemispherical grounded object is nonzero, but small, when the hemisphere is a good conductor. A floating object will also develop a potential across it when exposed to an electric field. Again, using the expressions for a sphere immersed in a uniform electric field surrounded by free space, it was shown that the field in the solid sphere is uniform and equal to

$$\vec{E}_{is} = 3E_e\left[\frac{j\omega\varepsilon_o}{\sigma_i + j\omega(2\varepsilon_o + \varepsilon_i)}\right]\hat{a}_z$$

This assumes sinusoidal steady-state conditions. Because the field is uniform, the voltage developed across the poles of the sphere is easily determined:

$$\begin{aligned} V &= -\int_R^{-R} \vec{E}_{is}\cdot d\vec{L} = -\int_R^{-R} 3E_e\left[\frac{j\omega\varepsilon_o}{\sigma_i + j\omega(2\varepsilon_o + \varepsilon_i)}\right]\hat{a}_z \cdot dz\hat{a}_z \\ &= 6RE_e\left[\frac{j\omega\varepsilon_o}{\sigma_i + j\omega(2\varepsilon_o + \varepsilon_i)}\right] \end{aligned} \quad (2.55)$$

(The voltage developed across the hemisphere would be half of this value.) The amplitude of the voltage developed across an object in a uniform electric field is *believed* by nescient individuals to be merely the

[11] Using antenna terminology, the effective length or height of the electrically-short cylindrical antenna is $0.5l_{th}$ for a triangular current distribution.

product of the field and longest length of the object. For example, a 0.5 m diameter sphere in a 10 kV/m rms field is incorrectly believed to develop a voltage of $0.5\,(10\times10^3)=5\,\text{kV}$ rms across it. Using the given expression, however, the voltage developed is ($\sigma_i = 0.2$ 1/Ω-m, $\varepsilon_{ri} = 81$, $f = 60$ Hz) in the mV range:

$$|V| = 6\left(\frac{0.5}{2}\right)(10\times10^3)\left|\frac{j(2\pi\times60)(8.854\times10^{-12})}{0.2+j(2\pi\times60)(2\times8.854\times10^{-12}+81\times8.854\times10^{-12})}\right| \approx 0.25\,\text{mV rms}$$

The voltage is quite small since the material is a good conductor and the frequency is low. At 1 kHz, a field of 300 V/m rms generates a voltage magnitude of about 0.13 mV rms. Conductive body suits are used by workers in high electric fields to further reduce this voltage drop or gradient. For an ungrounded conducting object, the voltage across the object is usually small. However, the open-circuit or unloaded voltage of the conducting object *relative to ground* (or some other reference) can reach a large value. The electric field in the radial direction outside of a solid sphere immersed in a uniform electric field was given as

$$E_{re} = \text{Re}\left[E_e R\cos\theta e^{j\omega t}\left\{\frac{1}{R} - 2\left[\frac{(\sigma_e-\sigma_i)+j\omega(\varepsilon_e-\varepsilon_i)}{(2\sigma_e+\sigma_i)+j\omega(2\varepsilon_e+\varepsilon_i)}\right]\frac{R^2}{r^3}\right\}\right] \quad \text{if } r > R$$

Assuming free space around a good-conducting sphere, the amplitude of this field directly below the pole of the sphere simplifies to

$$E_{re} \approx E_e\left(1+2\frac{R^3}{r^3}\right) \quad \text{if } r > R, \sigma_e = 0, \varepsilon_e = \varepsilon_o, \theta = 0°, \sigma_i \gg \omega\varepsilon_i \tag{2.56}$$

Referring to Figure 2.35, the voltage difference between the surface of the sphere R and some radial distance d from the center of the sphere is

$$\Delta V = -\int_R^d E_{re}dr = -\int_R^d E_e\left(1+2\frac{R^3}{r^3}\right)dr = -E_e\left(r-\frac{R^3}{r^2}\right)\Bigg|_R^d = -E_e\left(d-\frac{R^3}{d^2}\right) \tag{2.57}$$

A 600 V/m rms field would develop a voltage of about 470 V rms for a 2 meter sphere, which could crudely represent an ungrounded vehicle, at a distance of 0.3 m from the surface of the sphere:

$$|\Delta V| = 600\left[2.3-\frac{2^3}{(2.3)^2}\right] \approx 470\,\text{V rms}$$

The distance of 0.3 m could roughly represent the distance from the vehicle to the ground. The actual electric field distribution would be different from that given by E_{re} when the ground plane is this close

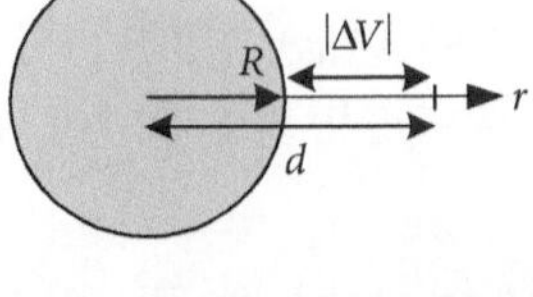

FIGURE 2.35 Potential difference between the sphere's surface and some radial distance $d - R$ from this point.

to the sphere. Measured open-circuit voltages for various size vehicles reached as high as several hundred volts when exposed to 632 V/m rms fields in the 0.3–3 MHz range.

For the electric field at the surface of a sphere (or hemisphere) immersed in a uniform field, there are several special cases worth examining. For sinusoidal steady-state conditions, the two components of the electric field external to the sphere are

$$E_{re} = \mathrm{Re}\left[E_e R\cos\theta e^{j\omega t}\left\{\frac{1}{R} - 2\left[\frac{(\sigma_e - \sigma_i) + j\omega(\varepsilon_e - \varepsilon_i)}{(2\sigma_e + \sigma_i) + j\omega(2\varepsilon_e + \varepsilon_i)}\right]\frac{R^2}{r^3}\right\}\right] \quad \text{if } r > R \tag{2.58}$$

$$\begin{aligned} E_{\theta e} &= -\frac{1}{r}\frac{d\Phi}{d\theta} \\ &= -\mathrm{Re}\left[E_e R\sin\theta e^{j\omega t}\left\{\frac{1}{R} + \left[\frac{(\sigma_e - \sigma_i) + j\omega(\varepsilon_e - \varepsilon_i)}{(2\sigma_e + \sigma_i) + j\omega(2\varepsilon_e + \varepsilon_i)}\right]\frac{R^2}{r^3}\right\}\right] \quad \text{if } r > R \end{aligned} \tag{2.59}$$

For free-space conditions around the sphere,

$$E_{re} = \mathrm{Re}\left[E_e R\cos\theta e^{j\omega t}\left\{\frac{1}{R} - 2\left[\frac{-\sigma_i + j\omega(\varepsilon_o - \varepsilon_i)}{\sigma_i + j\omega(2\varepsilon_o + \varepsilon_i)}\right]\frac{R^2}{r^3}\right\}\right] \quad \text{if } r > R, \sigma_e = 0, \varepsilon_e = \varepsilon_o \tag{2.60}$$

$$E_{\theta e} = -\mathrm{Re}\left[E_e R\sin\theta e^{j\omega t}\left\{\frac{1}{R} + \left[\frac{-\sigma_i + j\omega(\varepsilon_o - \varepsilon_i)}{\sigma_i + j\omega(2\varepsilon_o + \varepsilon_i)}\right]\frac{R^2}{r^3}\right\}\right] \quad \text{if } r > R, \sigma_e = 0, \varepsilon_e = \varepsilon_o \tag{2.61}$$

If the sphere is perfectly conducting, $\sigma_i \to \infty$, the electric field components at the surface of the sphere are

$$\lim_{\sigma_i \to \infty} E_{re} = \mathrm{Re}\left[E_e \cos\theta e^{j\omega t}\left(1 + 2\frac{R^3}{r^3}\right)\right] = 3E_e\cos\theta\cos(\omega t) \quad \text{if } r = R \tag{2.62}$$

$$\lim_{\sigma_i \to \infty} E_{\theta e} = -\mathrm{Re}\left[E_e \sin\theta e^{j\omega t}\left(1 - \frac{R^3}{r^3}\right)\right] = 0 \quad \text{if } r = R \tag{2.63}$$

As expected, the electric field is entirely normal (i.e., radial) to the perfectly conducting sphere for all frequencies. (The internal electric field is zero when the sphere is perfectly conducting.) For the other extreme condition, when the sphere is perfectly insulating and surrounded by free space, $\sigma_i = 0$,

$$\begin{aligned} E_{re} &= \mathrm{Re}\left[E_e \cos\theta e^{j\omega t}\left(\frac{3\varepsilon_i}{2\varepsilon_o + \varepsilon_i}\right)\right] \\ &= E_e \cos\theta \frac{3\varepsilon_i}{2\varepsilon_o + \varepsilon_i}\cos(\omega t) \quad \text{if } r = R, \sigma_i = 0, \sigma_e = 0, \varepsilon_e = \varepsilon_o \end{aligned} \tag{2.64}$$

$$E_{\theta e} = -\text{Re}\left[E_e \sin\theta e^{j\omega t}\left(\frac{3\varepsilon_o}{2\varepsilon_o+\varepsilon_i}\right)\right]$$
$$= -E_e \sin\theta \frac{3\varepsilon_o}{2\varepsilon_o+\varepsilon_i}\cos(\omega t) \quad \text{if } r=R, \sigma_i=0, \sigma_e=0, \varepsilon_e=\varepsilon_o \tag{2.65}$$

When both the sphere and its surroundings are perfectly insulating, the electric field along the sphere's surface has both tangential and radial components. For a lossy sphere (one that is neither perfectly conducting nor perfectly insulating), the electric field along the surface of the sphere has both tangential and normal components. (A lossy transmission line has electric field components both tangential and normal to the conductors.) Of course, if the conductor is a good conductor, the tangential component of the field is small compared to the normal component. Finally, when the sphere is perfectly insulating but the medium surrounding the sphere is conducting, the external electric field components are

$$E_{re} = \text{Re}\left[E_e R\cos\theta e^{j\omega t}\left\{\frac{1}{R} - 2\left[\frac{\sigma_e + j\omega(\varepsilon_e-\varepsilon_i)}{2\sigma_e + j\omega(2\varepsilon_e+\varepsilon_i)}\right]\frac{R^2}{r^3}\right\}\right] \quad \text{if } r>R, \sigma_i=0 \tag{2.66}$$

$$E_{\theta e} = -\text{Re}\left[E_e R\sin\theta e^{j\omega t}\left\{\frac{1}{R} + \left[\frac{\sigma_e + j\omega(\varepsilon_e-\varepsilon_i)}{2\sigma_e + j\omega(2\varepsilon_e+\varepsilon_i)}\right]\frac{R^2}{r^3}\right\}\right] \quad \text{if } r>R, \sigma_i=0 \tag{2.67}$$

Along the surface of the sphere,

$$E_{re} = \text{Re}\left\{E_e \cos\theta e^{j\omega t}\left[\frac{3j\omega\varepsilon_i}{2\sigma_e + j\omega(2\varepsilon_e+\varepsilon_i)}\right]\right\} \quad \text{if } r=R, \sigma_i=0 \tag{2.68}$$

$$E_{\theta e} = -\text{Re}\left\{E_e \sin\theta e^{j\omega t}\left[\frac{3\sigma_e + 3j\omega\varepsilon_e}{2\sigma_e + j\omega(2\varepsilon_e+\varepsilon_i)}\right]\right\} \quad \text{if } r=R, \sigma_i=0 \tag{2.69}$$

For ac conditions, both tangential and normal components can exist along the surface of the spherical cavity. At the poles of the cavity, $\theta = 0°$, the field is entirely radial to the surface. Along the "equator" of the sphere, $\theta = 90°$, the field is entirely tangential to the surface. For dc steady-state conditions, the normal or radial component is zero:

$$E_{re} = 0 \quad \text{if } r=R, \sigma_i=0, \omega=0 \tag{2.70}$$

$$E_{\theta e} = -\text{Re}\left(\frac{3}{2}E_e \sin\theta e^{j\omega t}\right) = -\frac{3}{2}E_e \sin\theta \quad \text{if } r=R, \sigma_i=0, \omega=0 \tag{2.71}$$

The normal component must be zero for dc steady-state conditions since the dc current through the surrounding medium must pass around the perfectly insulating spherical cavity (and $J_r = \sigma E_r$). The dc steady-state current into the spherical cavity must be zero. Hence, the current must pass tangentially around this spherical obstacle. For ac or nonzero frequencies, the charge along the surface of the cavity will vary in a sinusoidal fashion (increasing and decreasing with time). Although the current in the insulating cavity must still be zero, a normal electric field to the surface is required to "feed" the charge to this interface.

2.14 Insulating Spherical Shields

Referring to Figure 2.36, the electric field inside the cavity of an insulating spherical shell of inner radius a, outer radius b, and relative permittivity ε_r when placed in a uniform electric field of strength E_e is

$$E_i = \frac{9\varepsilon_r}{9\varepsilon_r - 2(\varepsilon_r - 1)^2\left(\frac{a^3}{b^3} - 1\right)} E_e \tag{2.72}$$

Determine the ratio E_i/E_e if the permittivity is very large. Free space exists outside the shell and inside the cavity. [Smythe; Weber; Jefimenko]

The conductivity of the shell is zero in this expression since it is insulating. However, the expression can be used for electroquasistatic situations for frequencies where the shell material is considered a good insulator (rather than a perfect insulator):

$$\frac{\sigma}{\omega\varepsilon_r\varepsilon_o} \ll 1$$

For very high shell permittivities, the ratio of the inner to outer field reduces to

$$\frac{E_i}{E_e} \approx \frac{9\varepsilon_r}{9\varepsilon_r - 2\varepsilon_r^2\left(\frac{a^3}{b^3} - 1\right)} = \frac{9}{9 + 2\varepsilon_r\left(1 - \frac{a^3}{b^3}\right)} \quad \text{if } \varepsilon_r \gg 1 \tag{2.73}$$

Furthermore, when the shell thickness is not small,

$$\frac{E_i}{E_e} \approx \frac{9}{9 + 2\varepsilon_r} \approx \frac{9}{2\varepsilon_r} \quad \text{if } \varepsilon_r \gg 1, b \gg a \tag{2.74}$$

The shielding effectiveness increases as the relative permittivity of the shell increases. For a thick shell of dielectric constant 10,

$$SE_{dB} = 20\log\left(\frac{E_e}{E_i}\right) = 20\log\left(\frac{9 + 2\varepsilon_r}{9}\right) \approx 10\text{ dB} \quad \text{if } b \gg a, \varepsilon_r = 10$$

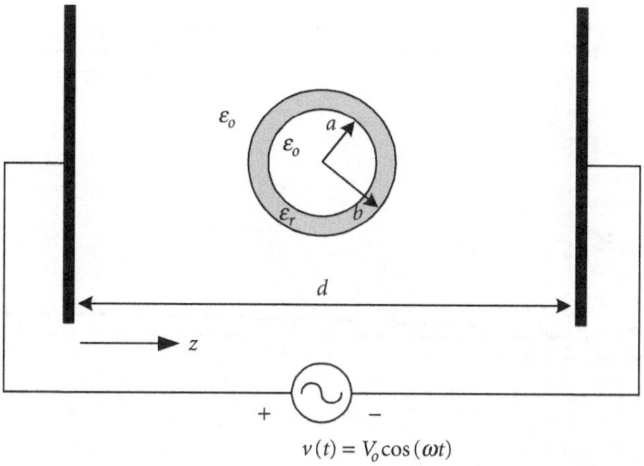

FIGURE 2.36 Spherical insulating shell between two charged parallel plates.

When the shell thickness, Δ, is small,

$$\frac{E_i}{E_e}=\frac{9}{9+2\varepsilon_r\left[1-\left(\frac{b-\Delta}{b}\right)^3\right]}=\frac{9}{9+2\varepsilon_r\left[3\frac{\Delta}{b}-3\left(\frac{\Delta}{b}\right)^2+\left(\frac{\Delta}{b}\right)^3\right]}\quad \text{if } \varepsilon_r \gg 1, \Delta=b-a$$

$$\approx\frac{9}{9+6\varepsilon_r\frac{\Delta}{b}}=\frac{3}{3+2\varepsilon_r\frac{\Delta}{b}}\quad \text{if } \varepsilon_r \gg 1, \Delta=b-a\ll b \tag{2.75}$$

The shielding effectiveness for a thin shell is

$$SE_{dB}=20\log\left(\frac{E_e}{E_i}\right)=20\log\left(\frac{3+2\varepsilon_r\frac{\Delta}{b}}{3}\right)=20\log\left(1+\frac{2\varepsilon_r}{3}\frac{\Delta}{b}\right)\quad \text{if } \varepsilon_r \gg 1, \Delta\ll b \tag{2.76}$$

Unless the relative permittivity is very large compared to b/Δ, the shielding effectiveness is small. The results in Mathcad 2.1 clearly show the low values in the shielding effectiveness (at least compared to conducting shields).

As seen by these expressions, if the relative permittivity approaches infinity, the electric field inside the cavity approaches zero. The electric field inside a perfectly conducting shield is also zero. When the shell's permittivity is infinite, the electric field distribution both inside the cavity and outside the shell is identical to a shell with infinite conductivity. Conductors, of course, have free charges while (ideal) insulators have no free charges only bound charges. In an insulator, the electric field from the bound

$b := 1$

$\Delta := 0.01, 0.02 .. 1$

$$SE(\varepsilon_r,\Delta) := 20\cdot\log\left[\frac{9\cdot\varepsilon_r-2\cdot(\varepsilon_r-1)^2\cdot\left[\frac{(b-\Delta)^3}{b^3}-1\right]}{9\cdot\varepsilon_r}\right]$$

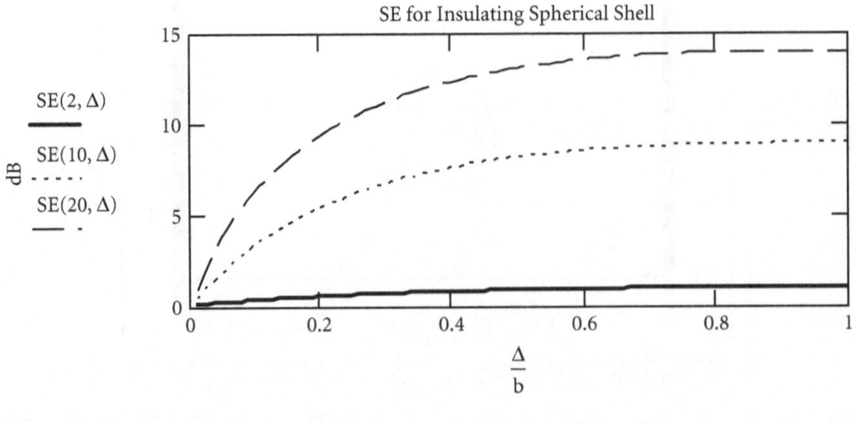

MATHCAD 2.1 SE vs. the ratio of the spherical shell thickness to outer radius.

charges along the surface of the shell generate a counter field (also referred to as a depolarizing field) that cancels some of the applied field. This is how an insulator can shield.

2.15 Insulating Cylindrical Shields

The electric field inside the cavity of an insulating cylindrical shell of inner radius a, outer radius b, and relative permittivity ε_r when placed in a uniform electric field of strength E_e is

$$E_i = \frac{4\varepsilon_r}{4\varepsilon_r - (\varepsilon_r - 1)^2\left(\dfrac{a^2}{b^2} - 1\right)} E_e \tag{2.77}$$

Determine the ratio E_i/E_e if the permittivity is very large. [Zahn]

This analysis is very similar to that presented for the insulating spherical shell. For large permittivities,

$$\frac{E_i}{E_e} \approx \frac{4}{4 + \varepsilon_r\left(1 - \dfrac{a^2}{b^2}\right)} \quad \text{if } \varepsilon_r \gg 1 \tag{2.78}$$

If the shell thickness is not small,

$$\frac{E_i}{E_e} \approx \frac{4}{4+\varepsilon_r} \approx \frac{4}{\varepsilon_r} \quad \text{if } \varepsilon_r \gg 1, b \gg a \tag{2.79}$$

For comparison to the spherical shell,

$$SE_{dB} = 20\log\left(\frac{E_e}{E_i}\right) = 20\log\left(\frac{4+\varepsilon_r}{4}\right) \approx 11\,\text{dB} \quad \text{if } b \gg a, \varepsilon_r = 10$$

The shielding effectiveness is slightly larger than for a spherical shell, but the cylindrical expressions assume that the cylindrical shell is very long (compared to the outer radius, b). The large permittivity approximation can easily be written in terms of the shell thickness, Δ:

$$\begin{aligned}\frac{E_i}{E_e} &= \frac{4}{4 + \varepsilon_r\left[1 - \dfrac{(b-\Delta)^2}{b^2}\right]} = \frac{4}{4 + \varepsilon_r\left[2\dfrac{\Delta}{b} - \left(\dfrac{\Delta}{b}\right)^2\right]} \quad \text{if } \varepsilon_r \gg 1, \Delta = b - a \\ &\approx \frac{2}{2 + \varepsilon_r \dfrac{\Delta}{b}} \quad \text{if } \varepsilon_r \gg 1, \Delta = b - a \ll b\end{aligned} \tag{2.80}$$

The shielding effectiveness in dB for this cylindrical shell is

$$SE_{dB} = 20\log\left(\frac{E_e}{E_i}\right) = 20\log\left(\frac{2 + \varepsilon_r \dfrac{\Delta}{b}}{2}\right) = 20\log\left(1 + \frac{\varepsilon_r}{2}\frac{\Delta}{b}\right) \quad \text{if } \varepsilon_r \gg 1, \Delta \ll b \tag{2.81}$$

2.16 EQS and Perfect Conductors

When is an electromagnetic system electroquasistatic? When can a real conductor or real insulator be considered perfect under electroquasistatic conditions? [Haus; Adler; Fano][12]

A pure electro*static* field is an electric field that does not vary with time. An electrostatic field, however, can vary in space. It is generated by nontime-varying stationary charge. For example, when a dc voltage is applied across two parallel plates separated by free space (and sufficient time is allowed to elapse so that the charge on the plates is in steady state), only an electric field exists across the plates. No magnetic field is generated by the constant charge on each plate.

Throughout this entire book, though, the fascinating consequences of time-varying electric and magnetic fields are presented. Sometimes, as with plane wave shielding, both the electric and magnetic fields are important, and their strong coupling is necessary to generate the electromagnetic wave. This coupling between time-varying electric and magnetic fields is evident by examining Maxwell's four equations. For immediate reference, Maxwell's equations, plus the charge conservation equation, are given in both differential and integral form in Table 2.3. For free space, $\vec{B} = \mu_o \vec{H}$ and $\vec{D} = \varepsilon_o \vec{E}$. Equation (2.82) indicates that a time-varying magnetic field (or magnetic flux density) will generate an electric field (or, more precisely, a nonzero curl of an electric field). Equation (2.83) indicates that both a current and a time-varying electric field (or electric flux density) will generate a magnetic field (or, more precisely, a nonzero curl of a magnetic field). It is thus commonly stated that time-varying electric and magnetic fields are coupled or linked together: a time-varying electric field implies the presence of a time-varying magnetic field and a time-varying magnetic field implies the presence of a time-varying electric field.

Fortunately, in many cases, it is *not* necessary to use Maxwell's equations in their complete unsimplified form to determine a reasonable approximation for the electric and magnetic field distribution. In the previous discussions, the emphasis has been on electric fields or, more specifically, electroquasistatic or EQS fields. In EQS situations, the electric field is usually the field variable of interest. The magnetic field is often not mentioned or is determined as if it has no effect on the electric field. However, EQS conditions do *not* imply that the electric and magnetic fields are not time varying. EQS conditions do imply, however, that the electric energy is much greater than the magnetic energy. In those cases where these energies can be determined, this energy relationship can be used as a check on whether EQS conditions exist.

TABLE 2.3 Basic Laws in Electromagnetics

Common Name	Differential Form	Integral Form	Eqn. #
Faraday's law	$\nabla \times \vec{E} = -\frac{\partial \vec{B}}{\partial t}$	$\oint_C \vec{E} \cdot d\vec{L} = -\frac{d}{dt} \iint_S \vec{B} \cdot d\vec{s}$	(2.82)
Ampère's law	$\nabla \times \vec{H} = \vec{J} + \frac{\partial \vec{D}}{\partial t}$	$\oint_C \vec{H} \cdot d\vec{L} = \iint_S \vec{J} \cdot d\vec{s} + \frac{d}{dt} \iint_S \vec{D} \cdot d\vec{s}$	(2.83)
Gauss' law	$\nabla \cdot \vec{D} = \rho_V$	$\oint\!\!\oint_S \vec{D} \cdot d\vec{s} = \iiint_V \rho_V dv$	(2.84)
Magnetic flux continuity	$\nabla \cdot \vec{B} = 0$	$\oint\!\!\oint_S \vec{B} \cdot d\vec{s} = 0$	(2.85)
Charge conservation	$\nabla \cdot \vec{J} + \frac{\partial \rho_V}{\partial t} = 0$	$\oint\!\!\oint_S \vec{J} \cdot d\vec{s} + \frac{d}{dt} \iiint_V \rho_V dv = 0$	(2.86)

[12]This section closely follows the presentation given in the text written by Professor Haus and Professor Melcher. This exceptional and insightful text is highly recommended for any serious student interested in the field of electromagnetics.

For EQS conditions, the field solution is generally simpler since (2.82) is approximated as

$$\nabla \times \vec{E} \approx 0 \quad (2.87)$$

[13]

The term $d\vec{B}/dt$ (or $\partial\vec{B}/\partial t$), which is sometimes referred to as the magnetic induction component, is neglected. The approximate expression given in (2.87) is quite "convenient" since it allows the electric field to be obtained from $\vec{E} = -\nabla\Phi$ and the voltage between two points to be obtained from $V = -\int_a^b \vec{E} \cdot d\vec{L}$. These two expressions are most safely used for situations that are EQS.

Before a specific example is presented, the general procedure for determining the electric and magnetic fields for EQS situations will be outlined. The electric field is first determined in the standard manner as with an electrostatic system using

$$\nabla \cdot (\varepsilon\vec{E}) = \rho_V \quad \text{and} \quad \nabla \times \vec{E} = 0 \quad (2.88)$$

If the volume charge density, ρ_V, is known, and the voltage is given at the boundaries such as along electrodes, then a zero-order estimate for the electric field can be determined without regard for the effects of the magnetic field. Note that the equations in (2.88) are not a function of *d*/*dt* (or the magnetic field). This does not imply that the electric field is not a function of time since the charge density or applied voltage can be time varying. Rather, the lack of *d*/*dt* in these two equations does imply that the present field distribution is not a function of any past field state (i.e., there is no field memory and no initial time conditions). The time variation of the current is then determined from

$$\nabla \cdot \vec{J} + \frac{\partial \rho_V}{\partial t} = 0$$

since it is a function of the charge density. Third, the magnetic field is determined from one or both of the remaining expressions

$$\nabla \times \vec{H} = \vec{J} + \frac{\partial \vec{D}}{\partial t} \quad \text{and} \quad \nabla \cdot \vec{B} = 0$$

since $\vec{J}$ and $\vec{D}$, both a function of the electric field, are known. Finally, if a better or more accurate expression for the electric field is desired, which considers the effects of the time-varying magnetic field, it can be obtained from (2.82).

Many situations have been discussed in this book that assume EQS or electrostatic conditions. Up to this point, however, there have been no examples where the magnetic field is determined from the EQS electric field and then an estimate of the error associated with assuming EQS conditions determined. Because of the relative simplicity of the analysis, the magnetic field and the error in the electric field will be determined for the classical, oppositely charged parallel plates. As shown in Figure 2.37, the two closely spaced, parallel good-conducting plates are separated by free space, and the plates are connected to many identical, time-varying voltage sources. These multiple sources help ensure that the field is uniform along the full depth of the parallel plates and that the current along the plates in the *x* direction is not a function of *y*. End effects or nonuniform fields along the outer edges are neglected. The standard result for the electric field everywhere between the plates is

$$\vec{E}(t) = -\frac{v(t)}{d}\hat{a}_z \quad (2.89)$$

[13]A field is said to be irrotational when its curl is equal to zero.

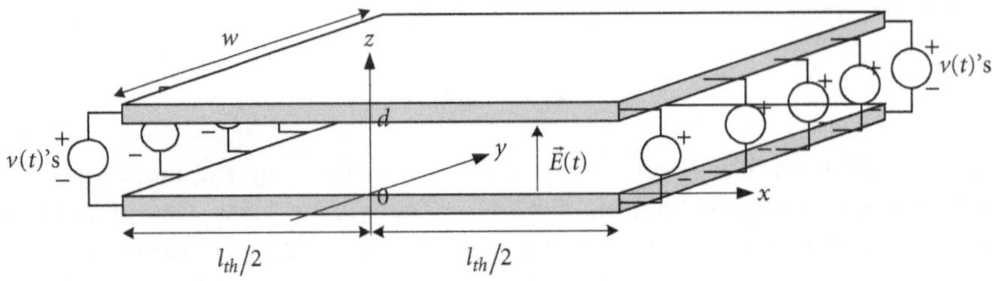

FIGURE 2.37 Many identical, parallel, time-varying voltage sources applied across two closely spaced parallel plates.

This result is based on Laplace's equation, which assumes zero volume charge density between the plates and electrostatic conditions (i.e., equations in (2.88)). It is considered the zero-order solution for the electric field. The boundary condition $D_{n2} - D_{n1} = \rho_S$, based on Gauss' law, allows the charge density on the inner surface of each plate to be immediately determined:

$$\rho_S = \begin{cases} -\varepsilon_o \dfrac{v(t)}{d} & z = 0 \\ \varepsilon_o \dfrac{v(t)}{d} & z = d \end{cases}$$

The application of this boundary condition along the upper plate is illustrated in Figure 2.38. When the voltage varies with time, the surface charge on each plate also varies with time. Furthermore, the surface charge is not presently a function of x. However, after improving the electrostatic-based electric field expression, it will be a function of x. The time change in the surface charge implies the existence of surface current along the inner surfaces of the two plates. The charge conservation equation, in integral form,

$$\oiint_S \vec{J} \cdot d\vec{s} + \frac{d}{dt} \iiint_V \rho_V dv = 0$$

can be used to determine this surface current. Briefly, this expression states that when charge is lost (or gained) in a closed region, current must pass outward from (or inward into) this region. Referring to Figure 2.39,

$$\underbrace{wK_x(x)}_{\substack{\text{total current}\\ \text{out of the right}\\ \text{side of the region}}} - \underbrace{wK_x(0)}_{\substack{\text{total current}\\ \text{out of the left}\\ \text{side of the region}}} + \frac{d}{dt}\left[\underbrace{(x-0)w\rho_S}_{\substack{\text{total charge}\\ \text{within the closed}\\ \text{region}}}\right] = 0 \tag{2.90}$$

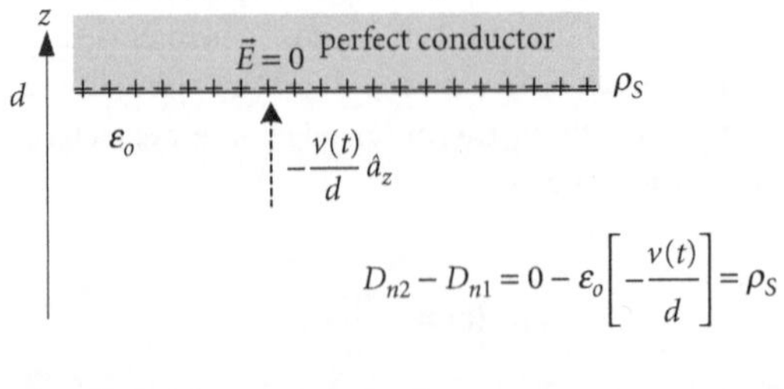

FIGURE 2.38 Lower surface of the upper plate. Although the surface charge is represented using "+" markers, the charge will change sign if $v(t)$ changes sign.

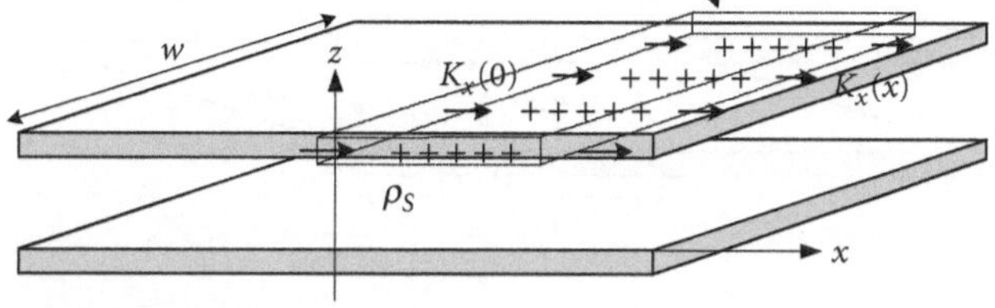

FIGURE 2.39 Closed surface passing through the upper plate. If the current into the left side is different from out the right side, the surface charge must be changing inside the region. (The voltage sources are not shown.)

With the push-push nature of the voltage supplies on the opposite sides of the plates, the current density at $x = 0$ is zero (symmetry arguments can also be used). Thus, (2.90) reduces to

$$K_x(x) = -x\frac{d\rho_S}{dt} = -\frac{x\varepsilon_o}{d}\frac{dv(t)}{dt} \tag{2.91}$$

As seen by (2.91), the surface current density is the greatest at the ends of the plates near the supplies and zero at the center where the opposing currents meet. If the voltage is increasing with time, the surface current $K_x(x)$ is negative: as the voltage increases, the charge on the upper plate increases. A current in the $-x$ direction, toward the z axis, is required to source this increasing charge. The magnetic field can be determined from the integral form in (2.83):

$$\oint_C \vec{H}\cdot d\vec{L} = \iint_S \vec{J}\cdot d\vec{s} + \frac{d}{dt}\iint_S \vec{D}\cdot d\vec{s}$$

The magnetic field generated by the current that is entirely in the x direction is in the y direction. Referring to Figure 2.40 and realizing that $\vec{J}$ is zero in the free-space region between the plates,

$$\underbrace{wH_y(x) - (x-0)(0) - w\,[H_y(0)] + (x-0)(0)}_{\text{line integral of the magnetic field around the closed path}} = \underbrace{0}_{\substack{\text{zero current}\\ \text{density normal to}\\ \text{the plates}}} + \frac{d}{dt}\left[\underbrace{-\varepsilon_o\frac{v(t)}{d}(xw)}_{\substack{\text{total electric flux over the}\\ \text{given surface}}}\right]$$

The direction of the surface vector is determined from the right-hand rule. Using symmetry arguments, $H_y(0)$ is zero. Therefore,

$$H_y(x) = -\frac{x\varepsilon_o}{d}\frac{dv(t)}{dt}$$

The electric field expression given in (2.89) assumed that $\nabla\times\vec{E} = 0$ or that $d\vec{B}/dt = 0$. With an estimate of the time-varying magnetic field between the plates now known, a correction term for the electric field

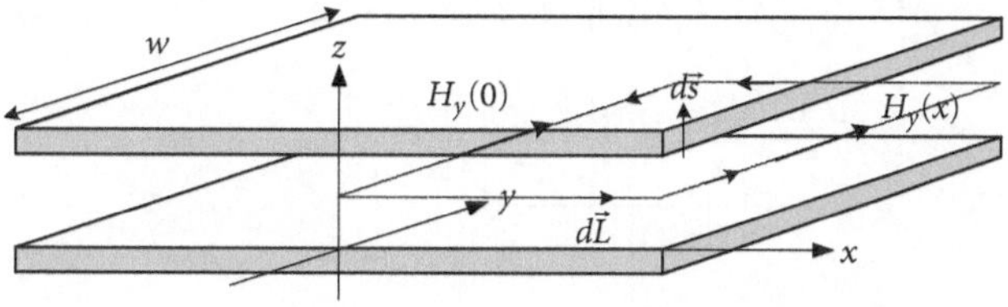

FIGURE 2.40 Open surface between and parallel to the two plates. (The voltage sources are not shown.)

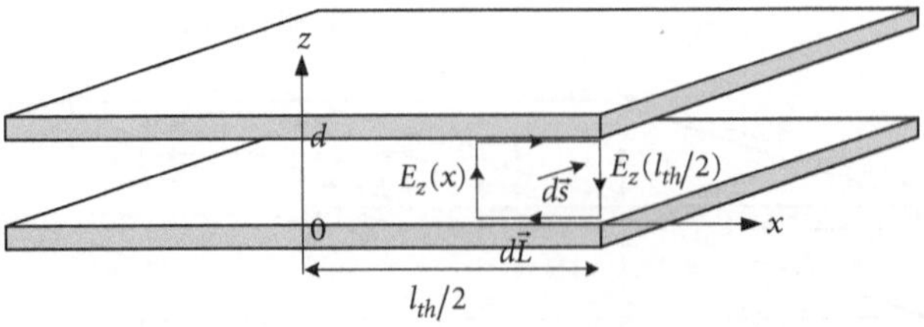

FIGURE 2.41 Closed contour for determining the electric field between two parallel plates. (The voltage sources are not shown.)

can be determined. The integral form in (2.82)

$$\oint_C \vec{E}\cdot d\vec{L} = -\frac{d}{dt}\iint_S \vec{B}\cdot d\vec{s}$$

can be employed to determine this correction term. Referring to the surface and its corresponding contour shown in Figure 2.41,

$$\underbrace{d[E_z(x)]+\left(\frac{l_{th}}{2}-x\right)(0)-d\left[E_z\left(\frac{l_{th}}{2}\right)\right]-\left(\frac{l_{th}}{2}-x\right)(0)}_{\text{line integral of the electric field around the closed path}} = -\frac{d}{dt}\int_0^d\int_x^{l_{th}/2}\mu_o\left[-\frac{x\varepsilon_o}{d}\frac{dv(t)}{dt}\hat{a}_y\right]\cdot dxdz\hat{a}_y$$

Evaluating the integral and solving for the electric field as a function of x,

$$E_z(x) = E_z\left(\frac{l_{th}}{2}\right)+\frac{\mu_o\varepsilon_o}{2d}\left(\frac{l_{th}^2}{4}-x^2\right)\frac{d^2v(t)}{dt^2} = -\frac{v(t)}{d}+\frac{\mu_o\varepsilon_o}{2d}\left(\frac{l_{th}^2}{4}-x^2\right)\frac{d^2v(t)}{dt^2} \tag{2.92}$$

When evaluating Faraday's law, the upper limit of integration was selected at $x = l_{th}/2$ since the electric field is forced to be equal to (2.89) by the voltage sources.[14] If an even better solution is desired for the electric field (and the magnetic field), the entire process can be repeated using (2.92) as the starting field instead of (2.89). What may be surprising is that the electric field between the two plates can vary even when the conductors are perfect with infinite conductivity! Probably, this variation is most easily understood by recalling that a voltage drop exists across a lossless inductor carrying a time-varying current and that the two parallel conductors in this example possess a nonzero inductance. This change in voltage from the ends of the plates, where the voltage supplies are located, to the center of the plates, at $x = 0$, results in the change in electric field with x. Probably the most interesting consequence of this work involving several of Maxwell's equations is the error associated with neglecting the correction term when estimating the electric field:

$$\left|\frac{E_{EQS}-E_{STATIC}}{E_{STATIC}}\right| = \left|\frac{E_z(x)-E_z\left(\frac{l_{th}}{2}\right)}{E_z\left(\frac{l_{th}}{2}\right)}\right| = \left|\frac{-\frac{v(t)}{d}+\frac{\mu_o\varepsilon_o}{2d}\left(\frac{l_{th}^2}{4}-x^2\right)\frac{d^2v(t)}{dt^2}-\left[-\frac{v(t)}{d}\right]}{-\frac{v(t)}{d}}\right|$$

$$= \frac{\frac{\mu_o\varepsilon_o}{2}\left(\frac{l_{th}^2}{4}-x^2\right)\left|\frac{d^2v(t)}{dt^2}\right|}{|v(t)|}$$

[14]If a single voltage source were applied across the two plates at the center of the plates, then the field would be constrained to $v(t)/d$ at the center.

If $v(t) = A\cos(\omega t)$, then this ratio is

$$\left|\frac{E_{EQS} - E_{STATIC}}{E_{STATIC}}\right| = \frac{\frac{\mu_o\varepsilon_o}{2}\left(\frac{l_{th}^2}{4} - x^2\right)|-A\omega^2\cos(\omega t)|}{|A\cos(\omega t)|} = \frac{\mu_o\varepsilon_o}{2}\left(\frac{l_{th}^2}{4} - x^2\right)\omega^2 \tag{2.93}$$

The maximum value of this error is at $x = 0$:

$$\left.\left|\frac{E_{EQS} - E_{STATIC}}{E_{STATIC}}\right|\right|_{\max} = \frac{\mu_o\varepsilon_o l_{th}^2\omega^2}{8} = \frac{1}{8}\left(\frac{l_{th}\omega}{c}\right)^2 \quad \text{where } c = \frac{1}{\sqrt{\mu_o\varepsilon_o}} \approx 3\times10^8 \text{ m/s},$$
$$= \frac{\pi^2}{2}\left(\frac{l_{th}}{\lambda_o}\right)^2 \quad \text{where } \lambda_o = \frac{c}{f} = \frac{2\pi c}{\omega} \text{ m} \tag{2.94}$$

This reassuring result indicates that when the length of the parallel plates is electrically small, $l_{th}/\lambda_o \ll 1$, the electric field expression obtained by neglecting the effects of the magnetic field is a good approximation to the higher order EQS approximation. For higher frequencies where the structure is not electrically small, the two terms given in (2.92) may be an inadequate description of the electric field.

In the previous situation, the two plates were constructed of perfect conductors and the space between the conductors was free space, a perfect insulator. When perfect conductors and perfect insulators are involved, usually the system can be classified as EQS or MQS by allowing the frequency of the applied signal to approach zero or dc. If the currents disappear leaving only voltage differences (or stationary charges), the system is likely EQS. (If the voltages disappear leaving only currents, the system is likely MQS.) These statements are reasonable. Many capacitors, for example, represented as two good-conducting plates separated by a good insulator, are (generally) electrostatic or electric energy-storage devices. At dc, capacitors are open-circuit zero-current devices. Even in crosstalk analysis, when the voltage is "large" and current is "small," the coupling can sometimes be thought of as mainly electric or capacitive.

Of course, no material is a perfect insulator or perfect conductor. All materials have nonzero and finite conductivities. It is only natural for an inquiring mind to ask under what conditions can a material be considered like a perfect insulator or perfect conductor for EQS conditions. After some analysis and reflection, it becomes clear that for a material to be considered like a perfect conductor

$$\tau \gg \tau_e \quad \text{good conductor, EQS} \tag{2.95}$$

where τ is the time of interest and τ_e is the electric time constant for the material. In other words, if the charges in the material can respond very quickly compared to the time of interest, then the material can be considered like a perfect conductor. Whether a material is a good conductor is relative. The overall electric time constant for a system containing regions with multiple permittivities and conductivities (or nonuniform electrical properties) will be a function of these electrical parameters, as well as the dimensions and shapes of these regions. However, for a region with uniform properties, $\tau_e = \varepsilon/\sigma$. Furthermore, if the applied signal were sinusoidal with a frequency ω rad/s, then a reasonable time of interest, τ, would be the reciprocal of ω since it is proportional to the period of the sinusoidal, $1/\omega = T/(2\pi)$. Therefore, for a uniform material to be considered a perfect conductor when the applied voltage is sinusoidal,

$$\frac{1}{\omega} \gg \frac{\varepsilon}{\sigma} \Rightarrow \frac{\sigma}{\omega\varepsilon} \gg 1 \quad \text{good conductor, EQS} \tag{2.96}$$

Equation (2.96) is the good-conductor criteria used in many places in this book. It is used for EQS (but not MQS) and far-field, plane wave situations. Traditional conductors such as copper obviously satisfy this inequality for even very high frequencies. Many organs and tissues in the human body also satisfy this inequality for lower frequencies. Clearly, when the time of interest is much shorter than the electrical time constant, the material is considered a good insulator:

$$\tau \ll \tau_e \quad \text{good insulator, EQS} \tag{2.97}$$

When a material is a good insulator, the response of the charges in the material is very slow compared to the time of interest. For sinusoidal signals,

$$\frac{1}{\omega} \ll \frac{\varepsilon}{\sigma} \Rightarrow \frac{\sigma}{\omega\varepsilon} \ll 1 \quad \text{good insulator, EQS} \tag{2.98}$$

With real materials, the good-conductor and good-insulator criteria can be used to determine, in the limit as the frequency approaches zero, whether a system is EQS. For example, in Figure 2.37 imagine that the two plates are not perfect conductors but instead copper, a material with a finite conductivity. Furthermore, imagine that the region between the plates is not a perfect insulator but instead a ceramic, a material with a nonzero conductivity. When a voltage is applied across the copper plates, voltage drops across the length of the plates and nonzero current passes through the insulation. Should this situation be considered EQS? For a sinusoidal voltage, the previous criteria can be used for both the copper conductor and ceramic insulator. If (2.96) is satisfied for the copper and (2.98) for the ceramic, then the system is EQS. Of course, the dimensions should still be electrically small.

When analyzing a system, it is not necessary that every part of the system be EQS to utilize the simplifying characteristics associated with EQS conditions. For example, a resistor pad at one end of an electrically-long transmission line can be analyzed using the traditional concept of voltage (i.e., $\nabla \times \vec{E} = 0$) even though the current distribution in the wires in the transmission line are described using MQS laws. EQS laws might describe the electrons and holes in a semiconductor device even though the relay connected to the device might be described by MQS laws. A high-voltage electrostatic painting process is probably adequately described using EQS laws unless strong arcing occurs, in which case, MQS laws might better describe at least the region near the arc.

3

Magnetic Field Shielding

Magnetic field shielding refers to the shielding of near-field magnetic fields. Near-field magnetic fields are frequently described as having a low impedance and, therefore, do not easily reflect off low-impedance metals. It is important to realize, however, that near-field magnetic fields are not plane waves, and their field distribution is typically complex and source dependent. Shielding of magnetic fields is usually difficult and expensive.

3.1 The "Magic" of Magnetic Field Shielding Revealed

Qualitatively explain how a shield "blocks out" or "diverts" near-field magnetic fields. [Hoburg, '95; Haus; Rikitake]

When analyzing and explaining how an object shields near-field magnetic fields, there are two "camps." One "camp" applies magnetoquasistatic concepts, assuming the object is electrically small. The other "camp" adapts methods from traditional plane wave shielding.

Through the magnetoquasistatic "camp" approach, magnetic shielding is accomplished by flux "shunting" and Eddy currents. To understand qualitatively how a magnetic material (a material with a relative permeability greater than one) can shield magnetic fields through flux "shunting," a simplistic atomic-level description of magnetic materials is helpful. As every young school kid knows, electrons orbit the nucleus of atoms. These circulating electrons generate a magnetic field, since current is the flow of charge and current generates a magnetic field. It is thus reasonable to model these orbiting electrons around their nucleus as a current loop. For nonmagnetic materials (such as most plastics and copper), the net magnetic field from all of the current loops, from each of the atoms, tends to cancel. However, for magnetic materials, the average magnetic field over some region may not cancel. When magnetic material is exposed to an external magnetic field, these regions or magnetic domains can partially or completely align resulting in larger overall magnetic fields. The total field is the sum of the induced field from the material and externally applied field. When the external field is removed, these domains of current loops may revert to their initial positions (or nearly so) or they may retain some of their "magnetization." For some magnetic materials, such as permanent magnets, even without an external field present, the domains do not revert to their initial positions, and a net magnetic field is generated by the medium. When magnetic material is placed in an external applied magnetic field, it is reasonable to accept that the material generally distorts the applied field. The permeability, μ, of a material is an important indicator of the field distorting ability of a material.

Simply stated, a shield of high permeability can "shunt," "divert," "attract," "channel," or "guide" (like a duct) a magnetic field. A magnetic field follows the path of least *reluctance* similar to how current follows the path of least resistance. High-permeability magnetic materials have low reluctance. The degree of magnetic shielding is determined by various factors including the shield's thickness, shape, and position relative to the applied magnetic field. This method of shielding is effective for dc and ac magnetic fields.[1]

In the magnetoquasistatic model, Eddy currents can also assist in magnetic shielding. As is known, a time-varying field will induce current in a conductor. This induced current then generates an induced magnetic field. This induced magnetic field, referred to as a "bucking field" or counter field, opposes the applied or incident field. For example, if an object to be shielded is surrounded by a thin highly conducting band "properly" oriented relative to the applied magnetic field, an induced circulating current will be produced in the band. The current in the conducting band generates a magnetic field that opposes the applied magnetic field. The sum of the applied and induced fields within the region enclosed by the band is less than the applied field alone. The degree of magnetic shielding is a function of the magnitude of the induced currents (and the frequency of the magnetic field) and the shield's conductivity, thickness, shape, and position relative to the applied magnetic field. This method of shielding an object is only effective for time-varying magnetic fields.

Through the plane-wave shielding "camp" approach, magnetic shielding is accomplished by absorption and reflections. Obviously, near-field magnetic fields are not plane waves. However, the plane-wave shielding concept can be adapted for near-field magnetic fields by using a modified near-field wave impedance. With thick shields (compared to the skin depth) of high conductivity, significant absorption loss of a magnetic field can occur in the shield. The electrons in a conductor move in response to the applied time-varying magnetic field. These moving electrons constitute current in the shield. Since all real conductors have a finite conductivity, energy is lost in the conductor with this charge flow. This ohmic loss reduces the strength of the field. In addition to absorption, reflections at the interfaces can assist in the shielding. The reflections are due to differences between the wave impedance of the near-field magnetic field on both sides of the shield and the wave impedance of the shielding material. (The wave impedance of the shield is a function of the frequency of the magnetic field and the permeability and conductivity of the shielding material.)

3.2 Magnetic Field from Simple Current Distributions

Tabulate the expressions for the magnetic field from simple line, surface, and volume current distributions. [Haus; Zahn]

The susceptibility of electrical devices (and living organisms) to magnetic fields is important. Engineers, though, are generally more comfortable with the circuit concepts of current and voltage than the electromagnetic concepts of magnetic field and electric field. The expressions in Table 3.1 allow the magnetic fields to be easily determined for some simple but important current distributions.

The meaning of line, surface, and volume current (or current density) must be clearly understood before these tabled results are applied. One idealized current distribution is the line or filamentary current. Its magnitude is nearly always described using the variable I with units of amperes (A). A line current has one nonzero dimension—length. Although a line current has a length, it has zero cross-sectional area. The total current in a line current of I is just I. Unlike charge, current has both a magnitude and direction and is, thus, technically a vector (e.g., $\vec{I}$). Probably, the most common use of the line current is the uniform current along a wire. At low frequencies (where the skin depth is large compared to the cross-sectional dimensions of the wire), the current distribution is essentially uniform over the cross section of the wire. When the cross-sectional dimensions of a current distribution are small compared to the distance from the current to the point of interest, the current is frequently modeled as a line current. In this case, the expression for the magnetic field from a straight, infinitely long line current is commonly used:

$$\vec{H} = \frac{I}{2\pi\rho}\hat{a}_\phi$$

[1]Sometimes the "shunting" due to high-permeability materials is undesirable. For example, with compasses in submarines and on the bridge of ships, magnetic materials can strongly reduce and disturb the reading.

TABLE 3.1 Current Distributions and Their Magnetic Fields

Current Distribution	Magnetic Field ($\vec{H}$)
Short-line current [Skilling; Demarest] 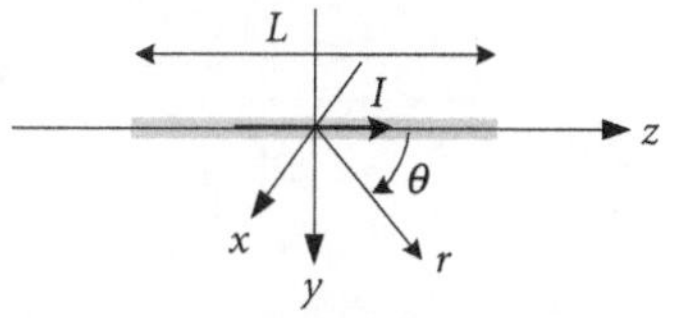	$\frac{IL\sin\theta}{4\pi r^2}\hat{a}_\phi$ if $r \gg L$ Magnetic field equivalent of the electric point charge, the total length of the line current is L, the short line current is along the z axis and centered about the xy plane
Infinite-line current [Demarest] 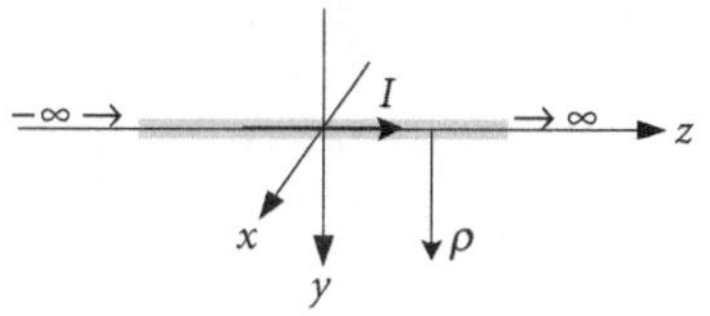	$\frac{I}{2\pi\rho}\hat{a}_\phi$ The field circulates around the current, direction is determined by the right-hand rule where the thumb is in the direction of the current, the infinite-length current is along the z axis
Parallel, oppositely directed infinite-line currents [Jefimenko] 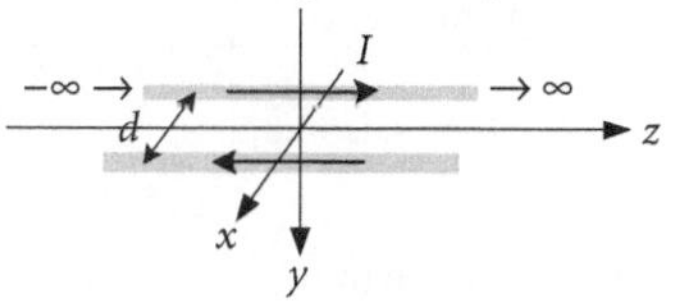	$-\frac{Iy}{2\pi}\left[\frac{1}{\left(x+\frac{d}{2}\right)^2+y^2}-\frac{1}{\left(x-\frac{d}{2}\right)^2+y^2}\right]\hat{a}_x$ $+\frac{I}{2\pi}\left[\frac{x+\frac{d}{2}}{\left(x+\frac{d}{2}\right)^2+y^2}-\frac{x-\frac{d}{2}}{\left(x-\frac{d}{2}\right)^2+y^2}\right]\hat{a}_y$ $=-\frac{I}{2\pi}\left[\frac{d}{x^2-\left(\frac{d}{2}\right)^2}\right]\hat{a}_y$ if $\lvert x\rvert \neq \frac{d}{2}$ and $y=0$ $=\frac{I}{2\pi}\left[\frac{d}{\left(\frac{d}{2}\right)^2+y^2}\right]\hat{a}_y$ if $x=0$ The two line currents in the $\pm z$ direction are in the xz plane, the distance between the currents is d and they are centered about the z axis
Parallel, similarly directed infinite-line currents [Jefimenko] 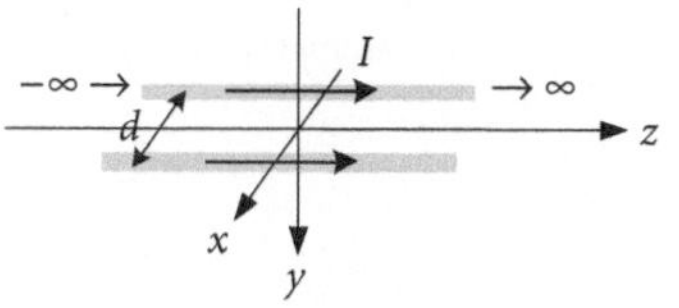	$-\frac{Iy}{2\pi}\left[\frac{1}{\left(x+\frac{d}{2}\right)^2+y^2}+\frac{1}{\left(x-\frac{d}{2}\right)^2+y^2}\right]\hat{a}_x$ $+\frac{I}{2\pi}\left[\frac{x+\frac{d}{2}}{\left(x+\frac{d}{2}\right)^2+y^2}+\frac{x-\frac{d}{2}}{\left(x-\frac{d}{2}\right)^2+y^2}\right]\hat{a}_y$ $=-\frac{Iy}{\pi}\left[\frac{1}{\left(\frac{d}{2}\right)^2+y^2}\right]\hat{a}_x$ if $x=0$ The two line currents in the z direction are in the xz plane, the distance between the currents is d and they are centered about the z axis

(Continued)

TABLE 3.1 Current Distributions and Their Magnetic Fields (Continued)

Current Distribution	Magnetic Field ($\vec{H}$)
Finite-length line current [Zahn; Inan] 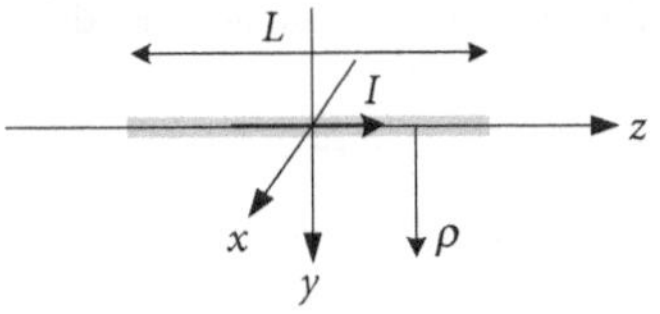	$\frac{I}{4\pi\rho}\left[\frac{\frac{L}{2}-z}{\sqrt{\rho^2+\left(z-\frac{L}{2}\right)^2}}+\frac{\frac{L}{2}+z}{\sqrt{\rho^2+\left(z+\frac{L}{2}\right)^2}}\right]\hat{a}_\phi$ $=0 \quad \text{if } \rho=0$ $=\frac{IL}{4\pi\rho\sqrt{\rho^2+\left(\frac{L}{2}\right)^2}}\hat{a}_\phi \quad \text{if } z=0$ The total length of the line current (or current "stick") is L, the current is along the z axis and centered about the xy plane
Magnetic dipole/small-loop line current [Demarest; Zahn] 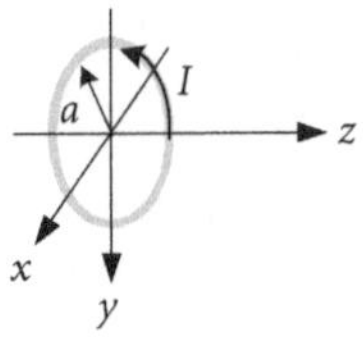	$\frac{Ia^2\cos\theta}{2r^3}\hat{a}_r+\frac{Ia^2\sin\theta}{4r^3}\hat{a}_\theta \quad \text{if } r\gg a$ $=\frac{Ia^2}{2r^3}\hat{a}_r \quad \text{if } r\gg a, \theta=0^\circ$ (along the $+z$ axis) $=\frac{Ia^2}{4r^3}\hat{a}_\theta \quad \text{if } r\gg a, \theta=90^\circ$ (along the $z=0$ plane) The loop of radius a is centered about the z axis in the xy plane, the field decreases rapidly with r, this field is identical in form to an electric dipole, for $r \gg a$ the shape of the loop is not important: let $S=\pi a^2$ where S is the area of any shape single-turn loop
Circular-loop line current [Demarest; Zahn] 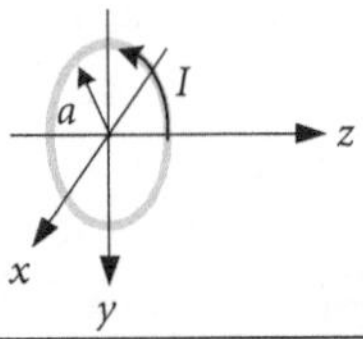	$\frac{Ia^2}{2(a^2+z^2)^{\frac{3}{2}}}\hat{a}_z$ along the z axis $(0,0,z)$ $=\frac{I}{2a}\hat{a}_z$ at the origin $(0,0,0)$ The loop of radius a is centered on the z axis in the xy plane
Square-loop line current [Marshall, '96; Watson; Lerner] 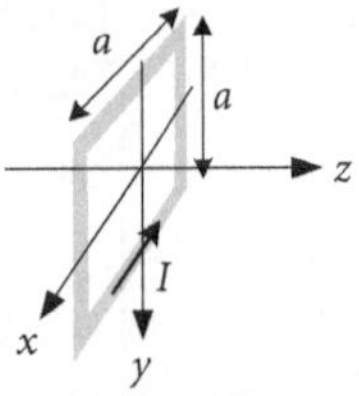	$\frac{4I}{\pi a\left(1+\frac{4z^2}{a^2}\right)\sqrt{2+\frac{4z^2}{a^2}}}\hat{a}_z$ along the z axis $(0,0,z)$ $=\frac{I2\sqrt{2}}{\pi a}\hat{a}_z$ at the origin $(0,0,0)$ The square loop has sides equal to a, the loop is located in the $z=0$ plane and centered on the z axis
Rectangular-loop line current [Zahn] 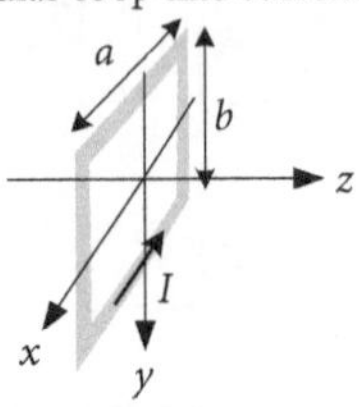	$\frac{2Iab(a^2+b^2+8z^2)}{\pi(a^2+4z^2)(b^2+4z^2)\sqrt{a^2+b^2+4z^2}}\hat{a}_z$ along the z axis $(0,0,z)$ $=\frac{2I\sqrt{a^2+b^2}}{\pi ab}\hat{a}_z$ at the origin $(0,0,0)$ $=\frac{4I}{\pi a\left(1+\frac{4z^2}{a^2}\right)\sqrt{2+\frac{4z^2}{a^2}}}\hat{a}_z$ if $a=b$ along the z axis $(0,0,z)$ $=\frac{I2\sqrt{2}}{\pi a}\hat{a}_z$ if $a=b$ at the origin $(0,0,0)$ The rectangular loop has sides equal to a and b, the loop is located in the $z=0$ plane and centered on the z axis

TABLE 3.1 Current Distributions and Their Magnetic Fields (Continued)

Current Distribution	Magnetic Field ($\vec{H}$)
N-sided polygon-loop line current [Zahn; Inan] 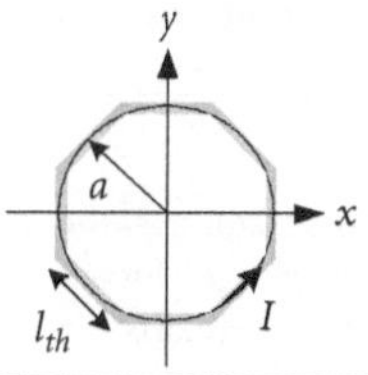	$\dfrac{NI\tan\left(\dfrac{\pi}{N}\right)}{2\pi a}\hat{a}_z$ at the origin (0,0,0) The polygon is located in the $z = 0$ plane and centered on the z axis, a is the radius of a circle passing through the corners of the polygon, the length of each of the N sides is $l_{th} = 2a\sin(\pi/N)$
Semicircular-loop line current [Lerner; Zahn] 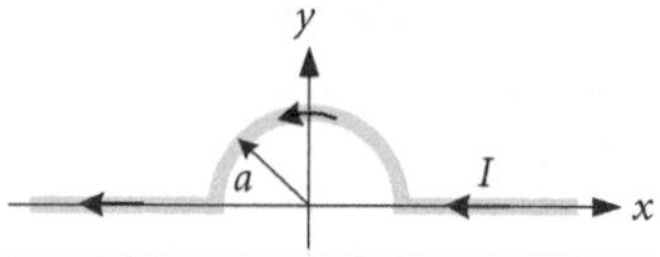	$\dfrac{I}{4a}\hat{a}_z$ at the origin (0,0,0) The straight line segments leading to the semicircle are along the x axis, the center of the semicircle of radius a is at the origin
Half-rectangular loop line current [Zahn] 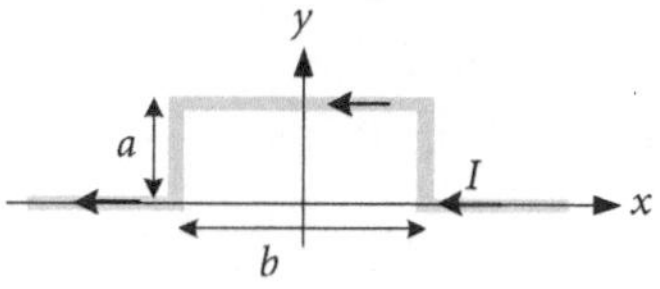	$\dfrac{I\sqrt{a^2+\left(\dfrac{b}{2}\right)^2}}{\pi ab}\hat{a}_z$ at the origin (0,0,0) The straight line segments leading to the rectangle are along the x axis, the rectangular loop has sides equal to a and b
Pie-shaped loop line current [Ulaby; Jefimenko] 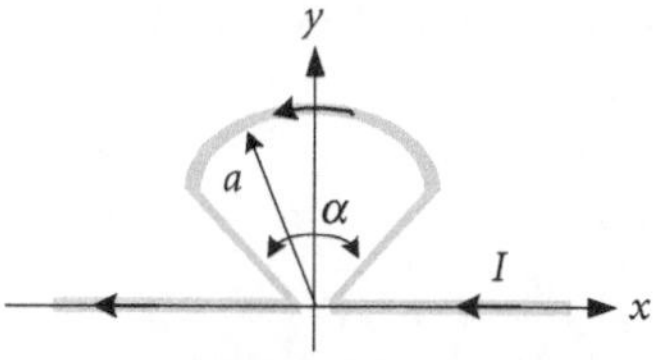	$\dfrac{\alpha I}{4\pi a}\hat{a}_z$ at the origin (0,0,0) a is the radius of the arc portion of the loop, α is the angle subtended by the arc in radians
Wedge-shaped loop line current 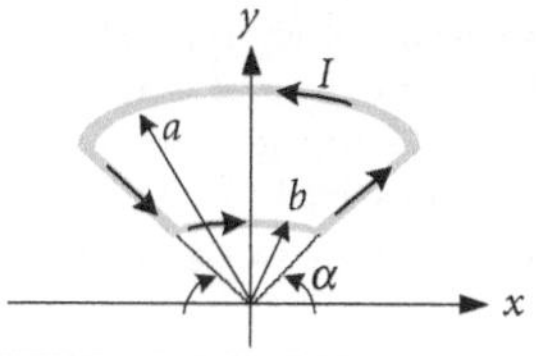	$\dfrac{\alpha I}{4\pi}\left(\dfrac{1}{a}-\dfrac{1}{b}\right)\hat{a}_z$ at the origin (0,0,0) a is the outer radius and b is the inner radius of the arc portions of the loop, α is the angle subtended by the arcs in radians
Hairpin line current [Lerner; Zahn] 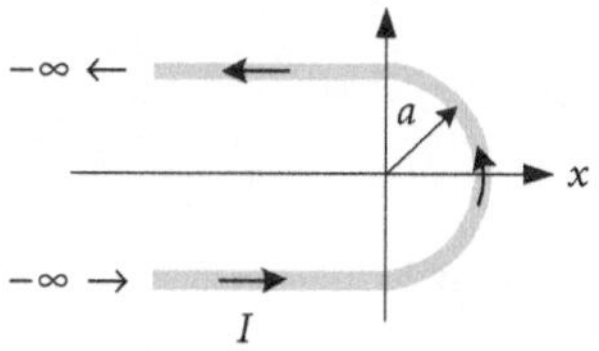	$\dfrac{I}{2a}\left(\dfrac{1}{2}+\dfrac{1}{\pi}\right)\hat{a}_z$ at the origin (0,0,0) The semi-infinite straight line segments leading to the semicircle are parallel to the x axis, the center of the semicircle of radius a is at the origin, the same field is obtained at the origin if the entire line current is reflected about the y axis

(Continued)

TABLE 3.1 Current Distributions and Their Magnetic Fields (Continued)

Current Distribution	Magnetic Field ($\vec{H}$)
Incomplete circular-loop line current [Nussbaum] 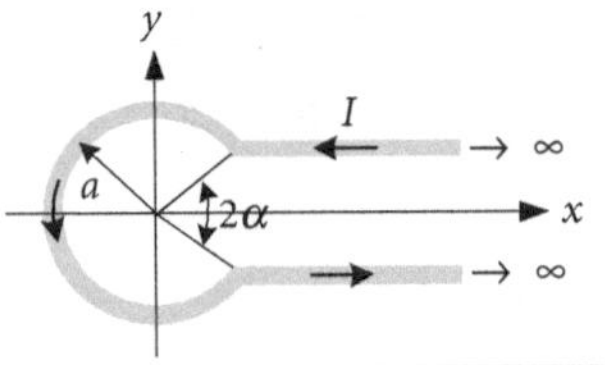 	$\frac{I}{2\pi a}\left[\pi-\alpha+\frac{(1-\cos\alpha)}{\sin\alpha}\right]\hat{a}_z$ at the origin (0,0,0) The semi-infinite straight line segments leading to the incomplete circle are parallel to the x axis, the center of the incomplete circle of radius a is at the origin, the angle α is in radians
Logarithmic-spiral line current [Nussbaum] 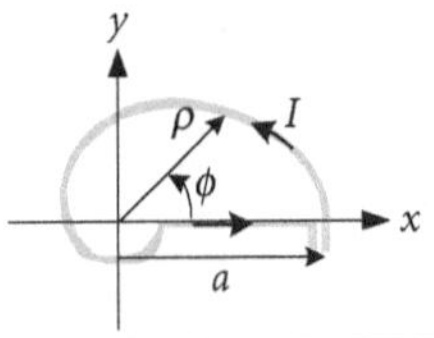 	$\frac{I}{4a}(e^2-1)\hat{a}_z$ at the origin (0,0,0) The line current follows the logarithmic spiral expression $\rho = ae^{-\frac{\phi}{\pi}}$
Infinite sheet of uniform surface current [Demarest; Zahn] 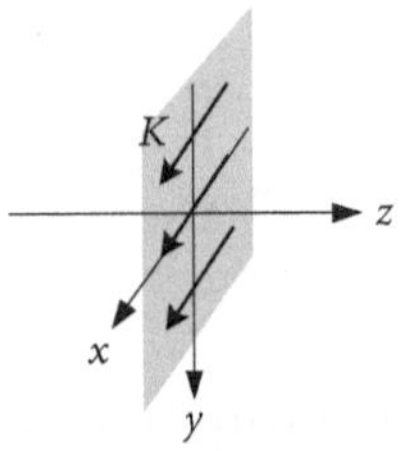 	$-\frac{K}{2}\hat{a}_y$ if $z>0$, $\frac{K}{2}\hat{a}_y$ if $z<0$ The uniform field is parallel to the sheet of current that is directed entirely in the x direction, it is independent of the distance from the sheet, the sheet extends to infinity in both the x and y directions and is in the $z = 0$ plane
Parallel infinite sheets of oppositely directed uniform surface currents [Zahn] 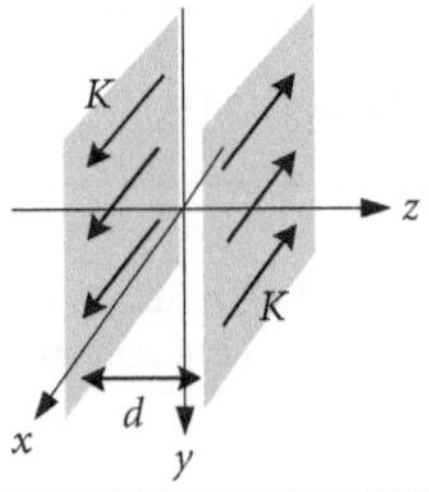 	$-K\hat{a}_y$ if $\lvert z\rvert<\frac{d}{2}$, 0 if $\lvert z\rvert>\frac{d}{2}$ The uniform field between the sheets of current is parallel to the sheets, the sheets are separated by a distance of d and are centered about the $z = 0$ plane, the sheets are located at $z = \pm d/2$, the sheet extends to infinity in both the x and y directions, the x directed surface currents in each sheet are of the same magnitude but opposite direction
Infinite-length strip, longitudinal, uniform surface current [Inan; Plonsey; Demarest] 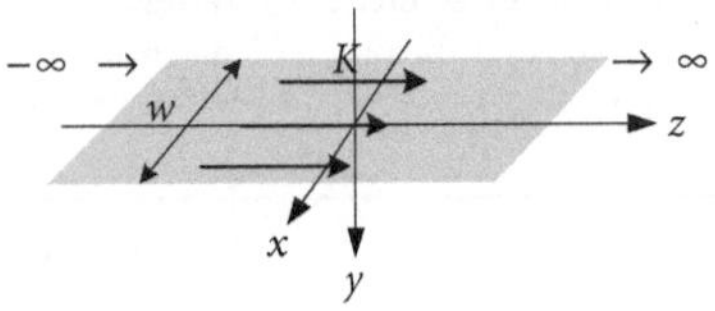 	$-\frac{K}{2\pi}\left[\tan^{-1}\left(\frac{x+\frac{w}{2}}{y}\right)-\tan^{-1}\left(\frac{x-\frac{w}{2}}{y}\right)\right]\hat{a}_x+\frac{K}{4\pi}\ln\left[\frac{\left(x+\frac{w}{2}\right)^2+y^2}{\left(x-\frac{w}{2}\right)^2+y^2}\right]\hat{a}_y$ $=-\frac{K}{\pi}\tan^{-1}\left(\frac{w}{2y}\right)\hat{a}_x$ if $y>0$ and in the yz plane $(0,y,z)$ The field is parallel to the surface at $x = 0$, the strip of width w is centered about $x = 0$, it is infinite in the z direction and located in the xz plane, the current on the strip is $I = Kw$ and is entirely in the z direction

TABLE 3.1 Current Distributions and Their Magnetic Fields (Continued)

Current Distribution	Magnetic Field $(\vec{H})$
Infinite-length strip, shunt, uniform surface current [Zahn] 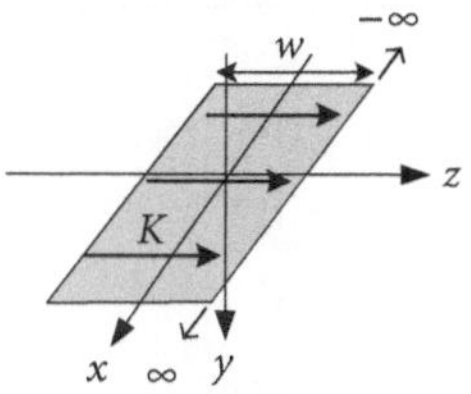	$-\dfrac{K}{2\pi}\left\{\dfrac{\pi}{2}+\sin^{-1}\left[\dfrac{\left(\frac{w}{2}\right)^2-y^2}{\left(\frac{w}{2}\right)^2+y^2}\right]\right\}\hat{a}_x$ if $y>0$ and in the xy plane $(x,y,0)$ The field is parallel to the surface at $z = 0$, the strip of width w is centered about $z = 0$, it is infinite in the x direction and located in the xz plane, the current on the strip is entirely in the z direction
Infinite-length cylindrical, uniform, circulating surface current [Inan; Jefimenko] 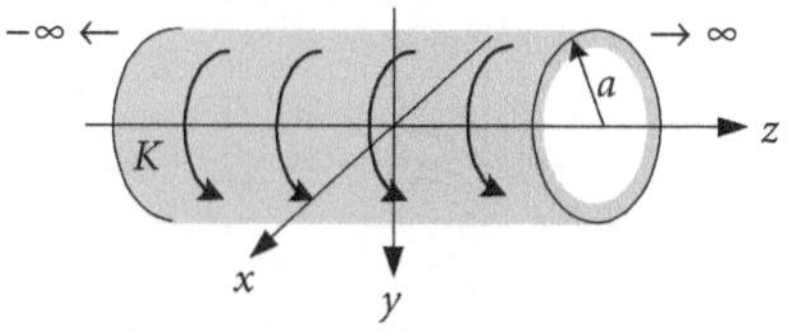	$K\hat{a}_z$ if $\rho<a$, 0 if $\rho>a$ The surface current is $\vec{K}=K\hat{a}_\phi$, the radius of the cylinder is a
Cylindrical, uniform, circulating surface current [Demarest; Zahn; Jefimenko] 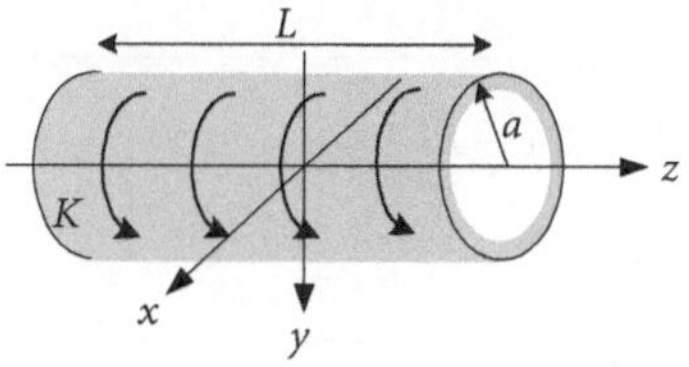	$\dfrac{K}{2}\left[\dfrac{\frac{L}{2}-z}{\sqrt{a^2+\left(z-\frac{L}{2}\right)^2}}+\dfrac{\frac{L}{2}+z}{\sqrt{a^2+\left(z+\frac{L}{2}\right)^2}}\right]\hat{a}_z$ along the z axis $(0,0,z)$ $=\dfrac{KL}{\sqrt{4a^2+L^2}}\hat{a}_z$ at the center $(0,0,0)$ $=\dfrac{KL}{2\sqrt{a^2+L^2}}\hat{a}_z$ at either end $\left(0,0,\pm\dfrac{L}{2}\right)$ $\dfrac{Ka^2L\cos\theta}{2r^3}\hat{a}_r+\dfrac{Ka^2L\sin\theta}{4r^3}\hat{a}_\theta$ if $r\gg L, r\gg a$ The axis of the cylinder of length L is along the z axis and centered about the xy plane, the surface current is $\vec{K}=K\hat{a}_\phi$, r and θ are spherical coordinate variables, the current on the cylinder of radius a is $I=KL$
Infinite-length cylindrical, uniform, longitudinal surface current [Zahn] 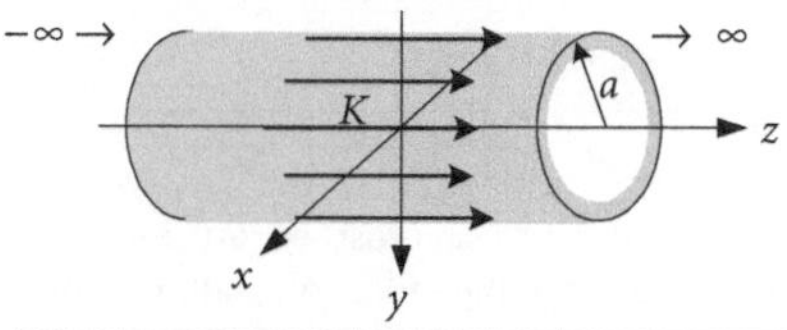	0 if $\rho<a$, $\dfrac{Ka}{\rho}\hat{a}_\phi$ if $\rho>a$ The surface current is $\vec{K}=K\hat{a}_z$, the current on the cylinder of radius a is $I=K2\pi a$

(Continued)

TABLE 3.1 Current Distributions and Their Magnetic Fields (Continued)

Current Distribution	Magnetic Field $(\vec{H})$
Infinite-length cylindrical, pitched, circulating surface current [Zahn] 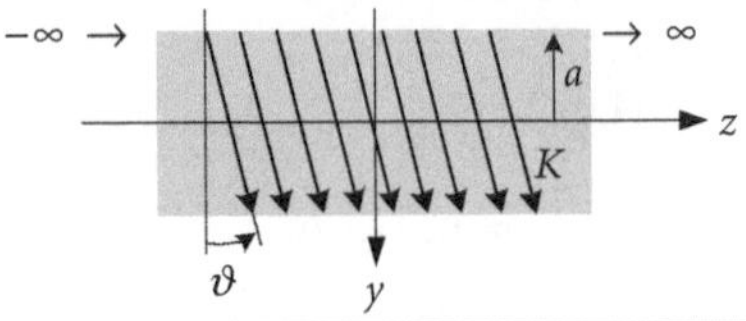	$K\cos\vartheta\,\hat{a}_z$ if $\rho<a$, $\dfrac{Ka\sin\vartheta}{\rho}\hat{a}_\phi$ if $\rho>a$ The surface current is $\vec{K}=K\cos\vartheta\,\hat{a}_\phi+K\sin\vartheta\,\hat{a}_z$, the radius of the cylinder is a
Spherical, uniform, circulating surface current [Zahn; Haus] 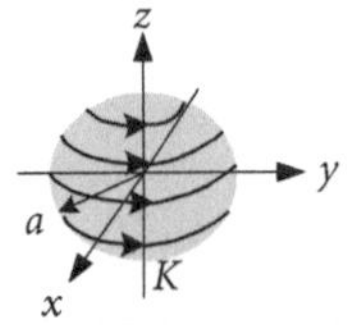	$\dfrac{K\pi}{4}\hat{a}_z$ at the origin (0,0,0) The surface current is $\vec{K}=K\hat{a}_\phi$, the current on the sphere of radius a is $I=K\pi a$
Spherical, nonuniform, circulating surface current [Haus] 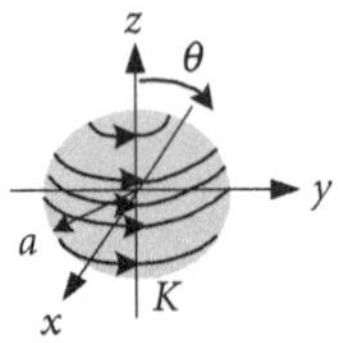	$\dfrac{NI\cos\theta}{3a}\hat{a}_r-\dfrac{NI\sin\theta}{3a}\hat{a}_\theta=\dfrac{NI}{3a}\hat{a}_z$ if $r<a$ $\dfrac{NIa^2\cos\theta}{3r^3}\hat{a}_r+\dfrac{NIa^2\sin\theta}{6r^3}\hat{a}_\theta$ if $r>a$ For this "flux" ball centered about the origin, the surface current is $\vec{K}=K\sin\theta\,\hat{a}_\phi=\frac{NI}{2a}\sin\theta\,\hat{a}_\phi$, where N is the total number of turns of thin wire tightly wound on the spherical form, I is the current in the wires, a is the radius of the sphere, and θ is the angle from the z axis, the turns density is sinusoidally distributed between the "north" and "south" poles of the sphere, notice that the field is uniform inside the sphere
Infinite-length cylindrical, nonuniform, longitudinal surface current [Haus] 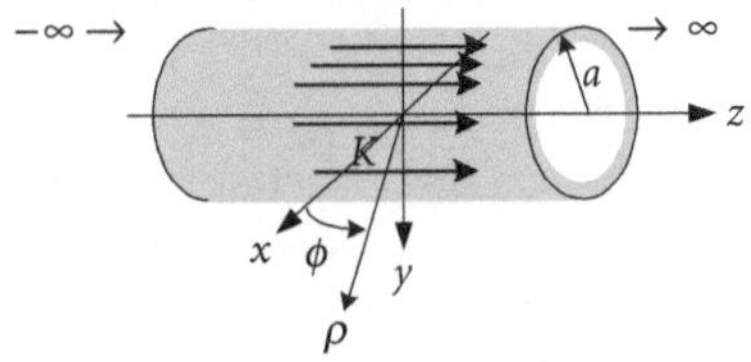	$\dfrac{NI\cos\phi}{4a}\hat{a}_\rho-\dfrac{NI\sin\phi}{4a}\hat{a}_\phi=\dfrac{NI}{4a}\hat{a}_x$ if $\rho<a$ $\dfrac{NIa\cos\phi}{4\rho^2}\hat{a}_\rho+\dfrac{NIa\sin\phi}{4\rho^2}\hat{a}_\phi$ if $\rho>a$ For this "flux" cylinder, the surface current is $\vec{K}=K\sin\phi\,\hat{a}_z=\frac{NI}{2a}\sin\phi\,\hat{a}_z$, where N is the total number of turns of thin wire tightly wound on the cylindrical form, I is the current in the wires, and a is the radius of the cylinder, the turns density is constant and equal to $N/(2a)$ when projected onto the $y=0$ plane, notice that the field is uniform inside the cylinder
Toroid-shaped uniform, circulating surface current [Demarest] 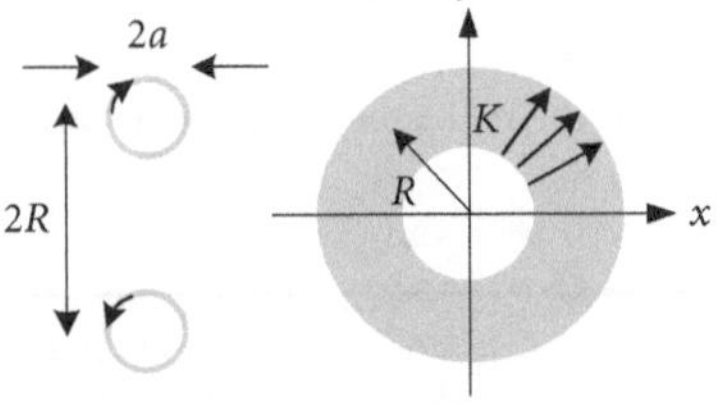	$\dfrac{K(R-a)}{\rho}\hat{a}_\phi$ $R-a<\rho<R+a$, 0 elsewhere $\approx K\hat{a}_\phi$ $R\gg a, R-a<\rho<R+a$ The toroid is doughnut shaped with a circular cross section, the cross section of the toroid has a radius of a, the radius of revolution of the toroid is R, the current on the toroid is $I=K2\pi(R-a)$

TABLE 3.1 Current Distributions and Their Magnetic Fields (Continued)

Current Distribution	Magnetic Field ($\vec{H}$)
Thin annular disk, uniform volume current [Haus] 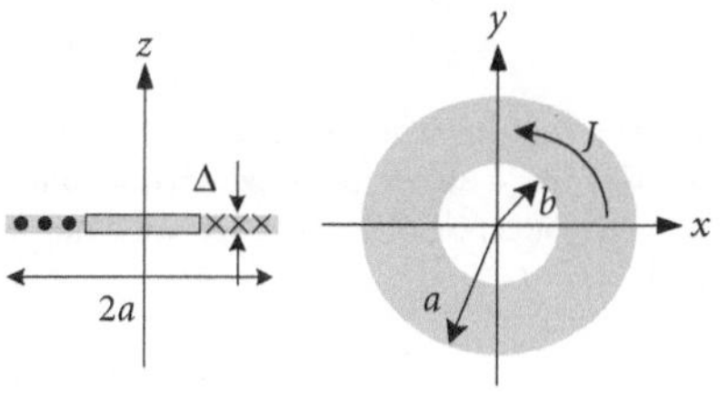 	$\frac{J\Delta}{2}\left[\frac{b}{\sqrt{b^2+z^2}}-\frac{a}{\sqrt{a^2+z^2}}+\ln\left(\frac{a+\sqrt{a^2+z^2}}{b+\sqrt{b^2+z^2}}\right)\right]\hat{a}_z$ along the z axis $(0,0,z)$ if $b \gg \Delta$ $=\frac{J\Delta\ln\left(\frac{a}{b}\right)}{2}\hat{a}_z$ at the origin $(0,0,0)$ if $b \gg \Delta$ The volume current is $\vec{J}=J\hat{a}_\phi$, the thin disk is centered about the z axis in the xy plane, the disk's outer radius is a and inner radius is b, its thickness is Δ, the current in the disk is $I=J\Delta(a-b)$
Finite-length cylindrical, uniform, circulating, volume current thin shell [Jefimenko; Jackson] 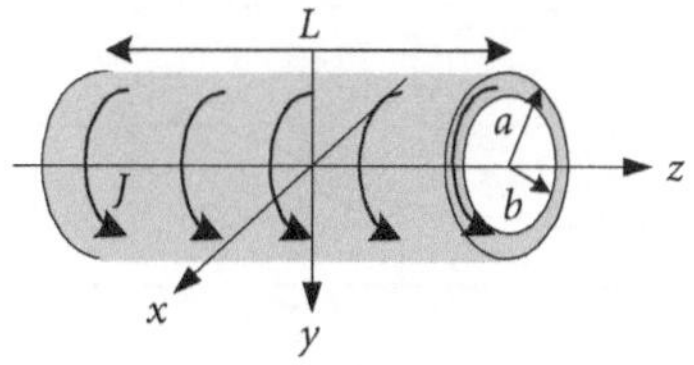 	the radial component is $\approx \pm\frac{J(a-b)\rho}{4a}\hat{a}_\rho$ if $a \gg a-b, L \gg a, a \gg \rho$, along the z axis near either face $\left(\rho,\phi,\pm\frac{L}{2}\right)$ the radial component is $\approx \frac{24J(a-b)a^2\rho z}{L^4}\hat{a}_\rho$ if $a \gg a-b, L \gg a, a \gg \rho, L \gg \lvert z\rvert$ along the z axis near the center (ρ,ϕ,z) The axis of the cylinder of length L is along the z axis and evenly spaced about the xy plane, the shell's outer radius is a and inner radius is b, the current in the cylinder is $I=JL(a-b)$
Infinite-length cylindrical, uniform, longitudinal volume current [Zahn; Demarest] 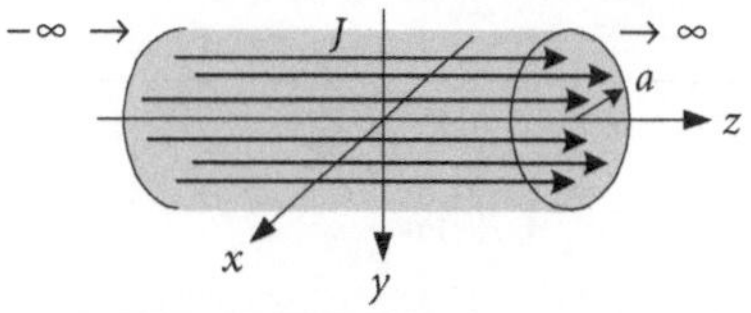 	$\frac{Ja^2}{2\rho}\hat{a}_\phi$ if $\rho \geq a$ The field circulates around the current, direction is determined by the right-hand rule where the thumb is in the direction of the current, the cylinder's axis is along the z axis, the cylinder's radius is a, the field is continuous at $\rho=a$ since no surface current is present, the current in the cylinder is $I=J\pi a^2$
Infinite-length cylindrical, uniform, longitudinal volume current shell [Demarest] 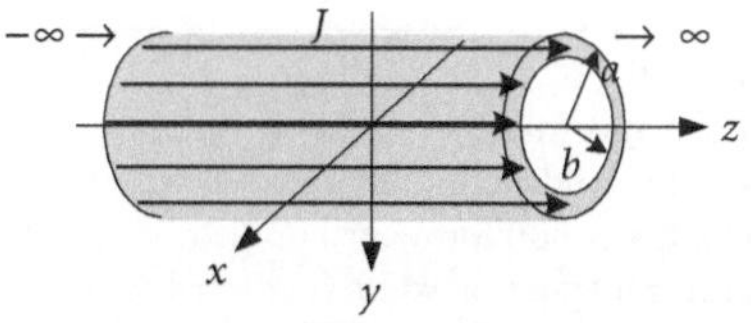 	$\frac{J(a^2-b^2)}{2\rho}\hat{a}_\phi$ if $\rho \geq a$ The cylinder's axis is along the z axis, the shell's outer radius is a and inner radius is b, the field is continuous at $\rho=a$ since no surface current is present, the current in the cylinder is $I=J\pi(a^2-b^2)$
Infinite-length, cylindrical, nonuniform, longitudinal volume current [Demarest] 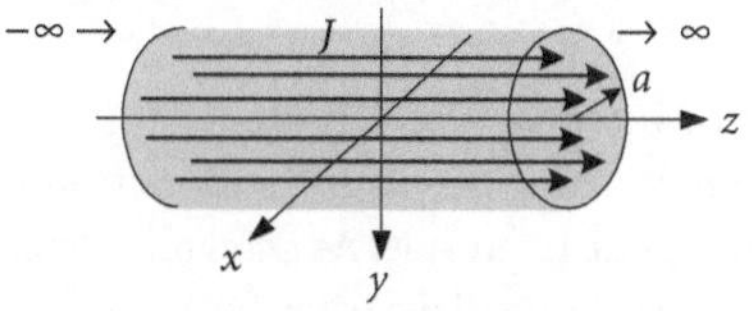 	$\frac{\int_0^a J(\rho)\rho d\rho}{\rho}\hat{a}_\phi$ if $\rho \geq a$ where $J(\rho)=0$ for $\rho > a$ The field outside this symmetrical cylindrical current distribution is only a function of the total current within the cylinder, the volume current is only a function of the radial (ρ) distance, the field is continuous at $\rho=a$ since no surface current is present

(Continued)

TABLE 3.1 Current Distributions and Their Magnetic Fields (Continued)

Current Distribution	Magnetic Field ($\vec{H}$)
Infinite-length longitudinal volume current with a parallel infinite-length cylindrical hole [Lerner; Inan] 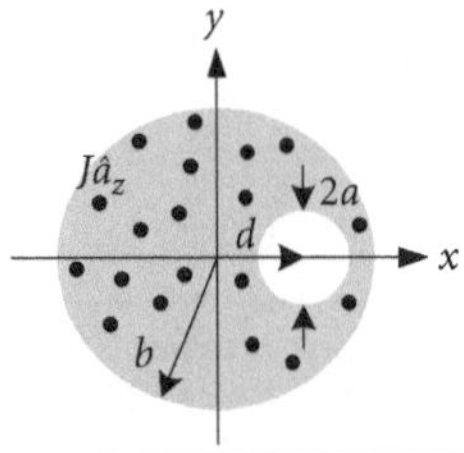	$\frac{Jd}{2}\hat{a}_y$ at the center of the cylindrical hole $(d,0,z)$ The cylinder's axis is along the z axis, the volume current is $\vec{J} = J\hat{a}_z$ everywhere in the cylinder of radius b except in the cylindrical hole of radius a, the distance between the center of the cylinder and center of the hole is d, the current is $I = J\pi(b^2 - a^2)$
Spherical, uniform, circulating, volume current shell [Zahn] 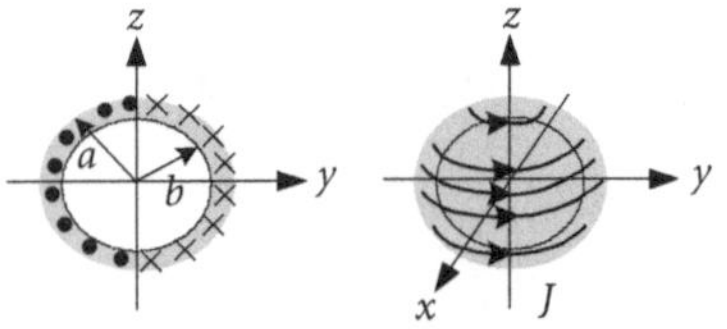	$\frac{J\pi(a-b)}{4}\hat{a}_z$ at the origin $(0,0,0)$ For this spherical shell centered about the origin the outer radius is a and inner radius is b, the volume current in the shell is $\vec{J} = J\hat{a}_\phi$, the current in the sphere is $I = (J\pi/2)(a^2 - b^2)$
Infinite-slab uniform volume current [Zahn; Jefimenko] 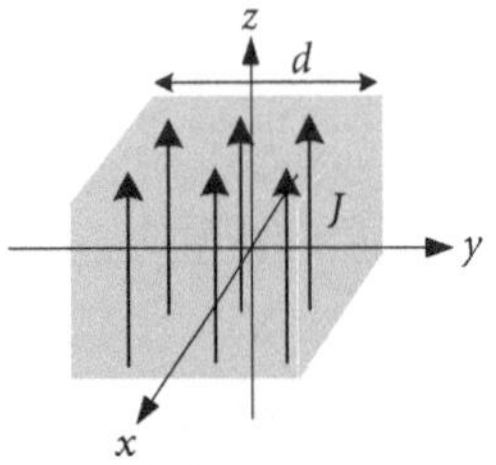	$-\frac{Jd}{2}\hat{a}_x \quad y \ge \frac{d}{2}, \quad \frac{Jd}{2}\hat{a}_x \quad y \le -\frac{d}{2}$ The uniform field is parallel to the two sides of the slab, the z-directed uniform volume current extends from $y = -d/2$ to $y = d/2$, the slab is centered about the z axis, it extends to infinity in both the x and z directions, the field is continuous at $y = \pm d/2$ since no surface current is present
Infinite-slab nonuniform volume current 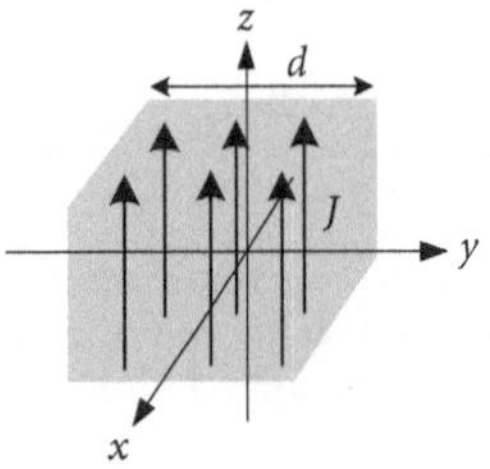	$-\frac{\int_{-\frac{d}{2}}^{\frac{d}{2}} J(y)dy}{2}\hat{a}_x \quad \text{if } y \ge \frac{d}{2}, \quad \frac{\int_{-\frac{d}{2}}^{\frac{d}{2}} J(y)dy}{2}\hat{a}_x \quad \text{if } y \le -\frac{d}{2}$ where $J(y) = 0 \quad \text{for } \lvert y \rvert > \frac{d}{2}$ The uniform field is parallel to the two sides of the slab, the z-directed volume current extends from $y = -d/2$ to $y = d/2$, the slab is centered about the z axis, it extends to infinity in both the x and z directions, the field outside this nonuniform current distribution is only a function of the total current per unit width (in the x direction), the volume current is only a function of y, the field is continuous at $y = \pm d/2$ since no surface current is present

where ρ is the smallest distance from the line current to the point of interest. The infinite line current is along the z axis in this expression. The magnetic field from a line current is *not* in the direction of the current. The direction of the magnetic field is determined using the right-hand rule. As shown in Figure 3.1, the right-hand thumb is pointed in the direction of the current. The remaining digits when curled around the line current indicate the direction of the magnetic field. The right-hand rule is invaluable for checking the approximate direction of the field from many line current distributions, including the field along the axis of circular current loop. The expression for the magnetic field from an infinite-length line current is frequently used for a *finite*-length line current when the smallest distance between the line current and point of interest is small compared to the actual length of the finite-length current: when the point of interest is close to the line current, the line current appears very long. The point of interest should also be far from either end of the finite-length line current to avoid the fringing fields. Straight finite-length line currents, appropriately summed, can be used to determine the magnetic field from multiple linear segments of current distributions, such as from a current-carrying square loop. They are also used for curved line currents by approximating the curve using a series of small straight segments and then summing the magnetic fields from each of these small straight segments.

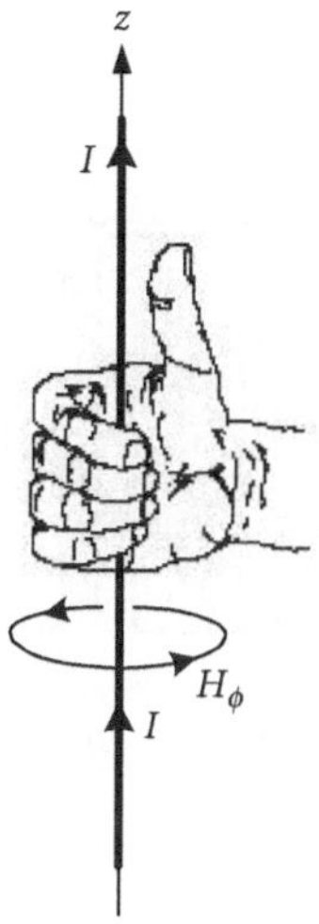

FIGURE 3.1 Right-hand rule to determine the direction of the magnetic field from a current-carrying wire.

A surface current has two nonzero dimensions — width and length. Its magnitude is described in Table 3.1 using the variable K and is measured in A/m. Surface current has zero depth and is the current per unit width. Surface current also has both a magnitude and direction and is, thus, a vector (e.g., $\vec{K}$). Surface current is initially the most "strange" to newcomers. If the surface current is not varying with position (i.e., the surface current is uniform), the total current over the surface can be determined by multiplying K by the width of the conductor that the current is carried. If it is varying along its width, the total current can be integrated over its width. For example, over the width from position a to b, the total current is

$$I = \int_a^b K dw \tag{3.1}$$

where K is the current normal to the width of interest. Surface current does not have to lie in a flat plane.[2] Surface current is frequently used to represent the current distribution along conductors at higher frequencies, where most of the current is contained within about one skin depth of the conductor's surface. For example, for a range of frequencies of interest, the current along a silver-coated waveguide is essentially along its surface. Representing a real current, which has nonzero depth, as a surface current with zero depth has certain mathematical advantages. When the depth of the current is small compared to the distance to the point of interest, the current is frequently modeled as a surface current. It is also used when the depth of the current is small compared to the other dimensions of the current. Often, the expression for the magnetic field from a flat, uniform surface current of infinite width and length is used. Assuming that the surface current is passing in the $+z$ direction and the surface current sheet is located in the yz plane,

$$\vec{H} = \frac{K}{2}\hat{a}_y \quad \text{if } x > 0, \quad \vec{H} = -\frac{K}{2}\hat{a}_y \quad \text{if } x < 0$$

The magnetic field is *not* in the direction of the current, but it is still parallel to the current sheet. The right-hand rule can again be used to determine this direction by pointing the thumb of the right hand in the direction of the surface current. The remaining digits of the hand when curled around the surface

---

[2]During a heavy rain, the water that passes down a curved children's slide is an approximation of a curved-surface water sheet.

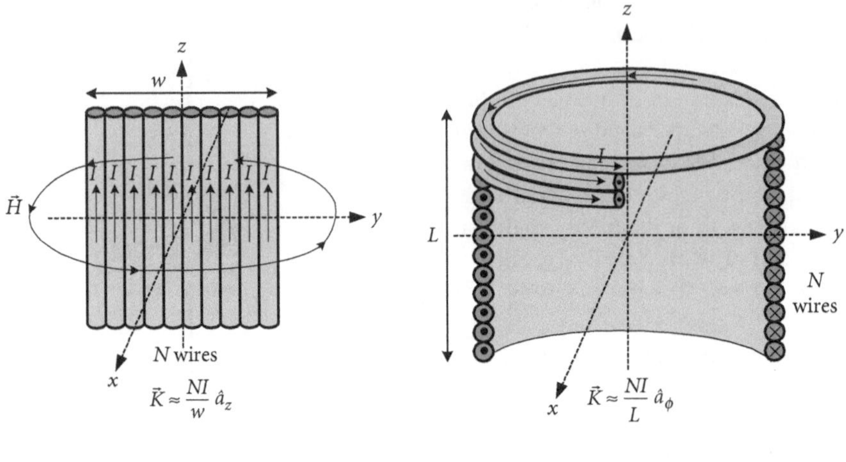

FIGURE 3.2 Surface currents approximated using closely spaced current-carrying wires.

current indicate the direction of the field (imagine either a very large hand or finite-width surface current). This expression for the field from an infinite[3] surface current is used for finite, uniform, surface current distributions when the smallest distance between the point of interest and surface current plane is small compared to the plane's nonzero dimensions. To avoid the fringing fields at the ends of the surface, the point of interest should be far from the edges. Notice that the magnetic field from this infinite current sheet is not a function of the distance from this surface current! Surface current is often used to represent the current in closely spaced insulated wires such as in a tightly wound solenoid or toroid. To generate a surface current, current-carrying wires are sometimes used. It is therefore important to be able to relate surface current (and sometimes volume current) to an equivalent N number of individually insulated wires in contact and in parallel. The relationships between NI and K for two important configurations are shown in Figure 3.2. These relationships for the surface currents are only approximations since real wires have a nonzero depth or diameter. The collection of parallel straight wires is approximating a flat, uniform surface current. Although w is not infinite, along the x axis and near the wires, the magnetic field is approximately the same as for an infinite current sheet. The solenoid or collection of N wire loops is approximating a circulating surface current. If the solenoid is constructed of a single piece of wire wound around an insulating cylindrical core, the wires will have a nonzero pitch to them. This will result in a slight pitch (from the horizontal or xy plane) in the surface current, which can be represented as

$$\vec{K} \approx \frac{NI}{L}\cos\vartheta\,\hat{a}_\phi + \frac{NI}{L}\sin\vartheta\,\hat{a}_z$$

where ϑ is the angle from no pitch. A nonzero value for ϑ will result in additional components in the field both inside and outside the coil. For a tightly wrapped group of small diameter wires, ϑ should be small. If a nonuniform surface current is desired, the spacing between the wires can be adjusted along the width of the surface.

The volume current, J, is measured in A/m^2. (Some sources use the variable J_S for surface current.) Both cross-sectional dimensions are nonzero for volume current. As with the other currents, volume current is actually a vector (e.g., $\vec{J}$). If the volume current is not changing over its cross section, the total current is simply AJ, where A is the cross-sectional area and J is the magnitude of the volume current. For example, if the uniform current density in a circular wire of radius a is J, then

$$I = \pi a^2 J$$

[3]The dimensions of the surface carrying the current are infinite (not the amplitude of the current).

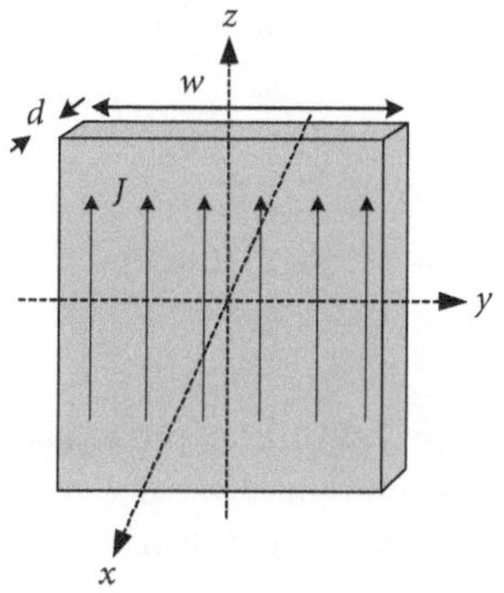

FIGURE 3.3 Relationship between the volume current and total current is given by (3.3) for this slab.

If the volume current is not constant with position, the total current can be obtained in the usual manner by integrating the normal component of the volume current over the cross section that it is passing:

$$I = \iint_S \vec{J} \cdot d\vec{s} \tag{3.2}$$

When volume current is confined to a region of small thickness (relative to the other dimensions) such as over a thin large slab, the volume current is sometimes approximated as a surface current. Referring to Figure 3.3, the volume current density, J, is simply related to the overall current, I:

$$I = wdJ \tag{3.3}$$

Since the surface current is the current per unit width, the surface current approximation for this volume current slab is

$$\vec{K} \approx \frac{I}{w}\hat{a}_z = \frac{wdJ}{w}\hat{a}_z = dJ\hat{a}_z \tag{3.4}$$

As a simple check, the unit of dJ is m(A/m^2) = A/m, which is the unit associated with surface current. Although the volume current was assumed constant or uniform in this example, this method also can be applied to nonuniform, volume current distributions. For example, imagine that the volume current in the previous thin slab varies according to the expression

$$\vec{J} = Je^{-x}\cos y\hat{a}_z$$

where $0 < x < d$ and $0 < y < w$. This volume current varies in both the x and y directions. A surface current "equivalent" will still vary in the y direction in the same sinusoidal manner as this volume current. However, there can be no x variation since surface current has no depth. The volume current in the x direction can be summed or integrated:

$$\vec{K} \approx \left(\int_0^d Je^{-x}dx \right) \cos y\hat{a}_z = J(1 - e^{-d})\cos y\hat{a}_z$$

If the total current is required over this slab, the magnitude of the surface current can be integrated over its width:

$$I = \int_0^w J(1-e^{-d})\cos y\, dy = J(1-e^{-d})\sin w$$

For this specific sinusoidal variation in the y direction, the total current could be zero (e.g., $w = \pi$), yet the surface current and volume current would not be zero.

The equations in Table 3.1 assume that the current distribution is known. Also, free space is assumed everywhere. Many of the currents shown are not physically realizable (i.e., cannot be built). For example, the current for the finite-length line current starts with a nonzero value (in free space) and ends with a nonzero value (in space). Nevertheless, they are quite valuable when used in conjunction with other current sources to generate a realizable distribution. (For example, four of these finite line currents can represent a square loop of current.) Unlike charge in electrostatic situations, current can exist inside perfect conductors in pure magnetostatic or dc situations. For the line, surface, and volume current expressions, the shaded regions shown can be assumed a conductor if needed to help visualize the currents. Other external conductors generally will affect the magnetic field if the field is time varying. The time-varying magnetic field will induce currents in these external conductors. These induced currents will then generate counter or induced magnetic fields, which should be summed with the original applied field. When the line, surface, and volume current densities are given in A, A/m, and A/m², respectively, and the distances are given in meters, the magnetic field is given in A/m. In the given table, sometimes the magnetic field is specified only along a certain direction (i.e., "along the z axis"), in a given region (i.e., "if $r > a$"), or at a specific location (i.e., "at the origin"). An absence of an expression in this table does not imply that the fields are zero in the unspecified region or location. In this table, a, b, d, L, l_{th}, w, Δ, α, ϑ, ρ_o, N, I, K, and J are constant with respect to position unless specifically stated otherwise, and the currents are considered in steady state. The directions shown for I, K, and J are important. Currents out of the page are sometimes represented using a "•" (i.e., the front or tip of the directional arrow) while currents into the page are sometimes represented using a "×" (i.e., the end or feathered end of the directional arrow). The Cartesian coordinate system variables are x, y, and z, the cylindrical coordinate system variables are ρ, ϕ, and z, and the spherical coordinate system variables are r, θ, and ϕ. Although Table 3.1 is not exhaustive,[4] it does contain many of the commonly seen and used expressions for the magnetic fields from simple current distributions.

If the desired current distribution is not given in Table 3.1, superposition of two or more of the provided currents might provide the desired distribution (at least at dc or low frequencies). However, the origins, coordinate systems, and field's directions for the various expressions should all be carefully noted. As an example, assume that the magnetic field along the z axis is desired for the two circular loops of line current shown in Figure 3.4. The distance between the loops is d, and the direction of the current

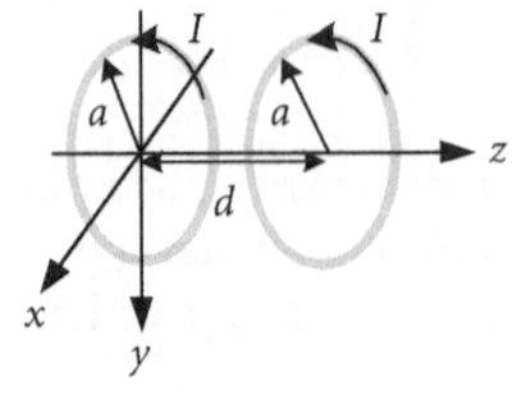

FIGURE 3.4 Two current-carrying circular loops centered on the z axis.

[4]Even though the author felt exhausted after compiling and checking these expressions.

is the same for both loops. For the first loop located at $z = 0$, the table result for the field along the axis for a circular loop of radius a is used directly:

$$H_{z,1}(0,0,z) = \frac{Ia^2}{2(a^2+z^2)^{\frac{3}{2}}}$$

For the second loop located at $z = d$, the magnetic field expression is obtained by a simple shift in the z variable:

$$H_{z,2}(0,0,z) = \frac{Ia^2}{2[a^2+(z-d)^2]^{\frac{3}{2}}}$$

Therefore, the magnetic field along the z axis when both loops are present is the sum of these two magnetic field expressions:

$$H_{z,total}(0,0,z) = H_{z,1}(0,0,z) + H_{z,2}(0,0,z) = \frac{Ia^2}{2(a^2+z^2)^{\frac{3}{2}}} + \frac{Ia^2}{2[a^2+(z-d)^2]^{\frac{3}{2}}} \tag{3.5}$$

The value of the z component of this total field midway between the loops at $z = d/2$ is given by the expression

$$H_{z,total}\left(0,0,\frac{d}{2}\right) = \frac{Ia^2}{\left(a^2+\frac{d^2}{4}\right)^{\frac{3}{2}}} \tag{3.6}$$

If $d = a$, then

$$H_{z,total}\left(0,0,\frac{d}{2}\right) \approx 0.72\frac{I}{a} \tag{3.7}$$

This field is larger than that at the center of a single loop, which is equal to $0.5I/a$. What is interesting about this two-loop current distribution is that the first derivative of the magnetic field with respect to z is zero at $z = d/2$:

$$\left.\frac{dH_{z,total}(0,0,z)}{dz}\right|_{z=\frac{d}{2}} = \left\{-\frac{3Iza^2}{2(a^2+z^2)^{\frac{5}{2}}} - \frac{3I(z-d)a^2}{2[a^2+(z-d)^2]^{\frac{5}{2}}}\right\}\Bigg|_{z=\frac{d}{2}} = 0$$

Furthermore, when the spacing between the loops is equal to their radius ($d = a$), the second derivative is also zero at $z = a/2$:

$$\left.\frac{d^2H_{z,total}(0,0,z)}{dz^2}\right|_{z=\frac{d}{2},\, d=a} = \left\{-\frac{3Ia^2}{2(a^2+z^2)^{\frac{5}{2}}} + \frac{15Iz^2a^2}{2(a^2+z^2)^{\frac{7}{2}}} - \frac{3Ia^2}{2[a^2+(z-a)^2]^{\frac{5}{2}}} + \frac{15I(z-a)^2a^2}{2[a^2+(z-a)^2]^{\frac{7}{2}}}\right\}\Bigg|_{z=\frac{a}{2}} = 0$$

This set of loops is referred to as a Helmholtz coil. Frequently, multiple-turn coils are used instead of single-turn loops. When the number of turns per coil is equal to N and the coils are tightly wound with a coil width small compared to d, then a good approximation for the magnetic field at the center is obtained by merely multiplying the previous field expressions by N. The zero first and second derivatives imply that the field is slowly varying along the z axis midway between the loops. Therefore, these loops provide a reasonably uniform magnetic field that can be used to determine the effect of magnetic fields on electronic devices and living matter. For example, these coils are used to test electronic compasses.

3.3 Magnetic Fields for Other Current Distributions

How can the magnetic field from a general current distribution be determined? What are the magnetic vector and scalar potentials? Can the magnetic vector potential at a point be used to determine the magnetic field at that same point? [Haus; Miner; Zahn; Hayt]

For many simple symmetrical current distributions, the magnetic field can be obtained quickly by the use of Ampère's law. Ampère's law, which is one of Maxwell's equations, in the point or differential form is

$$\nabla\times\vec{H}=\vec{J}+\frac{d\vec{D}}{dt} \tag{3.8}$$

where $\vec{J}$ is the volume current density vector and $\vec{D}=\varepsilon\vec{E}$ is electric flux density vector. The two sources of magnetic field, $\vec{H}$, are current and time-varying electric fields. However, the curl operation on $\vec{H}$ complicates the relationship between the current and time-varying electric flux density and the magnetic field. The integral form of Ampère's law is obtained by first integrating both sides of (3.8) over some surface S:

$$\iint\limits_S(\nabla\times\vec{H})\cdot d\vec{s}=\iint\limits_S\left(\vec{J}+\frac{d\vec{D}}{dt}\right)\cdot d\vec{s}=\iint\limits_S\vec{J}\cdot d\vec{s}+\iint\limits_S\frac{d\vec{D}}{dt}\cdot d\vec{s}$$

The normal to this surface has a direction and is thus a vector. The integration is over an open surface, so there is no volume associated with these integrations. Next, the powerful Stokes' theorem is used to eliminate the curling operation on the left side of this integral expression:

$$\oint\limits_C\vec{H}\cdot d\vec{L}=\iint\limits_S\vec{J}\cdot d\vec{s}+\iint\limits_S\frac{d\vec{D}}{dt}\cdot d\vec{s} \tag{3.9}$$

where C is the closed (or complete) contour or path around the perimeter of the surface, S. The relationship between the direction of the closed path and direction of $d\vec{s}$ is determined using the right-hand rule. In this case, when the fingers (except the thumb) of the right hand are in the direction of the path given for the contour, the thumb points in the direction of the normal to the surface and $d\vec{s}$. These directions are illustrated in Figure 3.5 for two current-carrying loops in the xy plane. The surface vectors, $d\vec{s}$, are in the $\pm z$ direction for these loops.

Equation (3.9) is Ampère's law in integral form. The first integral on the right side is the integration of the volume current density. The second integral involving $d\vec{D}/dt$ is the integration of the displacement current density. Sometimes, this displacement current contribution is negligible because the fields are magnetostatic or not time varying, or the contribution is zero because of the selection of S results in no

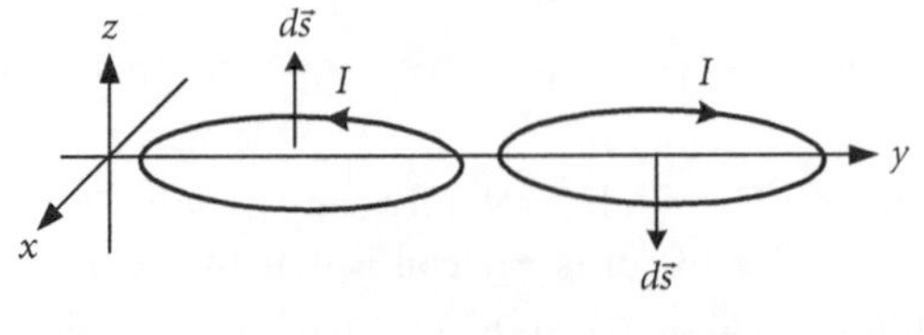

FIGURE 3.5 Relationship between the direction of the current and surface vector using the right-hand rule.

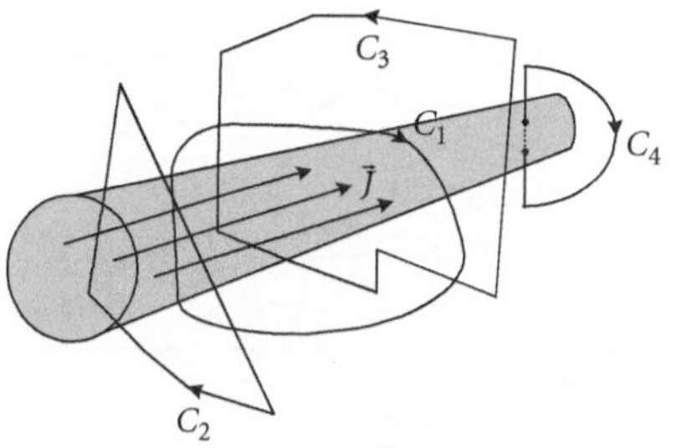

FIGURE 3.6 Only contour C_4 passes through the volume current.

displacement current passing through the surface. In these situations, the more common simplified form of Ampère's law is obtained:

$$\oint_C \vec{H}\cdot d\vec{L} = \iint_S \vec{J}\cdot d\vec{s} = I_{encl} \quad \text{if} \left|\iint_S \vec{J}\cdot d\vec{s}\right| \gg \left|\iint_S \frac{d\vec{D}}{dt}\cdot d\vec{s}\right| \tag{3.10}$$

where I_{encl} is the total current enclosed by the closed contour over the surface S. This important law states that the magnetic field summed around any closed path (in the same direction) is equal to the total current enclosed by this closed path. Any closed path selected (in the same direction) will result in the same result, regardless of the shape of the path, as long as the path encloses the same current. Several possible paths or contours enclosing a (unidirectional) volume current are shown in Figure 3.6. Contours C_1 and C_2 will result in the exact same result of

$$\iint_S \vec{J}\cdot d\vec{s} = I_{encl}$$

Contour C_3 will result in $-I_{encl}$ since a right-handed screw when advanced in the direction of the path shown by the arrow is in the opposite direction to the current density J. Contour C_4 will result in a value less than I_{encl} since the path cuts through the volume current and does not completely enclose it.

When the total current enclosed is zero, some students *incorrectly* believe that there is no magnetic field around this closed path since

$$\oint_C \vec{H}\cdot d\vec{L} = I_{encl} = 0$$

The magnetic field may be zero everywhere around the closed path in some situations. For example, when the closed contour encloses the outer cylinder of two concentric cylinders carrying equal but opposite currents (i.e., a coaxial cable), the magnetic field is zero outside the cylinders. However, the magnetic field outside two parallel conductors carrying equal but opposite currents (i.e., twin-lead line) is not zero even though the total current enclosed by a closed path containing both conductors is zero. If the net current enclosed by a closed path is zero, for some parts of the line integral the magnetic field can be positive, while for other parts the magnetic field can be negative.

There are many excellent textbooks in electromagnetics that clearly discuss how Ampère's law can sometimes be used to determine easily the magnetic field. Generally, to apply easily Ampère's law, symmetry is required and the proper Ampèrian closed path (closed contour surrounding the current distribution) must be selected. This usually implies that the contour selected be tangential to the magnetic field from the current. Ampère's law is wonderful for obtaining, in an elegant fashion, the magnetic field for many current distributions. However, unless an individual has a good understanding and "feel" for how the magnetic fields from a collection of currents add and subtract, it is recommended that Ampère's law not be used to determine specific field information. Many of the magnetic fields for line, surface,

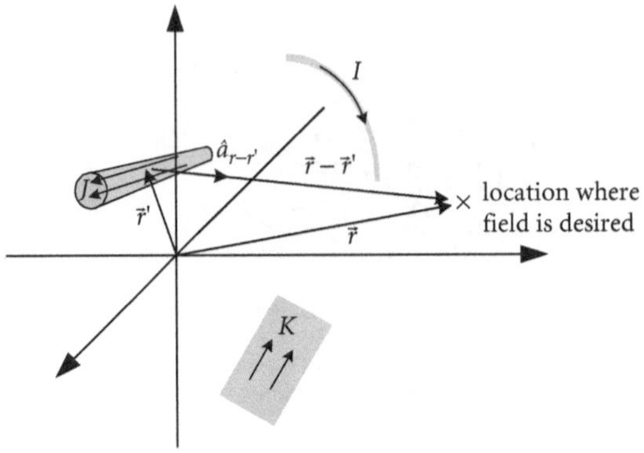

FIGURE 3.7 Magnetic fields generated by a line, surface, and volume current are desired at "×."

and volume currents that can be obtained simply using Ampère's law are contained in Table 3.1. Again, a field solution obtained via Ampère's law is frequently viewed as being elegant.

For a general current distribution, which does not necessarily possess symmetry or is not simple, the magnetic field can be determined directly or indirectly (via the magnetic potential). For any collection of line, surface, and volume currents, the magnetic field can be determined directly using the Biot-Savart law:

$$\vec{H}(r) = \int_{L'} \frac{\vec{I}'dL' \times \hat{a}_{r-r'}}{4\pi |\vec{r}-\vec{r}'|^2} + \iint_{S'} \frac{\vec{K}'dS' \times \hat{a}_{r-r'}}{4\pi |\vec{r}-\vec{r}'|^2} + \iiint_{V'} \frac{\vec{J}'dV' \times \hat{a}_{r-r'}}{4\pi |\vec{r}-\vec{r}'|^2} \tag{3.11}$$

These integrals can be evaluated numerically or analytically in some special cases. In some ways, the use of these integrals to determine the magnetic field may be considered the "brute force" approach. The primed variables represent the location of the actual currents so that $\vec{r}'$ is the vector from the origin to the location where the current(s) is located (the source of the magnetic field). The unprimed variable $\vec{r}$ is the vector from the origin to the location where the field is desired. Therefore, $\vec{r}-\vec{r}'$ is the vector from the current to the location where the field is desired. The quantity $|\vec{r}-\vec{r}'|$ is the distance between the source of the field (i.e., the current) and location where the field is desired. The vector $\hat{a}_{r-r'}$ is the unit vector from the current to this point of interest where the field is desired. In Figure 3.7, these vectors are shown for one location inside a volume current and for one point of interest. The vectors for other locations inside the volume current and from the line and surface currents are not shown. It is probably apparent that the field at any point of interest is the superposition or sum of the fields from each incremental current element, whether the element is a line, surface, or volume current. Initially, these integrals may seem confusing, but after a few examples their setup usually becomes mundane.

To illustrate the use of the Biot-Savart law, several worked problems will be provided. For the first current distribution, shown in Figure 3.8, the magnetic field will be determined along the z axis for a

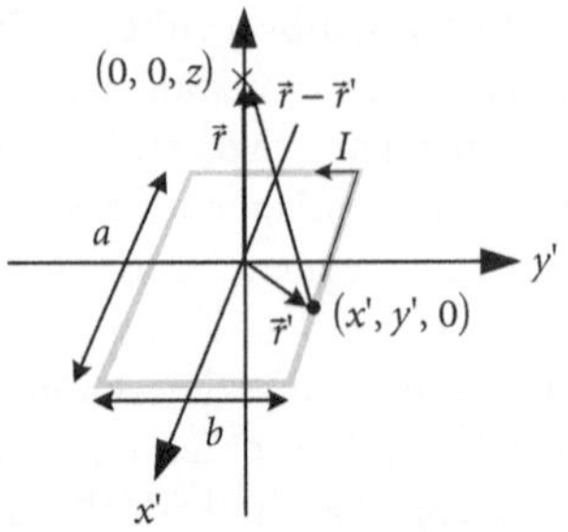

FIGURE 3.8 Magnetic field is desired along the z axis for this current-carrying rectangular loop.

rectangular current-carrying loop of dimensions *a* by *b* located in the *xy* plane and centered on the *z* axis. The current will be considered a line current. The integral is easily set up:

$$\vec{H}(0,0,z)=\int_{L'}\frac{\vec{I}'dL'\times\hat{a}_{r-r'}}{4\pi|\vec{r}-\vec{r}'|^2}$$

$$=\int_{L'}\frac{I(dx'\hat{a}_x+dy'\hat{a}_y)\times\dfrac{[(0-x')\hat{a}_x+(0-y')\hat{a}_y+(z-0)\hat{a}_z]}{\sqrt{(0-x')^2+(0-y')^2+(z-0)^2}}}{4\pi\left|\sqrt{(0-x')^2+(0-y')^2+(z-0)^2}\right|^2}$$

$$=\frac{I}{4\pi}\int_{L'}\frac{(dx'\hat{a}_x+dy'\hat{a}_y)\times(-x'\hat{a}_x-y'\hat{a}_y+z\hat{a}_z)}{[(x')^2+(y')^2+z^2]^{\frac{3}{2}}}$$

There are actually four separate integrals corresponding to each of the linear line segments of the rectangle. The first integral corresponds to the segment located at $y' = -b/2$. The second integral corresponds to the segment located at $x' = a/2$. The third integral corresponds to the segment located at $y' = b/2$. The fourth integral corresponds to the segment located at $x' = -a/2$. The sign of the current direction is determined by the order of the limits of integration not by the sign on the unit vectors $\hat{a}_x$ and $\hat{a}_y$:

$$\vec{H}(0,0,z)=\frac{I}{4\pi}\int_{-\frac{a}{2}}^{\frac{a}{2}}\frac{dx'\hat{a}_x\times\left(-x'\hat{a}_x+\frac{b}{2}\hat{a}_y+z\hat{a}_z\right)}{\left[(x')^2+\left(\frac{b}{2}\right)^2+z^2\right]^{\frac{3}{2}}}+\frac{I}{4\pi}\int_{-\frac{b}{2}}^{\frac{b}{2}}\frac{dy'\hat{a}_y\times\left(-\frac{a}{2}\hat{a}_x-y'\hat{a}_y+z\hat{a}_z\right)}{\left[\left(\frac{a}{2}\right)^2+(y')^2+z^2\right]^{\frac{3}{2}}}$$

$$+\frac{I}{4\pi}\int_{\frac{a}{2}}^{-\frac{a}{2}}\frac{dx'\hat{a}_x\times\left(-x'\hat{a}_x-\frac{b}{2}\hat{a}_y+z\hat{a}_z\right)}{\left[(x')^2+\left(\frac{b}{2}\right)^2+z^2\right]^{\frac{3}{2}}}+\frac{I}{4\pi}\int_{\frac{b}{2}}^{-\frac{b}{2}}\frac{dy'\hat{a}_y\times\left(\frac{a}{2}\hat{a}_x-y'\hat{a}_y+z\hat{a}_z\right)}{\left[\left(\frac{a}{2}\right)^2+(y')^2+z^2\right]^{\frac{3}{2}}}$$

The cross products are evaluated in the usual way:

$$\vec{H}(0,0,z)=\frac{I}{4\pi}\int_{-\frac{a}{2}}^{\frac{a}{2}}\frac{\left(\frac{b}{2}\hat{a}_z-z\hat{a}_y\right)dx'}{\left[(x')^2+\left(\frac{b}{2}\right)^2+z^2\right]^{\frac{3}{2}}}+\frac{I}{4\pi}\int_{-\frac{b}{2}}^{\frac{b}{2}}\frac{\left(\frac{a}{2}\hat{a}_z+z\hat{a}_x\right)dy'}{\left[\left(\frac{a}{2}\right)^2+(y')^2+z^2\right]^{\frac{3}{2}}}$$

$$+\frac{I}{4\pi}\int_{\frac{a}{2}}^{-\frac{a}{2}}\frac{\left(-\frac{b}{2}\hat{a}_z-z\hat{a}_y\right)dx'}{\left[(x')^2+\left(\frac{b}{2}\right)^2+z^2\right]^{\frac{3}{2}}}+\frac{I}{4\pi}\int_{\frac{b}{2}}^{-\frac{b}{2}}\frac{\left(-\frac{a}{2}\hat{a}_z+z\hat{a}_x\right)dy'}{\left[\left(\frac{a}{2}\right)^2+(y')^2+z^2\right]^{\frac{3}{2}}}$$

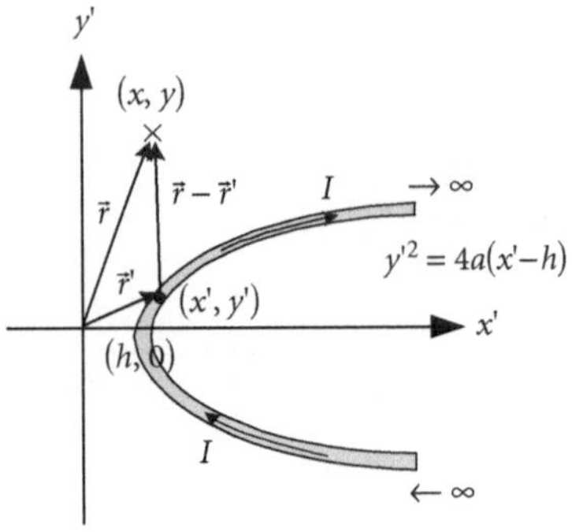

FIGURE 3.9 Integral expression for the magnetic field at "×" is desired for this line current.

Recognizing that $\int_c^d f(x)dx = -\int_d^c f(x)dx$, performing some simple manipulations, and then evaluating the integrals,

$$\vec{H}(0,0,z) = \frac{I}{4\pi}\hat{a}_z \int_{-\frac{a}{2}}^{\frac{a}{2}} \frac{b\,dx'}{\left[(x')^2 + \left(\frac{b}{2}\right)^2 + z^2\right]^{\frac{3}{2}}} + \frac{I}{4\pi}\hat{a}_z \int_{-\frac{b}{2}}^{\frac{b}{2}} \frac{a\,dy'}{\left[\left(\frac{a}{2}\right)^2 + (y')^2 + z^2\right]^{\frac{3}{2}}}$$

$$= \frac{2Iab(a^2+b^2+8z^2)}{\pi(a^2+4z^2)(b^2+4z^2)\sqrt{a^2+b^2+4z^2}}\hat{a}_z$$

For the second line current distribution, shown in Figure 3.9, the integrals will be set up to determine the magnetic field at the general point (x, y) in the xy plane. The current follows the parabolic curve

$$y^2 = 4a(x-h)$$

where $a > 0$ and $h > 0$. The vertex of the parabola is at $(h, 0)$ and its focus is at $(h + a, 0)$. There are no surface or volume currents. The magnetic field anywhere at the point (x, y) in the xy plane is described by

$$\vec{H}(x,y) = \int_{L'} \frac{\vec{I}'dL'\times\hat{a}_{r-r'}}{4\pi|\vec{r}-\vec{r}'|^2} = \int_{L'} \frac{I(dx'\hat{a}_x + dy'\hat{a}_y)\times\dfrac{[(x-x')\hat{a}_x + (y-y')\hat{a}_y]}{\sqrt{(x-x')^2+(y-y')^2}}}{4\pi\left|\sqrt{(x-x')^2+(y-y')^2}\right|^2}$$

$$= \frac{I}{4\pi}\int_{L'} \frac{[(y-y')dx' - (x-x')dy']\hat{a}_z}{[(x-x')^2+(y-y')^2]^{\frac{3}{2}}}$$

The primed variables represent the location of the line current. The integration is taken over the path of this line current. The previous integral is a function of both x' and y'. However, a line integration involves only dx' or dy'.[5] The integral can easily be written entirely in terms of either of these variables:

$$y' = \pm\sqrt{4a(x'-h)} \quad \Rightarrow dy' = \frac{\pm a}{\sqrt{a(x'-h)}}dx'$$

[5]When first exposed to curved line currents, some mathematically challenged students attempt to perform a $dx'dy'$ integration. This double differential is clearly nonsense since it would correspond to a surface integration not a line integration.

where the – multiplier corresponds to the portion of the parabola below the x axis. Substituting both of these expressions into the previous integral,

$$\vec{H}(x,y)=\frac{I}{4\pi}\int_{\infty}^{h}\frac{\left\{\left[y+\sqrt{4a(x'-h)}\right]dx'+(x-x')\frac{a}{\sqrt{a(x'-h)}}dx'\right\}\hat{a}_z}{\left\{(x-x')^2+\left[y+\sqrt{4a(x'-h)}\right]^2\right\}^{\frac{3}{2}}}$$

$$+\frac{I}{4\pi}\int_{h}^{\infty}\frac{\left\{\left[y-\sqrt{4a(x'-h)}\right]dx'-(x-x')\frac{a}{\sqrt{a(x'-h)}}dx'\right\}\hat{a}_z}{\left\{(x-x')^2+\left[y-\sqrt{4a(x'-h)}\right]^2\right\}^{\frac{3}{2}}}$$

The first integral represents the lower half of the parabolic current, and the second integral represents the upper half of the current. Since the unit vector $\hat{a}_z$ does not change direction, it may be taken past the integrals:

$$\vec{H}(x,y)=\frac{I}{4\pi}\hat{a}_z\int_{\infty}^{h}\frac{\left\{\left[y+\sqrt{4a(x'-h)}\right]+(x-x')\frac{a}{\sqrt{a(x'-h)}}\right\}dx'}{\left\{(x-x')^2+\left[y+\sqrt{4a(x'-h)}\right]^2\right\}^{\frac{3}{2}}}$$

$$+\frac{I}{4\pi}\hat{a}_z\int_{h}^{\infty}\frac{\left\{\left[y-\sqrt{4a(x'-h)}\right]-(x-x')\frac{a}{\sqrt{a(x'-h)}}\right\}dx'}{\left\{(x-x')^2+\left[y-\sqrt{4a(x'-h)}\right]^2\right\}^{\frac{3}{2}}}$$

The variables x, y, a, and h are constant with respect to x'. The magnetic field at any point (x, y) can be determined by evaluating these integrals. In Mathcad 3.1, these integrals are numerically evaluated along the x axis. The sign of the field can be checked using the right-hand rule for a current-carrying line conductor.

This same procedure can also be applied to surface and volume current distributions. For the third current distribution, shown in Figure 3.10, the magnetic field will be determined along the $+z$ axis from a disk carrying a circulating surface current equal to

$$\vec{K}=\rho\hat{a}_\phi$$

The disk is centered about the z axis and located in the xy plane. Obviously, this $\vec{K}$ is a nonuniform surface current that is zero at the center and equal to $\vec{K}(\rho=a)=a\hat{a}_\phi$ at its perimeter, where a is the radius of the disk. This surface current can be approximated by spiraling a current-carrying wire (with insulation) into the shape of a disk while decreasing the spacing of the wire with increasing radius. Working in the cylindrical coordinate system,

$$\vec{H}(0,0,z)=\iint_{S'}\frac{\vec{K}'dS'\times\hat{a}_{r-r'}}{4\pi|\vec{r}-\vec{r}'|^2}=\int_0^{2\pi}\int_0^{a}\frac{\rho'\hat{a}_\phi d\rho'\rho'd\phi'\times\frac{(-\rho'\hat{a}_\rho+z\hat{a}_z)}{\sqrt{(\rho')^2+z^2}}}{4\pi\left|\sqrt{(\rho')^2+z^2}\right|^2}$$

$$=\int_0^{2\pi}\int_0^{a}\frac{(\rho')^2d\rho'd\phi'\hat{a}_\phi\times(-\rho'\hat{a}_\rho+z\hat{a}_z)}{4\pi[(\rho')^2+z^2]^{\frac{3}{2}}}$$

$$=\int_0^{2\pi}\int_0^{a}\frac{(\rho')^3d\rho'd\phi'\hat{a}_z}{4\pi[(\rho')^2+z^2]^{\frac{3}{2}}}+\int_0^{2\pi}\int_0^{a}\frac{z(\rho')^2d\rho'd\phi'\hat{a}_\rho}{4\pi[(\rho')^2+z^2]^{\frac{3}{2}}}$$

$$I := 0.01 \qquad a := 0.025 \qquad h := 0.03$$

$$x := h - \frac{a}{10}, h - \frac{a}{10} + \frac{a}{500} .. h + \frac{a}{10}$$

$$H_z(x) := \frac{I}{4\cdot\pi}\cdot\int_{20\cdot h}^{h} \frac{\sqrt{4\cdot a\cdot(x_p - h)} + (x - x_p)\cdot\frac{a}{\sqrt{a\cdot(x_p - h)}}}{\left[(x - x_p)^2 + \left[\sqrt{4\cdot a\cdot(x_p - h)}\right]^2\right]^{\frac{3}{2}}} \, dx_p \ldots$$

$$+ \frac{I}{4\cdot\pi}\cdot\left[\int_{h}^{20\cdot h} \frac{-\sqrt{4\cdot a\cdot(x_p - h)} + -(x - x_p)\cdot\frac{a}{\sqrt{a\cdot(x_p - h)}}}{\left[(x - x_p)^2 + \left[\sqrt{4\cdot a\cdot(x_p - h)}\right]^2\right]^{\frac{3}{2}}} \, dx_p\right]$$

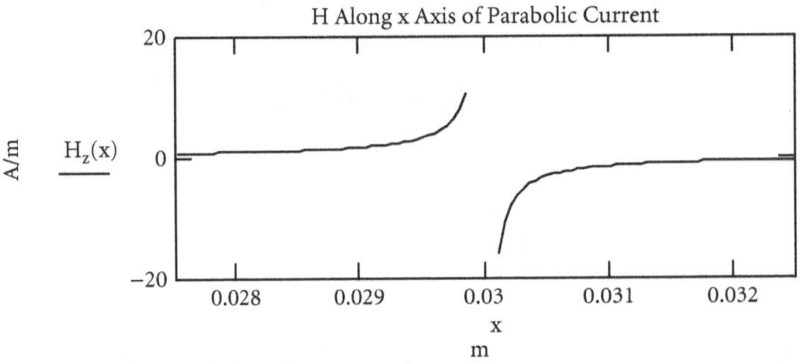

MATHCAD 3.1 Magnetic field vs. x for the current distribution in Figure 3.9.

One advantage of working in the Cartesian coordinate system is that the unit vectors, $\hat{a}_x$, $\hat{a}_y$, and $\hat{a}_z$, do not change their direction with integration. Unfortunately, the unit vector $\hat{a}_\rho$ cannot be pulled past the integral since its direction changes with ϕ' integration. By converting $\hat{a}_\rho$ to the Cartesian coordinate unit vector(s) equivalent, all of the unit vectors can be pulled past the integrals and the integrals

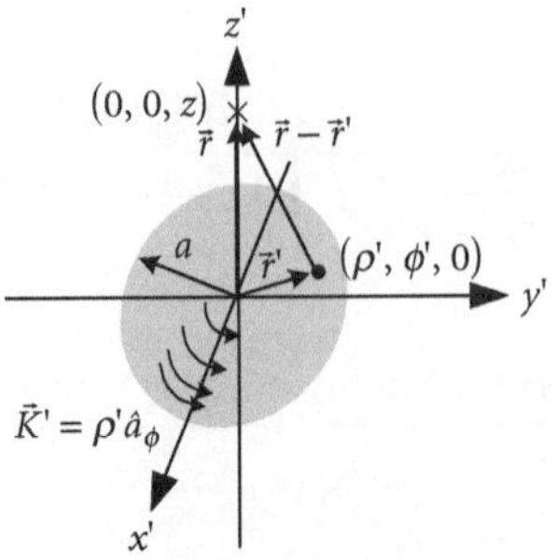

FIGURE 3.10 Magnetic field is desired at "×" along the z axis for this surface current.

evaluated:

$$\vec{H}(0,0,z)=\frac{1}{4\pi}\hat{a}_z\int_0^{2\pi}\int_0^{a}\frac{(\rho')^3 d\rho' d\phi'}{[(\rho')^2+z^2]^{\frac{3}{2}}}+\frac{z}{4\pi}\int_0^{2\pi}\int_0^{a}\frac{(\rho')^2 d\rho' d\phi'(\cos\phi'\hat{a}_x+\sin\phi'\hat{a}_y)}{[(\rho')^2+z^2]^{\frac{3}{2}}}$$

$$=\frac{2\pi}{4\pi}\hat{a}_z\int_0^{a}\frac{(\rho')^3 d\rho'}{[(\rho')^2+z^2]^{\frac{3}{2}}}+\frac{z}{4\pi}\hat{a}_x\int_0^{2\pi}\int_0^{a}\frac{(\rho')^2 d\rho'\cos\phi' d\phi'}{[(\rho')^2+z^2]^{\frac{3}{2}}}+\frac{z}{4\pi}\hat{a}_y\int_0^{2\pi}\int_0^{a}\frac{(\rho')^2 d\rho'\sin\phi' d\phi'}{[(\rho')^2+z^2]^{\frac{3}{2}}}$$

$$=\left[\frac{(a^2+2z^2)-2z\sqrt{a^2+z^2}}{2\sqrt{a^2+z^2}}\right]\hat{a}_z+0\hat{a}_x+0\hat{a}_y \quad z>0$$

The more experienced individual would usually not evaluate the integral involving $\hat{a}_\rho$ since by symmetry it is zero.

It is fairly obvious that determining the magnetic field for a given current distribution is more complicated than determining the electric field for a given charge distribution of "similar complexity." The vector nature of current and its cross product with the directional unit vector in the Biot-Savart law are responsible for this increased complexity. In electrostatics, the electric field can be obtained from the potential function. Since the potential function in electrostatics is a scalar, it can be simpler to determine than the electric field (at least for the tenderfoot). There are two types of potential functions for magnetic fields: the scalar magnetic potential and vector magnetic potential. However, most individuals find these magnetic potentials much less insightful than the electric potential, and they are often viewed as simply intermediate mathematical functions. Given the current distributions, the magnetic *vector* potential can be obtained from the integrals

$$\vec{A}(r)=\int_{L'}\frac{\mu_o\vec{I}'dL'}{4\pi|\vec{r}-\vec{r}'|}+\iint_{S'}\frac{\mu_o\vec{K}'dS'}{4\pi|\vec{r}-\vec{r}'|}+\iiint_{V'}\frac{\mu_o\vec{J}'dV'}{4\pi|\vec{r}-\vec{r}'|} \tag{3.12}$$

(The variables have the same meaning as with the Biot-Savart law.) These integrals are simpler than those in the Biot-Savart law. Furthermore, the magnetic vector potential is in the same direction as the current that is producing it since there are no cross products in the integrands. Therefore, if a current is entirely in the z direction, then $\vec{A}$ is also entirely in the z direction. The cost associated with the use of the magnetic vector potential is an additional operation to obtain the magnetic flux density:

$$\vec{B}=\mu_o\vec{H}=\nabla\times\vec{A} \tag{3.13}$$

The curl operation involves spatial differentiation. If the integrals for the magnetic vector potential are evaluated numerically, then to obtain the magnetic flux density, numerical differentiation of numerical data is required. This increases the error associated with the final field solution. The vector potential must be determined everywhere near the region of interest, not just at one location, to obtain these spatial derivatives. Also, since the curl operation involves differentiation, the magnetic vector potential is not unique. That is, each $\vec{B}$ does not have only one corresponding $\vec{A}$ vector function. For example, a constant vector can be added to $\vec{A}$ resulting in exactly the same $\vec{B}$.

When only volume currents are present, it can be shown that the magnetic vector potential satisfies the vector Poisson's equation:

$$\nabla^2\vec{A}=-\mu_o\vec{J} \tag{3.14}$$

This expression allows the magnetic vector potential to be determined *inside* a current distribution (and outside a current distribution if the right side of the equation is set to zero). For certain symmetrical volume current distributions, the magnetic vector potential can be evaluated quickly using (3.14). At first, $\nabla^2\vec{A}$ may seem meaningless. However, recall that ∇ is actually a vector operator and can be written as $\vec{\nabla}$. Therefore, $\vec{\nabla}\cdot\vec{\nabla}=\nabla^2$ is a scalar operator, and the vector $\vec{A}$ can be multiplied by a scalar. When working in the Cartesian coordinate system, the previous expression can be written as

$$\nabla^2 A_x = -\mu_o J_x,\ \nabla^2 A_y = -\mu_o J_y,\ \nabla^2 A_z = -\mu_o J_z \tag{3.15}$$

In other words, these three scalar equations state that each component of the magnetic vector potential is obtained from the corresponding component of the current density.

With help from Stokes' theorem, the expression for the total loop inductance can be written in terms of the magnetic vector potential:

$$L=\frac{\Phi}{I}=\frac{\iint_S \vec{B}\cdot d\vec{s}}{I}=\frac{\iint_S (\nabla\times\vec{A})\cdot d\vec{s}}{I}=\frac{\oint_C \vec{A}\cdot d\vec{l}}{I} \tag{3.16}$$

where C is the closed path along the conductor of the loop ($d\vec{l}$ is used instead of $d\vec{L}$ to help avoid confusion with the inductance variable, L). Instead of performing a surface integration of the magnetic flux density over the entire area of the loop, a closed line integration of the magnetic vector potential along the loop's perimeter can be determined. This can be helpful when determining inductance. When a current density is entirely in the z direction so that $\vec{A}$ is also entirely in the z direction, it can be shown that lines of constant A_z correspond to lines of magnetic flux. (The usefulness of this relationship is generally limited to two-dimensional configurations.) If the equipotential lines are determined for a two-dimensional electrostatic configuration of electrodes, these lines can be interpreted as lines of magnetic flux for the magnetostatic configuration of these same electrodes. For example, assume that the equipotentials between two infinitely long parallel conductors are determined for a given dc voltage across them. This situation is clearly electrostatic. When these conductors carry equal but opposite dc currents (clearly a magnetostatic situation), the magnetic flux lines for these two long conductors correspond to the equipotentials determined for the electrostatic situation.

Before a few examples are presented, the magnetostatic scalar potential will be defined. Like the electric potential, Φ, the magnetostatic scalar potential, ϑ, is a scalar and it obeys Laplace's equation:

$$\nabla^2\vartheta = 0 \quad \text{if } \vec{J}=0 \tag{3.17}$$

This expression allows the scalar potential to be determined *outside* a current distribution. Similar to the relationship between the electric field and electric potential, once ϑ is determined, the magnetic field is obtained via

$$\vec{H}=-\nabla\vartheta \tag{3.18}$$

The negative sign is included in this definition so that it is analogous to the electrostatic potential and electric field relationship.[6] The major restriction on the use of this scalar potential is that the current density must be zero throughout the region where the scalar potential is defined or used. Along the boundaries or other locations outside where the potential is defined, the current density need not be zero.

[6]Some authors define the magnetic scalar potential without the negative sign.

Like the electric potential, the scalar potential can be determined from the magnetic field:

$$\vartheta = -\int_a^b \vec{H} \cdot d\vec{L} \tag{3.19}$$

where $d\vec{L}$ is the differential path taken from a to b. Unlike electric potential, however, the value that is obtained for ϑ from this integral expression *is a function of the path taken.* If the path taken does not encircle any current, then the magnetic scalar potential is unique and single valued.

One example should be sufficient for illustrating how to obtain the magnetic vector potential directly from the integral definition. The magnetic vector potential will be determined everywhere for the rectangular current-carrying loop of dimensions a by b previously presented. The current will be considered a line current. There are four integrals corresponding to each of the linear line segments of the rectangle. Referring to Figure 3.11, along the z axis the integrals for magnetic vector potential are quickly set up:

$$\vec{A}(0,0,z) = \int_{L'} \frac{\mu_o \vec{I} dL'}{4\pi\sqrt{(0-x')^2+(0-y')^2+(z-0)^2}}$$

$$= \frac{\mu_o I}{4\pi}\hat{a}_x \int_{-\frac{a}{2}}^{\frac{a}{2}} \frac{dx'}{\sqrt{(x')^2+\left(\frac{b}{2}\right)^2+z^2}} + \frac{\mu_o I}{4\pi}\hat{a}_y \int_{-\frac{b}{2}}^{\frac{b}{2}} \frac{dy'}{\sqrt{\left(\frac{a}{2}\right)^2+(y')^2+z^2}}$$

$$+ \frac{\mu_o I}{4\pi}\hat{a}_x \int_{\frac{a}{2}}^{-\frac{a}{2}} \frac{dx'}{\sqrt{(x')^2+\left(\frac{b}{2}\right)^2+z^2}} + \frac{\mu_o I}{4\pi}\hat{a}_y \int_{\frac{b}{2}}^{-\frac{b}{2}} \frac{dy'}{\sqrt{\left(\frac{a}{2}\right)^2+(y')^2+z^2}} = 0$$

The sign of the current direction is determined by the order of the limits of integration, not by the sign on the unit vectors $\hat{a}_x$ and $\hat{a}_y$. Although the magnetic vector potential is zero along the z axis, this does not imply that the magnetic flux density is zero along the z axis. The curl of $\vec{A}$ is equal to $\vec{B}$, and the curl operation involves differentiation with respect to all of the coordinate variables (x, y, and z in the Cartesian coordinate system), not just the z variable. It is thus necessary (unless symmetry arguments are invoked) to determine $\vec{A}$ everywhere (near the region of interest) if $\vec{B}$ is to be determined. The magnetic

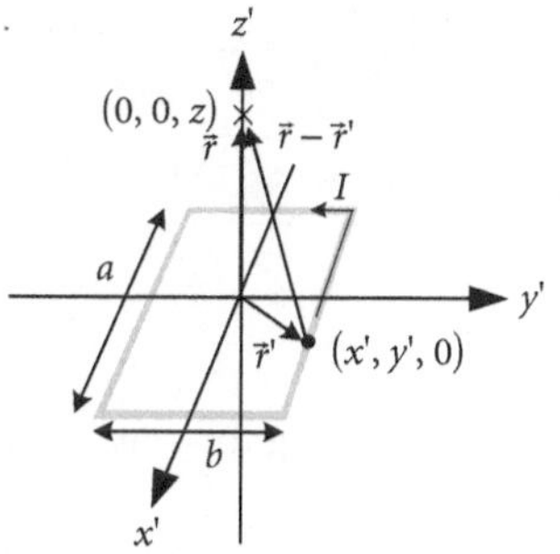

FIGURE 3.11 Magnetic vector potential is used to determine the magnetic flux density along the z axis for a current-carrying rectangular loop.

vector potential at any point (x, y, z) is given by

$$\vec{A}(x,y,z) = \int_{L'} \frac{\mu_o \vec{I} dL'}{4\pi\sqrt{(x-x')^2+(y-y')^2+(z-0)^2}}$$

$$= \frac{\mu_o I}{4\pi}\hat{a}_x \int_{-\frac{a}{2}}^{\frac{a}{2}} \frac{dx'}{\sqrt{(x-x')^2+\left(y+\frac{b}{2}\right)^2+z^2}} + \frac{\mu_o I}{4\pi}\hat{a}_y \int_{-\frac{b}{2}}^{\frac{b}{2}} \frac{dy'}{\sqrt{\left(x-\frac{a}{2}\right)^2+(y-y')^2+z^2}}$$

$$+ \frac{\mu_o I}{4\pi}\hat{a}_x \int_{\frac{a}{2}}^{-\frac{a}{2}} \frac{dx'}{\sqrt{(x-x')^2+\left(y-\frac{b}{2}\right)^2+z^2}} + \frac{\mu_o I}{4\pi}\hat{a}_y \int_{\frac{b}{2}}^{-\frac{b}{2}} \frac{dy'}{\sqrt{\left(x+\frac{a}{2}\right)^2+(y-y')^2+z^2}}$$

These integrals, which are very simple to set up, are evaluated in the standard manner (i.e., table, symbolic math package, or classical techniques). The expression for the magnetic vector potential, although byzantine, is

$$\vec{A}(x,y,z) = \frac{\mu_o I}{4\pi}\ln\left(\frac{a-2x+\sqrt{4x^2-4xa+4y^2+a^2+b^2+4z^2+4yb}}{-a-2x+\sqrt{4x^2+4xa+4y^2+a^2+b^2+4z^2+4yb}}\right)\hat{a}_x$$

$$+\frac{\mu_o I}{4\pi}\ln\left(\frac{b-2y+\sqrt{4x^2-4xa+4y^2+a^2+b^2+4z^2-4yb}}{-b-2y+\sqrt{4x^2-4xa+4y^2+a^2+b^2+4z^2+4yb}}\right)\hat{a}_y$$

$$+\frac{\mu_o I}{4\pi}\ln\left(\frac{-a-2x+\sqrt{4x^2+4xa+4y^2+a^2+b^2+4z^2-4yb}}{a-2x+\sqrt{4x^2-4xa+4y^2+a^2+b^2+4z^2-4yb}}\right)\hat{a}_x$$

$$+\frac{\mu_o I}{4\pi}\ln\left(\frac{-b-2y+\sqrt{4x^2+4xa+4y^2+a^2+b^2+4z^2+4yb}}{b-2y+\sqrt{4x^2+4xa+4y^2+a^2+b^2+4z^2-4yb}}\right)\hat{a}_y$$

The magnetic vector potential is entirely in the x and y directions since the line currents are entirely in the x and y directions. To determine the magnetic flux density anywhere, it is necessary to determine the curl of $\vec{A}$:

$$\vec{B} = \nabla \times \vec{A} = \left(\frac{\partial A_z}{\partial y} - \frac{\partial A_y}{\partial z}\right)\hat{a}_x + \left(\frac{\partial A_x}{\partial z} - \frac{\partial A_z}{\partial x}\right)\hat{a}_y + \left(\frac{\partial A_y}{\partial x} - \frac{\partial A_x}{\partial y}\right)\hat{a}_z$$

$$= -\frac{\partial A_y}{\partial z}\hat{a}_x + \frac{\partial A_x}{\partial z}\hat{a}_y + \left(\frac{\partial A_y}{\partial x} - \frac{\partial A_x}{\partial y}\right)\hat{a}_z$$

Without actually performing this differentiation, it is clear that the general expression for the magnetic flux density at any point is complex. However, along the z axis, the expression is relatively simple:

$$\vec{B} = \mu_o \frac{2Iab(a^2+b^2+8z^2)}{\pi(a^2+4z^2)(b^2+4z^2)\sqrt{a^2+b^2+4z^2}}\hat{a}_z \quad \text{along the } z \text{ axis } (0,0,z)$$

The tedious steps leading to this result from the magnetic vector potential will be left for the ambitious student.

Determining the magnetic vector potential via the vector Poisson's equation

$$\nabla^2 \vec{A} = -\mu_o \vec{J}$$

is relatively painless in some highly symmetrical configurations. Similar to when Poisson's equation was used to determine the electric potential, boundary conditions are required. In electrostatic problems, these boundaries frequently involve conducting surfaces (e.g., electrodes) where the potential is known. The boundary conditions, unfortunately, are not as intuitive for magnetic field problems. As a first example, the magnetic flux density will be determined everywhere via the magnetic potential for an infinitely long slab of thickness d carrying a uniform volume current of J entirely in the z direction. The slab extends from $y = -d/2$ to $y = d/2$ and is centered about the z axis as shown in Figure 3.12. It extends to infinity in both the x and z directions. Since the volume current is entirely in the z direction, the vector potential is entirely in the z direction. Outside the slab there is zero volume current. There are three regions, and Poisson's equation must be solved for each of these three regions:

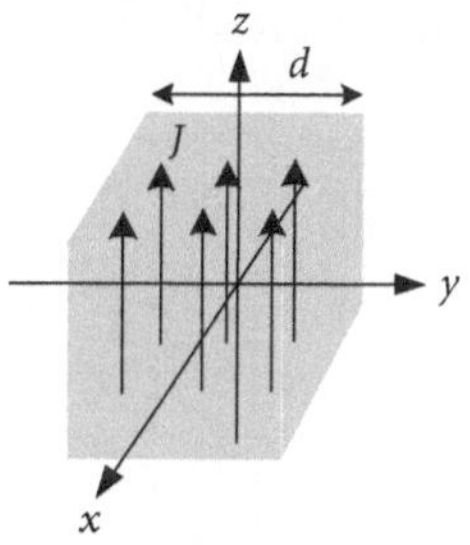

FIGURE 3.12 Volume current in a slab of thickness d.

$$\nabla^2 A_z = \begin{cases} -\mu_o J & -\frac{d}{2} < y < \frac{d}{2} \\ 0 & y > \frac{d}{2} \\ 0 & y < -\frac{d}{2} \end{cases}$$

Because of the symmetry of the situation, the vector potential should only vary in the y direction so that the previous equations reduce to

$$\frac{d^2 A_z}{dy^2} = \begin{cases} -\mu_o J & -\frac{d}{2} < y < \frac{d}{2} \\ 0 & y > \frac{d}{2} \\ 0 & y < -\frac{d}{2} \end{cases}$$

Integrating twice with respect to y for each region yields

$$A_z(y) = \begin{cases} -\frac{\mu_o J y^2}{2} + C_1 y + C_2 & -\frac{d}{2} < y < \frac{d}{2} \\ C_3 y + C_4 & y > \frac{d}{2} \\ C_5 y + C_6 & y < -\frac{d}{2} \end{cases}$$

Along the boundaries, both the normal and tangential components of $\vec{A}$ are continuous. This must be true since $\vec{B} = \nabla \times \vec{A}$ and a sudden jump or discontinuity in $\vec{A}$ would imply an infinite value in $\vec{B}$ since the ∇ operator involves differentiation (e.g., the derivative of the unit step $u(x)$ is an impulse function at $x = 0$). However, the *curl of* $\vec{A}$, $\nabla \times \vec{A}$, has discontinuous tangential components if surface current is present along the boundary. Rather than working with $\vec{A}$ boundary conditions to determine the constants of integration, the more familiar $\vec{B}$ and $\vec{H}$ boundary conditions will be applied. First, $\vec{B}$ must be determined via the curl of $\vec{A}$ for each of the three regions:

$$\vec{B} = \nabla \times \vec{A} = \left(\frac{\partial A_z}{\partial y} - \frac{\partial A_y}{\partial z} \right) \hat{a}_x + \left(\frac{\partial A_x}{\partial z} - \frac{\partial A_z}{\partial x} \right) \hat{a}_y + \left(\frac{\partial A_y}{\partial x} - \frac{\partial A_x}{\partial y} \right) \hat{a}_z = \frac{\partial A_z}{\partial y} \hat{a}_x$$

$$\vec{B} = \frac{dA_z}{dy} \hat{a}_x = \begin{cases} (-\mu_o J y + C_1)\hat{a}_x & -\frac{d}{2} < y < \frac{d}{2} \\ C_3 \hat{a}_x & y > \frac{d}{2} \\ C_5 \hat{a}_x & y < -\frac{d}{2} \end{cases}$$

Briefly, the two boundary conditions involving $\vec{B}$ and $\vec{H}$ are that the normal component of the flux density is continuous across a boundary consisting of two relative permeabilities μ_{r1} and μ_{r2}

$$B_{n1} = B_{n2} \quad \text{or} \quad \mu_{r1}\mu_o H_{n1} = \mu_{r2}\mu_o H_{n2}$$

and that the tangential component of the magnetic field not in the direction of K is discontinuous by the magnitude of the surface current along the boundary:

$$H_{t1} - H_{t2} = K$$

The tangential component that is in the same direction as the surface current is continuous across the boundary and not affected by the surface current. Both $\vec{B}$ and $\vec{H}$ are entirely in the x direction, which is tangential to the boundaries at $y = \pm d/2$. All regions are considered nonmagnetic and no surface currents are present so

$$H_{t1} - H_{t2} = 0 \quad \Rightarrow H_{x1} = H_{x2} \quad \Rightarrow B_{x1} = B_{x2} \quad \text{since } \mu_{r1} = \mu_{r2} = 1$$

Applying these boundary conditions at both boundaries,

$$-\mu_o J \frac{d}{2} + C_1 = C_3 \quad @\, y = \frac{d}{2}, \quad \mu_o J \frac{d}{2} + C_1 = C_5 \quad @\, y = -\frac{d}{2}$$

Finally, because of the symmetry of the slab about the z axis, the magnitude of the magnetic flux density on either side of the slab should be equal for equal distances from the slab. This implies that $C_1 = 0$; otherwise, the $\vec{B}$ field would not pass through the origin. Solving for the constants C_3 and C_5, the desired field expressions are obtained:

$$\vec{B} = \frac{dA_z}{dy} \hat{a}_x = \begin{cases} -\mu_o J y \hat{a}_x & -\frac{d}{2} \le y \le \frac{d}{2} \\ -\frac{\mu_o J d}{2} \hat{a}_x & y > \frac{d}{2} \\ \frac{\mu_o J d}{2} \hat{a}_x & y < -\frac{d}{2} \end{cases}$$

There are other "intuitive" boundary conditions that are used to determine the constants of integration for problems involving the vector Poisson's equation. For example, if the total current is zero (e.g., $Jy\hat{a}_z$ in the previous slab that exists from $-d/2 < y < d/2$), then the magnetic field outside the current distribution must eventually drop to zero far from the distribution. If the total current is not zero, but the configuration has finite dimensions (e.g., current around a finite-sized loop) and does not extend to infinity in any direction, then the magnetic field must also go to zero far from the distribution. If the magnetic field "blows up" or is infinite at certain locations, the corresponding constant of integration is eliminated so that the field is finite. Sometimes, Ampère's law is used to help determine a constant of integration.

As a second example involving the direct application of the vector Poisson's equation, the magnetic flux density will be determined everywhere for an infinitely long cylinder of radius a carrying a nonuniform volume current of $J\rho\hat{a}_z$. The axis of the cylinder is along the z axis. Since the current is entirely in the z direction, the vector potential is also entirely in the z direction:

$$\nabla^2 A_z = \begin{cases} -\mu_o J\rho & \rho < a \\ 0 & \rho > a \end{cases}$$

Working in the cylindrical coordinate system, Poisson's equation must be solved for two regions:

$$\frac{1}{\rho}\frac{d}{d\rho}\left(\rho\frac{dA_z}{d\rho}\right) = \begin{cases} -\mu_o J\rho & \rho < a \\ 0 & \rho > a \end{cases}$$

Integrating these equations in the standard way,

$$\frac{d}{d\rho}\left(\rho\frac{dA_z}{d\rho}\right) = \begin{cases} -\mu_o J\rho^2 & \rho < a \\ 0 & \rho > a \end{cases} \quad \Rightarrow \rho\frac{dA_z}{d\rho} = \begin{cases} -\dfrac{\mu_o J\rho^3}{3} + C_1 & \rho < a \\ C_3 & \rho > a \end{cases}$$

$$\frac{dA_z}{d\rho} = \begin{cases} -\dfrac{\mu_o J\rho^2}{3} + \dfrac{C_1}{\rho} & \rho < a \\ \dfrac{C_3}{\rho} & \rho > a \end{cases} \quad \Rightarrow A_z = \begin{cases} -\dfrac{\mu_o J\rho^3}{9} + C_1 \ln\rho + C_2 & \rho < a \\ C_3 \ln\rho + C_4 & \rho > a \end{cases}$$

The curl of the vector potential is equal to the magnetic flux density:

$$\vec{B} = \nabla\times\vec{A} = \left(\frac{1}{\rho}\frac{\partial A_z}{\partial\phi} - \frac{\partial A_\phi}{\partial z}\right)\hat{a}_\rho + \left(\frac{\partial A_\rho}{\partial z} - \frac{\partial A_z}{\partial\rho}\right)\hat{a}_\phi + \frac{1}{\rho}\left[\frac{\partial}{\partial\rho}(\rho A_\phi) - \frac{\partial A_\rho}{\partial\phi}\right]\hat{a}_z = -\frac{dA_z}{d\rho}\hat{a}_\phi$$

$$\vec{B} = -\frac{dA_z}{d\rho}\hat{a}_\phi = \begin{cases} \left(\dfrac{\mu_o J\rho^2}{3} - \dfrac{C_1}{\rho}\right)\hat{a}_\phi & \rho < a \\ -\dfrac{C_3}{\rho}\hat{a}_\phi & \rho > a \end{cases}$$

Since both regions are nonmagnetic and no surface current is present, the flux density must be continuous at $\rho = a$:

$$\frac{\mu_o Ja^2}{3} - \frac{C_1}{a} = -\frac{C_3}{a} \quad @\rho = a$$

The second condition required to solve for the magnetic flux density is that $C_1 = 0$; otherwise, the $\vec{B}$ field would be infinite at $\rho = 0$. Solving for C_3, the desired expressions are obtained:

$$\vec{B} = \begin{cases} \dfrac{\mu_o J \rho^2}{3} \hat{a}_\phi & \rho \le a \\ \dfrac{\mu_o J a^3}{3\rho} \hat{a}_\phi & \rho > a \end{cases}$$

These last examples will illustrate the determination and use of the scalar magnetic potential. If the scalar potential is provided or known, then it is generally a simple matter to determine the corresponding magnetic field since

$$\vec{H} = -\nabla \vartheta$$

For example, the magnetic scalar potentials both inside and outside a longitudinally wound, very long, cylindrical coil shown in Figure 3.13 are given by

$$\vartheta = \begin{cases} -\dfrac{NI\rho\cos\phi}{4R} & \rho < R \\ \dfrac{NIR\cos\phi}{4\rho} & \rho > R \end{cases}$$

where I is the current in the wire. As required when using the scalar magnetic potential, the current density is zero where the potential is defined. The current in the coil is modeled as a surface current, and it is in the $+z$ direction for $0 < \phi < \pi$ (and $-z$ direction for $\pi < \phi < 2\pi$). In the cylindrical coordinate system, the negative gradient of ϑ is equal to

$$\vec{H} = -\nabla\vartheta = -\frac{\partial\vartheta}{\partial\rho}\hat{a}_\rho - \frac{1}{\rho}\frac{\partial\vartheta}{\partial\phi}\hat{a}_\phi - \frac{\partial\vartheta}{\partial z}\hat{a}_z = -\frac{\partial\vartheta}{\partial\rho}\hat{a}_\rho - \frac{1}{\rho}\frac{\partial\vartheta}{\partial\phi}\hat{a}_\phi$$

Therefore,

$$\vec{H} = \begin{cases} \dfrac{NI\cos\phi}{4R}\hat{a}_\rho - \dfrac{NI\sin\phi}{4R}\hat{a}_\phi & \rho < R \\ \dfrac{NIR\cos\phi}{4\rho^2}\hat{a}_\rho + \dfrac{NIR\sin\phi}{4\rho^2}\hat{a}_\phi & \rho > R \end{cases}$$

For boundary conditions, the magnetic scalar potential, ϑ, is generally not continuous when surface current is present along the boundary. However, in free space, the normal derivative of ϑ is continuous

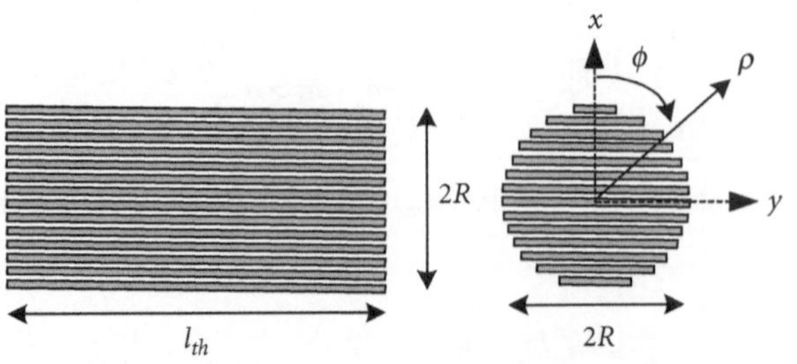

FIGURE 3.13 Longitudinally wound cylindrical coil.

$(d\vartheta_2/dn = d\vartheta_1/dn)$ even when a surface current is present. This relationship is easily verified for the cylindrical coil where the normal to the cylinder is in the ρ direction:

$$\frac{\partial}{\partial\rho}\left(-\frac{NI\rho\cos\phi}{4R}\right) = \frac{\partial}{\partial\rho}\left(\frac{NIR\cos\phi}{4\rho}\right)\bigg|_{\rho=R}$$

$$-\frac{NI\cos\phi}{4R} = -\frac{NIR\cos\phi}{4\rho^2}\bigg|_{\rho=R} = -\frac{NI\cos\phi}{4R}$$

However, the scalar potential is not continuous at the boundary:

$$-\frac{NI\rho\cos\phi}{4R}\bigg|_{\rho=R} \neq \frac{NIR\cos\phi}{4\rho}\bigg|_{\rho=R}$$

$$-\frac{NI\cos\phi}{4} \neq \frac{NI\cos\phi}{4}\bigg|_{\rho=R}$$

It is easily shown that the normal component of $\vec{B}$ is continuous and the tangential component of $\vec{H}$ is discontinuous (by the magnitude of the surface current) along the boundary:

$$\mu_o\frac{NI\cos\phi}{4R} = \mu_o\frac{NIR\cos\phi}{4\rho^2}\bigg|_{\rho=R}, \quad \frac{NIR\sin\phi}{4\rho^2} - \left(-\frac{NI\sin\phi}{4R}\right) = K\bigg|_{\rho=R}$$

$$\mu_o\frac{NI\cos\phi}{4R} = \mu_o\frac{NI\cos\phi}{4R}, \quad \frac{NI\sin\phi}{4R} + \frac{NI\sin\phi}{4R} = \frac{NI\sin\phi}{2R} = K$$

In the previous example, the scalar potential was provided. In some simple cases, the magnetic scalar potential can be determined via Laplace's equation:

$$\nabla^2\vartheta = 0 \quad \text{if } \vec{J} = 0$$

However, boundary conditions are needed. It seems as though the determination of the scalar magnetic potential is frequently not as straightforward or routine as the determination of the magnetic vector potential. Nevertheless, an example will be provided. The scalar potential is desired outside an infinitely long cylindrical wire carrying a current I in the z direction. The axis of the wire corresponds to the z axis. The scalar potential obeys the expression

$$\nabla^2\vartheta = \frac{1}{\rho}\frac{\partial}{\partial\rho}\left(\rho\frac{\partial\vartheta}{\partial\rho}\right) + \frac{1}{\rho^2}\frac{\partial^2\vartheta}{\partial\phi^2} + \frac{\partial^2\vartheta}{\partial z^2} = \frac{1}{\rho^2}\frac{\partial^2\vartheta}{\partial\phi^2} = 0$$

Outside the wire the scalar potential may be used since the current is zero. Because the magnetic scalar potential is not as commonly used as the electric potential, it may not be as immediately obvious how the magnetic potential varies. However, contours of ϑ are perpendicular to the $\vec{H}$ field lines. Integrating the previous expression twice,

$$\vartheta = C_1\phi + C_2$$

The magnetic field is the negative gradient of this scalar potential:

$$\vec{H} = -\nabla\vartheta = -\frac{\partial\vartheta}{\partial\rho}\hat{a}_\rho - \frac{1}{\rho}\frac{\partial\vartheta}{\partial\phi}\hat{a}_\phi - \frac{\partial\vartheta}{\partial z}\hat{a}_z = -\frac{1}{\rho}\frac{\partial\vartheta}{\partial\phi}\hat{a}_\phi = -\frac{C_1}{\rho}\hat{a}_\phi$$

To determine the constant of integration, realize that the magnetic field at the surface of a current-carrying infinitely long conductor is given by

$$H_\phi = \frac{I}{2\pi a}$$

The radius of the wire is equal to a. If no surface current is present, the tangential magnetic field is continuous. Therefore, C_1 is equal to $-I/2\pi$, and the magnetic scalar potential is

$$\vartheta = -\frac{I}{2\pi}\phi$$

The constant C_2 was set to zero, which is equivalent to setting the scalar potential reference at $\phi = 0$. As was previously stated, the magnetic scalar potential is not unique (if the path of interest encloses current). This nonuniqueness is obvious in this example since the potential at $\phi = 2\pi$ is equal to $-I$ not zero.

3.4 Magnetic Field Boundary Conditions

Discuss, but do not derive, the boundary conditions for magnetic fields. Clearly explain the effect of surface current on the magnetic fields. [Haus; Hayt; Melcher]

All realizable regions have boundaries. Depending on the electrical properties of the medium on both sides of the boundaries, and the shape and location of these boundaries, these boundaries can have a strong effect on the magnetic field distribution. The magnetic field, $\vec{H}$, is a fundamental quantity analogous to current. Steady dc and time-varying current, time-varying electric fields, and permanent magnets all generate magnetic fields. It thus should not be surprising that surface current, $\vec{K}$, which has no depth, can affect magnetic field boundary conditions, especially since it exists along interfaces or boundaries. With electric fields, the tangential components of the electric field are continuous across boundaries, even boundaries that have a single layer of surface charge. With magnetic fields, however, there is a jump or discontinuity in the tangential components of the magnetic field across a boundary containing surface current:

$$\hat{a}_n \times (\vec{H}_2 - \vec{H}_1) = \vec{K} \tag{3.20}$$

where $\hat{a}_n$ is the unit vector normal or perpendicular to the interface directed from region 1 to region 2. By taking the cross product of this normal vector with each of the $\vec{H}$ vectors, the tangential components are obtained. This boundary condition is valid for stationary, or magnetostatic, fields as well as time-varying magnetic fields. Sometimes, the scalar version of this expression is seen:

$$H_{t2} - H_{t1} = K \tag{3.21}$$

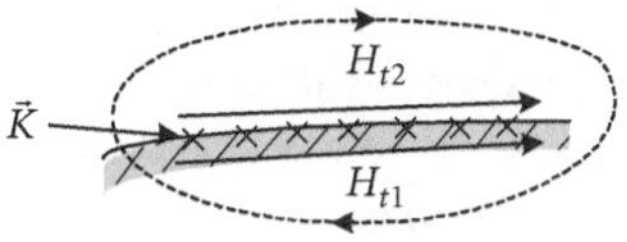

FIGURE 3.14 Magnetic fields along a surface and the fields generated by a surface current (directed into the page).

Students often eschew this scalar version, which requires insight into the directional nature of the surface current's magnetic field. Recall that an infinitely wide surface current directed entirely in the x direction, for example, generates a magnetic field entirely in the y direction. It does not generate a field perpendicular to the surface current in the z direction, and it does not generate a field in the direction of the current in the x direction. It is reasonable that there would be a discontinuity or jump in the tangential component of the magnetic field in the y direction across the surface current. Referring to Figure 3.14, the magnetic field generated by the surface current adds to H_{t2} but subtracts from H_{t1} as is indicated by (3.21). The direction of the magnetic field generated by the surface current is determined using the right-hand rule, where the thumb is in the direction of the current and the remaining curled digits are in the direction of the magnetic field generated by the current.

A vector at an interface of any shape can be separated into the sum of components tangential to the interface and normal to the interface. The only remaining boundary condition involves the normal component of the magnetic fields. At this point, a brief summary of the nature of magnetic materials would be helpful. As is well known, electrons (or an electron cloud) circulate around their nucleus in a material. Each of these circulating electron clouds, including the electrons spinning about their own axis, generates a magnetic field (since moving charge is the source of magnetic field). For most nonmagnetic materials, the net magnetic field from the orbiting electron clouds and spinning electrons tend to cancel resulting in no net external magnetic field. For magnetic materials, however, an applied magnetic field can cause a net torque on these circulating electrons in the material. The net magnetic field from the material can thus deviate from its "natural" state. As a result, the material itself can increase or decrease the magnetic field outside it.

In some cases, the field from the material, expressed in terms of the magnetization, $\vec{M}$, is linearly proportional to the applied magnetic field:

$$\vec{M} = \chi_m \vec{H} \tag{3.22}$$

where χ_m is the magnetic susceptibility. (Often for magnetic materials, though, the relationship between the magnetization and magnetic field is not linear.) As with electric fields, it is convenient to group the effects of the applied field and internally generated or induced field into one quantity. This quantity is referred to as the magnetic flux density, $\vec{B}$:

$$\vec{B} = \mu_o(\vec{H} + \vec{M}) = \mu_o(\vec{H} + \chi_m \vec{H}) = \mu_o(1 + \chi_m)\vec{H} \tag{3.23}$$

where μ_o is the permeability of free space, which is equal to $4\pi \times 10^{-7}$ H/m. Throughout most of this book and in most application literature, the commonly used and dimensionless quantity, the relative permeability, μ_r, is used:

$$\vec{B} = \mu_r \mu_o \vec{H} \quad \text{where } \mu_r = 1 + \chi_m \tag{3.24}$$

It is common to adapt the concept of relative permeability even when a material does not have a linear relationship between $\vec{M}$ and $\vec{H}$ (e.g., a differential relative permeability or an average relative permeability). The effect of the material on the applied magnetic field generally increases with μ_r. With this background, the final boundary condition can be stated:

$$B_{n2} = B_{n1} \tag{3.25}$$

In other words, the normal component of the magnetic flux density is continuous across a boundary. Surface current does not affect the normal component of the flux density. A difference in the permeability across a boundary is conveniently accounted for by using the $\vec{B}$ field in (3.25). An equivalent version of this boundary condition is

$$\hat{a}_n \cdot (\vec{B}_2 - \vec{B}_1) = 0 \tag{3.26}$$

where $\hat{a}_n$ is again the unit vector normal or perpendicular to the interface directed from region 1 to region 2. As will be discussed, there are also boundary conditions for the magnetic scalar and vector potentials. These conditions are discussed later when specific examples are given.

For a first example, the magnetic field is provided slightly below an interface containing a surface current as shown in Figure 3.15. The interface is located in the *xy* plane, and the surface current is entirely in the *y* direction. Above the interface, the relative permeability is equal to 100. Below the interface, the relative permeability is equal to 5. Both the magnetic field and magnetic flux density are desired slightly above the interface. The magnetic field slightly below the interface is given as

$$\vec{H}_5 = -2\hat{a}_x + 3\hat{a}_y + 4\hat{a}_z \quad \text{where } \mu_r = 5$$

Therefore, the flux density below the interface can be written down immediately:

$$\vec{B}_5 = 5\mu_o \vec{H}_5 = -10\mu_o\hat{a}_x + 15\mu_o\hat{a}_y + 20\mu_o\hat{a}_z \quad \text{where } \mu_r = 5$$

Before the fields are determined above the interface, realize that the surface current will only affect the magnetic field in the *x* direction since it does not generate a field in either the *y* or *z* directions. In the figure, only the components in the *yz* plane are shown. The boundary condition involving the surface current will first be applied:

$$\hat{a}_n \times (\vec{H}_2 - \vec{H}_1) = \vec{K}$$

where $\hat{a}_n$ is the unit vector normal to the interface directed from region 1 to region 2. If region 2 is selected as above the interface, then $\hat{a}_n = \hat{a}_z$. (If region 2 was selected as below the interface, then $\hat{a}_n = -\hat{a}_z$.) Therefore,

$$\hat{a}_z \times [H_x\hat{a}_x + H_y\hat{a}_y + H_z\hat{a}_z - (-2\hat{a}_x + 3\hat{a}_y + 4\hat{a}_z)] = K\hat{a}_y$$

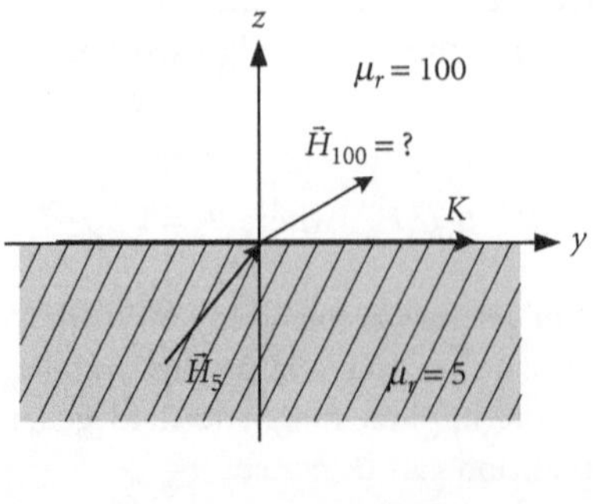

FIGURE 3.15 Boundary conditions are used to determine the magnetic field directly above the interface given the field directly below and surface current along the interface.

where $\vec{H}_{100} = H_x\hat{a}_x + H_y\hat{a}_y + H_z\hat{a}_z$. Taking the cross products,

$$(H_x + 2)\hat{a}_y - (H_y - 3)\hat{a}_x = K\hat{a}_y \quad \Rightarrow \begin{cases} H_x + 2 = K & \Rightarrow H_x = K - 2 \\ H_y - 3 = 0 & \Rightarrow H_y = 3 \end{cases}$$

As seen, only the x component of the magnetic field is affected by the surface current:

$$\vec{H}_{100} = (K-2)\hat{a}_x + 3\hat{a}_y + H_z\hat{a}_z$$

The y component is continuous across the boundary. The flux density vector above the interface is given by

$$\vec{B}_{100} = 100\mu_o\vec{H}_{100} = 100\mu_o(K-2)\hat{a}_x + 300\mu_o\hat{a}_y + 100\mu_o H_z\hat{a}_z$$

The continuity of the normal flux density, $B_{n2} = B_{n1}$, is the only relationship required to solve for the z component of the flux density. The normal to the surface is entirely in the z direction:

$$100\mu_o H_z\hat{a}_z = 20\mu_o\hat{a}_z \quad \Rightarrow H_z = \frac{1}{5}$$

The desired fields are then

$$\vec{H}_{100} = (K-2)\hat{a}_x + 3\hat{a}_y + \frac{1}{5}\hat{a}_z, \quad \vec{B}_{100} = 100\mu_o(K-2)\hat{a}_x + 300\mu_o\hat{a}_y + 20\mu_o\hat{a}_z$$

The next example is in the cylindrical coordinate system. Referring to Figure 3.16, the surface current is in both the radial and z directions:

$$\vec{K} = K_\rho\hat{a}_\rho + K_z\hat{a}_z$$

For $\phi > \pi/4$, the flux density is given as

$$\vec{B}_4 = 3\hat{a}_\rho - 2\hat{a}_\phi + 4\hat{a}_z$$

where the relative permeability is equal to 4. Near to the surface current for $\phi < \pi/4$, the magnetic field and flux density are desired. The relative permeability is equal to 10 for this region. Only the fields in the xy plane are shown in Figure 3.16. The magnetic field corresponding to $\vec{B}_4$ is

$$\vec{H}_4 = \frac{\vec{B}_4}{4\mu_o} = \frac{3}{4\mu_o}\hat{a}_\rho - \frac{1}{2\mu_o}\hat{a}_\phi + \frac{1}{\mu_o}\hat{a}_z$$

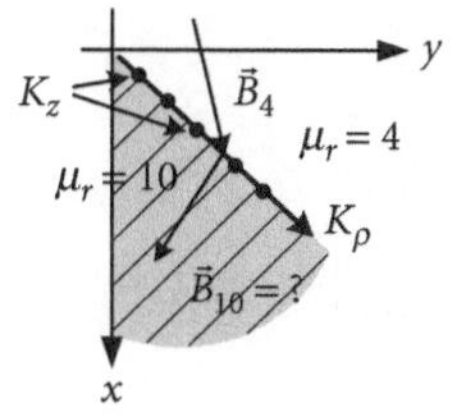

FIGURE 3.16 Interface is in the radial direction while the surface current is in both the radial and z directions.

The component of the surface current in the ρ direction will generate a magnetic field in the z direction, while the component of the surface current in the z direction will generate a magnetic field in the ρ direction. Therefore, neither tangential component of the magnetic field will be continuous. The boundary condition involving the surface current will first be applied:

$$\hat{a}_n \times (\vec{H}_2 - \vec{H}_1) = \vec{K}$$

where $\hat{a}_n$ is the unit vector normal to the interface directed from region 1 to region 2. If region 2 is selected as $\phi > \pi/4$, then $\hat{a}_n = \hat{a}_\phi$. (If region 2 was selected as $\phi < \pi/4$, then $\hat{a}_n = -\hat{a}_\phi$.) Letting $\vec{H}_{10} = H_\rho \hat{a}_\rho + H_\phi \hat{a}_\phi + H_z \hat{a}_z$,

$$\hat{a}_\phi \times \left[\frac{3}{4\mu_o}\hat{a}_\rho - \frac{1}{2\mu_o}\hat{a}_\phi + \frac{1}{\mu_o}\hat{a}_z - (H_\rho \hat{a}_\rho + H_\phi \hat{a}_\phi + H_z \hat{a}_z) \right] = K_\rho \hat{a}_\rho + K_z \hat{a}_z$$

Equating the components on the left side of the expression, after the cross products are taken, with the corresponding components on the right side of the expression,

$$-\left(\frac{3}{4\mu_o} - H_\rho\right)\hat{a}_z + \left(\frac{1}{\mu_o} - H_z\right)\hat{a}_\rho = K_\rho \hat{a}_\rho + K_z \hat{a}_z$$

$$\Rightarrow \begin{cases} -\left(\dfrac{3}{4\mu_o} - H_\rho\right) = K_z & \Rightarrow H_\rho = K_z + \dfrac{3}{4\mu_o} \\ \dfrac{1}{\mu_o} - H_z = K_\rho & \Rightarrow H_z = \dfrac{1}{\mu_o} - K_\rho \end{cases}$$

Therefore,

$$\vec{H}_{10} = \left(K_z + \frac{3}{4\mu_o}\right)\hat{a}_\rho + H_\phi \hat{a}_\phi + \left(\frac{1}{\mu_o} - K_\rho\right)\hat{a}_z$$

Since the relative permeability is 10 in this region,

$$\vec{B}_{10} = 10\mu_o \vec{H}_{10} = 10\mu_o\left(K_z + \frac{3}{4\mu_o}\right)\hat{a}_\rho + 10\mu_o H_\phi \hat{a}_\phi + 10\mu_o\left(\frac{1}{\mu_o} - K_\rho\right)\hat{a}_z$$

The last unknown, H_ϕ, is determined using the continuity of the normal component of the flux density:

$$-2\hat{a}_\phi = 10\mu_o H_\phi \hat{a}_\phi \quad \Rightarrow H_\phi = -\frac{1}{5\mu_o}$$

The desired fields are then

$$\vec{H}_{10} = \left(K_z + \frac{3}{4\mu_o}\right)\hat{a}_\rho - \frac{1}{5\mu_o}\hat{a}_\phi + \left(\frac{1}{\mu_o} - K_\rho\right)\hat{a}_z$$

$$\vec{B}_{10} = 10\mu_o\left(K_z + \frac{3}{4\mu_o}\right)\hat{a}_\rho - 2\hat{a}_\phi + 10\mu_o\left(\frac{1}{\mu_o} - K_\rho\right)\hat{a}_z$$

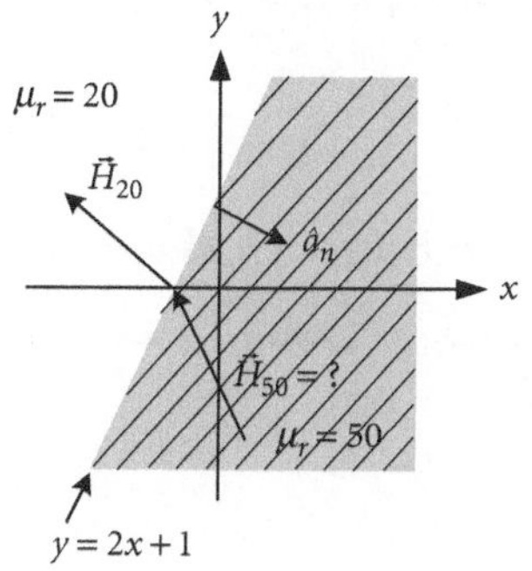

FIGURE 3.17 Interface is along the line $y = 2x + 1$. The unit normal to the interface has both x and y components.

The next example does not involve any surface current, but the interface is not along a constant coordinate surface. The interface between two magnetic regions is along $y = 2x+1$ as shown in Figure 3.17. For the region $y > 2x+1$ near the interface where $\mu_r = 20$, the magnetic field is

$$\vec{H}_{20} = -3\hat{a}_x + 2\hat{a}_y - 4\hat{a}_z \quad \text{where } \mu_r = 20$$

The fields are desired for the region near the interface for $y < 2x+1$ where $\mu_r = 50$. In Figure 3.17, only those components in the *xy* plane are shown. The flux density for $y > 2x+1$ is

$$\vec{B}_{20} = 20\mu_o\vec{H}_{20} = -60\mu_o\hat{a}_x + 40\mu_o\hat{a}_y - 80\mu_o\hat{a}_z \quad \text{where } \mu_r = 20$$

For this interface, the unit normal, $\hat{a}_n$, is in both the x and y directions. The unit normal to the interface is easily determined via the gradient of the function $f = 2x+1-y$:

$$\hat{a}_n = \frac{\nabla f}{|\nabla f|} = \frac{2\hat{a}_x - \hat{a}_y}{|2\hat{a}_x - \hat{a}_y|} = \frac{2\hat{a}_x - \hat{a}_y}{\sqrt{2^2+1^2}} = \frac{2\hat{a}_x - \hat{a}_y}{\sqrt{5}}$$

The gradient of f was divided by its magnitude to normalize the result. (The direction of this vector is as indicated in the figure. The unit vector in the opposite direction is easily determined by negating this result.) With the unit normal vector known, the boundary condition involving the surface current can be applied:

$$\hat{a}_n \times (\vec{H}_2 - \vec{H}_1) = \vec{K} = 0$$

where $\hat{a}_n$ is the unit vector normal to the interface directed from region 1 to region 2. Although it does not matter in this particular example because there is no surface current, region 2 is selected as $y < 2x + 1$. Letting $\vec{H}_{50} = H_x\hat{a}_x + H_y\hat{a}_y + H_z\hat{a}_z$,

$$\left(\frac{2\hat{a}_x - \hat{a}_y}{\sqrt{5}}\right) \times [H_x\hat{a}_x + H_y\hat{a}_y + H_z\hat{a}_z - (-3\hat{a}_x + 2\hat{a}_y - 4\hat{a}_z)] = 0$$

Taking the cross products and setting the individual components equal to zero,

$$2(H_y - 2)\hat{a}_z - 2(H_z + 4)\hat{a}_y + (H_x + 3)\hat{a}_z - (H_z + 4)\hat{a}_x = 0$$

$$\Rightarrow \begin{cases} -(H_z + 4) = 0 & \Rightarrow H_z = -4 \\ -2(H_z + 4) = 0 & \Rightarrow H_z = -4 \\ 2(H_y - 2) + (H_x + 3) = 0 & \Rightarrow H_x = 1 - 2H_y \end{cases}$$

The normal flux-density boundary condition is applied next:

$$\hat{a}_n \cdot (\vec{B}_2 - \vec{B}_1) = 0$$

$$\left(\frac{2\hat{a}_x - \hat{a}_y}{\sqrt{5}}\right) \cdot \begin{bmatrix} 50\mu_o H_x \hat{a}_x + 50\mu_o H_y \hat{a}_y + 50\mu_o H_z \hat{a}_z \\ -(-60\mu_o \hat{a}_x + 40\mu_o \hat{a}_y - 80\mu_o \hat{a}_z) \end{bmatrix} = 0$$

Taking the dot products and setting $H_x = 1 - 2H_y$,

$$2(50H_x + 60) - (50H_y - 40) = 100H_x + 160 - 50H_y = 100(1 - 2H_y) + 160 - 50H_y = 0$$

$$\Rightarrow H_y = \frac{26}{25}, \; H_x = 1 - \frac{52}{25} = -\frac{27}{25}$$

The desired fields are thus

$$\vec{H}_{50} = -\frac{27}{25}\hat{a}_x + \frac{26}{25}\hat{a}_y - 4\hat{a}_z, \quad \vec{B}_{50} = 50\mu_o \vec{H}_{50} = -54\mu_o \hat{a}_x + 52\mu_o \hat{a}_y - 200\mu_o \hat{a}_z$$

3.5 Flux Shunting Explained via Boundary Conditions

Using boundary conditions, explain how shields constructed of magnetic material "shunt" magnetic fields. [Zahn; Haus; Jackson]

In this discussion, the magnetic material will be assumed to have zero conductivity or the frequency of the magnetic field will be assumed sufficiently low so that the fields due to induced currents in the shield are negligible. Furthermore, the relative magnetic permeability of the shield, μ_s, will be assumed large compared to one, the free-space relative permeability surrounding the shield. In order to apply easily the boundary conditions, the shield will be considered flat. Two incident "rays" of magnetic field will be followed through and just past the shield slab. The following magnetic-field boundary conditions were previously discussed. The first boundary condition states that the normal component of the magnetic flux density, $\vec{B}$, must be continuous or equal across an interface:

$$B_{n1} = B_{n2} \tag{3.27}$$

The second boundary condition, assuming that no surface currents are present along the interface, states that the tangential components of the magnetic field, $\vec{H}$, must be continuous or equal across the interface:

$$H_{t1} = H_{t2} \tag{3.28}$$

The magnetic flux density and magnetic field in a permeable material are related through the permeability, $\vec{B} = \mu\vec{H} = \mu_r\mu_o\vec{H}$.

The slab of magnetic material is centered on the *xz* plane as shown in Figure 3.18. The source of the magnetic fields is to the left of the slab, $y < -d/2$. The region to the right of the slab, $y > d/2$, is to be shielded from the magnetic fields. The first "ray" of magnetic field to be analyzed is entirely perpendicular to the shield and is given as $\vec{H}_{e1} = H_e\hat{a}_y$. Using (3.27) at the left boundary, it immediately follows that

$$B_{ye} = B_{yi} \quad \Rightarrow \mu_o H_e = \mu_s \mu_o H_{yi} \quad \Rightarrow H_{yi} = \frac{H_e}{\mu_s}$$

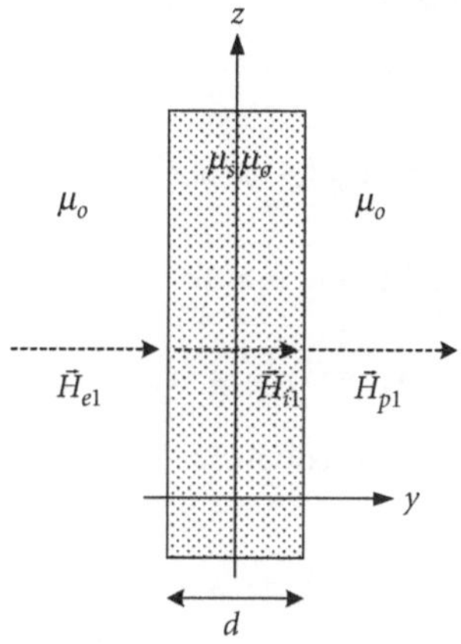

FIGURE 3.18 Magnetic field entirely normal to a slab of relative permeability μ_s.

Therefore, the magnetic field and flux density in the slab are

$$\vec{H}_{i1} = \frac{H_e}{\mu_s}\hat{a}_y, \quad \vec{B}_{i1} = \mu_s\mu_o\vec{H}_{i1} = \mu_o H_e \hat{a}_y \tag{3.29}$$

The magnetic field inside the slab is reduced by the magnetic nature of the slab. The direction of the field in the slab, however, is unchanged and is still perpendicular to the interface. Equation (3.27) can also be used to show that the magnetic field to the right of the slab is equal to $\vec{H}_{p1} = H_e\hat{a}_y$. For this perpendicular or transverse field, the slab provides no shielding.

Next, imagine that the "ray" of magnetic field to the left of the slab has both y and z components, $\vec{H}_{e2} = H_e\hat{a}_y + H_e\hat{a}_z$, as illustrated in Figure 3.19. Because the strength of both components is identical, the angle between this external field and left interface is 45°. Both the tangential and normal components

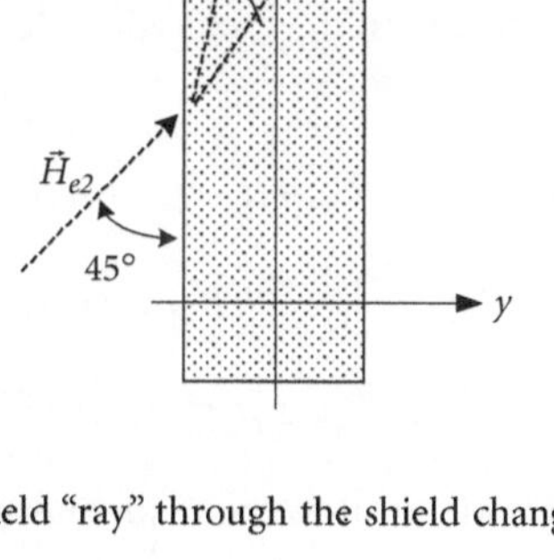

FIGURE 3.19 Direction of a magnetic field "ray" through the shield changes with shield permeability.

are required to determine the field inside the slab in this case. The application of the normal boundary condition is as before:

$$B_{ye} = B_{yi} \quad \Rightarrow \mu_o H_e = \mu_s \mu_o H_{yi} \quad \Rightarrow H_{yi} = \frac{H_e}{\mu_s}$$

The z component, obviously, is tangential to the interface. Applying (3.28),

$$H_{ze} = H_{zi} \quad \Rightarrow H_{zi} = H_e$$

Therefore, the magnetic field and magnetic flux density inside the slab are

$$\vec{H}_{i2} = \frac{H_e}{\mu_s}\hat{a}_y + H_e\hat{a}_z, \quad \vec{B}_{i2} = \mu_s\mu_o\vec{H}_{i2} = \mu_o H_e\hat{a}_y + \mu_s\mu_o H_e\hat{a}_z$$

If the relative permeability is very large (and the direction and strength of the external field remains the same), the field inside the slab is nearly entirely tangential to the interface:

$$\vec{H}_{i2} \approx H_e\hat{a}_z, \quad \vec{B}_{i2} \approx \mu_s\mu_o H_e\hat{a}_z \quad \text{if } \mu_s \gg 1$$

Two exiting locations for the field will be analyzed. One possible exiting location, assuming that the slab is sufficiently thick, is the top side of the slab. In this case, the normal component to the interface is now in the z direction and the tangent component is in the y direction. Applying the two boundary conditions to the unapproximated internal fields,

$$B_{zi} = B_{zp} \quad \Rightarrow \mu_s\mu_o H_e = \mu_o H_{zp} \quad \Rightarrow H_{zp} = \mu_s H_e$$

$$H_{yi} = H_{yp} \quad \Rightarrow H_{yp} = \frac{H_e}{\mu_s}$$

The field passing through the slab is therefore,

$$\vec{H}_{p2} = \frac{H_e}{\mu_s}\hat{a}_y + \mu_s H_e\hat{a}_z, \quad \vec{B}_{p2} = \mu_o\vec{H}_{p2} = \frac{\mu_o H_e}{\mu_s}\hat{a}_y + \mu_s\mu_o H_e\hat{a}_z$$

If the relative permeability is very large, the field exiting the slab is nearly entirely perpendicular to the slab:

$$\vec{H}_{p2} \approx \mu_s H_e\hat{a}_z, \quad \vec{B}_{p2} \approx \mu_s\mu_o H_e\hat{a}_z \quad \text{if } \mu_s \gg 1$$

In this case, the slab has shielded, diverted, or "shunted" the field to the top of the slab. The field exiting the slab, however, has been intensified by this "shunting." This slab has focused the magnetic field. Imagine that a continuous range of magnetic fields are incident to the slab over some entering area instead of one "ray" of the field. In this case, although the field exiting the slab into free space is stronger than that entering the slab, the exit area is smaller (and, thus, energy is conserved). Another possible exiting location, assuming that the slab is sufficiently thin, is the right side of the slab, as is also shown in Figure 3.19. Applying the boundary conditions as before,

$$B_{yi} = B_{yp} \quad \Rightarrow \mu_o H_e = \mu_o H_{yp} \quad \Rightarrow H_{yp} = H_e$$

$$H_{zi} = H_{zp} \quad \Rightarrow H_{zp} = H_e$$

$$\vec{H}_{p2} = H_e\hat{a}_y + H_e\hat{a}_z, \quad \vec{B}_{p2} = \mu_o\vec{H}_{p2} = \mu_o H_e\hat{a}_y + \mu_o H_e\hat{a}_z$$

The magnetic field and flux density are identical to the field entering the slab.

An external entering field such as $\vec{H}_{e2} = H_e\hat{a}_y + 10H_e\hat{a}_z$, which "hugs" the surface more closely, has a field in the slab that is even more tangential to the slab interface. As the relative permeability of the slab becomes very large, nearly all field components are "shunted" by this slab. Obviously, the relative permeability must be very large to "shunt" those field components that are nearly perpendicular to the slab. As either the relative permeability or thickness of the slab increases, more components of the field are diverted through the slab. In other words, the shielding effectiveness of the slab improves as the slab permeability and thickness increase. This redirection of the magnetic field increases the flux density in the slab, which can cause a real magnetic material to saturate at local "hot spots." For other shield shapes, such as spheres and cylinders, the principle is the same. Of course, a real magnetic field does not consist of a single "ray" of field, and the field direction can vary along the shield boundary.

The unperturbed field distribution (i.e., the field distribution without the slab present) will generally be affected by the addition of the magnetic shield. When a perfect conductor is placed in an electric field, the electric field is forced to be normal everywhere to the conductor's surface. Generally, the electric field distribution is modified or distorted by the conductor's presence. Analogously, when magnetic material with infinite permeability is placed in a magnetic field, the magnetic field external to the permeable material is forced to be normal to its surface. The magnetic field inside the permeable material, however, is tangential to the surface. Generally, the magnetic field distribution is modified or distorted by the addition of the magnetic material. Surfaces of high-permeability materials can be viewed as the magnetic equivalent of electric equipotentials.

3.6 Self Shielding Nature of Coax

Why is coaxial cable self shielding? Determine the general expressions for the magnetic fields everywhere in and around a coaxial cable assuming the currents in the two conductors are equal in magnitude but opposite in sign and uniformly distributed. Then, determine the magnetic fields assuming two different, more general current densities for both conductors. [Perry, '85; Boast; Anderson, '85]

An ideal coaxial cable (or coax) consists of two concentric cylindrical conductors. As will be shown, when the total current in the inner conductor is equal but opposite in sign to the total current in the outer conductor, the magnetic field outside of the outer conductor is zero. Thus, *when* the currents on the two conductors are balanced, the magnetic field emissions from the coax are ideally zero. In addition to performing the normal transmission line function of carrying the signal current, this outer conductor also acts as a shield: the outer conductor surrounds the inner conductor. Interestingly, as is discussed in this section, the outer and inner surfaces of the outer conductor almost appear decoupled from each other at high frequencies.

Fortunately, because the two conductors of a coaxial cable are both cylindrical and concentric, Ampère's law,

$$\oint_C \vec{H} \cdot d\vec{L} = \iint_S \vec{J} \cdot d\vec{s} = I_{encl} \tag{3.30}$$

where I_{encl} is the total current enclosed by the closed contour over the surface S can be used to determine easily the magnetic field everywhere. The length of the conductors, l_{th}, is assumed large compared to the outer radius of the outer conductor so that end effects can be neglected. The magnetic field expressions to be derived are valid away from the ends.

The axis of the two concentric conductors will be assumed along the z axis in the cylindrical coordinate system. The direction of increasing z is out of the page in Figure 3.20. In this first example, the currents in both the inner and outer conductors are assumed uniform: the current density does not vary with ρ, the radial position from the center (or with ϕ or z). Current uniformity is reasonable at low frequencies where the radius of the inner conductor and thickness of the outer conductor are less than the skin depth. Free space is assumed both between the two conductors and outside the outer conductor, but any lossless dielectric can be inserted between and outside them without affecting the magnetic field expressions.

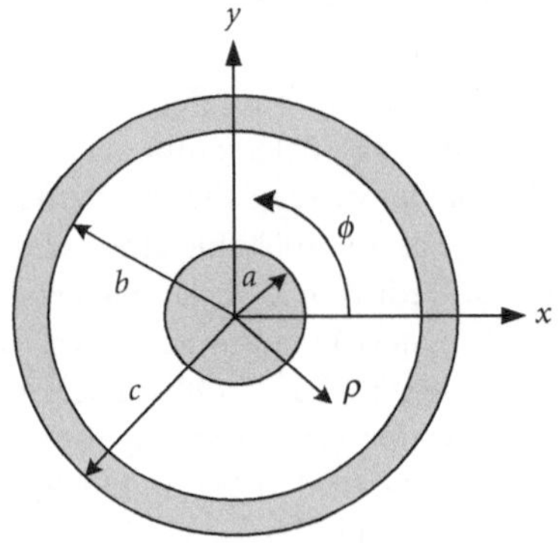

FIGURE 3.20 Two long, concentric cylindrical conductors centered about the z axis.

The total current in the inner conductor is I in the $+z$ direction, and the total current in the outer conductor is I in the $-z$ direction (or $-I$ in the $+z$ direction). The circular closed contour, C, when inside the inner conductor is shown in Figure 3.21. According to (3.30), this contour must be a closed contour: the contour must stop where it began. When this contour is at $\rho = a$, the total current enclosed by this contour is I. When this contour is at $\rho < a$, the total current enclosed by this contour is less than I. The direction of the magnetic field is as shown for current out of the page. The magnetic field from the current-carrying inner conductor is in the ϕ direction. The differential surface vector, $d\vec{s}$, is in the $+z$ direction if the direction of the closed contour integration is in the counterclockwise direction. (Using the right hand, the thumb is pointed in the direction $d\vec{s}$ when the remaining fingers are curled in the direction of C.) Therefore, using (3.30)

$$\int_0^{2\pi} \underbrace{H_\phi \hat{a}_\phi}_{\vec{H}} \cdot \underbrace{\rho d\phi \hat{a}_\phi}_{d\vec{L}} = \int_0^{\rho}\int_0^{2\pi} \underbrace{\frac{I}{\pi a^2}\hat{a}_z}_{\vec{J}} \cdot \underbrace{\rho d\phi d\rho \hat{a}_z}_{d\vec{s}} \quad \text{if } 0 \le \rho \le a$$

Since the current is uniformly distributed, the current density, $\vec{J}$, was easily obtained by dividing the total current, I, by the surface area of the inner conductor, πa^2. The right side of the previous expression is the total current contained within the closed contour located at ρ. Evaluating both sides of this expression,

$$\int_0^{2\pi} H_\phi \rho d\phi = 2\pi\rho H_\phi = \int_0^{\rho}\int_0^{2\pi} \frac{I}{\pi a^2}\rho d\phi d\rho = \frac{2I}{a^2}\int_0^{\rho}\rho d\rho = \frac{I\rho^2}{a^2}$$
$$\Rightarrow H_\phi = \frac{I\rho}{2\pi a^2} \quad \text{if } 0 \le \rho \le a \tag{3.31}$$

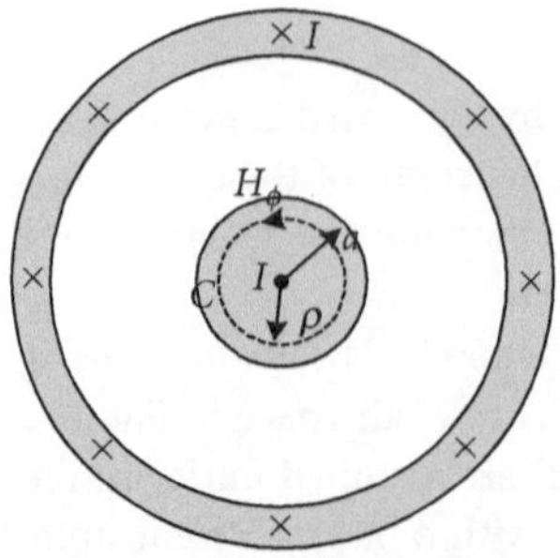

FIGURE 3.21 Total current in the inner conductor is in the opposite direction to the total current in the outer conductor. The contour, C, is shown inside the inner conductor.

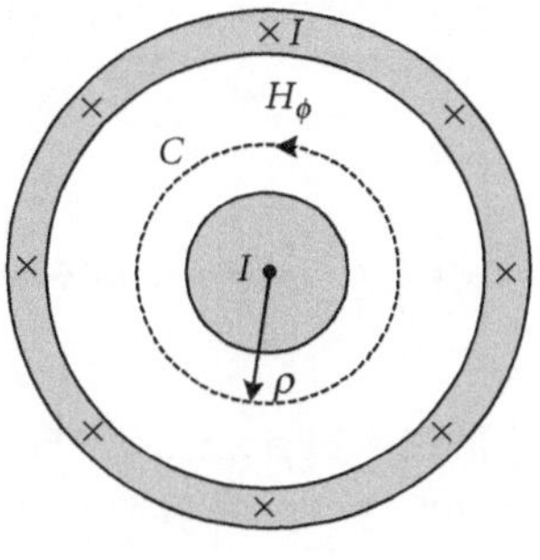

FIGURE 3.22 Contour C between the inner and outer conductor. Between the two conductors, the enclosed current by this contour is equal to I for all ρ.

At the center, $\rho = 0$, the magnetic field is zero since the total current enclosed is zero. As another check,

$$H_\phi(\rho = a) = \frac{I}{2\pi a}$$

which is the standard expression for the magnetic field at the surface of a current-carrying wire. For the region between the two conductors, the procedure is the same. A circular contour is shown in Figure 3.22. Between the conductors, in free space, the current density is zero and the magnetic field can be determined from

$$\int_0^{2\pi} H_\phi \hat{a}_\phi \cdot \rho d\phi \hat{a}_\phi = \int_0^a \int_0^{2\pi} \frac{I}{\pi a^2} \hat{a}_z \cdot \rho d\phi d\rho \hat{a}_z \quad \text{if } a \le \rho \le b$$

Carefully note that the upper limit of integration for ρ is a. The total current enclosed by the contour remains equal to I for this region, independent of ρ. Evaluating these integrals,

$$2\pi\rho H_\phi = I \quad \Rightarrow H_\phi = \frac{I}{2\pi\rho} \quad \text{if } a \le \rho \le b \tag{3.32}$$

which is the standard expression for the magnetic field outside of a very long current-carrying conductor. When the closed integration path, C, is inside the outer conductor as shown in Figure 3.23, the total

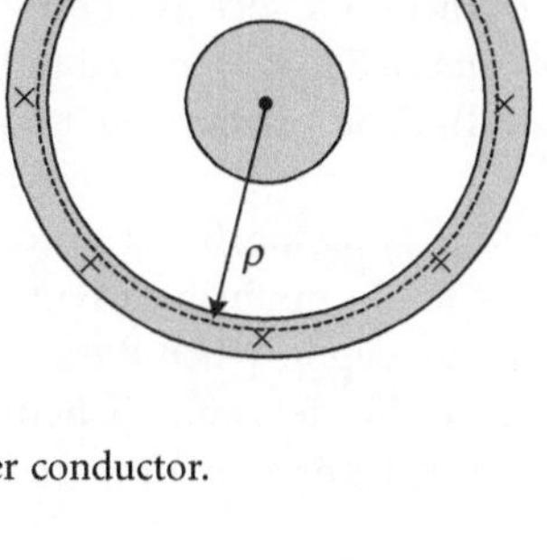

FIGURE 3.23 Contour C inside the outer conductor.

enclosed current is equal to the sum of I and the current enclosed between the contour and inner radius of the outer conductor:

$$\int_0^{2\pi} H_\phi \hat{a}_\phi \cdot \rho d\phi \hat{a}_\phi = I - \int_b^{\rho}\int_0^{2\pi} \frac{I}{\pi(c^2 - b^2)} \hat{a}_z \cdot \rho d\phi d\rho \hat{a}_z \quad \text{if } b \le \rho \le c$$

Again, carefully note the limits of integration for the ρ variable. The current density in the outer conductor is uniform and equal to I divided by the cross-sectional area of the outer conductor, $\pi(c^2 - b^2)$. Solving for the magnetic field,

$$2\pi\rho H_\phi = I - \frac{I(\rho^2 - b^2)}{c^2 - b^2} \quad \Rightarrow H_\phi = \frac{I(c^2 - \rho^2)}{2\pi(c^2 - b^2)\rho} \quad \text{if } b \le \rho \le c \tag{3.33}$$

Two simple checks can be performed on this result. First, the magnetic field must be continuous everywhere since there are no surface currents present and the fields are entirely tangential to the interfaces:

$$H_\phi(\rho = b) = \frac{I}{2\pi b}$$

Equation (3.32) evaluated at $\rho = b$ is equal to this result. Second, the magnetic field must be zero at the outer surface of the outer conductor since the total current enclosed is zero. Using (3.33), the field is indeed zero at the outer surface:

$$H_\phi(\rho = c) = 0$$

Obviously when the contour is anywhere outside the outer conductor, the total current enclosed is zero:

$$\int_0^{2\pi} H_\phi \hat{a}_\phi \cdot \rho d\phi \hat{a}_\phi = I - I = 0 \quad \Rightarrow H_\phi = 0 \quad \text{if } c \le \rho \tag{3.34}$$

This zero field is the important, zero (or low) magnetic-field emissions characteristic of coaxial cables. For two concentric cylindrical conductors carrying equal but oppositely directed currents, the total current enclosed by a contour surrounding the outside of the cable is zero. Thus, the magnetic field outside an ideal coax is zero without the need for costly shielding material. The magnetic field everywhere for a 75 Ω coax is shown in Mathcad 3.2.

When the return current in the outer conductor is not equal to the signal current in the inner conductor, the magnetic field is not zero outside the coax. At lower frequencies, the remaining current may actually be returning to the source through some other route such as a nearby ground plane. If the outer conductor of a coaxial cable were accidentally disconnected from the load or source, then the return current in the outer conductor would be zero (or small). (The current could, for example, return through a ground plane.)

Although a uniform current distribution is reasonable at lower frequencies, at higher frequencies it may not be. Frequently, to determine whether the uniform current assumption is reasonable for the inner conductor, its radius is compared to the skin depth. When the skin depth is greater than the radius, the uniform current assumption is reasonable. To determine whether the uniform current assumption is reasonable for the outer conductor, its thickness is compared to the skin depth in the same way.

$a := 0.0025 \qquad b := 0.0087 \qquad c := b + 0.001 \qquad I := 2$

$\rho := 0, \frac{a}{10} .. 1.1 \cdot c$

$$H_\phi(\rho) := \text{if}\left[\rho < a, \frac{I\cdot\rho}{2\cdot\pi\cdot a^2}, \text{if}\left[\rho < b, \frac{I}{2\cdot\pi\cdot\rho}, \text{if}\left[\rho < c, I\cdot\frac{(c^2-\rho^2)}{2\cdot\pi\cdot(c^2-b^2)\cdot\rho}, 0\right]\right]\right]$$

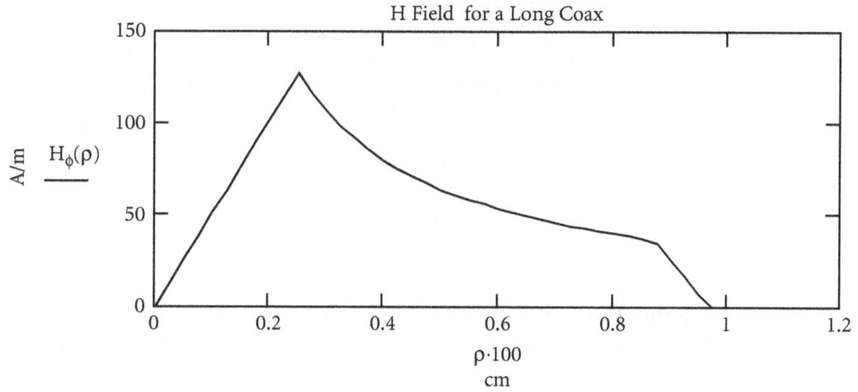

MATHCAD 3.2 Magnetic field vs. radial distance for a coaxial cable. The magnetic field is the greatest at the outer surface of the inner conductor. The magnetic field is continuous since no surface currents are present.

A nonuniform current distribution[7] inside the inner conductor will affect the magnetic field distribution inside the inner conductor, and a nonuniform current distribution inside the outer conductor will affect the magnetic field distribution in the outer conductor. However, the magnetic field outside the outer conductor and between the conductors are not affected. This assumes that the total current within the two conductors has not changed, just the current's distribution.

In this next example, the current densities in the inner and outer conductors are $\vec{J}_i(\rho) = J_i(\rho)\hat{a}_z$ and $\vec{J}_o(\rho) = J_o(\rho)\hat{a}_z$, respectively. These current densities are only a function of ρ so that the magnetic fields are still entirely in the ϕ direction and Ampère's law can still be easily applied. Since the actual variation of the current density with ρ is not given, the final magnetic field expressions involve integrals. For each of the four regions, the expressions are

$$\int_0^{2\pi} H_\phi \rho d\phi = 2\pi\rho H_\phi = \int_0^{\rho}\int_0^{2\pi} J_i(\rho)\rho d\phi d\rho \quad \Rightarrow H_\phi = \frac{\int_0^{\rho} \rho J_i(\rho) d\rho}{\rho} \quad \text{if } 0 \le \rho \le a \tag{3.35}$$

$$\int_0^{2\pi} H_\phi \rho d\phi = \int_0^{a}\int_0^{2\pi} J_i(\rho)\rho d\phi d\rho \quad \Rightarrow H_\phi = \frac{\int_0^{a} \rho J_i(\rho) d\rho}{\rho} \quad \text{if } a \le \rho \le b \tag{3.36}$$

[7]A current distribution that varies with ρ.

$$\int_0^{2\pi} H_\phi \rho d\phi = \int_0^a \int_0^{2\pi} J_i(\rho)\rho d\phi d\rho + \int_b^\rho \int_0^{2\pi} J_o(\rho)\rho d\phi d\rho \quad \text{if } b \le \rho \le c$$

$$\Rightarrow H_\phi = \frac{\int_0^a \rho J_i(\rho)d\rho + \int_b^\rho \rho J_o(\rho)d\rho}{\rho} \quad \text{if } b \le \rho \le c \tag{3.37}$$

$$\int_0^{2\pi} H_\phi \rho d\phi = \int_0^a \int_0^{2\pi} J_i(\rho)\rho d\phi d\rho + \int_b^c \int_0^{2\pi} J_o(\rho)\rho d\phi d\rho \quad \text{if } c \le \rho$$

$$\Rightarrow H_\phi = \frac{\int_0^a \rho J_i(\rho)d\rho + \int_b^c \rho J_o(\rho)d\rho}{\rho} \quad \text{if } c \le \rho \tag{3.38}$$

The magnetic field outside the coax is not necessarily zero unless the total current enclosed is zero. In terms of the current densities, the total current in the inner and outer conductors, respectively, are

$$I_i = \int_0^a \int_0^{2\pi} J_i(\rho)\rho d\phi d\rho, \quad I_o = \int_b^c \int_0^{2\pi} J_o(\rho)\rho d\phi d\rho \tag{3.39}$$

The total enclosed current for a closed contour outside the cable is $I_{encl} = I_i + I_o$.

Before other situations are studied, the external high-frequency inductance for a coaxial cable will be (painlessly) determined so that it can be compared to a later result. Equations (3.31) through (3.34) were derived for low frequencies where the currents are uniformly distributed over both conductors. As previously stated, the magnetic field in the space between the conductors is not affected by current distributions that vary with ρ. Thus, the high-frequency or external inductance, where the magnetic flux inside the conductors is neglected, is easily obtained:

$$L_e = \frac{\Phi}{I} = \frac{\iint_S \vec{B}\cdot d\vec{s}}{I} = \frac{\int_0^{l_{th}} \int_a^b \frac{\mu_o I}{2\pi\rho}\hat{a}_\phi \cdot d\rho dz \hat{a}_\phi}{I} = \frac{l_{th}\frac{\mu_o I}{2\pi}\int_a^b \frac{d\rho}{\rho}}{I} = 2\times 10^{-7} l_{th} \ln\left(\frac{b}{a}\right) \tag{3.40}$$

where l_{th} is the length of the coax, which is much greater than c. Equation (3.32) was used to determine the flux between the conductors. Equation (3.40) is referred to as the high-frequency inductance because at high frequencies the current is mostly along the outer periphery of the inner conductor and inner periphery of the outer conductor. The currents and fields are fairly small after a few skin depths into the conductors. At high frequencies, the skin depth is small and the flux contribution from fields inside the conductors to the total flux is negligible.

Since it seems that engineering students are always hungry for sugar (i.e., examples) even though protein (i.e., theory) will sustain them over a much longer period of time, a few additional examples involving current-carrying concentric conductors will be presented. In Figure 3.24, four concentric conductors carrying uniform currents are shown. Free space exists from $0 < \rho < \rho_0$ and $\rho_3 < \rho < \rho_4$. The center conductor actually consists of three conductors with insulation between the layers. Therefore, conduction current cannot pass between the center conductors once the current travels down the cable. Each conductor carries in the z direction a (potentially) different current density of J_1, J_2, J_3, and J_5, respectively, from the innermost to outermost conductor. The magnetic field outside this multiconductor

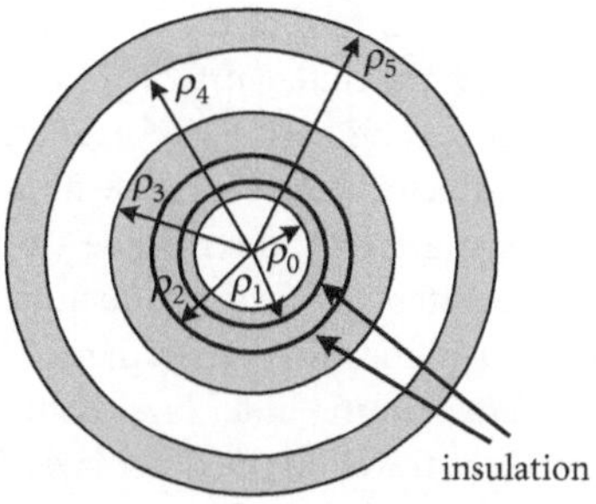

FIGURE 3.24 Multiconductor coaxial cable with a hollow center. The insulation is assumed thin compared to the thicknesses of the inner conductors.

cable is zero if the total current (not current density) in the cable is zero. Without explanation, the magnetic fields in terms of these current densities and the dimensions are

$$H_\phi = 0 \quad \text{if } 0 \le \rho \le \rho_0$$

$$H_\phi = \frac{\int_{\rho_0}^{\rho} \rho J_1(\rho)d\rho}{\rho} \quad \text{if } \rho_0 \le \rho \le \rho_1$$

$$H_\phi = \frac{\int_{\rho_0}^{\rho_1} \rho J_1(\rho)d\rho + \int_{\rho_1}^{\rho} \rho J_2(\rho)d\rho}{\rho} \quad \text{if } \rho_1 \le \rho \le \rho_2$$

$$H_\phi = \frac{\int_{\rho_0}^{\rho_1} \rho J_1(\rho)d\rho + \int_{\rho_1}^{\rho_2} \rho J_2(\rho)d\rho + \int_{\rho_2}^{\rho} \rho J_3(\rho)d\rho}{\rho} \quad \text{if } \rho_2 \le \rho \le \rho_3$$

$$H_\phi = \frac{\int_{\rho_0}^{\rho_1} \rho J_1(\rho)d\rho + \int_{\rho_1}^{\rho_2} \rho J_2(\rho)d\rho + \int_{\rho_2}^{\rho_3} \rho J_3(\rho)d\rho}{\rho} \quad \text{if } \rho_3 \le \rho \le \rho_4$$

$$H_\phi = \frac{\int_{\rho_0}^{\rho_1} \rho J_1(\rho)d\rho + \int_{\rho_1}^{\rho_2} \rho J_2(\rho)d\rho + \int_{\rho_2}^{\rho_3} \rho J_3(\rho)d\rho + \int_{\rho_4}^{\rho} \rho J_5(\rho)d\rho}{\rho} \quad \text{if } \rho_4 \le \rho \le \rho_5$$

$$H_\phi = \frac{\int_{\rho_0}^{\rho_1} \rho J_1(\rho)d\rho + \int_{\rho_1}^{\rho_2} \rho J_2(\rho)d\rho + \int_{\rho_2}^{\rho_3} \rho J_3(\rho)d\rho + \int_{\rho_4}^{\rho_5} \rho J_5(\rho)d\rho}{\rho} \quad \text{if } \rho_5 \le \rho$$

In high-Q circuits, Litz wire can be used at some frequencies to reduce ohmic losses. The ohmic losses and ac resistance of conductors increase with frequency due to the skin (and proximity) effect: the effective area that the current travels through the conductor decreases as the frequency increases. The insulated set of layered inner conductors shown in Figure 3.24 is an attempt to increase the effective surface area relative to a single, solid center conductor of radius ρ_3. Instead of the current just traveling along one outer surface of a single solid conductor, it can travel along the outer surfaces of three insulated conductors. (Of course, more than three insulated conductors can also be used.) Unfortunately, though, if a common voltage is applied to each of the inner insulated conductors, the current will not divide equally among them. The impedance of each conductor is not the same even if the thickness of each conductor is adjusted so that they have identical dc resistances. The ac resistances of the conductors are affected by the proximity effect. In addition, each conductor has a different inductance. Referring to the inductance table, the inductance of the outer layer conductor is typically less than the inner layer conductors. As a result, the current, even though the conductors are insulated, tends to travel on the outer layer. The impedance of each layer can be adjusted by inserting inductances in series with the conductors so as to equalize the impedances (and, of course, the ohmic losses of these inductances must be considered). Another solution, probably more reasonable, is to transpose the conductors at "cross-over" joints. For a two-conductor version containing only one "cross-over" joint, the outer layer conductor of the first half of the cable is connected to the inner layer conductor of the second half of the cable, and the inner layer conductor of the first half of the cable is connected to the outer layer conductor of the second half of the cable. This helps improve the symmetry of these inner conductors relative to the outermost return conductor similar to the transposing done with Litz wire. By transposing the inner conductors, the inductance (and capacitance) of these conductors relative to the outermost return conductor will be more similar.

The final example involving three concentric conductors is shown in Figure 3.25. The currents in the innermost and outermost conductors are both equal to $I/2$.[8] The current in the intermediate conductor has a magnitude of I and is in the opposite direction to the current in the other two conductors. The frequency is sufficiently low so that the currents can be assumed to be uniformly distributed. Without explanation, the magnetic fields are

$$H_\phi = \frac{\displaystyle\int_0^\rho \rho\left(\frac{-\frac{I}{2}}{\pi\rho_1^2}\right)d\rho}{\rho} = -\frac{I\rho}{4\pi\rho_1^2} \quad \text{if } 0 \le \rho \le \rho_1$$

$$H_\phi = \frac{\displaystyle\int_0^{\rho_1} \rho\left(\frac{-\frac{I}{2}}{\pi\rho_1^2}\right)d\rho}{\rho} = -\frac{I}{4\pi\rho} \quad \text{if } \rho_1 \le \rho \le \rho_2$$

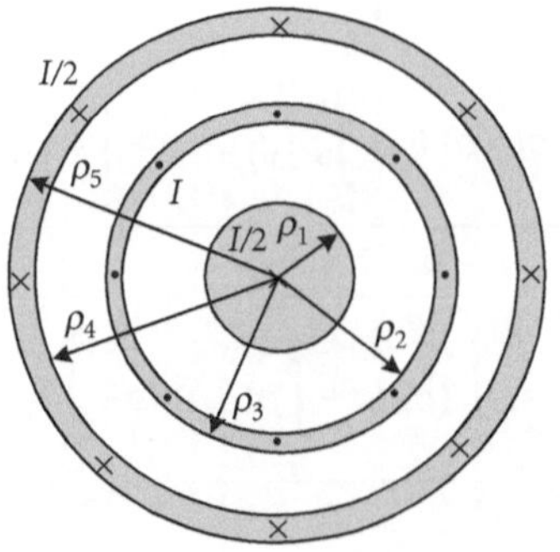

FIGURE 3.25 Three-conductor coaxial cable (sometimes referred to as a Triax).

[8]At low frequencies, this implies that the cross-sectional areas of the inner and outer conductors are the same.

$$H_\phi = \frac{\displaystyle\int_0^{\rho_1} \rho\left(\frac{-\frac{I}{2}}{\pi\rho_1^2}\right)d\rho + \int_{\rho_2}^{\rho} \rho\left[\frac{I}{\pi\left(\rho_3^2-\rho_2^2\right)}\right]d\rho}{\rho} = -\frac{I}{4\pi\rho} + \frac{I\left(\rho^2-\rho_2^2\right)}{2\pi\left(\rho_3^2-\rho_2^2\right)\rho} \quad \text{if } \rho_2 \le \rho \le \rho_3$$

$$H_\phi = \frac{\displaystyle\int_0^{\rho_1} \rho\left(\frac{-\frac{I}{2}}{\pi\rho_1^2}\right)d\rho + \int_{\rho_2}^{\rho_3} \rho\left[\frac{I}{\pi\left(\rho_3^2-\rho_2^2\right)}\right]d\rho}{\rho} = \frac{I}{4\pi\rho} \quad \text{if } \rho_3 \le \rho \le \rho_4$$

$$H_\phi = \frac{\displaystyle\int_0^{\rho_1} \rho\left(\frac{-\frac{I}{2}}{\pi\rho_1^2}\right)d\rho + \int_{\rho_2}^{\rho_3} \rho\left[\frac{I}{\pi\left(\rho_3^2-\rho_2^2\right)}\right]d\rho + \int_{\rho_4}^{\rho} \rho\left[\frac{-\frac{I}{2}}{\pi\left(\rho_5^2-\rho_4^2\right)}\right]d\rho}{\rho} \quad \text{if } \rho_4 \le \rho \le \rho_5$$

$$= \frac{I}{4\pi\rho} + \frac{-I\left(\rho^2-\rho_4^2\right)}{4\pi\left(\rho_5^2-\rho_4^2\right)\rho} \quad \text{if } \rho_4 \le \rho \le \rho_5$$

$$H_\phi = \frac{\displaystyle\int_0^{\rho_1} \rho\left(\frac{-\frac{I}{2}}{\pi\rho_1^2}\right)d\rho + \int_{\rho_2}^{\rho_3} \rho\left[\frac{I}{\pi\left(\rho_3^2-\rho_2^2\right)}\right]d\rho + \int_{\rho_4}^{\rho_5} \rho\left[\frac{-\frac{I}{2}}{\pi\left(\rho_5^2-\rho_4^2\right)}\right]d\rho}{\rho} = 0 \quad \text{if } \rho_5 \le \rho$$

It is interesting to determine the high-frequency inductance for this three-conductor cable. Although at high frequencies the current will be mainly along the surface of the three conductors (both surfaces for the intermediate conductor), this will not affect the magnetic fields between the conductors in free space. Because there are three conductors involved, and the magnitude of the current in the intermediate conductor is different from the inner and outer conductors, defining the inductance of this Triax is slightly more complicated. One method of defining the inductance is through the concept of differential-flux linkage. Briefly, the flux from $\rho_1 \le \rho \le \rho_2$ is multiplied by a factor of $-1/2$ since it is due to $-I/2$ from the inner conductor. The flux from $\rho_3 \le \rho \le \rho_4$ is multiplied by 1/2 since it is due to $I-(I/2)=I/2$ from both the inner and intermediate conductors. Thus, the inductance due to the flux outside of the conductors (i.e., the external inductance) is

$$L_e = \frac{\Phi}{I} = \frac{\displaystyle\iint_S \vec{B}\cdot d\vec{s}}{I} = \frac{\displaystyle\int_0^{l_{th}}\int_{\rho_1}^{\rho_2}\left(-\frac{1}{2}\right)\left(-\frac{\mu_o I}{4\pi\rho}\right)\hat{a}_\phi\cdot d\rho dz\hat{a}_\phi + \int_0^{l_{th}}\int_{\rho_3}^{\rho_4}\left(\frac{1}{2}\right)\frac{\mu_o I}{4\pi\rho}\hat{a}_\phi\cdot d\rho dz\hat{a}_\phi}{I}$$

$$= \frac{\mu_o l_{th}}{8\pi}\left[\ln\left(\frac{\rho_2}{\rho_1}\right)+\ln\left(\frac{\rho_4}{\rho_3}\right)\right] = \frac{10^{-7}l_{th}}{2}\left[\ln\left(\frac{\rho_2}{\rho_1}\right)+\ln\left(\frac{\rho_4}{\rho_3}\right)\right] = \frac{10^{-7}l_{th}}{2}\ln\left(\frac{\rho_2}{\rho_1}\times\frac{\rho_4}{\rho_3}\right)$$

When the thickness of the intermediate conductor is small, the inductance simplifies to

$$L_e \approx \frac{10^{-7}l_{th}}{2}\ln\left(\frac{\rho_4}{\rho_1}\right) \quad \text{if } \rho_2 \approx \rho_3$$

This inductance is one-fourth of the inductance of the standard two-conductor coax given in (3.40) (if $\rho_1 = a$ and $\rho_4 = b$). However, the capacitance between the intermediate conductor and the inner and outer conductors has increased (relative to the two-conductor coax). Since $Z_o = \sqrt{L/C}$ at high frequencies, the high-frequency characteristic impedance is potentially smaller for the three-conductor coax than a similar size two-conductor coax.

It should not always be assumed that the current distribution at higher frequencies is only along one surface of a conductor. Ideally for a two-conductor coax at higher frequencies, the current is along the outer surface of the inner conductor and inner surface of the outer conductor. However, frequently, the standard two-conductor coaxial cable is referred to as a three-wire line at higher frequencies. Current can also exist along the outer surface of the outer conductor. When connecting a coaxial cable directly to a dipole antenna, for example, significant currents can be induced from the antenna along the outer surface of the outer conductor of the cable. An external source (at a different frequency) can also induce currents along the outer surface of a coax. Even when the outer conductor is solid, the currents along the inner and outer surface of the outer conductor are isolated from each other when the thickness of the outer conductor is large compared to the skin depth (at the lowest frequency of interest). The current distributions are shown for a coaxial cable in Figure 3.26. For copper, the skin depth at 1 MHz is

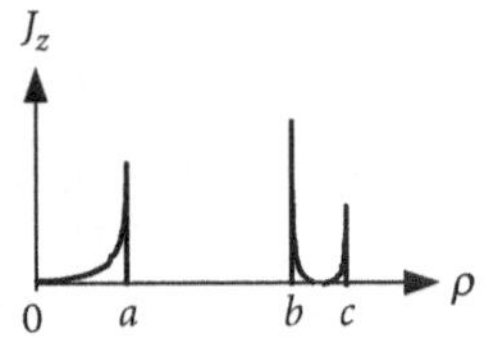

FIGURE 3.26 Surface current-like distributions along all three surfaces of a coaxial cable. The variables are defined in Figure 3.20.

$$\delta = \frac{1}{\sqrt{\pi f \mu_o \sigma}} = \frac{1}{\sqrt{\pi(10^6)(4\pi \times 10^{-7})(5.8\times 10^7)}} \approx 66\ \mu\text{m}$$

When modeling this three-wire effect, the standard transmission line model can be used for the currents along the outer surface of the inner conductor and inner surface of the outer conductor. Then, a third conductor of suitable electrical properties is added to represent the outside of the line.

The magnetic fields from concentric cylindrical conductors can be easily determined because the conductors have a common axis and because the current does not vary with the angle ϕ. An eccentric inner conductor complicates the analysis. If the symmetry is disturbed by shifting the inner conductor off center for a two-conductor coax, then it is more difficult to determine the magnetic field everywhere. Even though the enclosed current is still zero for a closed contour around the outer conductor, this does not imply that the magnetic field is zero everywhere outside this cable. If the current distribution is assumed uniform for both the outer conductor and shifted inner conductor, then a magnetic field will exist outside the coax. The proximity and skin effect will both affect the current distribution, further complicating the analysis.

3.7 Method of Images for Currents

Provide a brief table of images for currents near conducting and magnetic interfaces.

The method of images is sometimes used to determine the effect conducting and magnetic objects have on the magnetic field distribution from currents. With the method of images, conducting or magnetic objects are replaced with an image current or currents. As with charges and their images, the boundary conditions for the magnetic field and magnetic flux density must be satisfied along the surfaces of these conducting or magnetic objects. When the boundaries involve perfect conductors or materials of infinite permeability with zero conductivity, the boundary conditions are simple and can sometimes be easily qualitatively checked.

Even though the time-varying magnetic (and electric) field inside a perfect conductor is zero, surface current (and charge) can exist along the conductor's surface. Time-varying magnetic fields external to a conducting object can induce these currents along the surface. Unlike electric fields, which must be

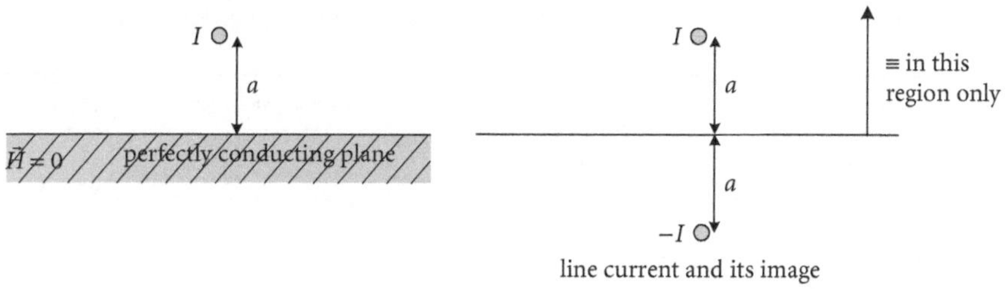

FIGURE 3.27 Line current above a perfectly conducting plane and its image current equivalent (without the ground plane).

entirely normal to a perfectly conducting surface, a tangential magnetic field can exist along a perfectly conducting surface. The normal component of the magnetic field, though, must be zero along a perfect conductor. The tangential components of the magnetic field can be discontinuous across a boundary containing surface current ($H_{t1} - H_{t2} = K$). The normal component of the magnetic field must be zero since the normal component of the magnetic flux density must be continuous across the boundary ($B_{n1} = B_{n2}$), and the time-varying fields are zero inside a perfect conductor.

In the right illustration in Figure 3.27, the image equivalent of a current-carrying conductor above a perfect ground plane is shown. As with electric images, the conducting plane does not need to be infinitely thick to use the method of images. (However, the general conditions required for the conducting plane to be perfect-like to magnetic fields are not the same as with electric fields.) One component of the magnetic field from both the original and image currents is shown in Figure 3.28. (Positive I is out of the page.) The normal components cancel along the line corresponding to the metal interface.

With magnetic objects where the permeability is infinite, $\vec{H}$ inside the object must be zero if the magnetic flux density, $\vec{B}$, is to remain finite since $\vec{B} = \mu\vec{H}$. However, assuming the conductivity of the magnetic object is not infinite, no surface current is present along the interface. Since $H_{t1} = H_{t2}$ along a magnetic object, when $\mu \to \infty$, the tangential component of $\vec{H}$ must be zero. The external magnetic field is therefore entirely normal to a magnetic material with infinite permeability. For the configuration shown in Figure 3.29, the image current is of the same magnitude and direction as the real current, I, for large values of μ_2. Along the interface, the tangential components of the magnetic field from the original current source and image current source cancel.

With the method of images, the location and strength of the image line current can be a function of the distance of the current from the surface or interface. Also, a single "real" current can generate more than one image current. To expand the size of Table 3.2, superposition can be applied to generate other

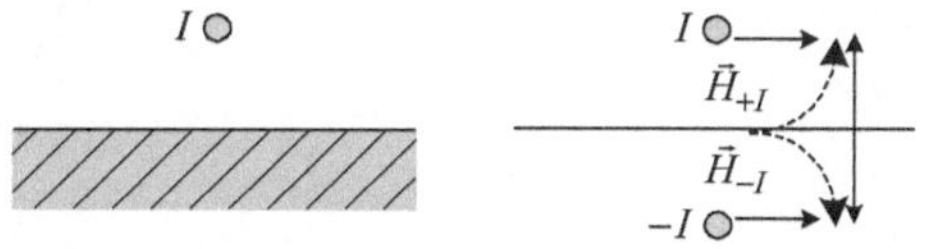

FIGURE 3.28 Normal components of the magnetic field from both the original real current and image current are equal but opposite along the interface corresponding to the surface of the perfect conductor.

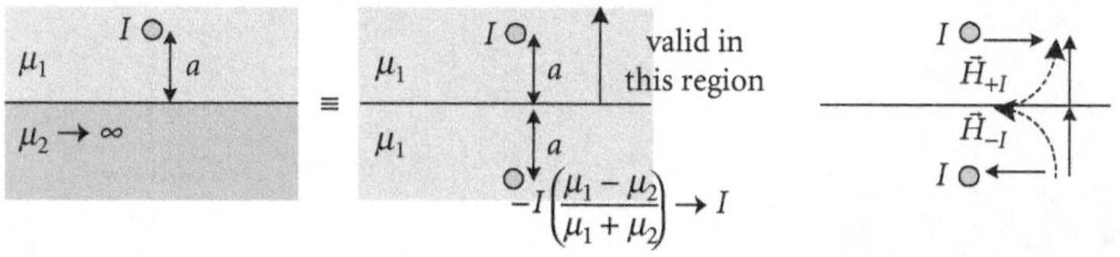

FIGURE 3.29 When the permeability of the lower region is very large, the image current approaches I.

TABLE 3.2 Line Current Near a Conductor or Magnetic Interface and its Equivalent Image Without the Conductor or Magnetic Interface

<table>
<tr><th>Actual Configuration</th><th>Equivalent Image Configuration</th></tr>
<tr><td>Horizontal line current above a grounded conducting plane [Zahn; Paul, ’92(b)]
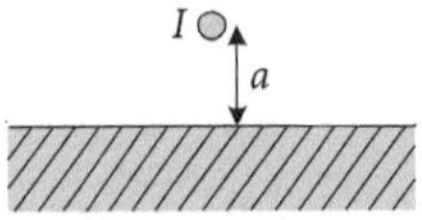</td><td>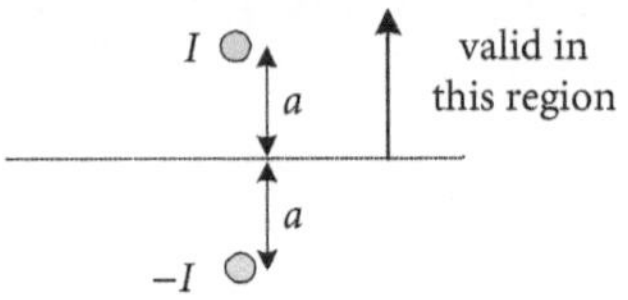</td></tr>
<tr><td>Vertical line current above a grounded conducting plane [Paul, ’92(b); Haus]
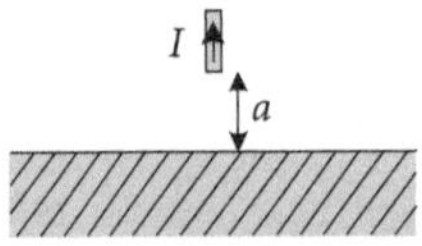</td><td>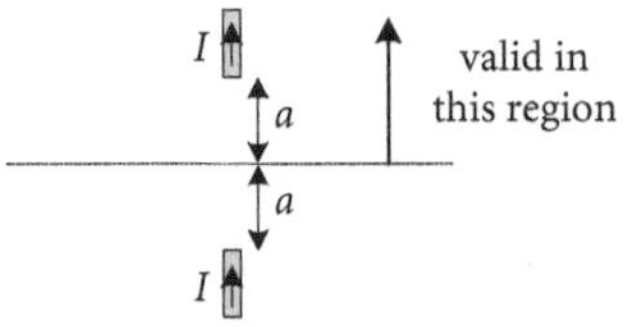</td></tr>
<tr><td>Line current above a magnetic plane [Weber; Jefimenko]
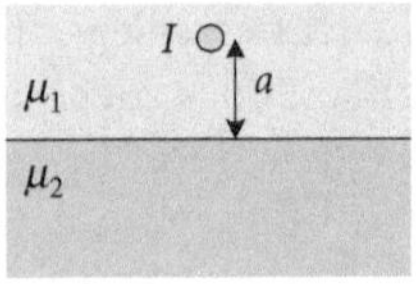</td><td>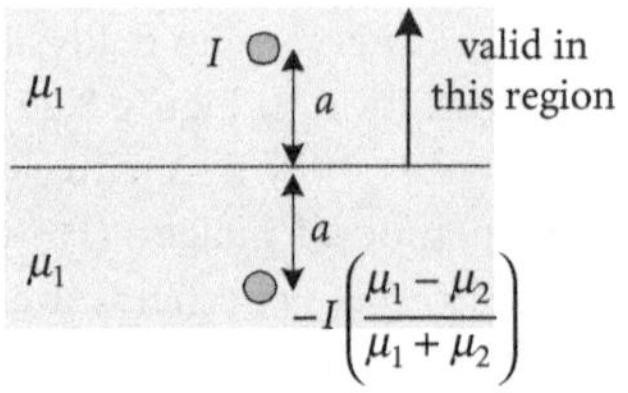</td></tr>
<tr><td>Line current above a magnetic plane [Weber; Jefimenko]
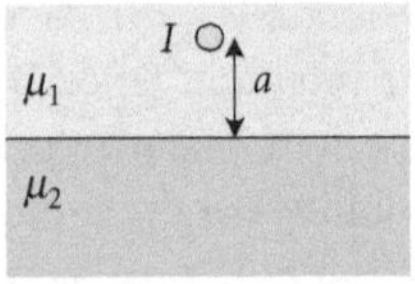</td><td>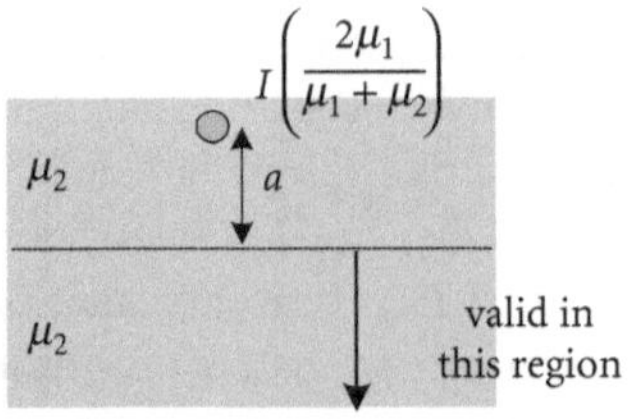</td></tr>
<tr><td>Line current between two perpendicular, grounded conducting planes [Lee, ’84(b); Wolff, ’67]
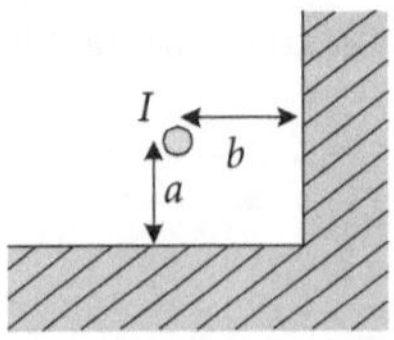</td><td>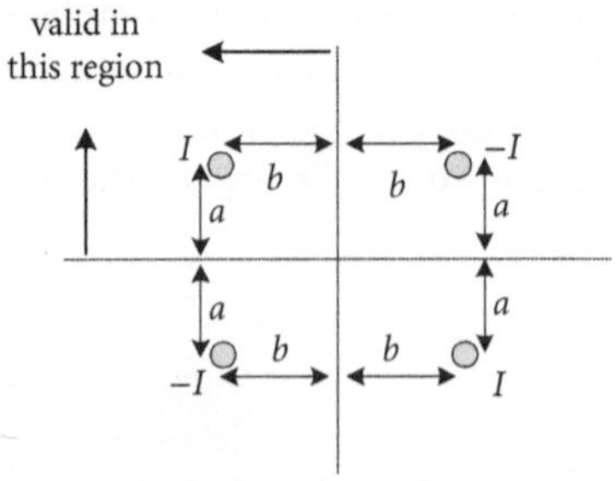</td></tr>
<tr><td>Line current between two grounded conducting planes at 45° [Lee, ’84(b); Wolff, ’67]
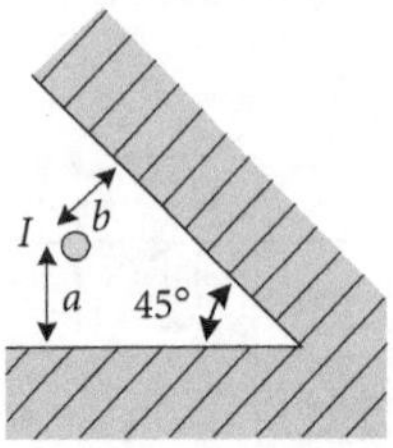</td><td>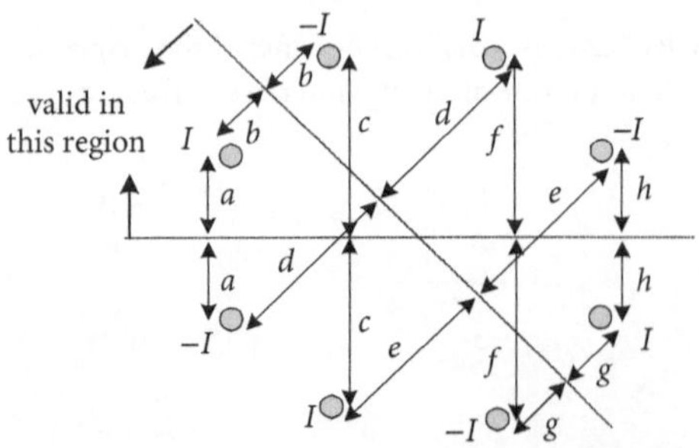</td></tr>
</table>

TABLE 3.2 Line Current Near a Conductor or Magnetic Interface and its Equivalent Image Without the Conductor or Magnetic Interface (Continued)

<table>
<tr><th>Actual Configuration</th><th>Equivalent Image Configuration</th></tr>
<tr><td>Line current between two grounded conducting planes at an angle $\beta = 180°/N$ [Lee, '84(b); Wolff, '67]
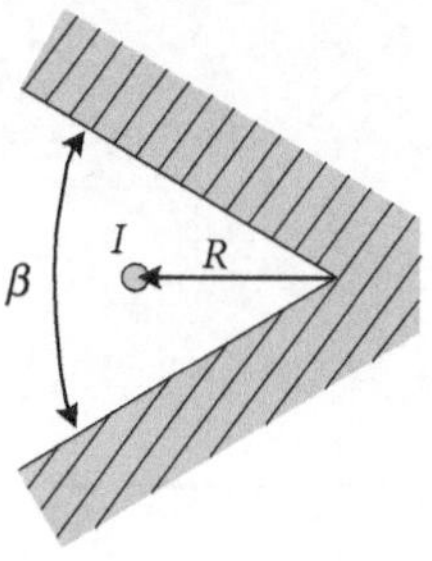
where N is an integer</td><td>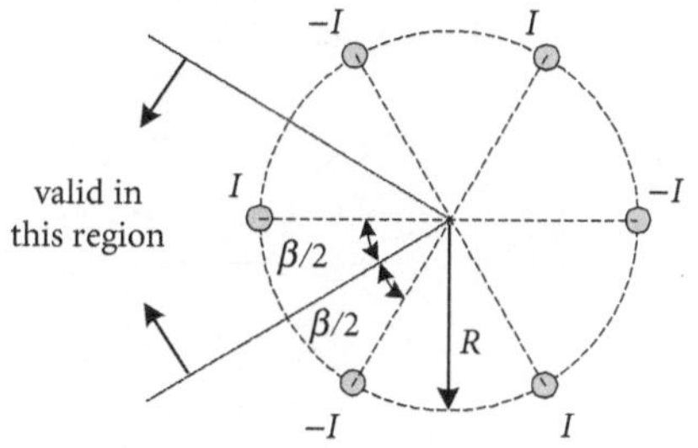
a total of $(2N-1)$ alternating sign images equally spaced along the circle of radius R</td></tr>
<tr><td>Line current outside a grounded, thin conducting cylinder [Paul, '92(b)]
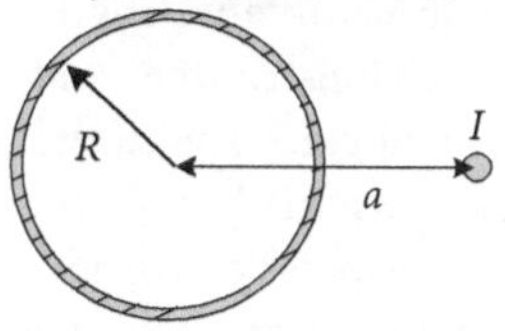</td><td>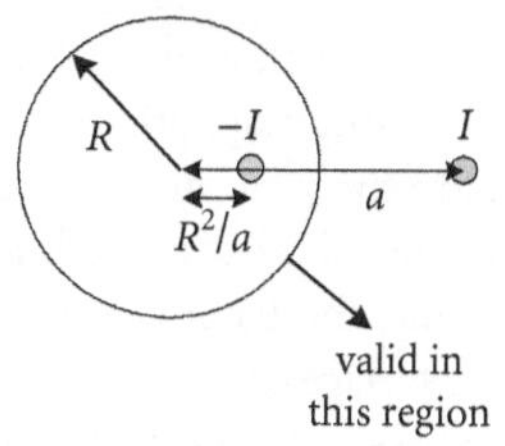</td></tr>
<tr><td>Line current inside a grounded, thin conducting cylinder [Paul, '92(b)]
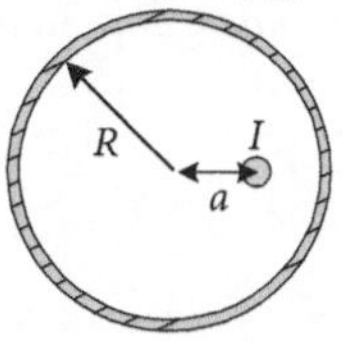</td><td>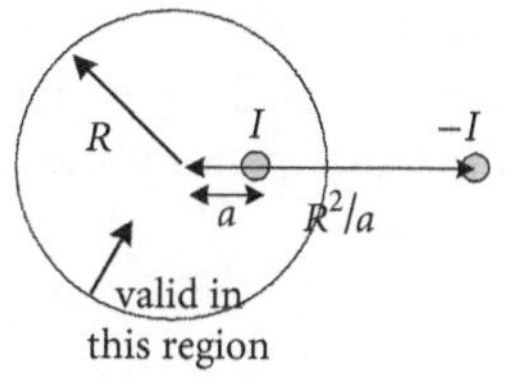</td></tr>
<tr><td>Line current outside a magnetic cylinder [Jefimenko; Weber]
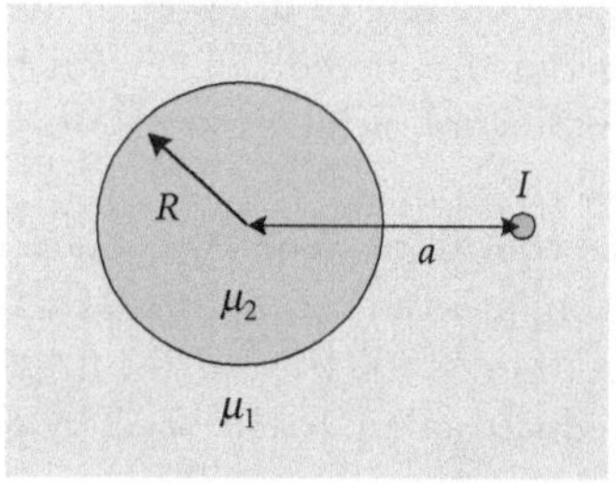</td><td>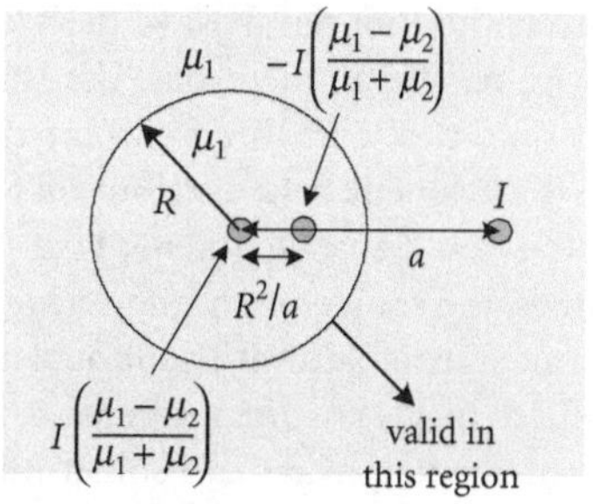</td></tr>
</table>

(Continued)

TABLE 3.2 Line Current Near a Conductor or Magnetic Interface and its Equivalent Image Without the Conductor or Magnetic Interface (Continued)

Actual Configuration	Equivalent Image Configuration
Line current outside a magnetic cylinder [Jefimenko; Weber] 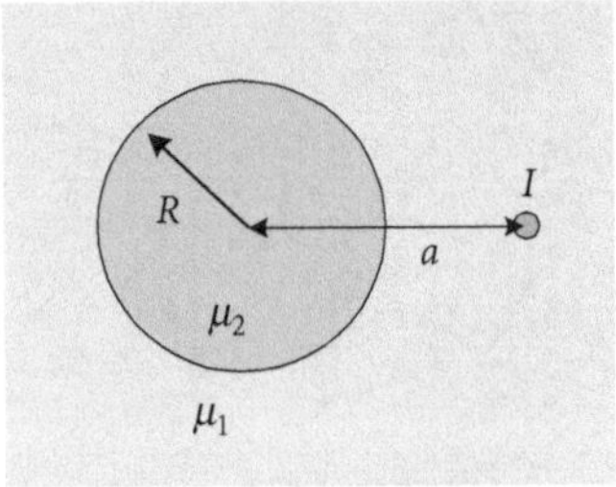	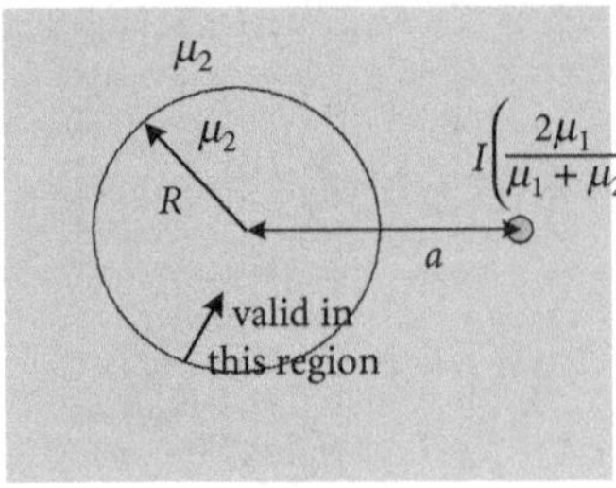

current distributions such as a current trace or current loop above a ground plane. In this table, the flat conducting planes and magnetic interfaces are theoretically infinite in width and length, and the cylindrical surfaces are theoretically infinite in length. However, the expressions can be used with excellent results if the distance between the line distribution and surface is small compared to width and length of the plane(s) and the length of the cylinder. The line currents are assumed long compared to their distance to the interface(s) and end effects are neglected. Although the expressions dealing with conductors were derived assuming perfect conductors, the results are quite good if they are just good conductors. Also, it is not necessary that the conducting planes be infinite in thickness. The method of images for currents can also be applied to thin good conductors (such as a twin-lead line surrounded by a thin circular conductor). Finally, for those configurations involving a *grounded* conductor, the grounded conductor is the return path for the line current.

3.8 Wire Partners Can Reduce Fields

Starting from the Biot-Savart law, derive the expression for the magnetic field from a long current-carrying wire parallel to the *z* axis. Use this expression to compare the rate of drop off of the magnetic field from one, two, and three current-carrying wires. When can the method of images be used to determine the magnetic field distribution for wires near other conductors, including earth ground? [Riley; Rauch; Zaffanella; Haus; Olsen, '83; Perz; Pettersson; Papakanellos]

Sometimes, the source of an interfering magnetic field can be controlled. When the source cannot be reduced to an acceptable level, traditional magnetic shielding techniques can be employed. Unfortunately, magnetic field shields are often expensive, heavy, large, and nonlinear. As will be seen, the magnetic fields from current-carrying wires can sometimes be inexpensively reduced to a tolerable level by placing the signal and return in close proximity.

The general expression for the magnetic field from a very long line current will be derived starting from the Biot-Savart law, (3.11):

$$\vec{H}(r) = \int_{L'} \frac{\vec{I}'dL' \times \hat{a}_{r-r'}}{4\pi|\vec{r} - \vec{r}'|^2}$$

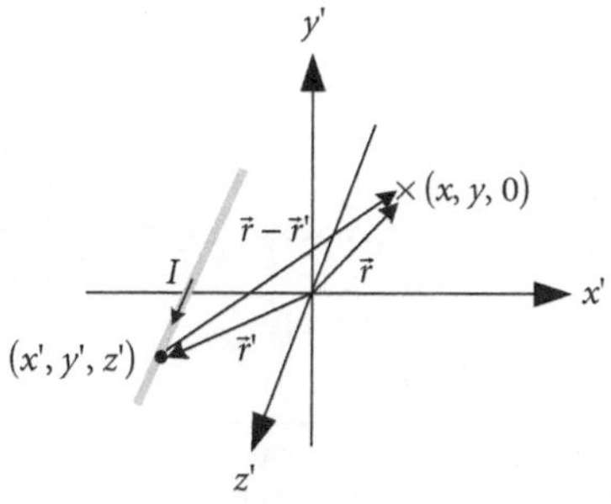

FIGURE 3.30 Magnetic field from an infinitely long line current parallel to the z axis is determined at (x, y) in the $z = 0$ plane.

Referring to Figure 3.30, the magnetic field at a point $(x, y, 0)$ from a single, infinitely long line current parallel to the z axis located at (x', y') is obtained by first setting up the integral

$$\vec{H}(x,y,0) = \int_{-\infty}^{\infty} \frac{I\hat{a}_z dz' \times \left[\dfrac{(x-x')\hat{a}_x + (y-y')\hat{a}_y + (0-z')\hat{a}_z}{\sqrt{(x-x')^2+(y-y')^2+(0-z')^2}}\right]}{4\pi\left[\sqrt{(x-x')^2+(y-y')^2+(0-z')^2}\right]^2}$$

$$= \int_{-\infty}^{\infty} \frac{I\hat{a}_z dz' \times [(x-x')\hat{a}_x + (y-y')\hat{a}_y - z'\hat{a}_z]}{4\pi[(x-x')^2+(y-y')^2+(z')^2]^{\frac{3}{2}}}$$

$$= \frac{I[(x-x')\hat{a}_y - (y-y')\hat{a}_x]}{4\pi} \int_{-\infty}^{\infty} \frac{dz'}{[(x-x')^2+(y-y')^2+(z')^2]^{\frac{3}{2}}}$$

and then evaluating it:

$$\vec{H}(x,y,0) = \frac{I}{2\pi}\left[\frac{-(y-y')\hat{a}_x}{(x-x')^2+(y-y')^2} + \frac{(x-x')\hat{a}_y}{(x-x')^2+(y-y')^2}\right] \tag{3.41}$$

where I is the magnitude of the current in the $+z$ direction.[9] The magnetic field from any number of very long parallel line currents (of different amplitudes and phases) located at different positions can be determined using this result and superposition. Of course, the line currents must not be too close, relative to their radius, so that they do not affect each other's current distributions.

In Figure 3.31, one very long current-carrying wire and two very long wires carrying equal but opposite directed currents (also referred to as a differential pair) are shown. In Mathcad 3.3, (3.41) is used to determine the magnetic field along the positive x axis for both of these configurations assuming

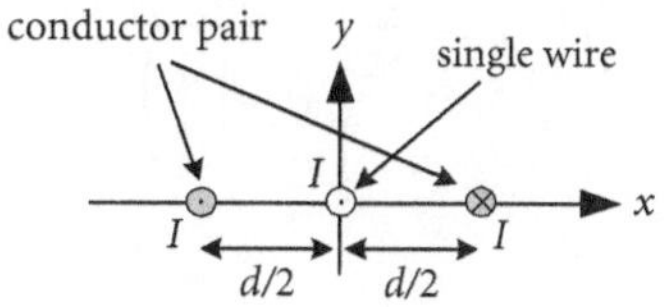

FIGURE 3.31 Single current-carrying wire along the z axis and two current-carrying wires centered about the z axis.

[9] This expression is in agreement with the relevant equations given in Table 3.1.

$$d := 4\cdot 1.03\cdot 10^{-3} \qquad K := 4\cdot\pi\cdot 10^{-7}\cdot 10^{7}$$

$$r := 4\cdot d,\ 4\cdot d + \frac{d}{100} .. 50\cdot d$$

$$H_x(x, y, x_p, y_p, I) := \frac{I}{2\cdot\pi}\cdot\left[\frac{-(y-y_p)}{(x-x_p)^2+(y-y_p)^2}\right] \qquad H_y(x, y, x_p, y_p, I) := \frac{I}{2\cdot\pi}\cdot\left[\frac{(x-x_p)}{(x-x_p)^2+(y-y_p)^2}\right]$$

$$B_s(x, y) := K\cdot\sqrt{H_x(x, y, 0, 0, 10)^2 + H_y(x, y, 0, 0, 10)^2}$$

$$B_d(x, y, d) := K\cdot\sqrt{\left(H_x\left(x, y, \frac{-d}{2}, 0, 10\right) + H_x\left(x, y, \frac{d}{2}, 0, -10\right)\right)^2 \ldots + \left(H_y\left(x, y, \frac{-d}{2}, 0, 10\right) + H_y\left(x, y, \frac{d}{2}, 0, -10\right)\right)^2}$$

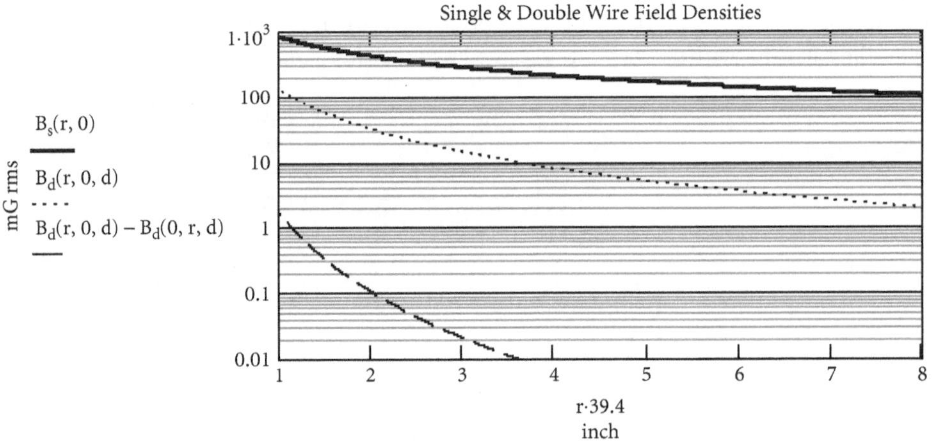

K converts a magnetic field in free space to a magnetic flux density in milligauss.

MATHCAD 3.3 Magnetic field vs. distance from the center of one current-carrying wire and two current-carrying wires with equal but opposite currents.

$I = 10$ A.[10] As is clearly seen, the magnetic field from the single wire is much greater than the magnetic field from the conductor pair. Also, the rate of drop off with distance is greater for the two wires. (Of course, if the currents were in the same direction for the conductor pair, the magnetic field would be greater than for the single wire.) For the differential pair, the difference between the fields along the x and y axis is also shown. Referring to Table 3.1, the magnetic field magnitudes on the side of the pair ($y = 0$) and directly above the pair ($x = 0$) for parallel conductors carrying equal but opposite currents with a magnitude of I are given by

$$|H_{side}| = \frac{I}{2\pi}\left[\frac{d}{x^2 - \left(\frac{d}{2}\right)^2}\right] \tag{3.42}$$

$$|H_{top}| = \frac{I}{2\pi}\left[\frac{d}{\left(\frac{d}{2}\right)^2 + y^2}\right] \tag{3.43}$$

[10]Although G or gauss is a common unit, it is not an SI unit.

Far from the two wires, these two field magnitudes are the same:

$$|H_{side}| \approx \frac{I}{2\pi}\left(\frac{d}{r^2}\right),\ |H_{top}| \approx \frac{I}{2\pi}\left(\frac{d}{r^2}\right) \quad \text{if } r = |x| = |y| \gg \frac{d}{2}$$

Although not shown in Mathcad 3.3, in all directions the magnitude of the field drops off at a rate of $1/r^2$ when distant from the wires for the differential pair. This should be compared to the drop off rate of $1/r$ for a single current-carrying wire. The magnitude of the field on the side of the pair is greater than on the top of the pair:

$$\frac{|H_{side}|}{|H_{top}|} = \frac{\dfrac{I}{2\pi}\left[\dfrac{d}{x^2 - \left(\dfrac{d}{2}\right)^2}\right]}{\dfrac{I}{2\pi}\left[\dfrac{d}{\left(\dfrac{d}{2}\right)^2 + y^2}\right]} = \frac{r^2 + \left(\dfrac{d}{2}\right)^2}{r^2 - \left(\dfrac{d}{2}\right)^2} > 1 \quad \text{if } r = |x| = |y| \tag{3.44}$$

This ratio is the greatest near the wires. Equation (3.44) should not be used too near the wires since this result is based on line currents, which have zero radius. At a distance of $r = 4d$, this ratio is nearly one:

$$\frac{|H_{side}|}{|H_{side}|} = \frac{16d^2 + \left(\dfrac{d}{2}\right)^2}{16d^2 - \left(\dfrac{d}{2}\right)^2} \approx 1.03 \quad \text{if } r = 4d$$

To understand why the fields drop off more rapidly for the differential pair, first study the sketch of the magnetic field around a single wire shown in Figure 3.32. For two parallel wires carrying currents of equal magnitude but opposite direction, superposition can be applied. For the conductor pair shown in Figure 3.31, a counter-clockwise field distribution is around the left conductor and a clockwise field distribution is around the right conductor. (This assumes that the conductors are sufficiently far apart so that the currents can be assumed constant and unaffected by their neighbor.) The magnetic fields from each of the two wires at one location along the y axis and x axis are shown in Figure 3.33. Simple vector analysis is used to determine the net field at each of these locations. As is clearly illustrated along the x axis and outside the wires, the fields are oppositely directed (but not of the same amplitude). The net current enclosed by a closed contour around these two wires is zero. Although this does not imply zero magnetic field around the wires, it does imply that the magnetic fields from the individual current distributions will be oppositely directed in some regions.

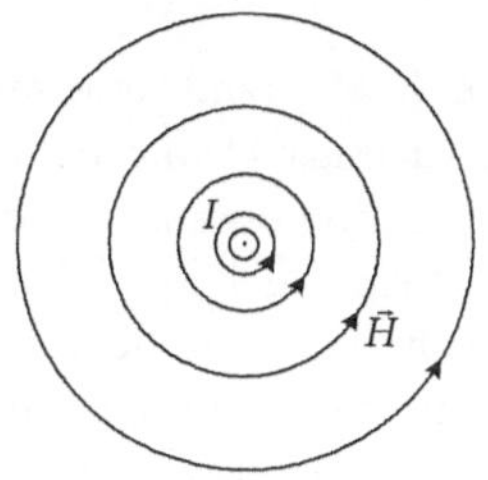

FIGURE 3.32 Circumferential magnetic field for a long current-carrying wire. The field drops off at a rate of 1/ r. (The current is out of the page.)

The net magnetic field outside the differential pair (at a given distance) decreases as the spacing, d, decreases. As the spacing approaches zero, the magnetic field approaches zero. (Although the wires cannot exist in the same space.) Using the expression for the magnetic field given in Table 3.1 for oppositely directed currents, as the spacing between the conductors approaches zero,

$$\vec{H}(x, y) \to 0 \quad \text{as } d \to 0 \tag{3.45}$$

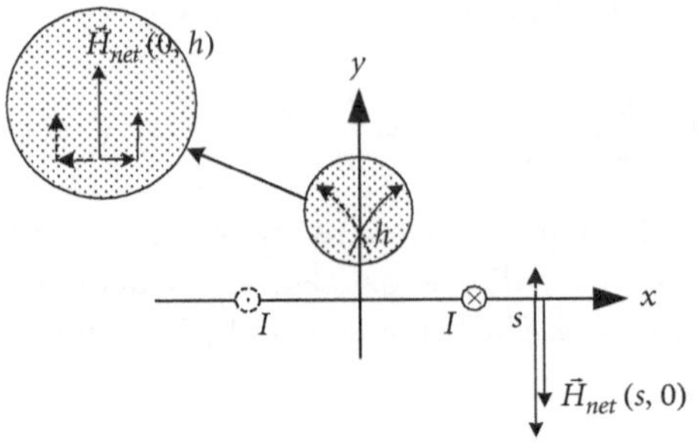

FIGURE 3.33 Using superposition and simple vector analysis, the magnetic field anywhere can be determined for two (or more) current-carrying wires.

Sometimes in the electrical wiring for power distribution, conductors are incorrectly wired resulting in larger-than-expected magnetic fields. Generally, if the "hot" wire and the neutral wire carrying the return current are separated, then the fields can drop off at a rate of $1/r$ rather than $1/r^2$. This can sometimes occur in the wiring of three-way switches when the "hot" and neutral are separated. It can also occur when using two different "hots" with a common neutral where one of the "hots" is not wired near the common neutral (e.g., not in the same cable or conduit). Connecting different grounds or grounding a single neutral at multiple locations can also generate strong magnetic fields. For example, metallic water pipes that may be touched by individuals (e.g., via a metal faucet) are required to be connected to the ground at the service panel. Where the water pipes are shared by many users, such as with a common metallic water main, the grounds of different users are indirectly connected. The grounds at the various service panels, which are connected to the pole/transformer grounds, are not perfect and not identical. Based on the potential between these different grounds, current can pass sneakily between users through the water lines. Since the ground is not a very effective magnetic field shield at low frequencies, the current-carrying water-pipe conductor can generate magnetic fields that drop off at a rate of $1/r$. Unexpected currents can also exist on other conductors connected to the ground system such as gas, phone, and cable TV lines. In some cases, insulated couplings can be used to reduce this surreptitious current.

In some older homes (and some museum villages), knob-and-tube wiring can be seen. The "hot" and neutral may be separated by one foot or more and could be attached to adjacent studs or rafters.[11] Because of this large separation distance, d, for a given distance from one of the conductors, the magnetic field is much greater than closely spaced conductors. As seen in Figure 3.33, along the side of the conductors, the strength of the magnetic field from the nearest conductor is much greater (and of opposite sign) than the magnetic field from the other conductor. This difference in strength is less dramatic when d is small (compared to the distance to the point of interest). If Δ is the distance to the nearest conductor, then (3.42) can be easily written in terms of this distance:

$$|H_{side}| = \frac{I}{2\pi}\left[\frac{d}{\left(\Delta+\frac{d}{2}\right)^2-\left(\frac{d}{2}\right)^2}\right] = \frac{I}{2\pi\Delta}\left(\frac{1}{1+\frac{\Delta}{d}}\right) \Rightarrow |H_{side}| \approx \begin{cases} \dfrac{Id}{2\pi\Delta^2} & \text{if } d \ll \Delta \\ \dfrac{I}{2\pi\Delta} & \text{if } d \gg \Delta \end{cases}$$

When $d \gg \Delta$ or the point of interest is close compared to the conductor separation, the magnetic field varies as a single current-carrying conductor. However, when $d \ll \Delta$ or the point of interest is far compared to the conductor separation, the field varies as a differential pair of current-carrying conductors.

[11]The conductors are periodically insulated from the studs (as well as supported and guided) by porcelain, doorknob shaped insulators. Inside the studs, the wire is passed through a porcelain *tube*.

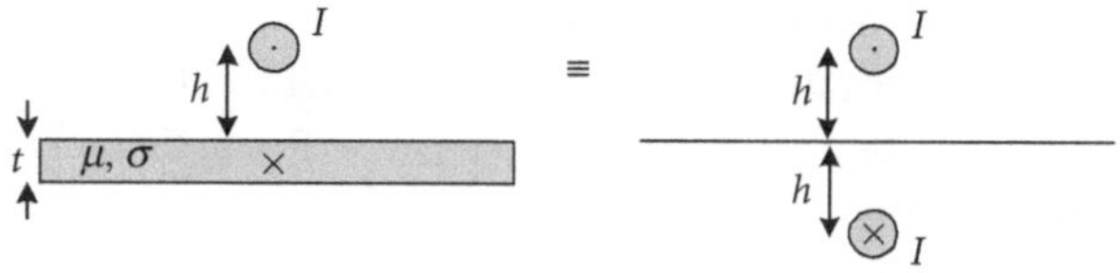

FIGURE 3.34 When the conducting plane can be considered perfect, it can be replaced with an image current.

The advantage of the differential pair in reducing the strength of the magnetic field can also be realized by using a good-conducting plane near a single conducting wire.[12] The ground plane is the return path for the current. When the plane can be considered perfect, the method of images can be used to replace the plane with an image current as shown in Figure 3.34. At dc, a nonmagnetic plane is transparent to magnetic fields, and the method of images cannot be (simply?) applied. (At low frequencies, the nonmagnetic plane is essentially transparent.) For magnetoquasistatic conditions and where h is large compared to the thickness t, the plane conductor can be considered perfect or good when (3.150) is satisfied:

$$\omega\mu\sigma l_1 l_2 = \omega\mu\sigma h t \gg 1 \tag{3.46}$$

When this condition is satisfied, the method of images can be applied. This does not imply that the current is entirely along the surface since the current will diffuse into the metal. However, the counter currents induced in the plane will be sufficiently large so that the metal will act like a perfect conductor. Then, the method of images can be applied (and the conductor will provide good shielding of ac magnetic fields). For a frequency of 60 Hz and a copper conducting plane,

$$(2\pi\times 60)(4\pi\times 10^{-7})(5.8\times 10^{7})ht \approx 2.7\times 10^{4} ht \gg 1$$

When this condition is satisfied, the magnetic fields normal to the plane conductor are small (as expected for very good conductors). If the distance h is equal to 5 cm, then t should be much greater than about 0.7 mm to satisfy this relationship. Although it is generally not necessary that the thickness of the plane conductor be greater than the skin depth to be considered a good conductor, for copper at 60 Hz $\delta = 1/\sqrt{\pi f \mu\sigma} \approx 8.5$ mm. For wires above the earth's surface, the analysis is much more complex. The conductivity of the earth is extremely poor compared to copper, and it is a function of the weather, season, and geology. Also, since the conductivity is not constant with depth, standard image theory does not apply. For a long conductor above a lossless earth, the electric field is normal to the conductor and earth. For a lossy earth, an electric field component also exists in the direction of the conductor. If the distance between the current-carrying conductor and earth is electrically small, then integral field expressions are available for a current-carrying wire above a large plane with uniform electrical parameters. Correction factors and "complex" images have been used to assist in the analysis.

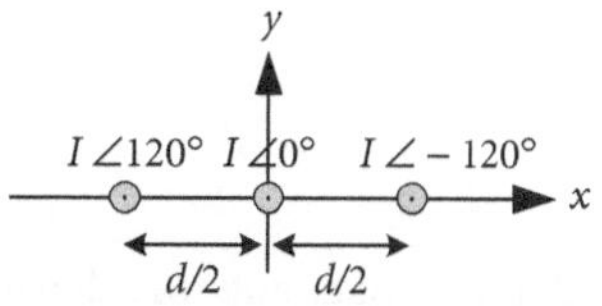

FIGURE 3.35 Three collinear conductors with currents 120° out phase.

The previous analysis of one current-carrying conductor and two parallel current-carrying conductors was relatively simple. In electric power transmission, two conductors carrying equal magnitude but oppositely directed currents are often referred to as a single-phase line. When there are three or more conductors, many possible physical layouts for the conductors exist. In addition, there are a number of possible combinations of current magnitudes and phases for these conductors. When other conducting objects, including the earth, are not close, (3.41) can be used with superposition to determine the field from any number of conductors. One example should be sufficient to illustrate the method. The three conductors shown in Figure 3.35

[12]The method of images can also be applied to other configurations such as a wire inside and parallel to a circular good-conducting cylinder.

are a three-phase system. The magnitudes of the current in each conductor are identical. Since the currents are out of phase, the magnetic fields from each of the current sources will also be out of phase. Sinusoidal steady-state conditions will be assumed and phasors will be used. Recall that a phasor with an amplitude of A at a phase angle θ can be written as

$$A\angle\theta = Ae^{j\theta} = A\cos\theta + jA\sin\theta \quad \text{where } j = \sqrt{-1}$$

As with the differential pair, the net current for this three-phase system is zero:

$$I\angle 0° + I\angle 120° + I\angle -120° = I + I[\cos(120°) + j\sin(120°)] + I[\cos(-120°) + j\sin(-120°)] = 0 \qquad (3.47)$$

The total x component of the magnetic field at some location (x, y) is the sum or superposition of the x component fields from each of the three current sources. Similarly, the total y component is the sum of the y components from each of the sources. When working with phasors in the frequency domain, both components of the total magnetic field will obviously be a function of their spatial location. Also, because of the phase difference between the currents, the field components will possess both an amplitude and a phase. In the frequency domain,

$$\vec{H}_s(x, y) = B(x, y)\angle\theta(x, y)\hat{a}_x + C(x, y)\angle\phi(x, y)\hat{a}_y$$

The time-domain magnetic field is given by

$$\vec{H}(x, y, t) = B(x, y)\cos[\omega t + \theta(x, y)]\hat{a}_x + C(x, y)\cos[\omega t + \phi(x, y)]\hat{a}_y$$

The magnitude of this vector is

$$|\vec{H}(x, y, t)| = \sqrt{\{B(x, y)\cos[\omega t + \theta(x, y)]\}^2 + \{C(x, y)\cos[\omega t + \phi(x, y)]\}^2} \qquad (3.48)$$

The upper limit on the magnitude for all time is

$$|\vec{H}(x, y, t)|_{\max} \le \sqrt{B^2(x, y) + C^2(x, y)} \qquad (3.49)$$

In Mathcad 3.4, the magnetic flux density is plotted as a function of distance from the origin for both the differential pair and the upper limit of the flux density for the three-conductor system shown in Figure 3.35. In Mathcad 3.5, the amplitudes of the x and y components are plotted for both the single and three-phase systems at a fixed radial distance $4d$ around the origin. The magnetic field from the three collinear conductors can actually exceed the field from the differential pair by changing one or more phases. In the extreme case where the phases of all three conductors are identical, at a large distance from the three conductors, they appear like a single conductor carrying a current of $3I$. Probably, for a given space containing two or more current-carrying conductors, the most important factor in reducing the strength of the magnetic fields from these conductors is having a net current of zero. When determining the net current, the amplitudes *and* phases of the currents must be considered.

$$d := 4\cdot 1.03\cdot 10^{-3} \qquad K := 4\cdot\pi\cdot 10^{-7}\cdot 10^{7} \qquad j := \sqrt{-1}$$

$$r := 4\cdot d, 4\cdot d + \frac{d}{100} .. 50\cdot d$$

$$H_x(x, y, x_p, y_p, I) := \frac{I}{2\cdot\pi}\cdot\left[\frac{-(y-y_p)}{(x-x_p)^2+(y-y_p)^2}\right] \qquad H_y(x, y, x_p, y_p, I) := \frac{I}{2\cdot\pi}\cdot\left[\frac{(x-x_p)}{(x-x_p)^2+(y-y_p)^2}\right]$$

$$B_{1p}(x, y, d) := k\cdot\sqrt{\left(\left|H_x\left(x, y, \frac{-d}{2}, 0, 10\right) + H_x\left(x, y, \frac{d}{2}, 0, 10\cdot e^{j\cdot\pi}\right)\right|\right)^2 \ldots + \left(\left|H_y\left(x, y, \frac{-d}{2}, 0, 10\right) + H_y\left(x, y, \frac{d}{2}, 0, 10\cdot e^{j\cdot\pi}\right)\right|\right)^2}$$

$$H_{3xs}(x, y, d) := H_x\left(x, y, \frac{-d}{2}, 0, 10\cdot e^{j\cdot\frac{2\cdot\pi}{3}}\right) + H_x\left(x, y, 0, 0, 10\cdot e^{j\cdot 0}\right) + H_x\left(x, y, \frac{d}{2}, 0, 10\cdot e^{-j\cdot\frac{2\cdot\pi}{3}}\right)$$

$$H_{3ys}(x, y, d) := H_y\left(x, y, \frac{-d}{2}, 0, 10\cdot e^{j\cdot\frac{2\cdot\pi}{3}}\right) + H_y\left(x, y, 0, 0, 10\cdot e^{j\cdot 0}\right) + H_y\left(x, y, \frac{d}{2}, 0, 10\cdot e^{-j\cdot\frac{2\cdot\pi}{3}}\right)$$

$$B_{3pu}(x, y, d) := K\cdot\sqrt{\left(\left|H_{3xs}(x, y, d)\right|\right)^2 + \left(\left|H_{3ys}(x, y, d)\right|\right)^2}$$

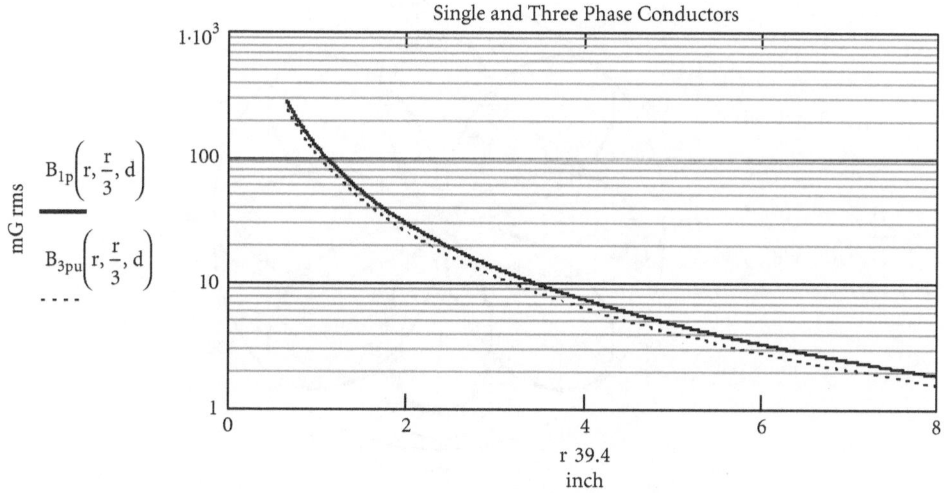

K converts a magnetic field in free space to a magnetic flux density in milligauss.

MATHCAD 3.4 Magnetic field as a function of distance from the origin along the line $y = x/3$ (in the first quadrant) for a single-phase system and the upper limit of the field for a three-phase system.

The magnetic field expressions given in this discussion are based on line currents, which have zero diameter. For multiple real conductors of nonzero diameters, the line current model is good if the conductors are not too close compared to the actual conductor's cross-sectional dimensions. When conductors of nonzero diameter are close, the proximity effect will disturb the current distribution on the wires and, hence, the magnetic field distribution. Also, the magnetic field expressions provided in this section, such as (3.41), are near-field expressions, and they should not be used to determine the value of the electric field in the far field. The magnetoquasistatic expressions for the magnetic field assume that both the length and spacing of the conductors are electrically short.

$$d := 4 \cdot 1.03 \cdot 10^{-3} \qquad K := 4 \cdot \pi \cdot 10^{-7} \cdot 10^{7} \qquad j := \sqrt{-1}$$

$$\theta := 0, \frac{\pi}{100} .. 2 \cdot \pi$$

$$x(\theta) := 4 \cdot d \cdot \cos(\theta) \qquad y(\theta) := 4 \cdot d \cdot \sin(\theta)$$

$$H_x(x, y, x_p, y_p, I) := \frac{I}{2 \cdot \pi} \cdot \left[\frac{-(y - y_p)}{\left[(x - x_p)^2 + (y - y_p)^2\right]} \right] \qquad H_y(x, y, x_p, y_p, I) := \frac{I}{2 \cdot \pi} \cdot \left[\frac{(x - x_p)}{\left[(x - x_p)^2 + (y - y_p)^2\right]} \right]$$

$$H_{1x}(x, y, d) := \left| H_x\left(x, y, \frac{-d}{2}, 0, 10\right) + H_x\left(x, y, \frac{d}{2}, 0, 10 \cdot e^{j \cdot \pi}\right) \right|$$

$$H_{1y}(x, y, d) := \left| H_y\left(x, y, \frac{-d}{2}, 0, 10\right) + H_y\left(x, y, \frac{d}{2}, 0, 10 \cdot e^{j \cdot \pi}\right) \right|$$

$$H_{3x}(x, y, d) := \left| H_x\left(x, y, \frac{-d}{2}, 0, 10 \cdot e^{j \cdot \frac{2 \cdot \pi}{3}}\right) + H_x\left(x, y, 0, 0, 10 \cdot e^{j \cdot 0}\right) + H_x\left(x, y, \frac{d}{2}, 0, 10 \cdot e^{-j \cdot \frac{2 \cdot \pi}{3}}\right) \right|$$

$$H_{3y}(x, y, d) := \left| H_y\left(x, y, \frac{-d}{2}, 0, 10 \cdot e^{j \cdot \frac{2 \cdot \pi}{3}}\right) + H_y\left(x, y, 0, 0, 10 \cdot e^{j \cdot 0}\right) + H_y\left(x, y, \frac{d}{2}, 0, 10 \cdot e^{-j \cdot \frac{2 \cdot \pi}{3}}\right) \right|$$

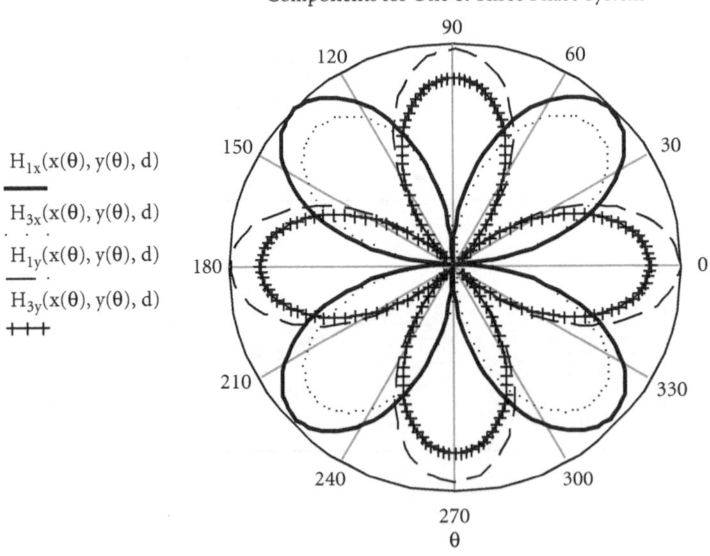

MATHCAD 3.5 Amplitude of the *x* and *y* components of the magnetic field for both the single-phase and three-phase systems. The +*x* axis corresponds to $\theta = 0°$ while the +*y* axis corresponds to $\theta = 90°$.

3.9 Thick Poor Conductors

Referring to Figure 3.36, a nonmagnetic solid cylinder of low conductivity, σ, and radius R is placed in a uniform magnetic field that is parallel to the axis of the cylinder. The magnetic field inside the cylinder is approximately the same as the magnetic field outside the cylinder if

$$\frac{\omega \mu_o \sigma R^2}{4} \ll 1 \tag{3.50}$$

If this cylinder is representing a human body with an average conductivity of 0.5 S/m, at what frequency is the magnetic field inside the body much different from the applied field? [Haus; Gandhi, '90]

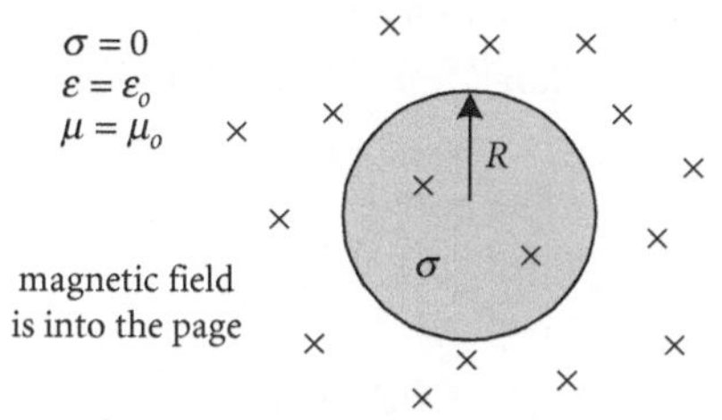

FIGURE 3.36 Conducting cylinder in a uniform parallel magnetic field.

It is well known that the direction of the magnetic field inside a long, tightly wound, current-carrying coil is along its axis. Thus, if a nonmagnetic solid cylinder is inserted inside a long current-carrying coil, and their axes are aligned, the cylinder will be exposed to a field parallel to its axis. Low-frequency magnetic fields (e.g., 1 kHz) can pass straight through many nonmagnetic conductors since the induced fields generated in the metal (from the "bucking" currents) are small compared to the applied magnetic field. As the conductivity of an object or the applied frequency decreases, the strength of these "bucking" currents decreases.

It was given that the induced magnetic fields inside a nonmagnetic low-conductivity body are small compared to the applied magnetic fields when (3.50) is satisfied. This equation involves not only the frequency of the applied field and electrical parameters of the cylinder but also the radius of the cylinder, a physical dimension. Since most biological materials including the human body are nonmagnetic ($\mu = \mu_o$),[13] and the average conductivity of the human body was given as 0.5 S/m, this inequality reduces to

$$\omega \ll \frac{6.4\times10^6}{R^2}$$

For a radius of 0.3 m, $\omega \ll 7.1\times10^7$ rad/s ~ 10 MHz. Therefore, the human body is transparent to magnetic fields with frequencies less than about 1 MHz. Near and above 10 MHz, inside the body the induced currents and their magnetic fields will be significant.

The skin depth is commonly used in describing the current distribution in a good conductor and in determining the effectiveness of a shield to a plane wave. The inequality (3.50) can also be written in terms of the skin depth for good conductors, $\delta = 1/\sqrt{\pi f \mu \sigma}$:

$$\frac{\omega \mu_o \sigma R^2}{4} = \frac{\pi f \mu_o \sigma R^2}{2} = \frac{1}{2}\left(\frac{R}{\delta}\right)^2 \ll 1$$

If the object's radius is small compared to the skin depth, the magnetic field passes through the object and little magnetic shielding is provided by the object. Using the previous conductivity for the human body, the skin depth at 60 Hz is

$$\delta = \frac{1}{\sqrt{\pi(60)(4\pi\times10^{-7})(0.5)}} \approx 92 \text{ m}$$

Even though the human body is considered a "good conductor" at 60 Hz, a magnetic field at this frequency passes through the human body. (See the later discussion on MQS.)

For frequencies below 1 MHz, a simple expression can be quickly obtained for the main component of the electric field inside the body. Given that the magnetic field inside the conductor is essentially the same as outside the conductor for these low frequencies, one of Maxwell's equations

$$\nabla \times \vec{E} = -\frac{\partial \vec{B}}{\partial t}$$

[13]Some bacteria, insects, and birds are weakly diamagnetic.

can be used to determine the induced electric field inside the conductor. Assuming that the applied external magnetic field is equal to $H_e \cos(\omega t)\hat{a}_z$, for sinusoidal steady-state conditions,

$$\nabla \times \vec{E}_s = -j\omega \vec{B}_s = -j\omega\mu_o H_e \hat{a}_z$$

In the cylindrical coordinate system,

$$\frac{1}{\rho}\left[\frac{\partial}{\partial \rho}(\rho E_{\phi s}) - \frac{\partial E_{\rho s}}{\partial \phi}\right]\hat{a}_z = -j\omega\mu_o H_e \hat{a}_z$$

The ρ component of the electric field will be assumed negligible compared to the ϕ component. This implies that most of the current is circulating in the ϕ direction since $\vec{J} = \sigma \vec{E}$. (The radial component of the current density must be zero at the surface, $\rho = R$, since free space surrounds the conductor.) Therefore,

$$\frac{1}{\rho}\frac{\partial}{\partial \rho}(\rho E_{\phi s}) = -j\omega\mu_o H_e \quad \Rightarrow \frac{\partial}{\partial \rho}(\rho E_{\phi s}) = -j\omega\mu_o H_e \rho \quad \Rightarrow \rho E_{\phi s} = \frac{-j\omega\mu_o H_e \rho^2}{2} + C$$

Solving for the electric field, realizing that the field must be finite even at $\rho = 0$ (i.e., $C = 0$), and using Euler's formula,

$$E_{\phi s} = \frac{-j\omega\mu_o H_e \rho}{2} \tag{3.51}$$

$$\begin{aligned}\vec{E} &= \mathrm{Re}(E_{\phi s}\hat{a}_\phi e^{j\omega t}) = \mathrm{Re}\left\{\frac{-j\omega\mu_o H_e \rho}{2}\hat{a}_\phi[\cos(\omega t) + j\sin(\omega t)]\right\} \\ &= \left(\frac{\omega\mu_o \rho}{2}\right) H_e \sin(\omega t)\hat{a}_\phi\end{aligned} \tag{3.52}$$

The electric field is greatest at the surface, as is expected. For a typical 60 Hz, 0.1 μT magnetic flux density inside a home (away from appliances), the maximum electric field at the surface of a 0.3 m radius, conducting cylinder is small:

$$|\vec{E}| = \left[\frac{(2\pi \times 60)\mu_o 0.3}{2}\right]\left(\frac{0.1 \times 10^{-6}}{\mu_o}\right) \approx 6\ \mu\text{V/m}$$

3.10 Thin Good Conductors

Referring to Figure 3.37, a uniform magnetic field is external and parallel to the axis of a nonmagnetic cylindrical shell of radius R. The cylinder has a conductivity of σ and a thickness Δ that is small compared to R. The differential equation that describes the relationship between the externally applied magnetic field, H_e, and resultant magnetic field inside of the cylinder, H_i, is given by

$$\frac{dH_i}{dt} + \frac{H_i}{\tau_m} = \frac{H_{ext}}{\tau_m} \quad \text{where } \tau_m = \frac{1}{2}\mu_o \sigma R \Delta \tag{3.53}$$

If $H_{ext}(t) = H_e \cos\omega t$, sketch the Bode magnitude plot of the ratio of the internal field to the external field. A *perfect* conductor will act as a perfect shield for near-field magnetic fields. For what frequencies can a nonideal shielding material be considered a good conductor for shielding near-field magnetic fields parallel to the cylinder's axis? [Haus; Zahn; Bridges, '88]

Working in the frequency domain, the sinusoidal steady-state solution for (3.53) can be readily obtained. The time-domain differential equation describing the relationship between the internal and

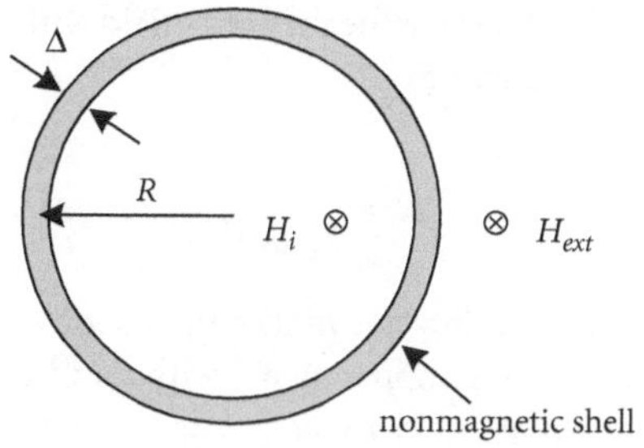

FIGURE 3.37 Applied external magnetic field parallel to the surface of a conducting cylindrical shell.

the external magnetic fields can be converted to the frequency domain by inspection (differentiation in the time domain corresponds to multiplication by $j\omega$ in the frequency domain):

$$j\omega H_{is} + \frac{H_{is}}{\tau_m} = \frac{H_{exts}}{\tau_m} \tag{3.54}$$

(It is assumed that the dimension R is electrically small.) Solving for the ratio of the inside to outside field,

$$G(\omega) = \frac{H_{is}}{H_{exts}} = \frac{1}{1 + j\omega\tau_m} = \frac{1}{1 + \dfrac{j\omega}{\dfrac{1}{\tau_m}}} = \frac{1}{1 + \dfrac{j\omega}{\dfrac{2}{\mu_o \sigma R \Delta}}} \tag{3.55}$$

This ratio is also referred to as the shielding factor. Perfect or complete shielding corresponds to a shielding factor of 0 (or $-\infty$ dB), while no shielding corresponds to a shielding factor of 1 (or 0 dB). This particular shield is acting like a low-pass filter. Low frequencies, including dc, are passed, while higher frequencies are attenuated. Interestingly, the shielding improves with increasing radius for this nonmagnetic shell. Using the circuits concept of inductance, as the cylinder's radius increases, its inductance increases (for circulating currents). Inductors oppose or resist changes in current and magnetic flux. For currents circulating around the cylinder in the ϕ direction, the resistance also increases with increasing radius but in proportion to its circumference not its cross-sectional area: $\approx 2\pi R/(\sigma \Delta l_{th})$, where l_{th} is the length of the cylinder, which is assumed large compared to R. Thus, for this particular shield, the relationship between the external and internal magnetic fields could be modeled as a current supply (representing the external magnetic field) in parallel with a parallel RL circuit (i.e., low-pass filter).

The Bode magnitude plot shown in Figure 3.38 is obtained directly from the shielding factor expression. For frequencies such that

$$\omega \gg \frac{2}{\mu_o \sigma R \Delta} \tag{3.56}$$

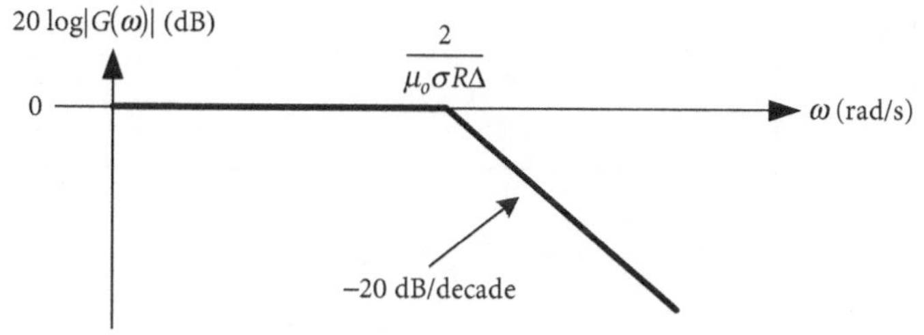

FIGURE 3.38 Bode magnitude plot of (3.55).

the thin conducting material is actually shielding the magnetic field without the use of expensive magnetic material. For copper, this expression reduces to

$$f \gg \frac{1}{230 R\Delta} \tag{3.57}$$

The lowest frequency for magnetic shielding vs. cylinder radius for various shield thicknesses is plotted in Mathcad 3.6. A 3.9 mil (≈ 0.1 mm) thick copper shield with a radius of at least 1 cm can be effective against near-field magnetic fields for frequencies much greater than about 4 kHz. Lower frequencies can only be shielded if the radius of the 3.9 mil shield is quite large (e.g., at 60 Hz, $R \approx 10$ m). A very thick 390 mil (≈ 1 cm) copper shield with a radius of at least 0.1 m can be effective at 60 Hz.

The second plot in Mathcad 3.6 is the maximum percent difference between the skin depth of copper at the lowest shielding frequency and the shield thickness. When the copper cylinder is not several skin-

$x := -30, -29.1 .. 50$

$$R(x) := \left(x + 1 - 10 \cdot \text{floor}\left(\frac{x}{10}\right)\right) \cdot 10^{\text{floor}\left(\frac{x}{10}\right)}$$

$$f(\Delta, R) := \frac{1}{230 \cdot \Delta \cdot R} \qquad \delta(\Delta, f, R) := \frac{1}{\sqrt{\pi \cdot f(\Delta, R) \cdot 4 \cdot \pi \cdot 10^{-7} \cdot 5.8 \cdot 10^{7}}} \qquad \%\text{diff}(\Delta, f, R) := 100 \cdot \frac{\delta(\Delta, f, R) - \Delta}{\delta(\Delta, f, R)}$$

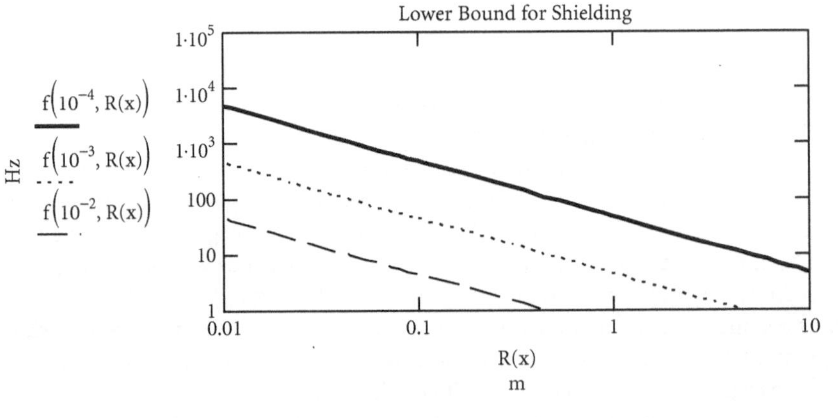

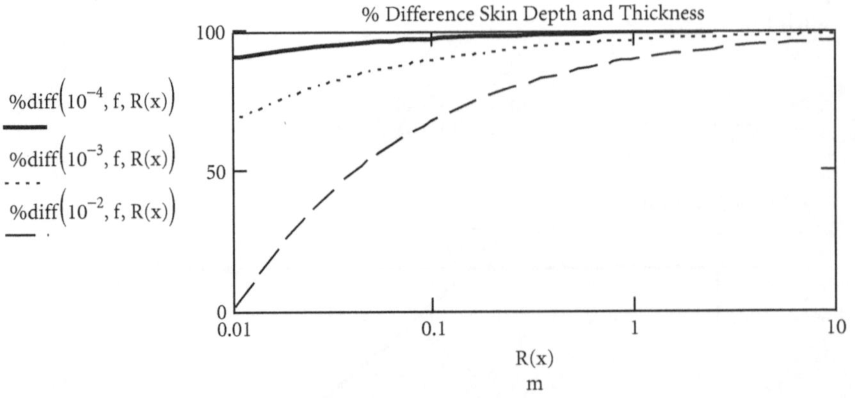

MATHCAD 3.6 Lowest frequency for shielding of parallel magnetic fields for a copper cylindrical shield of various thicknesses vs. shield radius. The percent difference between the skin depth and shield thickness is also plotted.

depths thick, then absorption of the field is not the likely mechanism of this form of shielding. Instead, Eddy currents are generated in the thin shield, and these currents produce a counter magnetic field that opposes the applied magnetic field. This method of shielding is not frequently (solely) utilized for demanding low-frequency magnetic shielding applications due to the size and thickness of the required shield. It is also true that an applied magnetic field transverse (perpendicular) to the axis of a thin cylindrical shell behaves in the same manner as a magnetic field parallel to the axis, with an identical time constant.

3.11 Spherical and Cylindrical Conducting and Magnetic Shields

The ratio of the magnetic field anywhere inside the cavity of a nonmagnetic spherical shield of outer radius a, inner radius b, wall thickness Δ (which is thin), and propagation constant γ to the applied, uniform, external magnetic field is

$$\frac{H_{is}}{H_{es}} = \frac{3\gamma a}{\gamma^2 b^2\left[\left(1+\dfrac{2}{(\gamma b)^2}\right)\sinh(\gamma\Delta)+\dfrac{3}{\gamma b}\cosh(\gamma\Delta)\right]} \tag{3.58}$$

Determine a low-frequency and high-frequency approximation for this expression, assuming a thin shield. Assume the shell is electrically small. Provide the analogous expressions for a cylindrical shell. [Shenfeld, '68; Bridges, '88; Hasselgren; Cooley; Hoburg, '95; Mager, '70; Chatterton; Gustafson]

There are many expressions for the shielding effectiveness of spherical and long cylindrical shells. As will be seen, most of these expressions are actually just special cases for low, high, or higher frequencies and for nonmagnetic or magnetic conductors. The expressions given in this section are for spherical and infinitely long cylindrical shields. Although most real shields are not spherical or cylindrical with infinite length, these analytical expressions provide an excellent indication (and, frequently, an upper limit) of what can be expected for shields of other shapes. Mathematical derivations are followed by a discussion of the results. The general expressions for dc or low-frequency purely magnetic (i.e., negligible conductivity) shields are provided later.

The propagation constant of the nonmagnetic shell is defined as

$$\gamma = \alpha + j\beta = \sqrt{j\omega\mu_o(\sigma + j\omega\varepsilon)} \tag{3.59}$$

where μ_o is the free-space permeability, σ is the conductivity of the shell, and ε is the permittivity of the shell. Equation (3.59) is identical to the propagation constant expression used for plane waves. Assuming that the shell is a good conductor,

$$\sigma \gg \omega\varepsilon$$

the propagation constant reduces to

$$\begin{aligned}\gamma &\approx \sqrt{j\omega\mu_o\sigma} = \sqrt{\omega\mu_o\sigma}\angle 45^\circ = \sqrt{2\pi f\mu_o\sigma}\left(\frac{1}{\sqrt{2}}+j\frac{1}{\sqrt{2}}\right)\\ &= \sqrt{\pi f\mu_o\sigma}(1+j) = \frac{1}{\delta}+j\frac{1}{\delta}\end{aligned} \tag{3.60}$$

where δ is the standard expression for the skin depth of a good conductor. To obtain the low-frequency approximation, the product of the propagation constant and shell thickness is considered small (i.e., low-operating frequency and/or thin shell)

$$|\gamma\Delta| \ll 1 \quad \Rightarrow \left|\left(\frac{1+j}{\delta}\right)\Delta\right| \ll 1 \quad \Rightarrow \delta \gg \Delta$$

which results in the following approximations for the hyperbolic functions:

$$\sinh(\gamma\Delta) \approx \gamma\Delta \quad \left(\sinh x = x + \frac{x^3}{3!} + \frac{x^5}{5!} + \cdots\right)$$

$$\cosh(\gamma\Delta) \approx 1 \quad \left(\cosh x = 1 + \frac{x^2}{2!} + \frac{x^4}{4!} + \cdots\right)$$

The final approximation is that the shell thickness is small compared to the shell radius $(a \approx b \gg \Delta)$. Utilizing all these approximations, the expression for the ratio of the internal to unperturbed external magnetic field is obtained:

$$\begin{aligned}\frac{H_{is}}{H_{es}} &\approx \frac{3\gamma b}{\gamma^2 b^2 \left\{\left[1+\frac{2}{(\gamma b)^2}\right]\gamma\Delta + \frac{3}{\gamma b}\right\}} = \frac{3}{\gamma^2 b\Delta + 3 + 2\frac{\Delta}{b}} \approx \frac{3}{\gamma^2 b\Delta + 3} \quad \text{if } \delta \gg \Delta \\ &\approx \frac{1}{1+\frac{jb\omega\mu_o\sigma\Delta}{3}} = \frac{1}{1+\frac{j\omega}{\frac{3}{b\mu_o\sigma\Delta}}} \quad \text{if } \delta \gg \Delta\end{aligned} \tag{3.61}$$

This equation is considered the low-frequency approximation valid for good-conducting thin shields where $\delta \gg \Delta$. This ratio is also referred to as the shielding factor or shielding gain. The smaller the shielding factor, the better the shielding. For the low-frequency approximation, the internal field is small compared to the external field when

$$\left|\frac{1}{1+\frac{j\omega}{\frac{3}{b\mu_o\sigma\Delta}}}\right| \ll 1 \quad \Rightarrow \frac{3}{b\mu_o\sigma\Delta} \ll \omega \tag{3.62}$$

For nonmagnetic good conductors, the product $\mu_o\sigma$ ranges from about 20 to 70. It is possible to obtain some shielding of low-frequency magnetic fields with nonmagnetic material.

To obtain the high-frequency expression, again, several approximations are required. The material of the shell is once more considered a good conductor so that the propagation constant is given by the approximation in (3.60). The next approximation is that the frequency is sufficiently high so that the following two inequalities involving both the shell thickness and radius are satisfied:

$$|\gamma\Delta| \gg 1 \text{ (or } \Delta \gg \delta) \quad \text{and} \quad \left|\frac{3}{\gamma b}\right| \ll 1 \text{ (or } b \gg \delta)$$

In other words, the skin depth should be much less than the shell thickness. This implies that the skin depth is also much less than the shell radius. The final approximation is that the shell is sufficiently thin so that the internal radius of the shell is about equal to the external radius of the shell ($a \approx b$). Using the exponential definitions for the hyperbolic functions (instead of the series definitions) and the previous high-frequency approximations, it is seen that the hyperbolic sine is about equal to the hyperbolic cosine:

$$\sinh(\gamma\Delta) = \frac{e^{\gamma\Delta} - e^{-\gamma\Delta}}{2} \approx \frac{e^{\gamma\Delta}}{2} \quad \text{if } |\gamma\Delta| \gg 1$$

$$\cosh(\gamma\Delta) = \frac{e^{\gamma\Delta} + e^{-\gamma\Delta}}{2} \approx \frac{e^{\gamma\Delta}}{2} \quad \text{if } |\gamma\Delta| \gg 1$$

The ratio of the internal magnetic field to the external magnetic field reduces to

$$\begin{aligned} \frac{H_{is}}{H_{es}} &\approx \frac{3\gamma a}{\gamma^2 b^2 \left[\sinh(\gamma\Delta) + \frac{3}{\gamma b}\cosh(\gamma\Delta)\right]} \\ &\approx \frac{3\gamma a}{\gamma^2 b^2 \sinh(\gamma\Delta)} \approx \frac{3}{\gamma b \sinh(\gamma\Delta)} \quad \text{if } b \gg \delta \qquad (3.63) \\ &\approx \frac{3}{\gamma b \frac{e^{\gamma\Delta}}{2}} \quad \text{if } b \gg \delta, \Delta \gg \delta \end{aligned}$$

Using the definition for the propagation constant, the magnitude of this ratio for high frequencies is

$$\begin{aligned} \left|\frac{H_{is}}{H_{es}}\right| &= \left|\frac{6}{\gamma b e^{\gamma\Delta}}\right| \approx \left|\frac{6\delta}{(1+j)be^{(1+j)\frac{\Delta}{\delta}}}\right| = \left|\frac{6\delta}{(1+j)be^{\frac{\Delta}{\delta}}e^{j\frac{\Delta}{\delta}}}\right| \quad \text{if } b \gg \delta, \Delta \gg \delta \\ &= \frac{6\delta}{\sqrt{2}be^{\frac{\Delta}{\delta}}} = \frac{3\sqrt{2}\delta e^{-\frac{\Delta}{\delta}}}{b} \quad \text{if } b \gg \delta, \Delta \gg \delta \end{aligned} \qquad (3.64)$$

where the relationship $|e^{jx}| = 1$ was used. The internal magnetic field is small compared to the external magnetic field when

$$\frac{3\sqrt{2}\delta e^{-\frac{\Delta}{\delta}}}{b} \ll 1 \qquad (3.65)$$

At high frequencies, where the skin depth is small compared to the shell's thickness, a conducting shell is effective in shielding magnetic fields. Although (3.64) is a high-frequency approximation, it (and all the other expressions in this section) should only be used when the dimensions of the shield are electrically small.

The previous results are valid for nonmagnetic materials. The equation for a thin, magnetic spherical shell at low frequencies is

$$\frac{H_{is}}{H_{es}} = \frac{1}{1 + \frac{j\omega}{\frac{3}{b\Delta\mu_o\sigma}} + \frac{2}{3}\frac{(\mu_r - 1)^2\Delta}{\mu_r b}} \quad \text{if } \delta \gg \Delta \qquad (3.66)$$

Even at very low frequencies, the expression for the shielding factor is still a function of the relative permeability. The equation at high frequencies is

$$\left|\frac{H_{is}}{H_{es}}\right| = \frac{3\sqrt{2}\mu_r \delta e^{-\frac{\Delta}{\delta}}}{b} \quad \text{if } b \gg \mu_r\delta, \Delta \gg \delta \tag{3.67}$$

where μ_r is the relative permeability of the shell. The shielding factor is a function of both the permeability and conductivity (via the skin depth) at high frequencies. When $\Delta \gg \delta$, the magnetic field will vary significantly in the shield material from the outer surface to the inner surface of the shell. Since for real magnetic materials the permeability will vary with the intensity of the magnetic field (and possibly saturate), this high-frequency expression should be used with caution. An expression for "not too high frequencies" is given by

$$\left|\frac{H_{is}}{H_{es}}\right| \approx \frac{3\sqrt{2}be^{-\frac{\Delta}{\delta}}}{\mu_r\delta} \quad \text{if } \Delta \gg \delta, b \gg \Delta, \mu_r\delta \gg b \tag{3.68}$$

The analogous expressions for cylindrical shields will be derived shortly.

For long, thin cylindrical shells the equations are quite similar. The shielding effectiveness expressions will be provided for applied fields that are both parallel and transverse to the axis of the cylinder.[14] (A magnetic field at any angle to the shell's surface can be decomposed into components tangential and normal to the surface.) Referring to Figure 3.39, the general expression for a magnetic field parallel to the cylinder's axis is

$$\frac{H_{is}}{H_{es}} = \frac{1}{\cosh(\gamma\Delta) + \frac{\gamma a}{2\mu_r}\sinh(\gamma\Delta)} \tag{3.69}$$

The corresponding low-frequency approximation for magnetic, conducting, thin cylindrical $(a \approx b)$ shells is

$$\frac{H_{is}}{H_{es}} = \frac{1}{1 + \frac{\frac{j\omega}{2}}{b\mu_o\sigma\Delta}} \quad \text{if } \delta \gg \Delta \tag{3.70}$$

which is the same equation given previously for nonmagnetic shells. This equation is not a function of the relative permeability. At dc, or $\omega = 0$, there is no shielding of magnetic fields parallel to the cylinder's surface.

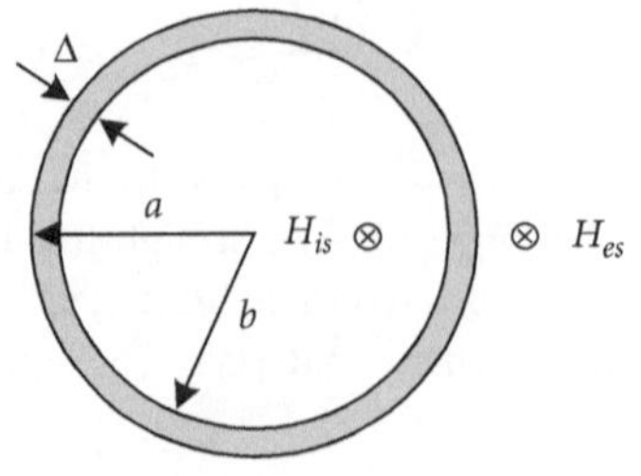

FIGURE 3.39 Applied external magnetic field parallel to the surface of a conducting, magnetic cylindrical shell.

---

[14] The applied transverse field is perpendicular to a plane passing through the axis of the cylinder.

This implies that there is no field distortion (or magnetic "ducting") at dc and, obviously, there are no Eddy currents at dc to produce a counter field.[15] The high-frequency approximation for magnetic conducting shells is

$$\left|\frac{H_{is}}{H_{es}}\right| = \frac{2\sqrt{2}\mu_r \delta e^{-\frac{\Delta}{\delta}}}{b} \quad \text{if } b \gg \mu_r \delta, \Delta \gg \delta \tag{3.71}$$

For frequencies that are not too large, alternative "lower" high-frequency expressions can be derived. First, assume that $|\gamma\Delta| \gg 1$ or equivalently $\Delta \gg \delta$:

$$\begin{aligned}\frac{H_{is}}{H_{es}} &= \frac{1}{\cosh(\gamma\Delta) + \frac{\gamma a}{2\mu_r}\sinh(\gamma\Delta)} \approx \frac{1}{\frac{e^{\gamma\Delta}}{2} + \frac{\gamma b}{2\mu_r}\frac{e^{\gamma\Delta}}{2}} \\ &= \frac{2e^{-\gamma\Delta}}{1 + \frac{\gamma b}{2\mu_r}} \quad \text{if } \Delta \gg \delta, a \approx b \gg \Delta\end{aligned} \tag{3.72}$$

$$\left|\frac{H_{is}}{H_{es}}\right| = \left|\frac{2e^{-\left(\frac{1+j}{\delta}\right)\Delta}}{1 + \left(\frac{1+j}{\delta}\right)\frac{b}{2\mu_r}}\right| = \frac{2e^{-\frac{\Delta}{\delta}}}{\sqrt{\left(1 + \frac{b}{2\mu_r \delta}\right)^2 + \left(\frac{b}{2\mu_r \delta}\right)^2}} \tag{3.73}$$

Second, rather than assuming $b \gg \mu_r \delta$, assume that $\mu_r \delta \gg b$:

$$\left|\frac{H_{is}}{H_{es}}\right| \approx \frac{2e^{-\frac{\Delta}{\delta}}}{\sqrt{1 + \left(\frac{b}{2\mu_r \delta}\right)^2}} \approx 2e^{-\frac{\Delta}{\delta}} \quad \text{if } \Delta \gg \delta, b \gg \Delta, \mu_r \delta \gg b \tag{3.74}$$

This expression can be used for situations such as when $b = 50$ mm, $\Delta = 2$ mm, $\delta = 0.2$ mm, and $\mu_r =$ 20,000 where $\mu_r \delta = 4 \gg b = 0.05$. To use (3.71), it would be necessary that $b \gg \mu_r \delta$. If b and μ_r were fixed, the skin depth would have to be much smaller to satisfy this inequality, which would imply a much higher frequency or conductivity.

The shielding effectiveness of a cylindrical shield to a parallel magnetic field is shown in Mathcad 3.7. The thickness of the 0.3 m radius cylindrical shell is 0.15 mm. For frequencies less than about 100 kHz, the low-frequency magnetic shielding approximation is quite good, while for frequencies greater than about 100 kHz, the high-frequency approximation is quite good. For a material with a relative permeability of 10 and conductivity of 10^7 1/Ω-m, the skin depth at 100 kHz is

$$\delta = \frac{1}{\sqrt{\pi f \mu \sigma}} = \frac{1}{\sqrt{\pi(100 \times 10^3)(10 \times 4\pi \times 10^{-7})(10^7)}} \approx 0.16 \text{ mm}$$

Even though $\Delta \approx \delta$ at 100 kHz, both the low-frequency and high-frequency expressions provide good results near this frequency. Referring to Figure 3.40, when the magnetic field is transverse (i.e., perpendicular)

---

[15]Furthermore, using the fundamental boundary condition, $H_{t1} = H_{t2}$, the components of the magnetic field tangential to the inner (or outer) surface of the cylindrical shell must be equal since no surface currents are present.

$$\mu_o := 4\cdot\pi\cdot10^{-7} \quad \varepsilon_o := 8.854\cdot10^{-12} \quad \sigma := 1\cdot10^{7} \quad j := \sqrt{-1}$$

$$x := 10, 11..\,70$$

$$\omega(x) := \left(x + 1 - 10\cdot\mathrm{floor}\left(\frac{x}{10}\right)\right)\cdot 10^{\mathrm{floor}\left(\frac{x}{10}\right)}$$

$$\gamma(\omega, \mu_r) := \sqrt{j\cdot\omega\cdot\mu_r\cdot\mu_o\cdot(\sigma + j\cdot\omega\cdot\varepsilon_o)} \qquad \delta(\omega, \mu_r) := \frac{1}{\sqrt{\pi\cdot\frac{\omega}{2\cdot\pi}\cdot\mu_r\cdot\mu_o\cdot\sigma}}$$

$$SE_{pLF}(\omega, b, \Delta, \mu_r) := 20\cdot\log\left(\left|1 + \frac{j\cdot\omega}{\frac{2}{b\cdot\mu_o\cdot\sigma\cdot\Delta}}\right|\right)$$

$$SE_{pHF}(\omega, b, \Delta, \mu_r) := 20\cdot\log\left[\left|\frac{b}{\left(2\cdot\sqrt{2}\cdot\mu_r\cdot\delta(\omega, \mu_r)\cdot e^{\frac{-\Delta}{\delta(\omega, \mu_r)}}\right)}\right|\right]$$

$$SE_p(\omega, b, \Delta, \mu_r) := 20\cdot\log\left[\left|\cosh(\gamma(\omega, \mu_r)\cdot\Delta) + \frac{\gamma(\omega, \mu_r)\cdot(b + \Delta)}{2\cdot\mu_r}\cdot\sinh(\gamma(\omega, \mu_r)\cdot\Delta)\right|\right]$$

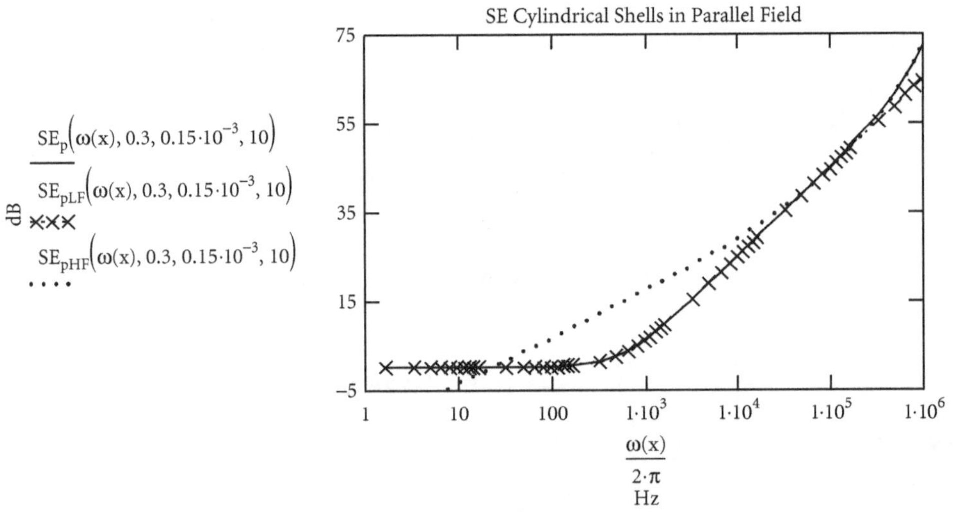

MATHCAD 3.7 Comparison of the low-frequency and high-frequency approximations to the exact result for a thin cylindrical shell exposed to a parallel magnetic field.

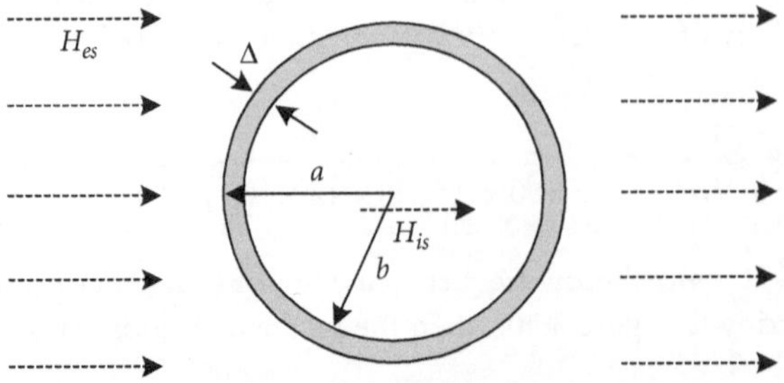

FIGURE 3.40 Applied external magnetic "perpendicular" to the surface of a conducting, magnetic cylindrical shell.

to a plane passing through the axis of the thin cylindrical shell, which is referred to in this book as "perpendicular" to the surface, the general expression is

$$\frac{H_{is}}{H_{es}} = \frac{1}{\cosh(\gamma\Delta) + \frac{1}{2}\left(\frac{\gamma a}{\mu_r} + \frac{\mu_r}{\gamma a}\right)\sinh(\gamma\Delta)} \tag{3.75}$$

The low-frequency approximation for this expression, which is different from (3.70), is

$$\frac{H_{is}}{H_{es}} = \frac{1}{1 + \frac{j\omega}{\frac{2}{b\mu_0\sigma\Delta}} + \frac{(\mu_r - 1)^2\Delta}{2\mu_r b}} \tag{3.76}$$

while the high-frequency approximation is the same as (3.71):

$$\left|\frac{H_{is}}{H_{es}}\right| = \frac{2\sqrt{2}\mu_r\delta e^{-\frac{\Delta}{\delta}}}{b} \tag{3.77}$$

The corresponding, "lower" high-frequency approximation is obtained in a manner similar to that given for the parallel magnetic field equation:

$$\begin{aligned}
\left|\frac{H_{is}}{H_{es}}\right| &\approx \left|\frac{1}{\frac{e^{\gamma\Delta}}{2} + \frac{1}{2}\left(\frac{\gamma b}{\mu_r} + \frac{\mu_r}{\gamma b}\right)\frac{e^{\gamma\Delta}}{2}}\right| = \left|\frac{2e^{-\gamma\Delta}}{1 + \frac{1}{2}\left(\frac{\gamma b}{\mu_r} + \frac{\mu_r}{\gamma b}\right)}\right| \quad \text{if } \Delta \gg \delta,\, a \approx b \gg \Delta \\
&= \frac{2e^{-\frac{\Delta}{\delta}}}{\left|1 + \frac{1}{2}\left[\frac{(1+j)b}{\mu_r\delta} + \frac{\mu_r\delta}{(1+j)b}\right]\right|} \quad \text{if } \Delta \gg \delta,\, b \gg \Delta \\
&\approx \frac{2e^{-\frac{\Delta}{\delta}}}{\left|1 + \frac{\mu_r\delta}{2(1+j)b}\right|} = \frac{2e^{-\frac{\Delta}{\delta}}}{\left|1 + \frac{\mu_r\delta(1-j)}{4b}\right|} \\
&\approx \frac{2e^{-\frac{\Delta}{\delta}}}{\left|\frac{\mu_r\delta(1-j)}{4b}\right|} = \frac{8be^{-\frac{\Delta}{\delta}}}{\sqrt{2}\mu_r\delta} = \frac{4\sqrt{2}be^{-\frac{\Delta}{\delta}}}{\mu_r\delta} \quad \text{if } \Delta \gg \delta,\, b \gg \Delta,\, \mu_r\delta \gg b
\end{aligned} \tag{3.78}$$

For transverse fields, a magnetic shell does distort the field at dc ($\omega = 0$) and, thus, the shell provides some shielding of dc fields:

$$\frac{H_{is}}{H_{es}} = \frac{1}{1 + \frac{(\mu_r - 1)^2\Delta}{2\mu_r b}} \quad \text{if } \omega = 0 \tag{3.79}$$

General shielding effectiveness expressions are available that include the effect of the conductivity and permeability for spherical shells and infinitely long cylindrical shells of any thickness. These MQS-based expressions assume that the dimensions of the shell are electrically small. As before, the inner radius of the shell is b and outer radius of the shell is a. For spherical shields, the general daunting expression derived by Hoburg is

$$\frac{H_{is}}{H_{es}} = \frac{3}{2(R_{11} - R_{12}) + (R_{22} - R_{21})} \tag{3.80}$$

where

$$R_{11} = \frac{1}{(\gamma a)^3}\{[(\gamma a)^2 + 1 - (\gamma b)(\gamma a)]\sinh[\gamma(a-b)] + \{(\gamma b)[(\gamma a)^2 + 1] - (\gamma a)\}\cosh[\gamma(a-b)]\} \tag{3.81}$$

$$R_{12} = \frac{\mu_o}{\mu}\frac{1}{2(\gamma a)^3}\left\{\begin{aligned}&[(\gamma b)(\gamma a) - [(\gamma b)^2 + 1][(\gamma a)^2 + 1]]\sinh[\gamma(a-b)]\\&+[(\gamma a) - (\gamma b)][1 - (\gamma b)(\gamma a)]\cosh[\gamma(a-b)]\end{aligned}\right\} \tag{3.82}$$

$$R_{21} = \frac{\mu}{\mu_o}\frac{2}{(\gamma a)^3}\{[1 - (\gamma b)(\gamma a)]\sinh[\gamma(a-b)] - [(\gamma a) - (\gamma b)]\cosh[\gamma(a-b)]\} \tag{3.83}$$

$$R_{22} = \frac{1}{(\gamma a)^3}\{[(\gamma b)(\gamma a) - (\gamma b)^2 - 1]\sinh[\gamma(a-b)] + \{(\gamma a)[(\gamma b)^2 + 1] - (\gamma b)\}\cosh[\gamma(a-b)]\} \tag{3.84}$$

$$\gamma = \sqrt{\pi f \mu_r \mu_o \sigma}(1 + j) = \frac{1}{\delta} + j\frac{1}{\delta}$$

For cylindrical shells,

$$\frac{H_{is}}{H_{es}} = \frac{2\mu_r \dfrac{a}{b}}{\left\{\begin{aligned}&[\mu_r K_1(\gamma b) - \gamma b\dot{K}_1(\gamma b)][\mu_r I_1(\gamma a) + \gamma a\dot{I}_1(\gamma a)]\\&-[\mu_r I_1(\gamma b) - \gamma b\dot{I}_1(\gamma b)][\mu_r K_1(\gamma a) + \gamma a\dot{K}_1(\gamma a)]\end{aligned}\right\}} \tag{3.85}$$

where $I_1(x)$ is the modified Bessel function of the first kind of order one and $\dot{I}_1(x)$ is its derivative, $K_1(x)$ is the modified Bessel function of the second kind of order one and $\dot{K}_1(x)$ is its derivative, and γ is as previously defined for the spherical shell.

The shielding gain or shielding factor is plotted in Figure 3.41 as a function of both the conductivity and permeability for a spherical shell at a frequency of 60 Hz. The inner radius of the shell is 0.1 m and

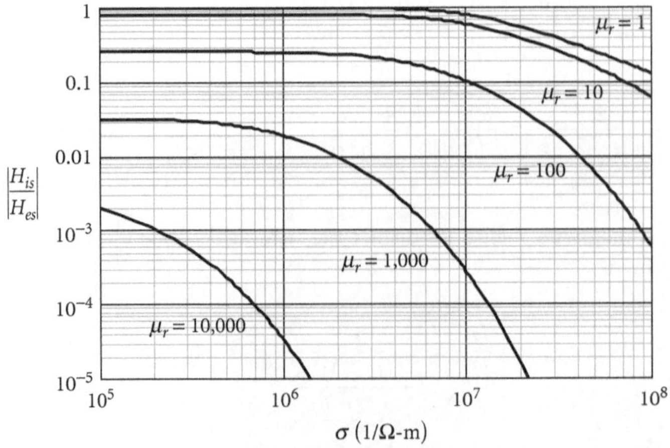

FIGURE 3.41 Shielding gain (or factor) of a spherical shell at 60 Hz as a function of the conductivity and permeability of the shell. The inner radius of the shell is 0.1 m, and its thickness is 5 mm.

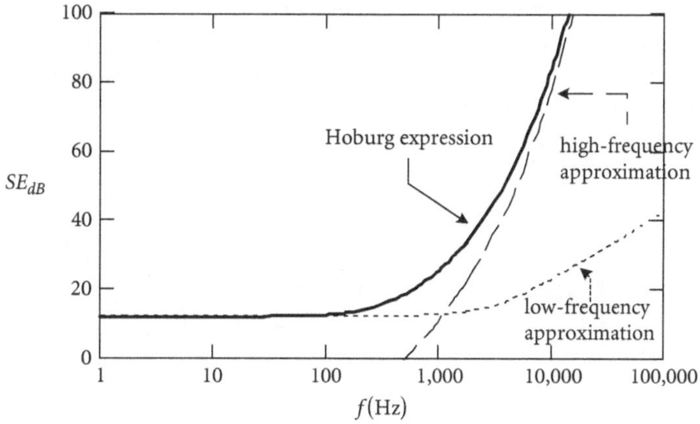

FIGURE 3.42 Low-frequency and high-frequency approximations compared to Hoburg's expression for a spherical shell as a function of the frequency. The conductivity and relative permeability of the shell are 10^6 1/Ω-m and 100, respectively. The inner radius of the shell is 0.1 m, and its thickness is 5 mm.

its thickness is 5 mm. Although not shown, the shielding factor curves for a very long cylindrical shell, with the same inner radius and thickness, are very similar. The shielding improves, corresponding to a smaller gain ratio, as the conductivity or the permeability increases. To obtain very small shielding factors for a single layer shield at low frequencies, it is necessary to use high-relative permeability materials. In Figure 3.42, Hoburg's expression is compared to the following low-frequency and high-frequency approximations for a thin, magnetic spherical shell:

$$SE_{dB,LF} = 20\log\left|1+\frac{j\omega}{\dfrac{3}{b\Delta\mu_o\sigma}}+\frac{2}{3}\frac{(\mu_r-1)^2\Delta}{\mu_r b}\right| \quad \delta \gg \Delta \tag{3.86}$$

$$SE_{dB,HF} = 20\log\left|\frac{b}{3\sqrt{2}\mu_r\delta e^{-\frac{\Delta}{\delta}}}\right| \quad \Delta \gg \delta \tag{3.87}$$

For a conductivity of 10^6 1/Ω-m, relative permeability of 100, and thickness of 5 mm, the skin depth is equal to the thickness at a frequency of about 100 Hz:

$$\delta = \frac{1}{\sqrt{\pi f\mu\sigma}} = \frac{1}{\sqrt{\pi f(100\times 4\pi\times 10^{-7})10^6}} = \Delta = 5\times 10^{-3} \quad \Rightarrow f \approx 100\text{ Hz}$$

At a frequency of 10 kHz, the skin depth is about 0.5 mm. Considering the simplicity of the low-frequency and high-frequency approximations, they are surprisingly good.

A shield that is both conducting and magnetic has two possible mechanisms for shielding. The conducting nature of the shield allows for counter currents to be induced and counter magnetic fields to be generated reducing the net magnetic field. The magnetic nature of the shield allows for magnetic "shunting" or redirecting of the magnetic field away from the area to be protected. Frequently, for magnetic conducting shields, at low frequencies, the magnetic "shunting" effect is the most important.

At higher frequencies, both mechanisms are important. For both the spherical and cylindrical shells, the degree of shielding throughout the inside of the cavity is constant for a uniform applied field. At low frequencies for nonmagnetic conducting materials, the shielding effectiveness of the long cylindrical shell is slightly better than a spherical shell of the same thickness and radius:

$$\left|\frac{s_{cylinder,LF}}{s_{sphere,LF}}\right| = \left|\frac{\dfrac{1}{1+\dfrac{\dfrac{j\omega}{2}}{b\mu_o\sigma\Delta}}}{\dfrac{1}{1+\dfrac{\dfrac{j\omega}{3}}{b\mu_o\sigma\Delta}}}\right| = \left|\frac{6+j2\omega b\mu_o\sigma\Delta}{6+j3\omega b\mu_o\sigma\Delta}\right| < 1 \tag{3.88}$$

(The shielding effectiveness increases as the shielding factor decreases.) For magnetic or nonmagnetic conducting materials, at high frequencies, the shielding effectiveness of the cylinder is about 3.5 dB better than a sphere:

$$\left|\frac{s_{cylinder,HF}}{s_{sphere,HF}}\right| = \frac{\dfrac{2\sqrt{2}\mu_r\delta e^{-\frac{\Delta}{\delta}}}{b}}{\dfrac{3\sqrt{2}\mu_r\delta e^{-\frac{\Delta}{\delta}}}{b}} = \frac{2}{3} < 1$$

However, for magnetic materials where the flux "shunting" dominates over Eddy current effects, the sphere provides better shielding than the cylinder. For large permeabilities, the ratio of shielding factors is

$$\frac{s_{cylinder}}{s_{sphere}} \approx \frac{\dfrac{1}{1+\dfrac{1}{2}\dfrac{\mu_r\Delta}{b}}}{\dfrac{1}{1+\dfrac{2}{3}\dfrac{\mu_r\Delta}{b}}} = \frac{6+4\dfrac{\mu_r\Delta}{b}}{6+3\dfrac{\mu_r\Delta}{b}} > 1$$

3.12 Pure Magnetic Spherical Shell

The magnetic field inside the cavity of a magnetic spherical shell of inner radius b, outer radius a, and relative permeability μ_r when placed in a uniform magnetic field of strength H_e is

$$H_i = \frac{9\mu_r}{9\mu_r - 2(\mu_r-1)^2\left(\dfrac{b^3}{a^3}-1\right)} H_e \tag{3.89}$$

Determine H_i if the permeability is very large. Determine the necessary thickness of a steel body suit around an individual using an electric drill if 1 mG of field is tolerable (a typical value inside a house). Assume the external magnetic field near the drill is 2.5 mT at 60 Hz. [Thomas; Jackson; Gandhi, '90]

Because (3.89) does not involve the conductivity, this shielding expression does not consider counter currents and the fields they produce. Hence, the shielding is only due to the "magnetic duct" effect. When the relative permeability of the shell is one, corresponding to nonmagnetic material, the shielding effectiveness of the shell is 0 dB: the shell has no effect on the external field. Also, if the relative permeability is not large and shell thickness is small ($a \approx b$), the shielding effectiveness is poor. However, when the permeability is large ($\mu_r \gg 1$), this equation reduces to

$$H_i = \frac{9\mu_r}{9\mu_r - 2(\mu_r - 1)^2\left(\dfrac{b^3}{a^3} - 1\right)} H_e \approx \frac{9\mu_r}{9\mu_r - 2\mu_r^2\left(\dfrac{b^3}{a^3} - 1\right)} H_e = \frac{9}{9 - 2\mu_r\left(\dfrac{b^3}{a^3} - 1\right)} H_e \tag{3.90}$$

Furthermore, when the shield's thickness, $\Delta = a - b$, is small compared to the radius, *and* the relative permeability is large, this equation further simplifies to

$$\begin{aligned} H_i &= \frac{9}{9 - 2\mu_r\left[\dfrac{b^3}{(b+\Delta)^3} - 1\right]} H_e = \frac{9}{9 - 2\mu_r\left[\dfrac{b^3}{b^3\left(1 + \dfrac{\Delta}{b}\right)^3} - 1\right]} H_e \\ &= \frac{9}{9 - 2\mu_r\left[\left(1 + \dfrac{\Delta}{b}\right)^{-3} - 1\right]} H_e \approx \frac{9}{9 - 2\mu_r\left[\left(1 - \dfrac{3\Delta}{b}\right) - 1\right]} H_e = \frac{1}{1 + \dfrac{2}{3}\dfrac{\mu_r \Delta}{b}} H_e \end{aligned} \tag{3.91}$$

Although the shield is thin, sometimes a is substituted for b in the last expression in (3.91) to provide a more conservative result. The shielding effectiveness increases with the permeability and shield thickness but decreases with the shield radius. As the radius increases, the reluctance increases for a given shell thickness. The magnetic reluctance is a measure of the "resistance" of magnetic material to magnetic flux. Recall that the shielding effectiveness improved with increasing radius for nonmagnetic shields. Shields of high permeability provide a low-reluctance path for magnetic fields. The permittivity of the shell has no effect on the magnetic fields; however, the electric field is affected by the permittivity. Insulators, which are usually nonmagnetic, do not affect the magnetic field.

Note that (3.89) through (3.91) do not involve the frequency of the magnetic field. Equation (3.89) was derived assuming the magnetic field was nontime varying (i.e., magnetostatic). For time-varying fields, Eddy currents are induced in a shield when the conductivity is not zero. In (3.66), it was given that for a magnetic, conducting, spherical thin shell at low frequencies ($\Delta \ll \delta$)

$$\frac{H_{is}}{H_{es}} = \frac{1}{1 + \dfrac{j\omega}{\dfrac{3}{b\Delta\mu_o\sigma}} + \dfrac{2}{3}\dfrac{(\mu_r - 1)^2\Delta}{\mu_r b}}$$

When the permeability is large, this equation reduces to

$$\frac{H_{is}}{H_{es}} \approx \frac{1}{1 + \dfrac{j\omega}{\dfrac{3}{b\Delta\mu_o\sigma}} + \dfrac{2}{3}\dfrac{\mu_r\Delta}{b}} \tag{3.92}$$

where σ is the conductivity of the shell. Furthermore, if the frequency is sufficiently low so that

$$\omega \ll \frac{3}{b\Delta\mu_o\sigma} \tag{3.93}$$

the equation further reduces to

$$\frac{H_{is}}{H_{es}} \approx \frac{1}{1+\frac{2}{3}\frac{\mu_r\Delta}{b}} \tag{3.94}$$

assuming that $\mu_r \gg 1$. The two frequency-related conditions imposed on this final result are then

$$2\pi f \ll \frac{3}{b\Delta\mu_o\sigma} \quad \text{or} \quad \sqrt{\frac{2}{3}\frac{b\Delta}{\mu_r}} \ll \delta = \frac{1}{\sqrt{\pi f\mu_r\mu_o\sigma}} \tag{3.95}$$

and

$$\Delta \ll \delta \tag{3.96}$$

It is sometimes stated that (3.89), a magnetostatic-based equation, is valid if the skin depth is "large."

Now, referring back to the original body-suit problem statement, the shielding required in this situation is, converting 1 mG to T,

$$20\log\left(\frac{1\times10^{-3}\times10^{-4}}{2.5\times10^{-3}}\right) \approx -88\text{ dB}$$

Even though steel is a conductor with about one-tenth the conductivity of copper, (3.89) can still be used to determine the potential usefulness of steel in guiding magnetic fields around the drill operator. (In this case, at 60 Hz, $\delta \approx 0.85$ mm.) As is seen in Mathcad 3.8, there are no body suit sizes (of 0.1 mm

$\mu_r := 1000$

$b := 0.1, 0.11 .. 0.5$

$$G(b,\Delta) := \frac{9\cdot\mu_r}{9\cdot\mu_r - 2\cdot(\mu_r - 1)^2\cdot\left[\left(\frac{b}{b+\Delta}\right)^3 - 1\right]}$$

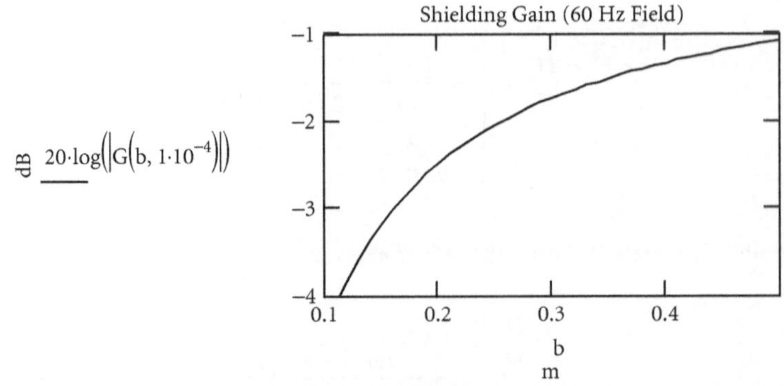

MATHCAD 3.8 Shielding gain of a 0.1 mm thick steel spherical shield vs. shield radius.

$$\Delta := 0.01$$

$$x := -20, -19.1 .. 10$$

$$b(x) := \left(x + 1 - 10 \cdot \text{floor}\left(\frac{x}{10}\right)\right) \cdot 10^{\text{floor}\left(\frac{x}{10}\right)}$$

$$G(b, \Delta, \mu_r) := \frac{9 \cdot \mu_r}{9 \cdot \mu_r - 2 \cdot (\mu_r - 1)^2 \cdot \left[\left(\frac{b}{b + \Delta}\right)^3 - 1\right]}$$

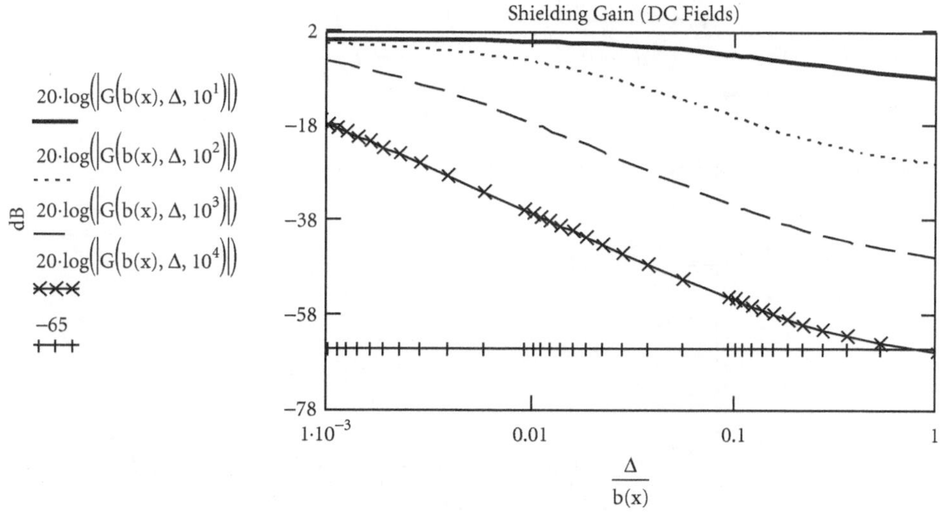

MATHCAD 3.9 Shielding gain at dc vs. the ratio of the spherical shield's thickness to radius. The permeability is varied from 10 to 10,000. As expected, the shielding effectiveness increases as the permeability increases.

thickness) that satisfy this shielding requirement. (This thickness satisfies the frequency restrictions placed on this equation.) Near-field magnetic field shielding of this degree is extremely difficult when using only steel.

Multiple thin shells of high permeability can provide greater shielding levels. When the layers are close, the shielding from each layer is additive. However, when the shields are well separated, the layers are decoupled from each other and the shielding from each layer is multiplicative.

In Mathcad 3.9, a dc applied field is assumed so there are no restrictions on the thickness of the shield. It can be seen that even when the relative permeability is 10,000, the shielding gain saturates at a value of about –65 dB. This occurs when the shielding thickness is equal to the radius of the magnetic shell. (The shielding gain in dB is the negative of the shielding effectiveness in dB.)

3.13 Pure Magnetic Cylindrical Shell

The magnetic field inside the cavity of a magnetic cylindrical shell of inner radius b, outer radius a, and relative permeability μ_r when placed in a uniform magnetic field of strength H_e at right angle to it (i.e., transverse to the axis of the cylinder) is

$$H_i = \frac{4\mu_r}{4\mu_r - (\mu_r - 1)^2\left(\frac{b^2}{a^2} - 1\right)} H_e \tag{3.97}$$

Determine H_i if the permeability is very large. Are there any assumptions related to the frequency inherent in this equation? A field of 0.25 G can produce a one character shift on a display screen. If a 60 Hz 1 G signal is present, determine the necessary thickness and permeability of a practical shield to be placed around the cathode ray tube to avoid this character shift. State specifically the material to be employed. [Thomas; Jackson]

When the shell is nonmagnetic, $\mu_r = 1$, there is no difference between the field inside and outside the shell. When the relative permeability is large, this equation reduces to

$$H_i \approx \frac{4\mu_r}{4\mu_r - (\mu_r)^2\left(\frac{b^2}{a^2}-1\right)} H_e = \frac{4}{4-\mu_r\left(\frac{b^2}{a^2}-1\right)} H_e \tag{3.98}$$

If the relative permeability is large *and* shield thickness (Δ) is small, then

$$\begin{aligned} H_i &= \frac{4}{4-\mu_r\left[\frac{b^2}{(b+\Delta)^2}-1\right]} H_e = \frac{4}{4-\mu_r\left[\frac{b^2}{b^2\left(1+\frac{\Delta}{b}\right)^2}-1\right]} H_e \\ &= \frac{4}{4-\mu_r\left[\left(1+\frac{\Delta}{b}\right)^{-2}-1\right]} H_e \approx \frac{4}{4-\mu_r\left[\left(1-\frac{2\Delta}{b}\right)-1\right]} H_e = \frac{1}{1+\frac{1}{2}\frac{\mu_r\Delta}{b}} H_e \end{aligned} \tag{3.99}$$

where $a = b + \Delta$. Although the shield is thin, sometimes a is substituted for b in the last expression in (3.99) to provide a more conservative result. The shielding effectiveness for spherical shields (of equal thickness, radius, and permeability) is slightly greater than for long cylindrical shields.

As with the spherical shields, (3.97) was derived assuming that the applied field is static (dc). To use this equation at nonzero frequencies for shells of nonzero conductivity, the skin depth of the material should be "large." Generally, this implies that the skin depth should be much greater than the dimensions of the system (such as the shell thickness).

It will be assumed that if the field is reduced to one-tenth of 0.25 G that the character shifting will be negligible or acceptable. (Actually, a 0.01 G magnetic field can cause a wiggle.) The shielding required in this situation is

$$20\log\left(\frac{0.025}{1}\right) \approx -32 \text{ dB}$$

The analysis in Mathcad 3.10 indicates that a cylindrical shield of 5 cm radius and 0.1 mm ($\approx$ 4 mils) thickness, with a relative permeability of 40,000, will satisfy this shielding requirement. (The skin depth, δ, is about 0.25 mm for one particular magnetic material.) Some interesting results for dc applied fields are given in Mathcad 3.11.

Typically, as the relative permeability increases, the saturation level of the material decreases as is shown in Table 3.3. Ferrimagnetic materials saturate around 2,000–5,000 G (0.2–0.5 T). A ferrimagnetic material should be adequate for a 1 G (100 μT) magnetic field.

$$\Delta := 0.1 \cdot 10^{-3}$$

$$b := 0.01, 0.011 .. 0.1$$

$$G(b, \Delta, \mu_r) := \frac{4 \cdot \mu_r}{4 \cdot \mu_r - (\mu_r - 1)^2 \cdot \left[\left(\frac{b}{b + \Delta} \right)^2 - 1 \right]}$$

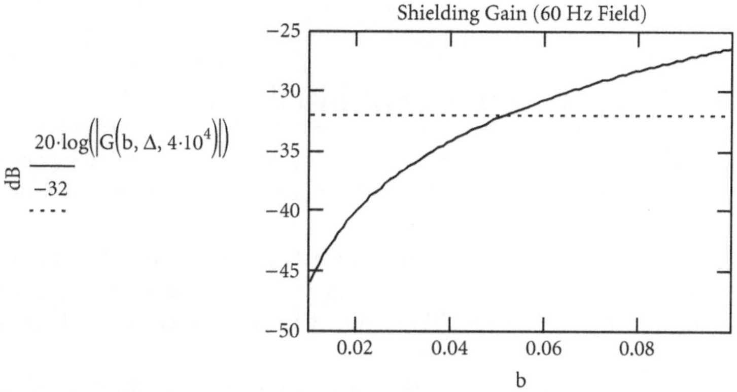

MATHCAD 3.10 Shielding gain of a 0.1 mm thick high-permeability cylindrical shell vs. shield radius.

$$\Delta := 0.01$$

$$x := -20, -19.1 .. 10$$

$$b(x) := \left(x + 1 - 10 \cdot \mathrm{floor}\left(\frac{x}{10} \right) \right) \cdot 10^{\mathrm{floor}\left(\frac{x}{10} \right)}$$

$$G(b, \Delta, \mu_r) := \frac{4 \cdot \mu_r}{4 \cdot \mu_r - (\mu_r - 1)^2 \cdot \left[\left(\frac{b}{b + \Delta} \right)^2 - 1 \right]}$$

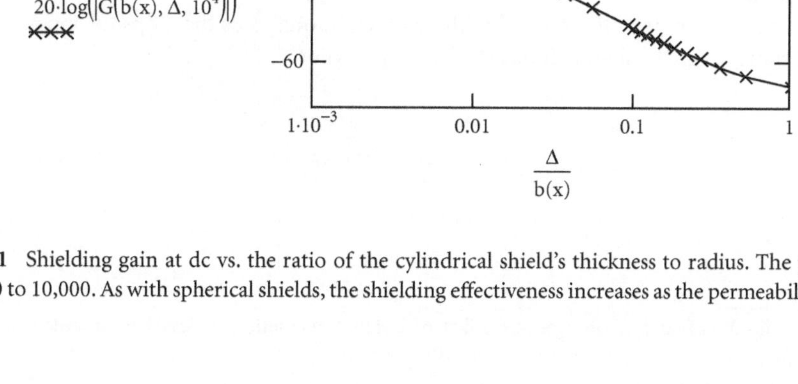

MATHCAD 3.11 Shielding gain at dc vs. the ratio of the cylindrical shield's thickness to radius. The permeability is varied from 10 to 10,000. As with spherical shields, the shielding effectiveness increases as the permeability increases.

TABLE 3.3 Typical Relationship Between the Relative Permeability and Saturation Level

μ_r	Saturation Level
300	2 T
10,000	1.5 T
35,000	0.8 T

3.14 Finite-Length Cylindrical Shell

A cylindrical shield is finite in length. After reviewing the relevant literature, explain qualitatively and quantitatively the effects of this finite length on the shielding effectiveness. [Mager, '70; Mager, '68; Watson; Haus; Jackson]

A shielded device nearly always requires one or more external connections. For this reason, most shields have openings. For some shielding cans, for example, the top or bottom of the can is left open so that the internal device can be adjusted or accessed. There are a limited number of simple analytical expressions that model these openings.

Before the modified shielding expressions are presented, the concepts of magnetic charge and demagnetization will be discussed. When a magnetic material is placed in a magnetic field, the field is usually disturbed or distorted by this magnetic material. It is helpful to model this effect via induced, hypothetical magnetic "charges" inside and along the surface of the magnetic material. With dielectrics, it is easy to visualize induced, real electric charges along the surface of the dielectric. With magnetic materials, it is more difficult to visualize magnetic charges especially since they are not actually "charge."[16] Nevertheless, the concept of magnetic charge is helpful. Actually, this concept is probably familiar to most educated children since they probably have experimented with magnets and the attraction of the north and south poles. An ideal magnet is a magnetic material that is permanently magnetized. For an ideal magnet, the magnetic charges are always present, with or without an external magnetic field. These poles are a source of magnetic field. With magnetic materials that are not permanently magnetized or do not have any residual magnetization, poles can only appear when exposed to an external magnetic field. These poles of magnetic charge are a source of magnetic field.

When an infinitely long cylinder is placed in a parallel magnetic field, no magnetic surface charges are induced along the surface since the field is entirely tangential to its surface. For either volume or surface magnetic charge to be induced, the magnetic field must have some nonzero component in the direction of the gradient of the permeability (i.e., in the direction that the permeability is changing). Assuming the frequency is dc or sufficiently low so that any conductivity-related effects are negligible, the parallel magnetic field passes straight through the cylinder. However, when this long cylinder is placed in a transverse field, magnetic surface charges are induced on both sides of the cylinder. These magnetic charges generate a field that opposes the applied field. The shielding action can also be explained in terms of magnetic field "shunting."

The shielding effectiveness of a finite-length cylinder (modeled as the ellipsoid shown in Figure 3.43) in a parallel static magnetic field is indicated by Mager to be

$$SE_p = \frac{H_e}{H_i} \approx (4N)SE_t + 1 \tag{3.100}$$

[16]Some theories discuss the theoretical possibility of isolated magnetic charges or monopoles.

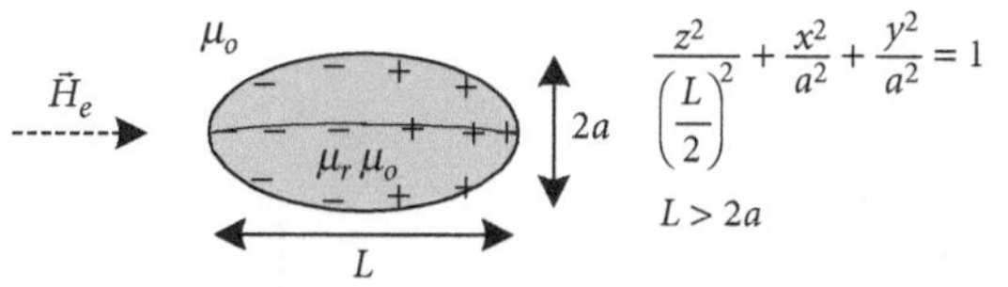

FIGURE 3.43 Magnetic field parallel to the longer axis of a prolate spheroid. A crude sketch of the magnetic surface charges is shown.

where SE_t is the shielding effectiveness for a thin-wall, high-relative permeability cylinder of infinite length in a transverse magnetic field. The variable N is the demagnetization factor for a solid prolate spheroid:

$$N = \frac{1}{k^2 - 1}\left[\frac{k}{\sqrt{k^2-1}}\ln(k + \sqrt{k^2-1}) - 1\right] \quad \text{where } k = \frac{L}{2a} \tag{3.101}$$

The length-to-diameter ratio of the solid spheroid is the same as the finite-length cylinder (where the diameter corresponds to the outer diameter of the cylinder). The relative permeability is the average value over the cross section of the shell and interior of the thin-walled cylinder. Using the expression given in (3.100),

$$SE_p \approx (4N)\left(1 + \frac{1}{2}\frac{\mu_r \Delta}{a}\right) + 1 \approx \frac{2N\mu_r \Delta}{a} + 1 \quad \text{if } \mu_r \Delta \gg a \tag{3.102}$$

where Δ is the thickness and a is the outer radius of the cylinder. For very long cylinders modeled as a prolate spheroid, $k \gg 1$ and N is small. For an infinite-length spheroid, N is zero:

$$N \approx \frac{1}{k^2}\left[\frac{k}{\sqrt{k^2}}\ln(k + \sqrt{k^2}) - 1\right] \approx \frac{1}{k^2}\ln(2k) \approx 0 \quad \text{if } k \gg 1 \tag{3.103}$$

In this case, $SE_p = 1$ for an infinite-length cylinder in a parallel static field. For finite-length cylinders, the magnetic charges induced on the two ends of the prolate spheroid model of the cylinder increase SE_p. The magnetic charges generate a demagnetization field as discussed previously. Another way of viewing this increase in shielding effectiveness is that the rounded ends of the spheroid are guiding or "shunting" the magnetic field around the interior. Mager indicates that for a cylinder with end caps,

$$SE_p = \frac{(4N)SE_t + 1}{1 + \frac{a}{L}} \approx \frac{(4N)\left(1 + \frac{1}{2}\frac{\mu_r \Delta}{a}\right) + 1}{1 + \frac{a}{L}} \tag{3.104}$$

As the length increases, the shielding effectiveness according to (3.104) decreases. This relationship is verified in Mathcad 3.12. The magnetic fields that are parallel to the sides of the finite-length cylinder are actually perpendicular to the end caps or to the top and bottom of the closed cylinder. These *perpendicular* fields pass through to the interior of the cylinder. For large relative permeabilities and some curvature to these end caps, the fields are partially shunted past the interior of the cylinder.

Mager also provides some expressions for the variation of the fields within open-ended cylinders in static magnetic fields. For magnetic fields that are transverse to a finite-length open-ended cylinder, the interior magnetic field along the axis of the cylinder is (due to the open end only)

$$H_{i,oe} \approx \frac{H_{et}}{3} e^{-\frac{k_t z}{b}} \quad \text{where } k_t = \begin{cases} 3.832 & \text{(theoretical)} \\ 3.52 & \text{(measured mean)} \end{cases} \tag{3.105}$$

$$a := 0.01$$

$$L := 0.02, 0.021..\,0.3$$

$$N(a, L) := \frac{1}{\left(\frac{L}{2\cdot a}\right)^2 - 1}\cdot\left[\frac{\frac{L}{2\cdot a}}{\sqrt{\left(\frac{L}{2\cdot a}\right)^2 - 1}}\cdot \ln\left[\frac{L}{2\cdot a} + \sqrt{\left(\frac{L}{2\cdot a}\right)^2 - 1}\right] - 1\right]$$

$$SE_p(\mu_r, \Delta, a, L) := \frac{\left[4\cdot N(a, L)\cdot\left(1 + \frac{1}{2}\cdot\frac{\mu_r\cdot\Delta}{a}\right)\right] + 1}{1 + \frac{a}{L}}$$

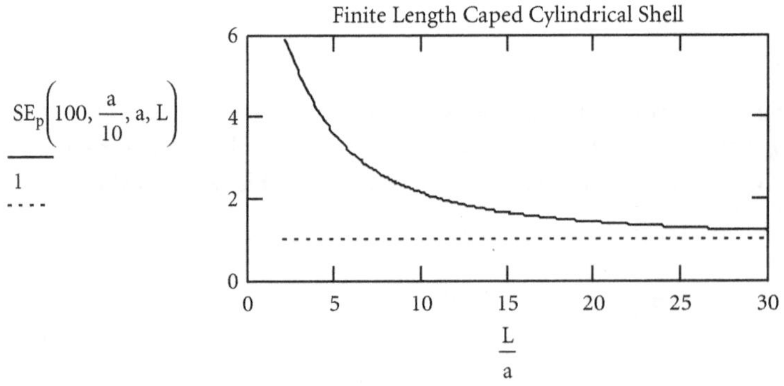

MATHCAD 3.12 Shielding effectiveness of a finite-length caped cylindrical shell as a function of L/a. The relative permeability of the shell is 100.

where H_{et} is the external, transverse magnetic field. For distances of about one diameter into the cylinder, the interior magnetic field due to the open end is small:

$$H_{i,oe} \approx \frac{H_{et}}{3} e^{-\frac{(3.832)(2b)}{b}} \approx 0.16\times 10^{-3} H_{et}$$

For fields parallel to the cylinder, the interior magnetic field is (due to the open end only)

$$H_{i,oe} \approx 1.3 H_{ep}\sqrt{\frac{L}{2a}}e^{-\frac{k_p z}{b}} \quad \text{where } k_p = \begin{cases} 2.405 & \text{(theoretical)} \\ 2.26 & \text{(measured mean)} \end{cases} \tag{3.106}$$

where H_{ep} is the external, parallel magnetic field. As shown in Figure 3.44, a is the outer radius, b is the inner radius, and z is along the axis of the cylinder. The fields decay exponentially into the cylinder. As expected, as the inner radius of the cylinder increases, the field decreases more slowly with z. However, for parallel fields, increasing the outer radius decreases the strength of the interior field since more

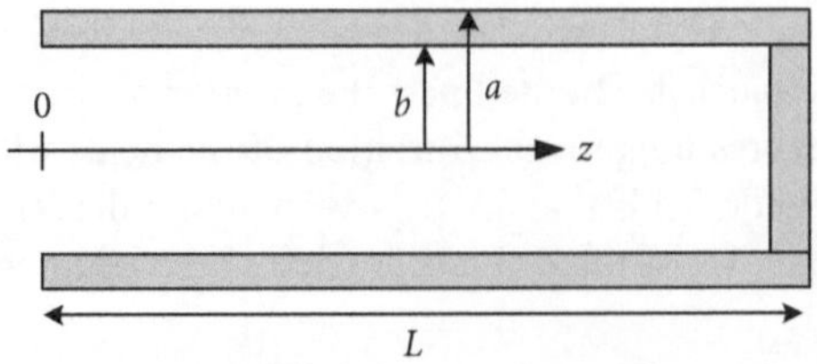

FIGURE 3.44 Side view of an open-ended cylindrical shell.

material is available for flux channeling. At the middle of the cylinder, $z = L/2$, the corresponding shielding effectiveness expressions are

$$SE_{oe,t} = \frac{H_{et}}{2H_{i,oe}} = \frac{3}{2}e^{\frac{k_t L}{2b}} \quad \text{if } z = L/2 \tag{3.107}$$

$$SE_{oe,p} = \frac{H_{ep}}{2H_{i,oe}} = \frac{1}{2.6}\sqrt{\frac{2a}{L}}e^{\frac{k_p L}{2b}} \quad \text{if } z = L/2 \tag{3.108}$$

(At the center, with *both* ends of the cylinder open, the internal magnetic field, $H_{i,oe}$, was multiplied by two.) Mager states that the total shielding effectiveness is essentially the parallel combination of SE_{oe} with the corresponding *SE* without the openings:

$$SE_{eff,t} = \frac{SE_{oe,t}SE_t}{SE_{oe,t} + SE_t}, \quad SE_{eff,p} = \frac{SE_{oe,p}SE_p}{SE_{oe,p} + SE_p} \tag{3.109}$$

This expression is assuming the *SE*'s are much greater than one (similar to assuming the *Q*'s of a lossy *L* and *C* are large to obtain an effective *Q* for a circuit).

3.15 Shielding the Source and Shielding Reciprocity

Frequently, magnetic materials such as μ-metal and steel are used in magnetic field shielding, but copper can also be used. A time-varying magnetic field induces currents, Eddy currents, in the copper, generating a "scattering" field that opposes the external magnetic field. Determine the shielding effectiveness of a 0.15 mm thick, 0.3 m radius spherical copper shield around a 1 kHz magnetic field source. [Hoburg, '95; Hasselgren; Rikitake; Gustafson; Wait, '77]

In this situation, the source of the magnetic field is inside the shield. The shielding equations given previously assumed that the source of the magnetic field was outside the shell or shield. Although, reciprocity can be applied to shields (with linear properties), it must be carefully applied.[17] Generally, a shield that is effective in reducing magnetic fields originating on one side of the shield (e.g., its interior) is also effective in reducing magnetic fields originating on the other side of the shield (e.g., its exterior). However, the degree of shielding, and whether the magnetic material saturates, can be different.

The shielding expressions previously given for the spherical shell and infinitely long cylindrical shell can also be applied for specific source types inside the shield. Recall that the shielding factors for the spherical and cylindrical shell were determined for external fields, which were uniform before being perturbed or distorted by the shield. The fields inside the shielded region or cavity were uniform. These same shielding factor expressions also apply when the magnetic field source inside the shield is a magnetic dipole centered about the origin (for the sphere) or axis (for the cylinder). An electric dipole consists of two equal but opposite sign charges spaced very close together. For the spherical shield, the magnetic dipole source is a small current-carrying loop centered about the sphere's origin. For the long cylindrical shield, the magnetic dipole source is two long, closely spaced conductors carrying equal but opposite sign currents that are centered about the cylinder's axis.[18] The magnetic field outside the shield for these two magnetic dipole sources has the same field distribution as without the shield but attenuated everywhere by the amount given by the shielding factor. Referring to Figure 3.45, for the current sources to

[17]Even the author, during his informative years, has accidentally misused reciprocity.

[18]Some students ponder why a single line current was not used instead of a pair of line currents. The source of the magnetic field cannot be a single line current for the cylindrical shield in Figure 3.45 since the return current would have to be at infinity outside the shield.

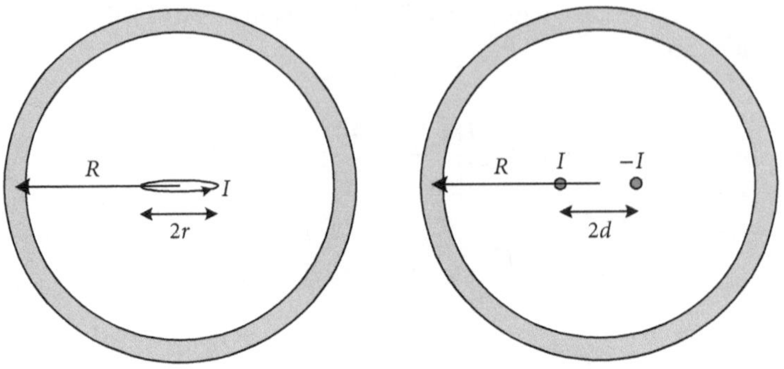

FIGURE 3.45 Magnetic dipole sources inside a spherical and cylindrical shell.

be considered small, $R \gg r$ for the sphere and $R \gg d$ for the cylinder. Many magnetic field sources can be (at least roughly) modeled as a magnetic dipole if the point of interest is far from the source. For example, the magnetic field outside a steel tube enclosing a pair of current-carrying conductors can sometimes be modeled in this way.

The magnetic field from the 1 kHz source will be modeled as a magnetic dipole. With this major assumption, and by permitting the shield around the source to be spherical in shape, the equations given previously may be employed. Using the low-frequency approximation for a thin spherical shield, magnetic shielding occurs when (3.62) is satisfied:

$$3 \ll b\omega\mu_o\sigma\Delta$$

This low-frequency approximation is valid when

$$\sigma \gg \omega\varepsilon,\ b \gg \Delta,\ \gamma\Delta \ll 1,$$

Copper is definitely a good conductor at 1 kHz. If the radius of this shield, b, is about 0.3 m, then the thickness of the shield, Δ, should be less than 3 cm. Finally,

$$|\gamma\Delta| \approx |\sqrt{j\omega\mu_o\sigma}\,\Delta| = \Delta\sqrt{\omega\mu_o\sigma} = \Delta\sqrt{(2\pi\times10^3)(4\pi\times10^{-7})(5.8\times10^7)}$$

$$\approx 680\Delta \ll 1 \quad \Rightarrow \Delta \ll 1.5\,\text{mm}$$

The shield should be less than 0.15 mm thick. If the shield is 0.15 mm ($\approx$ 6 mils) thick with a radius of 0.3 m, then the inequality is satisfied:

$$3 \ll b\omega\mu_o\sigma\Delta \approx 21$$

A shell thickness of 0.15 mm is much less than the skin depth of copper at 1 kHz:

$$\delta = \frac{1}{\sqrt{\pi f \mu_o \sigma}} = 2.1\,\text{mm} \approx 82\,\text{mil}$$

Because the skin depth is large compared to 0.15 mm, absorption has only a minor role in the magnetic shielding for these small thicknesses. The "bucking" currents are generating a counter magnetic field that opposes the applied magnetic field. There is 17 dB of magnetic shielding:

$$20\log\left|\frac{1}{1+\dfrac{jb\omega\mu_o\sigma\Delta}{3}}\right| \approx 20\log\left|\frac{1}{1+j6.9}\right| \approx -17\,\text{dB}$$

$\mu_o := 4\cdot\pi\cdot10^{-7}$ $\quad\varepsilon_o := 8.854\cdot10^{-12}$ $\quad\sigma := 5.8\cdot10^{7}$ $\quad j := \sqrt{-1}$

$x := 10, 11..\ 60$

$$\omega(x) := \left(x + 1 - 10\cdot\text{floor}\left(\frac{x}{10}\right)\right)\cdot10^{\text{floor}\left(\frac{x}{10}\right)}$$

$$\gamma(\omega) := \sqrt{j\cdot\omega\cdot\mu_o\cdot(\sigma + j\cdot\omega\cdot\varepsilon_o)}$$

$$SE(\omega, b, \Delta) := 20\cdot\log\left[\left|\frac{\gamma(\omega)^2\cdot b^2\cdot\left[\left[1+\frac{2}{(\gamma(\omega)\cdot b)^2}\right]\cdot\sinh(\gamma(\omega)\cdot\Delta) + \frac{3}{\gamma(\omega)\cdot b}\cdot\cosh(\gamma(\omega)\cdot\Delta)\right]}{[3\cdot\gamma(\omega)\cdot(b+\Delta)]}\right|\right]$$

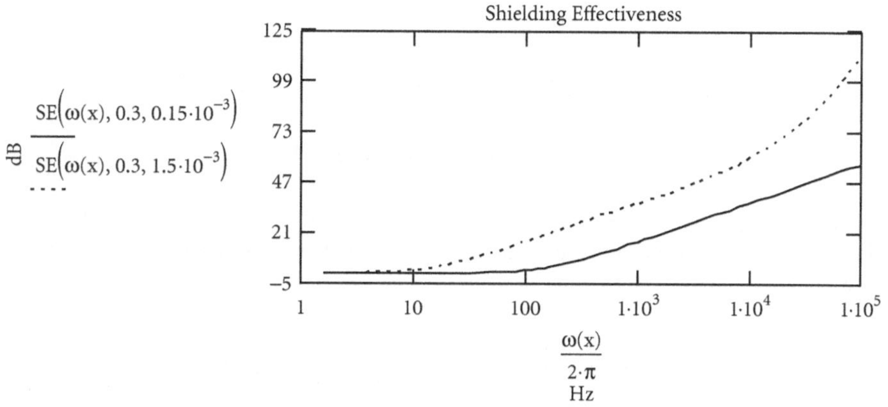

MATHCAD 3.13 SE is plotted vs. frequency for a 0.3 m radius spherical shield of 0.15 mm and 1.5 mm thickness. As expected, the shielding effectiveness increases with thickness.

Generally speaking, if a shield is effective against near-field magnetic fields, it is also effective against near-field electric fields and plane waves since most magnetic materials are conductive.

The inverse of Equation (3.58) is used to determine the shielding effectiveness vs. frequency for a 0.15 mm and 1.5 mm thick, 0.3 m radius shield. As seen in Mathcad 3.13, as the frequency increases, the shielding effectiveness increases.

3.16 Shielding a Cosmetologist with a Body Suit

Determine the necessary thickness of an aluminum body suit around a cosmetologist using a hair dryer if 0.1 μT rms of field is acceptable. Assume that the external magnetic field near the source, a hair dryer, is 10 μT rms at 60 Hz. [Hotte; Chou; Guy; Joyner; Gustafson]

An experienced individual will immediately realize the futility of shielding against low-frequency magnetic fields using (reasonably thin) aluminum, a nonmagnetic conducting material with $\mu_r = 1$. However, materials with $\mu_r > 1$ can provide some guiding of magnetic fields.[19] A shield constructed of magnetic material, at even low frequencies, can "collect" the flux density thereby distorting the magnetic field. This distortion is a function of the shape of the shield and its permeability. Furthermore, if the shield is

[19]If a magnetic field is entirely normal to a large slab of magnetic material, the magnetic field exiting the slab will be equal to that entering the slab. One of the classical boundary conditions in electromagnetics states that the normal flux density, B_n, is continuous across a boundary. Thus, magnetic material does not always provide shielding.

conducting and the frequency of the magnetic field is sufficiently high, a reasonable counter magnetic field is generated to further assist in shielding.

Before the properties of real body suits are presented, an idealized long, thin, cylindrical, conducting nonmagnetic shield will be analyzed. The approximate low-frequency expression for the magnitude of the ratio of the internal to external magnetic field (for either parallel or transverse fields) is

$$\left|\frac{H_{is}}{H_{es}}\right| = \left|\frac{1}{1+\dfrac{j\omega}{\dfrac{2}{b\mu_o\sigma\Delta}}}\right| = \frac{1}{\sqrt{1+\left(\dfrac{\omega b\mu_o\sigma\Delta}{2}\right)^2}} \tag{3.110}$$

where b is the radius and Δ is the thickness of the conductive cylindrical shield. This ratio should be less than or equal to $0.1/10 = 0.01$ or -40 dB. This approximation is valid when

$$\sigma \gg \omega\varepsilon,\, b \gg \Delta,\, \gamma\Delta \ll 1,$$

Aluminum is definitely a good conductor at 60 Hz, hence, the first inequality is satisfied. If the radius of this shield, b, is about 0.3 m, then the thickness of the shield, Δ, should be less than 3 cm to satisfy the second inequality. Finally,

$$|\gamma\Delta| \approx |\sqrt{j\omega\mu_o\sigma}\Delta| = \Delta\sqrt{\omega\mu_o\sigma} = \Delta\sqrt{(2\pi\times 60)(4\pi\times 10^{-7})(0.61\times 5.8\times 10^7)}$$
$$\approx 130\Delta \ll 1 \quad \Rightarrow \Delta \ll 7.7\text{ mm}$$

The shield should be less than 0.77 mm thick. The required cylinder thickness can be solved using the low-frequency approximation:

$$\frac{1}{\sqrt{1+\left[\dfrac{(2\pi\times 60)(0.3)(4\pi\times 10^{-7})(0.61\times 5.8\times 10^7)\Delta}{2}\right]^2}} < 0.01 \quad \Rightarrow \Delta > 4\text{ cm}$$

Unfortunately, this thickness is much greater than 0.77 mm and, thus, the low-frequency approximation should not be used. The high-frequency approximation can be used if the thickness of the shell is much greater than the skin depth of aluminum at 60 Hz:

$$\Delta \gg \delta = \frac{1}{\sqrt{\pi f\mu_o\sigma}} = 1.1\text{ cm}$$

Using the high-frequency approximation and solving for Δ,

$$\left|\frac{H_{is}}{H_{es}}\right| = \frac{2\sqrt{2}\mu_r\delta e^{-\frac{\Delta}{\delta}}}{b} < 0.01 \quad \Rightarrow \Delta > 2.5\text{ cm}$$

Although $\Delta > \delta$, it is not much greater than δ. The exact expressions for both the parallel and transverse shielding effectiveness of the cylindrical shell are plotted in Mathcad 3.14. The aluminum shell of radius 0.3 m should have a thickness of 2.5 cm ($\approx$ 1 inch) to have a shielding effectiveness of 40 dB at 60 Hz.

$\mu_o := 4\cdot\pi\cdot 10^{-7} \qquad \varepsilon_o := 8.854\cdot 10^{-12} \qquad \sigma := 0.61\cdot 5.8\cdot 10^{7} \qquad j := \sqrt{-1}$

$x := 10, 11 .. 30$

$$\omega(x) := \left(x + 1 - 10\cdot \text{floor}\left(\frac{x}{10}\right)\right)\cdot 10^{\text{floor}\left(\frac{x}{10}\right)}$$

$$\gamma(\omega, \mu_r) := \sqrt{j\cdot\omega\cdot\mu_r\cdot\mu_o\cdot(\sigma + j\cdot\omega\cdot\varepsilon_o)}$$

$$SE_p(\omega, b, \Delta, \mu_r) := 20\cdot\log\left[\left|\cosh(\gamma(\omega,\mu_r)\cdot\Delta) + \frac{\gamma(\omega,\mu_r)\cdot(b+\Delta)}{2\cdot\mu_r}\cdot\sinh(\gamma(\omega,\mu_r)\cdot\Delta)\right|\right]$$

$$SE_t(\omega, b, \Delta, \mu_r) := 20\cdot\log\left[\left|\cosh(\gamma(\omega,\mu_r)\cdot\Delta) \ldots + \frac{1}{2}\cdot\left[\frac{\gamma(\omega,\mu_r)\cdot(b+\Delta)}{\mu_r} + \frac{\mu_r}{[\gamma(\omega,\mu_r)\cdot(b+\Delta)]}\right]\cdot\sinh(\gamma(\omega,\mu_r)\cdot\Delta)\right|\right]$$

$\Delta := 2.5\cdot 10^{-2}$

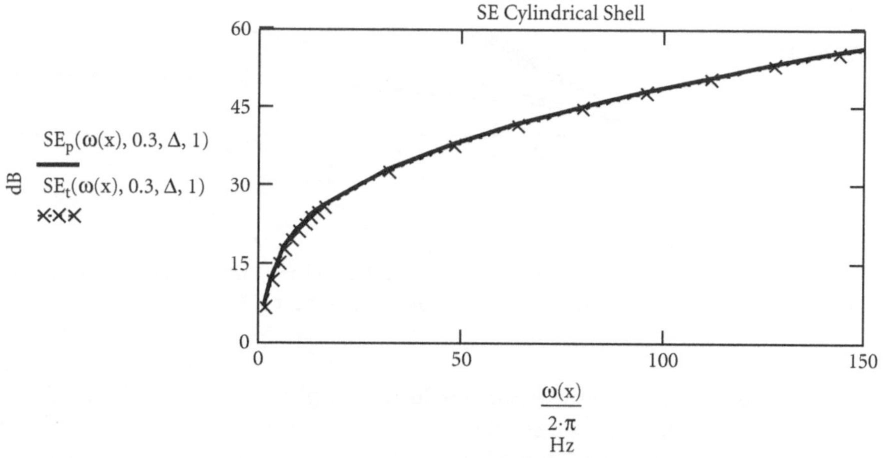

$SE_p(2\cdot\pi\cdot 60, 0.3, \Delta, 1) = 40.627 \qquad SE_p(2\cdot\pi\cdot 60, 0.3, \Delta, 1) - SE_t(2\cdot\pi\cdot 60, 0.3, \Delta, 1) = -1.517 \times 10^{-4}$

MATHCAD 3.14 Shielding effectiveness of a cylindrical shell for both parallel and transverse magnetic fields.

For the particular set of parameters used, the shielding effectiveness for either parallel or transverse fields is essentially the same.

An aluminum body suit that is about 2 cm thick is not practical. If a magnetic material such as steel is used (μ_r about 200 or greater), the thickness is reduced to the mm range as is shown in Mathcad 3.15 (for an SE of about 40 dB). For finite-length open-ended cylindrical shields, such as with some shielding cans, the SE is sometimes much less than predicted by the theoretical equations. As with most shields, openings are a major source of loss.

The expressions used for the cylindrical body suit assumed a constant permeability and infinite length. For distances of about one diameter inside a finite-length cylinder from its open end, the shielding is about that of an infinite-length cylindrical shield. Furthermore, the field intensity will vary within the shell material and, hence, the permeability will also vary. In some locations, the permeability may saturate while in other locations it may not.

Metal whole-body suits, gloves, and hoods are commercially available. Workers use them when the power source of the field cannot be de-energized or other forms of shielding are inadequate. High fields cannot be avoided, for example, when maintenance personnel are working on live, high-voltage, power line conductors and when medical workers are operating near high-power field sources used in some

$\mu_o := 4\cdot\pi\cdot10^{-7}$ $\quad \varepsilon_o := 8.854\cdot10^{-12}$ $\quad \sigma := 0.5\cdot5.8\cdot10^{7}$ $\quad \omega := 2\cdot\pi\cdot60$ $\quad b := 0.3$ $\quad j := \sqrt{-1}$

$\gamma(\mu_r) := \sqrt{j\cdot\omega\cdot\mu_r\cdot\mu_o\cdot(\sigma + j\cdot\omega\cdot\varepsilon_o)}$

$\Delta := 0.0001, 0.0002..\ 0.01$

$$SE_t(\Delta, \mu_r) := 20\cdot\log\left[\left|\cosh(\gamma(\mu_r)\cdot\Delta) + \frac{1}{2}\cdot\left[\frac{\gamma(\mu_r)\cdot(b+\Delta)}{\mu_r} + \frac{\mu_r}{[\gamma(\mu_r)\cdot(b+\Delta)]}\right]\cdot\sinh(\gamma(\mu_r)\cdot\Delta)\right|\right]$$

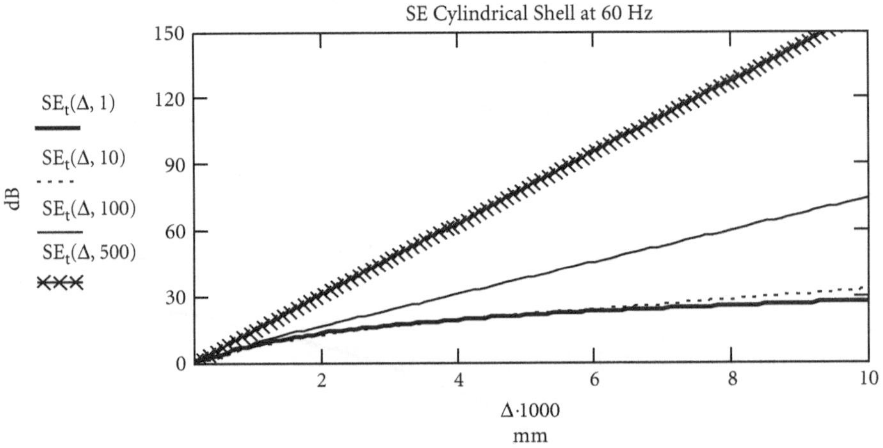

MATHCAD 3.15 SE vs. thickness for a 0.3 m radius cylindrical shield for various relative permeabilities. In these plots, the relative conductivity (with respect to copper) is assumed to be 0.5.

cancer treatments. Body suits can provide some reduction in plane waves and electric fields, but their effectiveness against low-frequency magnetic fields (e.g., < few MHz) is not good. To provide a relatively light and flexible suit, the suits are composed of metallized fabric such as fiberglass cloth with carbon threads, woven silver-coated nylon yarn, and woven stainless steel fibers. (Although magnetic material, such as steel and μ-metal, when used to construct a typical solid shield can be effective, the properties of magnetic materials can change dramatically with processing and when in fiber form.) It is also important to realize that face openings in hoods, zippers, and loose overlaps in the suit can allow electromagnetic fields to pass directly to the user. These openings can even act like waveguides to these signals. For example, a 15 cm wide opening is about one-half of a wavelength at 1 GHz.

As with ESD coatings, a parameter that is used to characterize the goodness of a body suit is its resistance. Unfortunately, the resistance, partially because of the mesh nature of the material, can change with washing, wear, time, corrosion, and movement. The method used to measure the resistance, including the contact resistance and location of the test electrodes, will also affect the results. Furthermore, the resistance can change with the applied test voltage and is thus nonlinear.

In the presence of a strong electric field, a large voltage can appear across an individual. Although the electric field may be approximately known when an individual is not present, with or without a suit, a person will perturb or distort the electric field. The position of the individual relative to the field and equipotentials will partially determine the strength of the voltage difference. For example, when an individual is oriented mostly along an equipotential, the voltage across the individual will be small compared to when the individual is oriented perpendicular to this equipotential. The degree of distortion generated by introducing a conductive (or dielectric) body into an electric field is a function of the shape of the body and its electric properties. For an object with finite resistance, a voltage difference across the object implies current through the object. Depending on this voltage difference, the resultant current can be "imperceptible to annoying." Although more complex models are available, a simple low-frequency

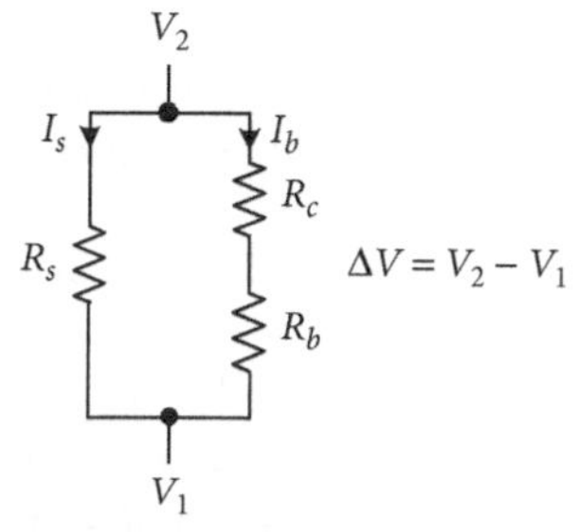

FIGURE 3.46 Simple model of a body in a conductive suit.

model of an individual in a body suit consists of three resistances: (i) resistance of the body suit, R_s, (ii) resistance of the body, R_b, and (iii) contact resistance between the body and suit, R_c. The resistance of the body and suit can be quite variable, and the contact resistance is probably the most difficult to predict and measure. The contact resistance is a function of the condition and area of the skin and the inner surface of the suit. It can also vary with body movement. (If the inner surface of the suit has an insulating liner, this contact resistance can be very large.) The model shown in Figure 3.46 assumes that the contact resistance is in series with the body resistance and that the suit resistance is in parallel with this series combination. When there is a voltage difference across this parallel combination, current exists in both the body and suit. If the resistance of the suit is small compared to the contact and body resistances, then the current in the body suit is large compared to the current in the body. If a constant current is injected into this circuit (instead of a constant voltage across the circuit), then using current division, the smaller the suit resistance, the smaller the body current. Again, this model is extremely simplistic since it does not consider factors such as skin effect, capacitance, inductance, multiple contact points along the length of the body, break points along the suit, and multiple current injection points due to, for example, openings in the shield. As a numerical example, if $R_s = 400\ \Omega$, $R_c = 3.2\ \text{k}\Omega$, $R_b = 500\ \Omega$, and $\Delta V = 4$ V rms, then

$$I_s = \frac{\Delta V}{R_s} = \frac{4}{400} = 10\ \text{mA rms},\quad I_b = \frac{\Delta V}{R_c + R_b} = \frac{4}{3{,}700} \approx 1.1\ \text{mA rms}$$

A suit current of 10 mA rms could correspond to barehand work on a 765 kV rms line. A current of 1.1 mA rms at 60 Hz generates a "warmth/slight tingling" sensation. Suit resistances as low as 10 Ω have been manufactured and 5–100 Ω is specified by national and international standards.

The sparking and flammability of the suit is also a factor that should be considered. A low suit resistance will divert more current through the suit. However, if the resistance is too low, the likelihood of discharging and sparking to other charged bodies is greater. If the suit does not contain an insulating inner liner, the user can feel this sparking.

3.17 Power Line Shielding via Burying

How effective is typical medium-dry soil at 60 Hz in shielding electric and magnetic fields? [Haus; Wait, '71; Zaborszky; Whitaker; Wait, '77; Xu, '95; Xu, '96; Wait, '97; Collin, '69; Heiland]

Over the years, there has been considerable (but sporadic) interest in reducing the strength of electromagnetic fields from power lines due to their perceived harmful effects. There are various methods of reducing the fields including lowering the current levels in the conductors, raising the height of the conductors, transposing the conductors, changing the spacing and phasing of the conductors, shielding the conductors, and burying the conductors. To reduce the general population's exposure to these fields, the power lines can also be rerouted and the right-of-way widened.

There are obvious aesthetic advantages in burying power lines, and sometimes it is a necessity in dense urban environments. Buried lines, however, are currently still more costly than overhead lines and possibly more difficult to maintain (but they do not have to support their own weight nor endure wind and ice as is the case with overhead lines).

Conductors used in overhead power lines are bare since they are separated by air, a very good insulator, and their physical separation from each other is usually large so voltage breakdown between conductors is not usually an issue (but corona discharge around the conductors can be a problem). Because of the nonzero conductivity of the earth, when conductors are buried, they are insulated from each other and grouped together. When bundled together, the conductors are referred to as a cable. The cable can be buried directly or carried in a buried conduit. Since the conductors in an underground cable are often close, voltage breakdown between the conductors can be an issue. With the high breakdown strength of certain oils, the conductors can be contained within an oil-filled pipe to reduce the likelihood of breakdown. The oil can also assist in cooling. For most voltages encountered, heating (due to I^2R losses) is a major problem.

As was previously discussed, probably the two most important factors in reducing the near-field magnetic field from a collection of current-carrying conductors are a net current of zero and close spacing of the conductors. The phasor sum of all of the currents (including both their amplitudes and phases) at a given position along their length should sum to zero. When there is an imbalance and the net current is not zero, the magnetic field from the group of conductors drops off more like $1/r$, when not surrounded by any metallic or magnetic material. This net current is sometimes referred to as a common-mode current. Even when the net current is zero, magnetic fields still exist around the conductors. To reduce these fields, cables can be shielded. In Figure 3.47, a three-phase line is shown inside a steel pipe. The induced currents in the pipe and their directions are also shown at one instant of time. Steel, for its cost, is a decent material for magnetic shielding with a relatively good conductivity, permeability, and saturation level. Since steel is both conductive and magnetic, substantial magnetic shielding can occur due to both induced currents and magnetic shunting.

Both electric and magnetic fields still penetrate through the steel surrounding the current-carrying conductors. (The references should be consulted for the complex set of equations for the fields outside a shielded set of conductors.) The surrounding earth will have an effect on both the time-varying electric and magnetic fields. Both the conductivity and permittivity of the earth will vary with frequency. They will also vary with factors such as geology, season, and depth. Surprisingly, the relative permeability of earth is generally within 1% of one (or of free-space's value). For this analysis, it will be assumed that the electrical parameters of the earth are $\sigma = 10^{-3}$ mhos/m, $\varepsilon_r = 20$, and $\mu_r = 1$ for all frequencies below a few kHz. For near-field electric field shielding of sinusoidal signals, the ratio of the conduction to displacement current determines whether the earth is a good shield. For a sinusoidal signal in steady state at a frequency of 60 Hz, this ratio is

$$\frac{\sigma}{\omega\varepsilon} = \frac{10^{-3}}{2\pi(60)(8.854\times10^{-12})(20)} \approx 1.5\times10^{4}$$

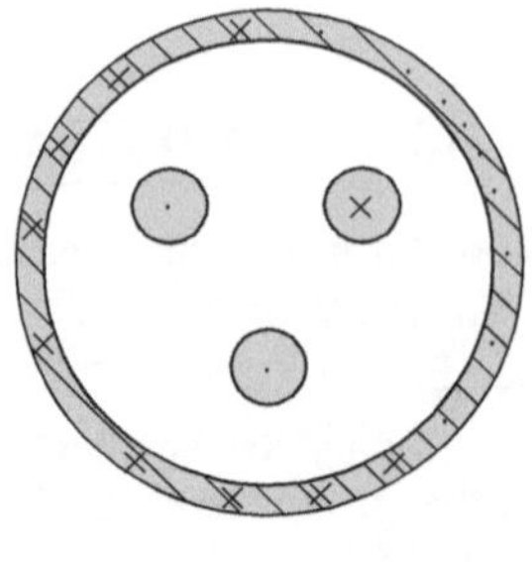

FIGURE 3.47 Rough sketch of the distribution of the induced currents inside a conducting pipe surrounding a three-phase line.

Since this ratio is much greater than one, earth is a very good conductor at lower frequencies and, hence, an excellent attenuator of the electric fields from the underground cable. For that matter, the electric fields from an overhead cable above the earth are also greatly attenuated when passing through the earth. Even at the tenth harmonic of 60 Hz, or 600 Hz, the earth is a good conductor.

The skin depth is a good indicator of magnetic field shielding when the dimension of interest is much greater than the skin depth. The skin depth at 60 Hz, using the previous parameters for earth, is given by

$$\delta = \frac{1}{\sqrt{\pi f \mu \sigma}} = \frac{1}{\sqrt{\pi(60)(4\pi \times 10^{-7})(10^{-3})}} \approx 2.1 \times 10^{3} \text{ m}$$

If the dimension of interest is the depth of the cable below the surface, it is fairly obvious that δ is much greater than the depth. If δ were of the same order or smaller than the depth, then the earth would provide some significant attenuation of the magnetic field. As discussed later in the MQS section, if the induced currents in the earth are sufficiently large, then it is not necessary that the depth be of the same order or larger than the skin depth. Using (3.150), if

$$\omega\mu\sigma l_1 l_2 = 2\pi(60)(4\pi \times 10^{-7})(10^{-3}) l_1 l_2 \approx 4.7 \times 10^{-7} l_1 l_2 \gg 1$$

then the induced current in the earth can generate a sufficient counter field to provide some magnetic field shielding. In this expression, the l's are the characteristic lengths. Even if both of these lengths are taken as the depth of a typical underground cable parallel to the earth-air interface, this expression is not satisfied.

An overhead power line does not necessarily provide a greater magnetic field level along the earth's surface than an underground cable. It is true that closely spaced conductors generally generate a smaller magnetic field at a fixed distance than widely spaced conductors. However, underground conductors that are buried at a shallow depth without a magnetic shield surrounding them can produce stronger magnetic fields than an overhead line at the same current level.

Occasionally, positive image currents are used in the analysis of underground conductors. The basis for positive images will be presented since they can be easily misapplied. Referring to Figure 3.48, imagine that a long conductor carrying a line current I, parallel to the z axis, is located at $x = -h$, $y = 0$. For $x < 0$, representing the earth, the conductivity σ_e is much greater than the conductivity σ_a for $x > 0$, representing the air. The earth-air interface is at the $x = 0$ plane, and the bare conductor is located in the earth. The depth h is electrically small, the length of the conductor is large compared to h, and h is large compared to the wire radius. Before this two-region problem is analyzed, the current density from a line current along the z axis in a region of uniform conductivity σ_e will be determined. For steady-state conditions, the conservation of charge equation reduces to

$$\oiint_S \vec{J} \cdot d\vec{s} = 0$$

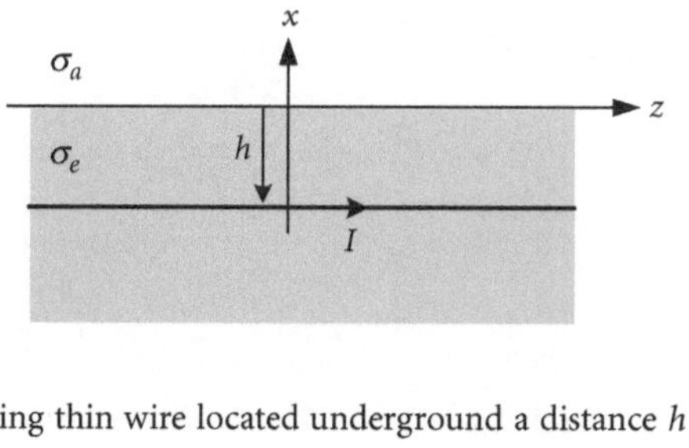

FIGURE 3.48 Bare, current-carrying thin wire located underground a distance h from the surface of the earth.

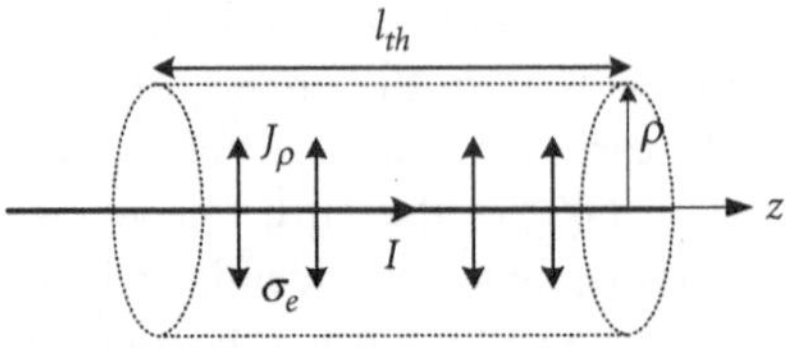

FIGURE 3.49 Current density from a line current I passes radially outward into the surrounding region of conductivity σ_e.

In other words, the total current into the wire must equal the total current passing outward from the bare wire.[20] Referring to Figure 3.49,

$$I = \int_0^{l_{th}}\int_0^{2\pi} J_\rho \hat{a}_\rho \cdot \rho d\phi dz \hat{a}_\rho = 2\pi\rho l_{th} J_\rho \quad \Rightarrow J_\rho = \frac{I}{2\pi\rho l_{th}} = \sigma_e E_\rho \quad \Rightarrow E_\rho = \frac{I}{2\pi\rho l_{th}\sigma_e} \tag{3.111}$$

If it is difficult to visualize a *constant* line source that is losing current along its length, refer to Figure 3.50. (Similarly, a point source can be visualized as a single insulated wire with an uninsulated end or tip.) The corresponding electric potential is given by

$$\Phi = -\frac{I}{2\pi l_{th}\sigma_e}\ln(\rho) + C = -\frac{I}{2\pi l_{th}\sigma_e}\ln\sqrt{x^2+y^2} + C \tag{3.112}$$

where C is a constant. Recall that $\vec{E} = -\nabla\Phi$.

Returning to the situation shown in Figure 3.48, at the earth-air interface, the normal component of the current density must be zero since the conductivity of the air is zero or at least much less than that of the earth. Since the normal current density is zero, the normal component of the electric field must also be zero (if $\sigma_e \neq 0$). The zero, normal, electric field condition is satisfied by inserting a positive image (not a negative image current as used for current sources near perfect conductors) equally spaced from the interface. As shown in Figure 3.51, in addition to the positive image current, when determining the potential for $x < 0$, the conductivity of the air region is replaced with the conductivity of the earth. The potential is, referring to the form of the solution given in (3.112), the sum of the potential from the line current at $x = -h$ and its positive image current at $x = h$:

$$\Phi = -\frac{I}{2\pi l_{th}\sigma_e}\left[\ln\sqrt{(x+h)^2+y^2} + \ln\sqrt{(x-h)^2+y^2}\right] \quad \text{if } x \le 0 \tag{3.113}$$

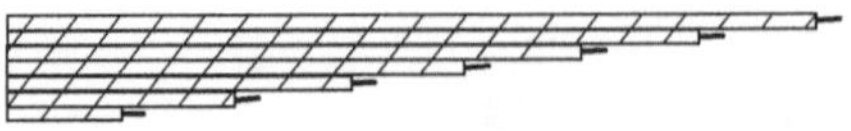

FIGURE 3.50 Line current modeled as many insulated wires with their bare ends spaced throughout the length of the line.

[20]Of course, rarely would a current-carrying bare wire be buried in the ground.

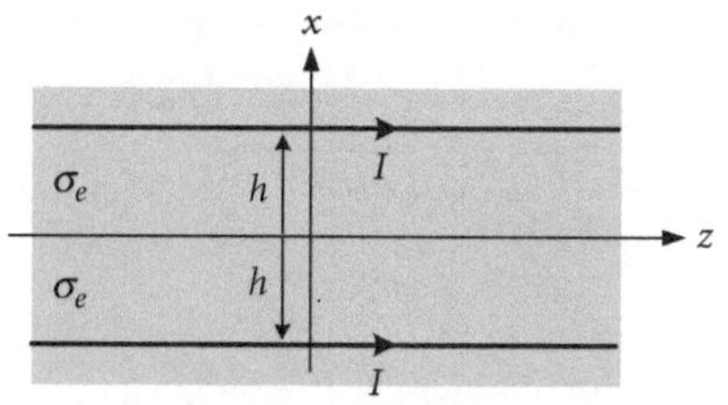

FIGURE 3.51 Original line current and its positive image.

The potential must be a function of y as well as x if the normal component of the electric field is to drop to zero at $x = 0$. The corresponding electric field is

$$\begin{aligned}\vec{E} &= -\nabla\Phi = -\frac{\partial\Phi}{\partial x}\hat{a}_x - \frac{\partial\Phi}{\partial y}\hat{a}_y \\ &= \frac{I}{2\pi l_{th}\sigma_e}\left\{\left[\frac{x+h}{(x+h)^2+y^2}+\frac{x-h}{(x-h)^2+y^2}\right]\hat{a}_x+\left[\frac{y}{(x+h)^2+y^2}+\frac{y}{(x-h)^2+y^2}\right]\hat{a}_y\right\} \quad \text{if } x<0\end{aligned} \tag{3.114}$$

In the earth, the current density can be obtained from this expression since for a linear conductor $\vec{J} = \sigma\vec{E}$. Evaluating this expression at $x = 0^-$, it is seen that the electric field is indeed entirely tangential to the interface slightly below the surface:

$$\vec{E} = \frac{Iy}{\pi l_{th}\sigma_e(h^2+y^2)}\hat{a}_y \quad \text{if } x = 0^- \tag{3.115}$$

In other words, the conduction current, which is proportional to the electric field, is not passing into the air region. Also, as another check on this result, at a distance of Δ in the x direction near the actual wire (and in the xz plane),

$$\begin{aligned}\vec{E} &= \frac{I}{2\pi l_{th}\sigma_e}\left(\frac{1}{\Delta}+\frac{1}{-2h+\Delta}\right)\hat{a}_x \quad \text{if } x=-h+\Delta,\ y=0 \\ &\approx \frac{I}{2\pi l_{th}\sigma_e}\left(\frac{1}{\Delta}+\frac{1}{-2h}\right)\hat{a}_x \approx \frac{I}{2\pi\Delta l_{th}\sigma_e}\hat{a}_x \quad \text{if } x=-h+\Delta,\ y=0,\ h \gg \Delta\end{aligned} \tag{3.116}$$

When Δ is small compared to h, the effect of the image current is negligible, and the resultant electric field is of the form given in (3.111). The potential along the interface is obtained from (3.113)

$$\Phi = -\frac{I}{\pi l_{th}\sigma_e}\ln\sqrt{h^2+y^2} \quad \text{if } x = 0 \tag{3.117}$$

The potential cannot change instantaneously across the interface (unless a doublet of charge is present). This potential along the interface can be used to determine the potential anywhere above the interface in the air. For the air region, above the interface, the potential is equal to

$$\Phi = -\frac{I}{\pi l_{th}\sigma_e}\ln\sqrt{(x+h)^2+y^2} \quad \text{if } x>0 \tag{3.118}$$

At $x = 0$ along the interface, (3.118) is equal to (3.117). Since there is also no net charge in the air region, it must satisfy Laplace's equation or $\nabla^2\Phi = 0$ (which it does). The corresponding electric field is

$$\begin{aligned}\vec{E} &= -\nabla\Phi = -\frac{\partial\Phi}{\partial x}\hat{a}_x - \frac{\partial\Phi}{\partial y}\hat{a}_y \\ &= \frac{I}{\pi l_{th}\sigma_e}\left[\frac{x+h}{(x+h)^2+y^2}\hat{a}_x + \frac{y}{(x+h)^2+y^2}\hat{a}_y\right] \quad \text{if } x>0\end{aligned} \tag{3.119}$$

Just above the interface, the electric field is

$$\vec{E} = \frac{I}{\pi l_{th}\sigma_e}\left(\frac{h}{h^2+y^2}\hat{a}_x + \frac{y}{h^2+y^2}\hat{a}_y\right) \quad \text{if } x=0^+ \tag{3.120}$$

As a simple check on this result, notice that the $\hat{a}_y$ component in (3.120), which is tangential to the interface, is equal to (3.115). The tangential components along the interface must be equal from the boundary condition $E_{ta} = E_{te}$. With the electric field known on both sides of the interface, the charge along the interface can be determined. Even in steady-state conditions, a surface charge can exist along the earth-air interface since there is a change in the conductivity. This surface charge is equal to

$$\rho_S = D_{na} - D_{ne} = \varepsilon_a E_{na} - \varepsilon_e E_{ne} = \varepsilon_a E_{na} - \varepsilon_e \frac{\sigma_a}{\sigma_e} E_{na} = \left(1 - \frac{\varepsilon_e\sigma_a}{\varepsilon_a\sigma_e}\right)\varepsilon_a E_{na} \quad @x=0 \tag{3.121}$$

where the n subscript refers to the component normal to the interface, which is in the x direction. Since the conductivity of the air region is zero and permittivity of the earth is finite, (3.121) simplifies to

$$\rho_S = \varepsilon_a E_{na} = \varepsilon_o E_{xa} = \frac{I\varepsilon_o}{\pi l_{th}\sigma_e}\left(\frac{h}{h^2+y^2}\right) \quad @x=0 \tag{3.122}$$

The surface charge is greatest directly above the current-carrying wire. A rough sketch of the current density and charge density are shown in Figure 3.52. It should be fairly obvious that there are many assumptions in the use of a positive, real image current for a bare, long uniform line current electrically close to an interface between a high-conductivity and low-conductivity region in steady-state conditions!

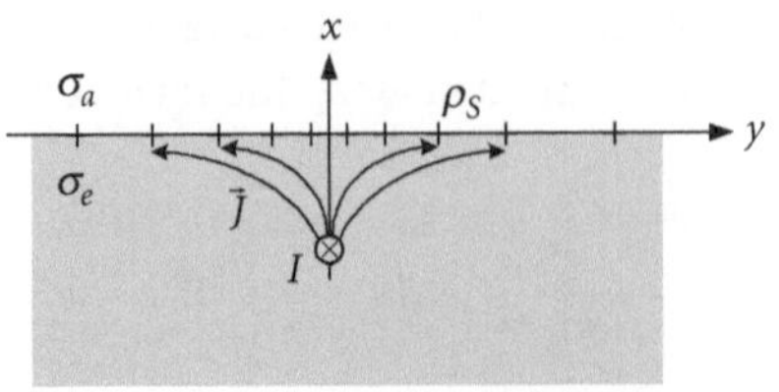

FIGURE 3.52 Rough sketch of the current density and surface charge distribution along the earth-air interface. The current density curves as it approaches the interface so that it is entirely tangential to the interface.

Inherent in the use of the electric field and surface charge expressions is the assumption of EQS conditions. If the electric energy is much greater than the magnetic energy, an estimate for the magnetic field can be obtained from the electric field. However, if the magnetic energy is much greater than the electric energy and the frequency is sufficiently low so that MQS conditions can be assumed, then the magnetic field can be obtained from the current distribution. (A point current source also has a positive image equal distance from the interface.)

It might be tempting to use the concept of a real, positive image current to analyze other situations such as a long, uniform line current insulated from its surroundings. For the situation given in Figure 3.48, for example, imagine that the current-carrying conductor is coated with a good insulator. In this case, assuming again that the frequency is sufficiently low so that all displacement currents are negligible, the electric field is entirely in the z direction (i.e., in the direction of the current). The magnetic field has both x and y components and can be obtained from Faraday's law:

$$\nabla \times \vec{E} = -\frac{\partial \vec{B}}{\partial t}$$

Assuming a free-space permeability and sinusoidal steady-state conditions,

$$\nabla \times \vec{E}_s = \left(\frac{\partial E_{zs}}{\partial y} - \frac{\partial E_{ys}}{\partial z}\right)\hat{a}_x + \left(\frac{\partial E_{xs}}{\partial z} - \frac{\partial E_{zs}}{\partial x}\right)\hat{a}_y + \left(\frac{\partial E_{ys}}{\partial x} - \frac{\partial E_{xs}}{\partial y}\right)\hat{a}_z = -j\omega\mu_o \vec{H}_s$$

$$\frac{\partial E_{zs}}{\partial y}\hat{a}_x - \frac{\partial E_{zs}}{\partial x}\hat{a}_y = -j\omega\mu_o \vec{H}_s \quad \Rightarrow H_{xs} = \frac{j}{\omega\mu_o}\frac{\partial E_{zs}}{\partial y}, \; H_{ys} = \frac{-j}{\omega\mu_o}\frac{\partial E_{zs}}{\partial x}$$

Unfortunately, although a positive, real image current can be used for some limiting conditions and situations, it generally should not be used. Instead, more complex analytical expressions or a numerical program should be used.

Finally, the magnetic field from a buried magnetic dipole (i.e., a small current-carrying loop) in a nonmagnetic ground with a uniform conductivity will be examined. All distances are electrically small. Referring to Figure 3.53, the integral expressions for the magnetic field at (ρ, z) above the interface in the air region are given by

$$H_\rho(\rho, z) = \frac{IA}{2\pi h^3}\int_0^\infty \frac{x^3 e^{-x\left(\frac{z}{h}\right)} e^{-\sqrt{x^2 + j\sigma\mu_o\omega}} J_1\left(x\frac{\rho}{h}\right)}{x + \sqrt{x^2 + j\sigma\mu_o\omega}} dx \quad \text{if } z > 0 \tag{3.123}$$

$$H_z(\rho, z) = \frac{IA}{2\pi h^3}\int_0^\infty \frac{x^3 e^{-x\left(\frac{z}{h}\right)} e^{-\sqrt{x^2 + j\sigma\mu_o\omega}} J_0\left(x\frac{\rho}{h}\right)}{x + \sqrt{x^2 + j\sigma\mu_o\omega}} dx \quad \text{if } z > 0 \tag{3.124}$$

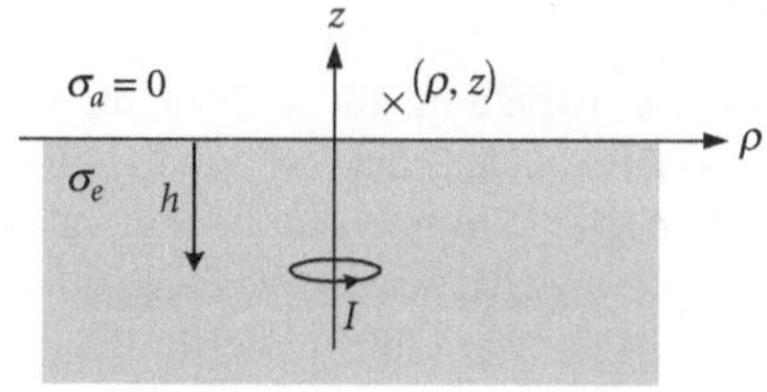

FIGURE 3.53 Small current-carrying loop of area A located a distance h below the ground.

$$\mu_o := 4\cdot\pi\cdot10^{-7} \quad I := 1 \quad A := \pi\cdot0.1^2 \quad \omega_o := 2\cdot\pi\cdot60 \quad h_{max} := 1 \quad j := \sqrt{-1}$$

$$h := 0, \frac{h_{max}}{100} .. h_{max}$$

$$H_z(h, \omega, \sigma, z, \rho) := \frac{I\cdot A}{2\cdot\pi\cdot h^3}\cdot\int_0^{50\cdot h_{max}} \frac{x^3\cdot e^{-x\cdot\frac{z}{h}}}{x+\sqrt{x^2+j\cdot(\sigma\cdot\mu_o\cdot\omega)}}\cdot e^{-\sqrt{x^2+j\cdot(\sigma\cdot\mu_o\cdot\omega)}}\cdot J0\left(x\cdot\frac{\rho}{h}\right)dx$$

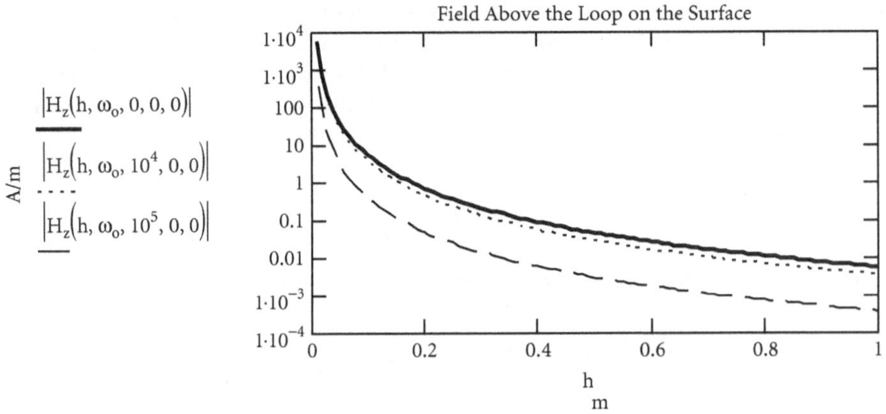

$$\delta(\omega, \sigma) := \frac{1}{\sqrt{\pi\cdot\frac{\omega}{2\cdot\pi}\cdot\mu_o\cdot\sigma}} \qquad \delta(\omega_o, 10^4) = 0.65 \qquad \delta(\omega_o, 10^6) = 0.065$$

MATHCAD 3.16 Magnetic field at the surface directly above a small current-carrying loop vs. the depth of loop as the conductivity of the medium is varied.

where $J_0()$ and $J_1()$ are the Bessel functions of the first kind of order zero and one, respectively. The variable I is the current in the loop of area A. Directly above the loop, the magnetic field is entirely in the z direction as expected. The magnetic field at the surface of the earth directly above the loop is plotted vs. the depth of the loop in Mathcad 3.16. As is clearly seen, the conductivity of the earth must be unrealistically large to have a significant effect on the magnetic field. For a more realistic conductivity of the earth (e.g., 0.01 1/Ω-m), the skin depth is very large. For the surrounding earth to have a strong effect on the magnetic field, the depth should be comparable or larger than a skin depth. This usually implies frequencies much greater than 60 Hz for reasonable depths. In Mathcad 3.17, the horizontal and vertical components of the magnetic field are plotted vs. distance along the surface. As is seen, the *shape* of the horizontal component vs. distance curve does not seem to be affected much by the change in the conductivity.

3.18 Wave Impedance Concept

The wave impedance model is sometimes used to determine the shielding effectiveness of materials to near-field magnetic sources. When are the reflective losses important for a single-layer shield? [Schulz, '68; Sjögren; Whitehouse]

Although there are many sources of near-field magnetic fields, there are only a few that can be simply described starting from Maxwell's equations. The field expressions for a magnetic dipole, though, are well known and relatively simple. A magnetic dipole is a current-carrying electrically-small loop. It is analogous to an electric dipole, which is two equal but opposite sign charges separated by a small distance. For a magnetic dipole, the magnitude of the wave impedance in the near field is

$$|Z_w| = \eta_o\beta_o r = 377\frac{2\pi}{\lambda_o}r\ \Omega \quad \text{when } \beta_o r \ll 1 \tag{3.125}$$

$$\mu_o := 4\cdot\pi\cdot 10^{-7} \quad I := 1 \quad A := \pi\cdot 0.1^2 \quad \omega_o := 2\cdot\pi\cdot 60 \quad \rho_{max} := 2 \quad h := 1 \quad j := \sqrt{-1}$$

$$\rho := 0, \frac{\rho_{max}}{100} .. \rho_{max}$$

$$H_\rho(h,\omega,\sigma,z,\rho) := \frac{I\cdot A}{2\cdot\pi\cdot h^3}\cdot\int_0^{20\cdot\rho_{max}} \frac{x^3\cdot e^{-x\cdot\frac{z}{h}}}{x+\sqrt{x^2+j\cdot(\sigma\cdot\mu_o\cdot\omega)}}\cdot e^{-\sqrt{x^2+j\cdot(\sigma\cdot\mu_o\cdot\omega)}}\cdot J1\left(x\cdot\frac{\rho}{h}\right)dx$$

$$H_z(h,\omega,\sigma,z,\rho) := \frac{I\cdot A}{2\cdot\pi\cdot h^3}\cdot\int_0^{50\cdot\rho_{max}} \frac{x^3\cdot e^{-x\cdot\frac{z}{h}}}{x+\sqrt{x^2+j\cdot(\sigma\cdot\mu_o\cdot\omega)}}\cdot e^{-\sqrt{x^2+j\cdot(\sigma\cdot\mu_o\cdot\omega)}}\cdot J0\left(x\cdot\frac{\rho}{h}\right)dx$$

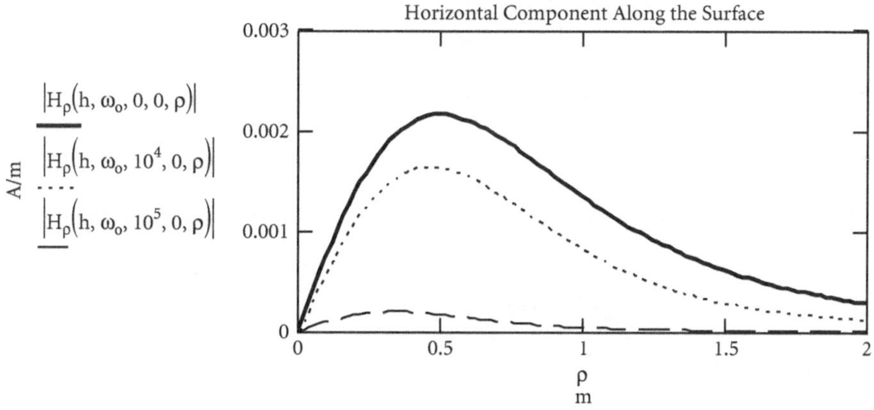

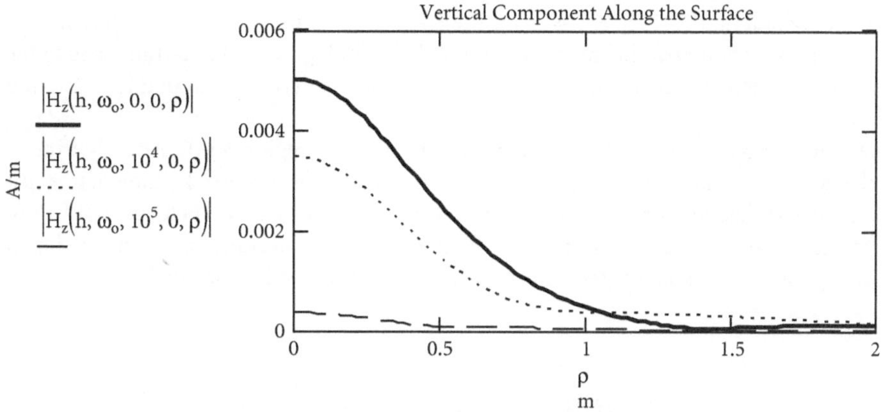

MATHCAD 3.17 Horizontal and vertical components of the magnetic field from a buried current-carrying loop vs. the distance along the surface as the conductivity of the medium is varied.

where r is the distance from the magnetic dipole. This impedance is less than 377 Ω when r is less than $\lambda_o/2\pi$. This is probably why magnetic field sources are referred to as low-impedance sources.[21] Unfortunately, from a magnetic shielding viewpoint, the ratio of r/λ_o is often very small for magnetic field sources (in the near field) resulting in small wave impedances. For example, at a distance of 5 cm from a 1 kHz small

[21]Usually, the current through the loop is large "compared" to the voltage across the terminals of the loop. The magnitude of the ratio of the voltage to the current, or impedance, is thus small.

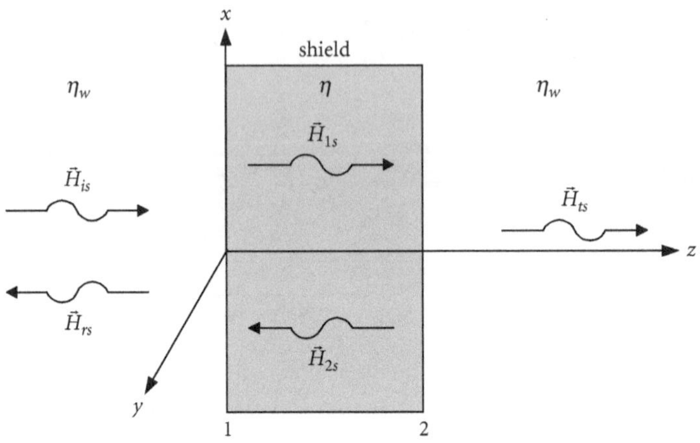

FIGURE 3.54 Magnetic field "waves" in and near a flat shield.

current-carrying loop in free space, the magnitude of the wave impedance is about 0.4 mΩ:

$$|Z_w| = 377\frac{\dfrac{2\pi}{\dfrac{3\times 10^8}{10^3}}}{}\left(\frac{5}{100}\right) \approx 0.4\ \text{m}\Omega$$

The wave impedance model must be carefully used in the near field since the field is *not* a plane wave in the near field. Equation (3.125) is based on the transverse components of the electric and magnetic fields: these components are parallel to the flat shield shown in Figure 3.54 and transverse to the direction of propagation. Nevertheless, in some situations, this approach can be helpful in understanding magnetic field shielding.

Imagine that the source of the electromagnetic field is in the region $z < 0$. Although the expressions for the reflection and transmission coefficients for plane waves are commonly known, it is instructive to determine these coefficients using the magnetic field instead of the electric field as the working variable. In terms of the magnetic field terms shown in Figure 3.54, the reflection and transmission coefficients for a positive-going wave incident upon the first interface (air-to-shield, labeled 1) are

$$\rho_1 = \frac{H_{rs}}{H_{is}} = \frac{-\dfrac{E_{rs}}{\eta_w}}{\dfrac{E_{is}}{\eta_w}} = -\frac{E_{rs}}{E_{is}} = -\frac{\eta-\eta_w}{\eta+\eta_w} = \frac{\eta_w-\eta}{\eta+\eta_w} \tag{3.126}$$

$$\mathrm{T}_1 = \frac{H_{1s}}{H_{is}} = \frac{\dfrac{E_{1s}}{\eta}}{\dfrac{E_{is}}{\eta_w}} = \frac{\eta_w}{\eta}\frac{E_{1s}}{E_{is}} = \frac{\eta_w}{\eta}\frac{2\eta}{\eta+\eta_w} = \frac{2\eta_w}{\eta+\eta_w} \tag{3.127}$$

Note that these coefficients are different from the coefficients for electric fields. For example, the reflection coefficient for the magnetic field is the negative of the reflection coefficient for the electric field.

The coefficients at the second interface (shield-to-air, labeled 2) for the magnetic fields are

$$\rho_2 = \frac{H_{2s}}{H_{1s}} = \frac{-\frac{E_{2s}}{\eta}}{\frac{E_{1s}}{\eta}} = -\frac{E_{2s}}{E_{1s}} = -\frac{\eta_w - \eta}{\eta + \eta_w} = \frac{\eta - \eta_w}{\eta + \eta_w} \tag{3.128}$$

$$T_2 = \frac{H_{ts}}{H_{1s}} = \frac{\frac{E_{ts}}{\eta_w}}{\frac{E_{1s}}{\eta}} = \frac{\eta}{\eta_w}\frac{E_{ts}}{E_{1s}} = \frac{\eta}{\eta_w}\frac{2\eta_w}{\eta + \eta_w} = \frac{2\eta}{\eta + \eta_w} \tag{3.129}$$

For electric fields, the field is mostly reflected at the first interface since the reflection coefficient

$$\frac{E_{rs}}{E_{is}} = \frac{\eta - \eta_w}{\eta + \eta_w}$$

is close to –1. (The impedance, η, of a metallic shield at low frequencies is very small, and the wave impedance, η_w, for electric fields is high, greater than 377 Ω, in the near field.) However, the wave impedance for magnetic fields is less than 377 Ω. The reflection coefficient for magnetic fields at the first interface is

$$\rho_1 = \frac{\eta_w - \eta}{\eta + \eta_w}$$

To obtain a reflection coefficient close to –1 would require a large shield impedance relative to the wave impedance:

$$\eta_w \ll \eta = \sqrt{\frac{j\omega\mu}{\sigma + j\omega\varepsilon}} \tag{3.130}$$

Materials with high permeability, low conductivity, and low permittivity satisfy this inequality. The total loss is also a function of the reflection from the second interface, absorption, and multiple reflections. The magnitude of the shield impedance of a high-permeability material, such as μ-metal, is shown in Mathcad 3.18. The shield impedance is not necessarily much greater than the wave impedance of a magnetic field source. Real magnetic materials have nonzero conductivity.

As just stated, for electric field sources, most of the total reflection occurs at the first interface.[22] The reflection coefficient for magnetic fields at the second interface is

$$\rho_2 = \frac{\eta - \eta_w}{\eta + \eta_w}$$

The wave impedance for magnetic fields is often greater than the shield impedance of good conductors. Although this reflection coefficient is often negative, it is not necessarily near –1. For that matter, as the distance between the source and good-conducting shield decreases, the magnitude of this ratio decreases. The reflection losses are not as great for a magnetic field as with an electric field.

The shielding effectiveness of the shield for magnetic fields can be easily determined if the multiple-reflective loss term is neglected:

$$\frac{H_{ts}}{H_{is}} = \frac{H_{ts}}{H_{1s}}\frac{H_{1s}}{H_{is}} = \frac{2\eta}{\eta + \eta_w}\frac{2\eta_w}{\eta + \eta_w} = \frac{4\eta\eta_w}{(\eta + \eta_w)^2} \tag{3.131}$$

[22]For electric fields, the reflection coefficient is about +1 at the second interface for a good conducting shield. However, most of the electric field has already reflected off the first interface. After absorption, the electric field at the second interface is often not important.

$$\mu(\mu_r) := \mu_r \cdot 4 \cdot \pi \cdot 10^{-7} \qquad \varepsilon_r := 1 \qquad \varepsilon := \varepsilon_r \cdot 8.854 \cdot 10^{-12} \qquad \sigma(\sigma_r) := \sigma_r \cdot 5.8 \cdot 10^{7} \qquad j := \sqrt{-1}$$

$$\omega := 10^2, 2 \cdot 10^2 .. 10^5$$

$$\eta(\omega, \mu_r, \sigma_r) := \sqrt{\frac{j \cdot \omega \cdot \mu(\mu_r)}{\sigma(\sigma_r) + j \cdot \omega \cdot \varepsilon}}$$

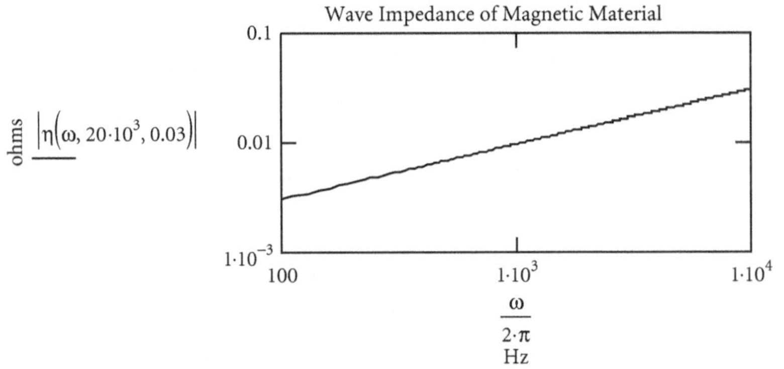

MATHCAD 3.18 Wave impedance vs. frequency for a magnetic material with a relative permeability of 20,000.

The ratio for electric fields is the same:

$$\frac{E_{ts}}{E_{is}} = \frac{4\eta\eta_w}{(\eta+\eta_w)^2} \approx \frac{4\eta\eta_w}{(\eta_w)^2} = \frac{4\eta}{\eta_w}$$

which is a small quantity. Since η is not necessarily much less than η_w for magnetic fields, this same approximation would not necessarily be valid for (3.131). For magnetic fields, (3.131) is about one if $\eta \approx \eta_w$. Again, the reflective loss for magnetic shields is not large.

Equation (3.131) is valid if the multiple-reflective loss is negligible, which is true if the absorption loss is large. (For thin shields, the multiple-reflective losses can be important.) For magnetic fields, the absorption loss is

$$A_{dB} = 20\log\left(e^{\frac{d}{\delta}}\right) \approx 8.69\frac{d}{\delta} = 8.69d\sqrt{\pi\mu f\sigma} \tag{3.132}$$

For electric fields, the absorption loss is less critical because most of the signal is reflected off the first interface. For magnetic fields, this loss is important. At low frequencies, the product of the conductivity and permeability should be maximized. This product is large for high-μ materials.

It is actually quite easy to use improperly the wave impedance model. In the previous analysis, it was assumed that the magnetic fields were parallel to the shield (or transverse to the direction of propagation). Unlike far fields, near fields are generally complex with several components that may be equally important and not parallel to the shield's surface.[23] For example, the external magnetic field to a high-permeability shield is mostly normal to its surface. These other components cannot be easily handled using a plane wave-like, one-dimensional model. It was also assumed in this analysis that the wave impedance at the input of the shield was identical to the wave impedance at the output of the shield. Unfortunately, because the near-field wave impedance generally varies with position, this also may not be true. Furthermore, the wave impedance of the actual source of interest may not be known or the source may not be accurately modeled as a magnetic dipole. Sometimes the wave impedance for a larger current-carrying loop may

[23] A shield does not need to be flat for a magnetic field to be parallel to its surface.

be more appropriate. Finally, holes and seams in the shield and conductors passing through the shield can cause the measured performance of the shield to deviate significantly from the idealized performance.

3.19 Flat Shielding of Current-Carrying Loops

How effective is a flat conductor in shielding the magnetic field generated by a current-carrying small loop? [Moser, '67; Hasselgren; Hansen; Olsen, '96]

Simple expressions are available for the shielding effectiveness of a large conducting plate when both the source of the magnetic field and receiver of the magnetic field are coaxial loops parallel to the plate. A single loop source and single loop receiver are shown in Figure 3.55. For the plate to be considered large, its length and width should be large compared to the dimensions *a*, *b*, and *c*.[24] The loop radii are electrically small (so that the current along the loops is constant), and the plate is in the near field of the source.

This first expression for the shielding effectiveness assumes that the voltage induced on both sides of the plate is the same. This would be true if the thickness, *t*, is less than the skin depth, δ, of the plate material. The shielding effectiveness defined as the ratio of the voltage induced at the receiver loop without the shield to the voltage induced at the receiver loop with the shield is given by (in dB)

$$SE_{dB} = 20\log\sqrt{(2.2\times10^{-6}at\sigma f)^2+1} \quad \text{if } 0.15<\frac{c}{a}<0.9 \tag{3.133}$$

where σ is the conductivity of the plate and *f* is the frequency of the source. This thin-plate shielding is due to Eddy currents in the shield. As is expected, the effectiveness increases with the thickness and conductivity of the shield, as well as the frequency of the source. Over the specified range for c/a, the limited experimental work available has shown only a slight variation in the *SE* with loop spacing, 2*c*. As an example, if $a = 3.5$ cm, $c = 1.5$ cm, $f = 10$ kHz, $t = 3.2$ mm, and $\sigma = 3.5\times10^7$ 1/Ω-m (aluminum), $SE_{dB} \approx 39$ dB.

When the receiver loop radius is very small compared to the source loop radius, a comparatively simple formula that includes the spacing between the loops is given by

$$SE_{dB} = 20\log\left|\frac{\sinh\left[(1+j)\frac{t}{\delta}\right]}{(1+j)\frac{t}{\delta}}\frac{jt}{\mu_r\delta^2}\frac{a^2+c^2}{3c}\right| \quad \text{where } \delta=\frac{1}{\sqrt{\pi f\mu_r\mu_o\sigma}} \tag{3.134}$$

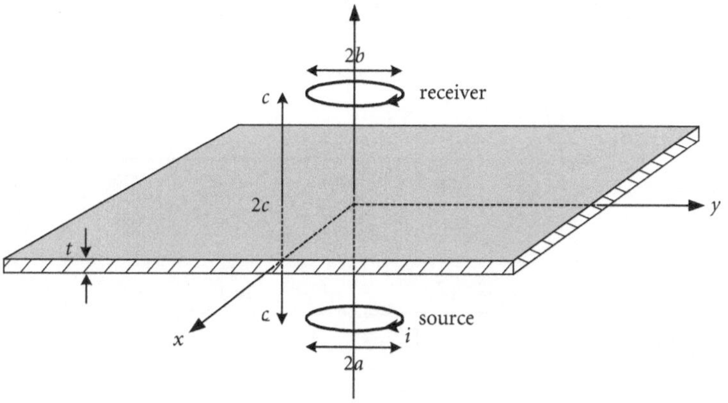

FIGURE 3.55 Infinitely large conducting plate between two loops.

[24]As expected, if the plate is not large compared to *c* and the radii of the loops, leakage around the plate occurs.

This expression is valid if both of the following conditions are satisfied:

$$\frac{L}{\mu_r \delta} \gg 1 \quad \text{and} \quad \frac{tL}{\mu_r \delta^2} \gg 1 \tag{3.135}$$

where L is a characteristic length that is a function of the shield parameters. A possible value for this length is $L = \sqrt{a^2 + b^2 + c^2}$. The source loop should be sufficiently far from the plate so that the magnetic field is mostly parallel to the plate's surface. The normal component of the magnetic field at the plate is assumed negligible. For spacings, c, that are too small (e.g., $c < a$), the SE_{dB} obtained from (3.134) implies that the effectiveness decreases with increasing spacing from the shield! The accuracy of the expression seems to improve with increased spacing.

The shielding effectiveness of a large aluminum plate with a thickness of 0.32 cm ≈ 1/8" is shown in Mathcad 3.19 as a function of frequency. Both (3.133) and (3.134) are plotted in Mathcad 3.19. Although both expressions are plotted from 100 Hz to 10 kHz, to satisfy the restrictions on the equations, (3.133)

$\mu_o := 4 \cdot \pi \cdot 10^{-7}$ $\quad \mu_r := 1$ $\quad \sigma := 0.6 \cdot 5.8 \cdot 10^7$ $\quad a := 0.035$ $\quad b := 0.0001$ $\quad c := 0.015$

$t := 0.0032$ $\quad j := \sqrt{-1}$

$f := 10^2, 1.1 \cdot 10^2 .. 10^4$

$$\delta(f) := \frac{1}{\sqrt{\pi \cdot f \cdot \mu_r \cdot \mu_o \cdot \sigma}}$$

$$SE_1(f) := 20 \cdot \log\left(\left| \frac{\sinh\left[(1+j) \cdot \frac{t}{\delta(f)}\right]}{\left[(1+j) \cdot \frac{t}{\delta(f)}\right]} \cdot \frac{j \cdot t}{\mu_r \cdot \delta(f)^2} \cdot \frac{a^2 + c^2}{3 \cdot c} \right|\right)$$

$$SE_2(f) := 20 \cdot \log\left[\sqrt{\left(2.2 \cdot 10^{-6} \cdot a \cdot t \cdot \sigma \cdot f\right)^2 + 1}\right]$$

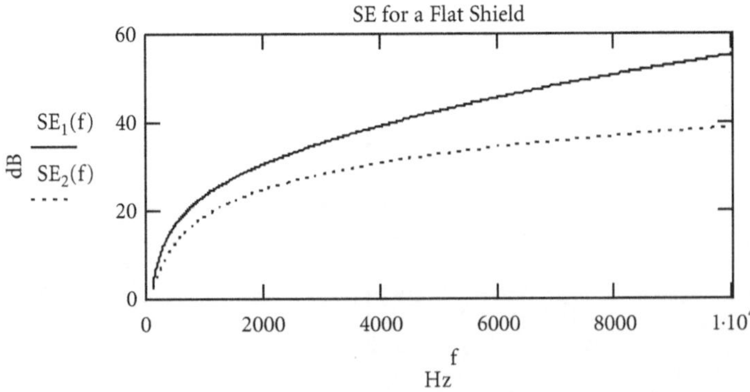

$$\frac{\sqrt{a^2 + b^2 + c^2}}{\mu_r \cdot \delta\left(10^3\right)} = 14.114 \qquad \frac{\sqrt{a^2 + b^2 + c^2}}{\mu_r \cdot \delta\left(10^2\right)} = 4.463 \qquad \frac{\delta\left(10^3\right)}{t} = 0.843$$

$$\frac{t \cdot \sqrt{a^2 + b^2 + c^2}}{\mu_r \cdot \delta\left(10^3\right)^2} = 16.741 \qquad \frac{t \cdot \sqrt{a^2 + b^2 + c^2}}{\mu_r \cdot \delta\left(10^2\right)^2} = 1.674 \qquad \frac{\delta\left(10^1\right)}{t} = 8.431$$

MATHCAD 3.19 Shielding effectiveness vs. frequency using (3.133) and (3.134).

$\mu_o := 4\cdot\pi\cdot 10^{-7}$ $\quad \mu_r := 1$ $\quad \sigma := 0.6\cdot 5.8\cdot 10^{7}$ $\quad a := 0.035$ $\quad b := 0.0001$ $\quad t := 0.0032$

$f := 10^3$ $\quad j := \sqrt{-1}$

$c := 0.05, 0.06..\, 0.1$

$$\delta := \frac{1}{\sqrt{\pi\cdot f\cdot \mu_r\cdot \mu_o\cdot \sigma}} \qquad SE_1(c) := 20\cdot\log\left(\left|\frac{\sinh\left[(1+j)\cdot\frac{t}{\delta}\right]}{\left[(1+j)\cdot\frac{t}{\delta}\right]}\cdot\frac{j\cdot t}{\mu_r\cdot\delta^2}\cdot\frac{a^2+c^2}{3\cdot c}\right|\right)$$

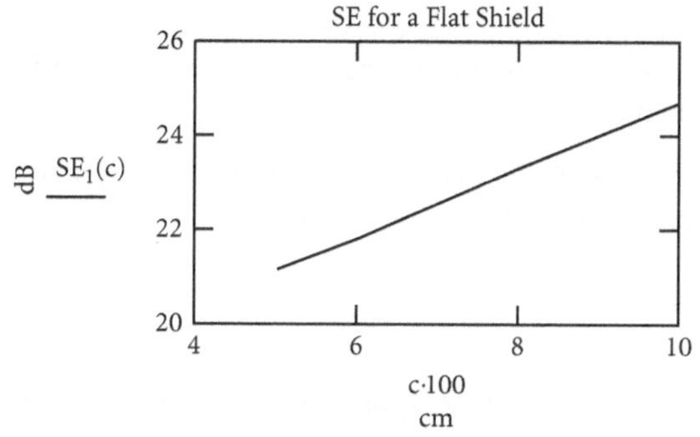

$$\frac{\sqrt{a^2+b^2+0.05^2}}{\mu_r\cdot\delta} = 22.622$$

$$\frac{t\cdot\sqrt{a^2+b^2+0.05^2}}{\mu_r\cdot\delta^2} = 26.832$$

MATHCAD 3.20 Shielding effectiveness vs. loop-to-shield spacing.

should only be used at lower frequencies (e.g., 10 Hz for the given set of parameters) and (3.134) at higher frequencies (e.g., 1 kHz for the given set of parameters). Equation (3.134) is plotted in Mathcad 3.20 as a function of the loop-to-shield spacing for a fixed frequency of 1 kHz.

The previous expressions were for conducting plates where Eddy or induced currents were the main mechanism for shielding. However, shielding can also occur through flux shunting when the plate is magnetic (i.e., relative permeability greater than one). The plane-wave approach to shielding discussed previously can be applied (with caution). Other much more complicated expressions are also available.

3.20 Grounding Shields

Is it necessary to ground magnetic shields? Explain.

The answer to this question may surprise EMC neophytes: grounding a shield has no effect on magnetic shielding. Magnetic shields either redirect magnetic fields via their small "resistance" (reluctance) to magnetic fields or produce a counter field that opposes the applied magnetic field. However, to dissipate static charge, for safety reasons, or for convenience (e.g., mechanical considerations), a magnetic shield may be grounded. (The author can imagine a situation where the number of grounding points on a magnetic shield are so numerous that the ground becomes a part of the magnetic shield and, hence, affects the shielding effectiveness. Also, ground loops can be produced if a shield is grounded at more than one location.)

3.21 Cheap Shielding

Provide several disadvantages of magnetic shielding. Name several alternative, less expensive methods of "shielding" against magnetic fields.

Earth, motors, transformers, permanent magnets, power lines, and switching power supplies are only a few examples of magnetic field sources. Significant magnetic field reduction is usually difficult to achieve.

Magnetic shields are often

1. expensive
2. sensitive to machining and working operations
3. frequency dependent
4. nonlinear
5. saturatable
6. heavy
7. large.

Before using the brute-force method of surrounding an object with magnetic material, it is highly recommended that the following alternative methods be tried:

1. increase the distance between the source of the magnetic field and the receiver or victim
2. use a shorted turn around the object to produce a counter field
3. rotate the source or receiver
4. move the position of the cables, including the power lines
5. redesign.

Since the fields from many products, including household appliances, vary as $1/r^3$ in the near field, the first method can be quite effective.

3.22 Nonideal Shapes

How does the shape of a magnetic shield affect its shielding performance? [Rikitake; Thomas; Cooley]

The equations normally seen for magnetic shields correspond to spheres and infinitely long cylinders. Constructing a spherical shield is frequently not practical and constructing an infinitely long cylinder is not possible. There is some information available for shields that are neither of these idealized shapes.

When a magnetic shell, which does not have either of these idealized shapes, is placed in a uniform magnetic field, the internal field in the cavity deviates from the ideal uniform field, influencing the shield's performance. For shield shapes such as cubes, a first-order estimate of the shielding effectiveness can be obtained by using the spherical formula. The radius in the spherical formula is set equal to the radius of a sphere with the same volume as the cube. For a long rectangular shield of square cross section, the infinitely long cylinder formula can be used if the radius is set equal to the radius of a circle with the same area as the rectangular shield's cross section. If the length of the rectangular shield is less than three times its width, the infinitely long cylinder equations should not be used (or used with caution). The field at its center will approach that of a cubical shield of the same cross section.

To improve upon the previous first-order estimates, the distortion of the field should be considered. Unfortunately, either numerical or experimental work is usually required. There is one important trend, however, that can provide insight into the distortion produced by nonideal shapes. At the corners of cubical and rectangular shields, the field tends to concentrate. To reduce these field gradients, large scale rounding of the edges is required. Also, since magnetic fields tend to "leak" from sharp corners, sharp corners should be avoided.

The direction of the external field for cylindrical shields does have an impact on the shield's performance. For cylinders of finite length, the use of a lid or cap at one end can reduce the effectiveness of the shield at low frequencies if the magnetic field is parallel to the axis of the cylinder. However, if the external field is perpendicular to the surface of the cylindrical shield, the performance of the shield can be improved by capping the end of the cylinder. For a finite-length cylindrical shield of relative permeability from 100 to 1,000 to have a shielding effectiveness (for perpendicular low-frequency magnetic fields) similar to an infinite cylinder, the length-to-radius ratio should be greater than or equal to ten.

3.23 Reducing the Magnetic Coupling Between Inductors

The T-equivalent circuit is used in place of an ordinary transformer. How should the coils be oriented to reduce the coupling between the inductors in the T-equivalent circuit? [Weber; Haus]

An interesting application of cheap shielding involves the orientation of the three inductors in a T-shaped equivalent circuit. Before this shielding technique is discussed, it is helpful to understand qualitatively the distribution of the magnetic fields around a current-carrying loop and solenoid. The magnetic field distribution around a current-carrying loop is shown in Figure 3.56. Although only the fields in one plane are actually sketched, by symmetry, the fields are identical in the other planes leading from the loop's axis. (When the fields from the leads are not neglected, the field distribution would not be symmetrical.) Along the axis of the loop, the field is entirely perpendicular to the plane of the loop. Unlike electric fields, which can terminate on charge, magnetic field lines must be closed. Therefore, this field line along the axis must return somewhere far from the loop.

A solenoid or long coil consists of (usually) many connected loops of wire with the same axis as shown in Figure 3.57. Each current-carrying loop or turn generates a magnetic field, and the sum of these fields can generate a strong field within the coil. Furthermore, if the loops are closely spaced, each with approximately the same radius, and the total length of the solenoid is large compared to the radius of the loops, then the field is relatively uniform inside the solenoid. As seen in the figure, the fields near

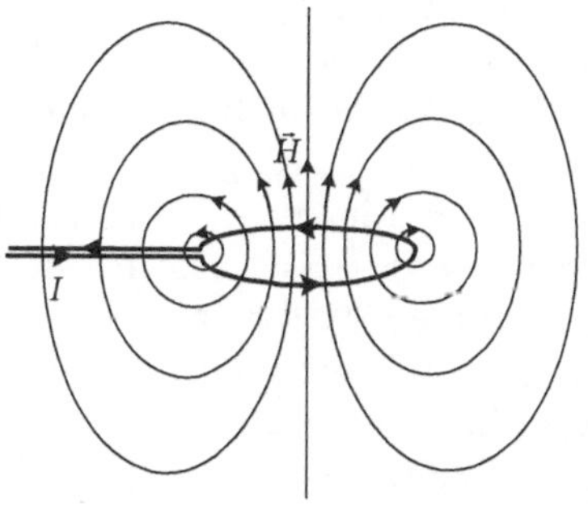

FIGURE 3.56 Magnetic field distribution around a current-carrying loop (for the fields in the plane of the paper). The fields from the two conductors leading to the loop are not shown.

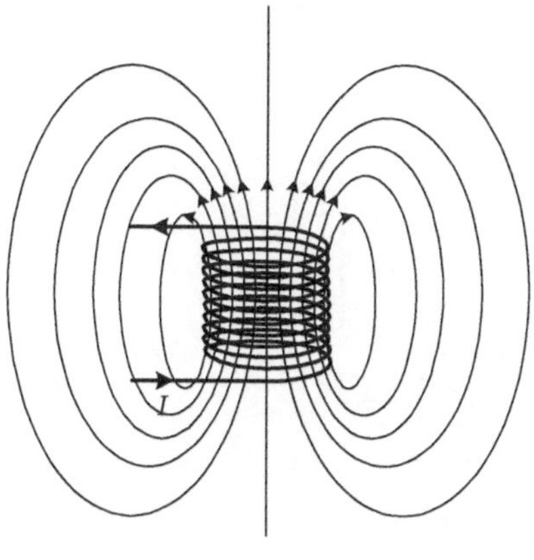

FIGURE 3.57 Magnetic field distribution around a current-carrying coil (for the fields in the plane of the paper). The fields from the two conductors leading to the loop are not shown.

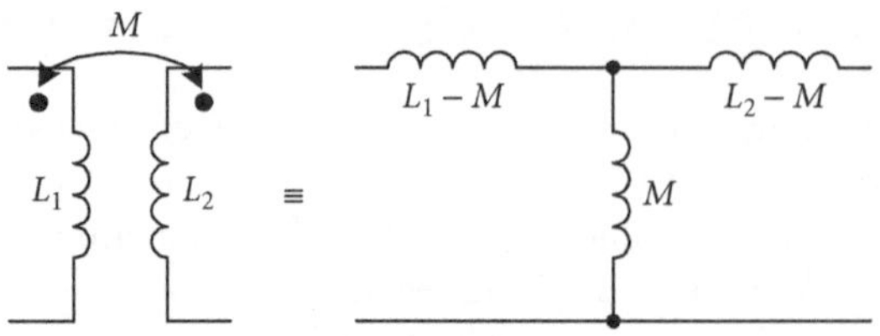

FIGURE 3.58 Linear transformer and its T-equivalent.

the ends are not as uniform. As the spacing between the loops decreases and the solenoid's length increases, the magnetic field outside the solenoid decreases. The flux density is concentrated inside the coil and diluted outside the coil (since the cross-sectional area that the field passes through is much greater outside the coil). For a very long coil with closely spaced turns, the flux density outside the coil is often assumed small or zero. When the spacing between the loops (or the pitch) increases, the field leakage near the outer surface of the solenoid increases.

To have greater control over the bandwidth of a transformer (and to have possibly a greater Q), its T-equivalent can be used. The T-equivalent of a transformer is shown in Figure 3.58. With the transformer shown on the left, magnetic coupling exists between L_1 and L_2, and the input and output are conductively isolated. With the T-equivalent, there is no magnetic coupling between any of the three inductors, but the input and output are not conductively isolated. When fine control of the mutual coupling, M, is desired, the T-equivalent circuit can sometimes be used since it does not rely on the less-than-predictable nature of magnetic coupling. However, there should be no magnetic coupling between the three inductors when using the T-equivalent. Because of the directional nature of the magnetic fields emanating from the ends of a solenoid, the three inductors can be placed mutually orthogonal to each other as shown in Figure 3.59. The magnetic coupling between the three inductors should be small for this arrangement, and this layout does not require high-permeability shields. The coupling between the three inductors will also decrease as the distance between them increases, but the space the three inductors will occupy will increase.

If the three inductors were in close proximity with parallel axes, the magnetic coupling between the inductors could be strong. This would somewhat defeat the purpose of using the T-equivalent for the transformer. To reduce the electric field or capacitive coupling between each of the three inductors, each inductor can be enclosed in a metallic can and appropriately grounded. (To decrease the induced currents in the can from the time-varying magnetic fields, a thin slot can be cut in each metal can.)

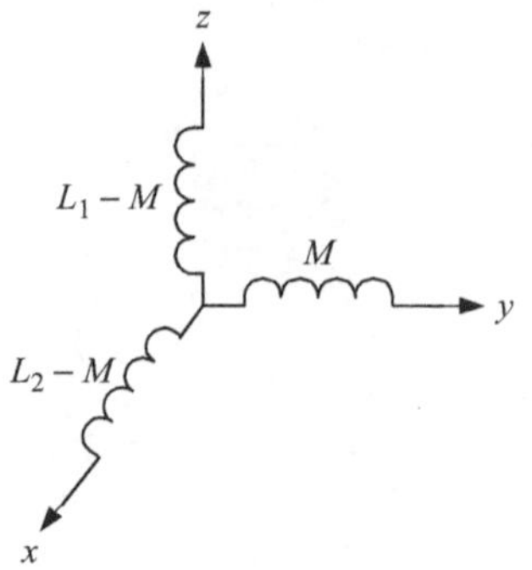

FIGURE 3.59 Three inductors with axes mutually orthogonal to minimize mutual coupling.

Sometimes, to increase the operating frequency of an inductor, several smaller value inductances, each with higher operating frequencies, are placed in series.[25] If there is no magnetic coupling between the three inductors, the sum of the smaller value inductances can be set equal to the desired inductance. If the coupling is not negligible, the mutual coupling between the inductors must be considered. For example, the equivalent inductance for three equal inductors in series (each with an inductance of L) is

$$L_{eq} = 3(L + 2M)$$

This expression assumes that the mutual inductances between each of the inductors are equal to M. If the series inductors can be oriented mutually perpendicular to each other, then the mutual coupling should be small.[26]

3.24 Typical Magnetic Flux Densities

Tabulate typical, magnetic flux density magnitudes. [Gauger; COMAR; IPCS, '87; Cleveland, '99; Gandhi, '90; Geddes; Reilly; Hiziroğlu; Caola; Hansson; Maruvada; Perry, '94]

A competent engineer should "know" whether a number is reasonable. The data contained in Table 3.4 can be used to facilitate this "feeling" of reasonableness. For those sources where the magnetic field is provided at multiple distances, the rapid drop off in the magnetic field with increasing distance is apparent. For free-space conditions, the relationship between the magnitude of the magnetic flux density and magnetic field is $|\vec{B}| = \mu_o |\vec{H}|$.

If possible, when performing a magnetic field measurement at 60 Hz, the magnetic field probe should be close (e.g., 1 cm) rather than far away (e.g., 1 m) from the source. The strength of near-field magnetic fields generally decreases rapidly from the source varying as $1/r^2$or $1/r^3$, where r is the distance from the source. Background noise and other magnetic field sources can strongly interfere with the measurement if the magnetic field probe is not close to the source. Sometimes, interfering 60 Hz sources can be turned off or turned on and off and their effect noted. Magnetic field sources are directional, and the magnetic field measured at different locations at the same distance from the source can vary. Furthermore, the orientation of the probe itself can affect the measurement. However, low-frequency magnetic field measurements are not significantly influenced by the (nonmagnetic) individual holding the magnetic field probe.

Although the interaction of electromagnetic fields with tissues and cells is not completely understood, it is considered unlikely that exposure to 1 μT 60 Hz frequencies can result in health problems.[27] However, ELF magnetic fields, either sinusoidal or pulsed, from 100 μT to 10 mT can produce biological effects. Some of these effects can be beneficial such as in bone healing. In magnetic resonance imaging (MRI) systems, static and time-varying magnetic fields (and radio frequency fields) are used to generate an image of the internal structure of the body: during these tests the subject is exposed to strong 1.5–2 T magnetic fields. For static magnetic fields, subjects exposed to 4 T showed a statistical difference in nausea, vertigo, and metallic taste compared to subjects exposed to 1.5 T. Fields as high as 0.75 mT applied longitudinally to the whole body, or 1.5 mT just to the head, could not be detected by the (human) subjects. When performing these tests, extreme precautions were taken to prevent the subject from using external cues such as acoustic vibrations and dimming of the lights in detecting these fields. Some subjects can detect (via sensations in the lower back and face) pulsed 1.4 kHz sinusoidal fields with a repetition frequency of 14 Hz and a peak value of the dB/dt equal to 86 T/s. At a frequency of 1.27 kHz, some subjects can detect (via mild sensations in the lower back) a 60 T/s peak magnetic field.

---

[25]Sometimes, several lower value, higher resonant frequency, capacitors are used in parallel instead of a single, higher value, lower resonant frequency, capacitor.

[26]The use of three orthogonal inductors has utility in optimizing the noise figure and bandwidth of an amplifier.

[27]Even if some future research showed that there was some (miniscule?) consequence of these low-level magnetic fields, it must be compared against other risks in life such as driving an automobile or riding a buggy.

TABLE 3.4 Typical Magnetic Flux Densities

Source	Approximate $\|\vec{B}\|$ Field
Magnet (ceramic, cylindrical, 2 cm diameter, 1 cm long)	130,000 μT (dc) @ pole surface 20,000 μT (dc) @ 1 cm from pole surface 10,000 μT (dc) @ 3 cm from pole surface
Magnet (rare earth, samarium-cobalt, cylindrical, 2 cm diameter, 1 cm long)	350,000 μT (dc) @ pole surface 60,000 μT (dc) @ 1 cm from pole surface 20,000 μT (dc) @ 3 cm from pole surface
Earth's field near surface of equator	30 μT (dc)
Earth's field near surface of poles	60 μT (dc)
FCC limit for controlled/occupational exposure	2.05 μT rms (0.3–3 MHz) (6.15/*f*)[a] μT rms (3–30 MHz) 0.205 μT rms (30–300 MHz)
FCC limit for uncontrolled/general population exposure	2.05 μT rms (0.3–1.34 MHz) (2.75/*f*)[a] μT rms (1.34–30 MHz) 0.0917 μT rms (30–300 MHz)
765 kV rms transmission lines (1 kA per phase)	15 μT rms (60 Hz) (directly underneath the lines near the surface) 1–10 μT rms (60 Hz) (20–60 m along the surface from the base of the transmission lines)
138 kV rms transmission lines	1–1.4 μT rms (60 Hz) (10 m along the surface from the base of the transmission lines)
500 kV rms transmission lines	0.24 μT rms (60 Hz) (90 m along the surface from the base of the transmission lines)
Residential building	0.07 μT rms (60 Hz) U.S. houses, away from appliances 0.04 μT rms (50 Hz) Swedish houses, away from appliances 0.013 μT rms (50 Hz) Norwegian houses, away from appliances
House electrical wiring (10 A rms, #12 AWG)	5 μT rms (60 Hz) @ 1 inch 0.005 μT rms (60 Hz) @ 3 ft
Welding arcs (industrial)	100-5,800 μT rms (0–50 Hz) @ 0-0.8 m
Can openers	1,000–2,000 μT rms (60 Hz) @ 1 inch 0.07–0.9 μT rms (60 Hz) @ 3 ft
Electric shavers	15–1,500 μT rms (60 Hz) @ 1 inch < 0.01–0.35 μT rms (60 Hz) @ 3 ft
Sabre and circular saws	210–1,000 μT rms (60 Hz) @ 1 inch < 0.05–1.0 μT rms (60 Hz) @ 3 ft
Library or shop security/theft deterrent system	< 1,000 μT rms (0.1–10 kHz)
Drills	400–800 μT rms (60 Hz) @ 1 inch 0.08–0.2 μT rms (60 Hz) @ 3 ft
Vacuum cleaners	200–800 μT rms (60 Hz) @ 1 inch 0.12–1.7 μT rms (60 Hz) @ 3 ft
Mixers	50–700 μT rms (60 Hz) @ 1 inch 0.01–0.23 μT rms (60 Hz) @ 3 ft

TABLE 3.4 Typical Magnetic Flux Densities (Continued)

Source	Approximate $|\vec{B}|$ Field
Fluorescent desk lamps	40–400 μT rms (60 Hz) @ 1 inch 0.02–0.21 μT rms (60 Hz) @ 3 ft
Induction stirrer (industrial)	200–300 μT rms (10 Hz) @ 2 m
Garbage disposals	80–250 μT rms (60 Hz) @ 1 inch 0.02–0.1 μT rms (60 Hz) @ 3 ft
Microwave ovens	75–200 μT rms (60 Hz) @ 1 inch 0.3–0.8 μT rms (60 Hz) @ 3 ft
Fluorescent fixtures (workroom)	13–200 μT rms (60 Hz) @ 1 inch < 0.01–0.28 μT rms (60 Hz) @ 3 ft
Electric ranges	5.5–200 μT rms (60 Hz) @ 1 inch < 0.01–0.07 μT rms (60 Hz) @ 3 ft
Portable heaters	10–170 μT rms (60 Hz) @ 1 inch 0.01–0.25 μT rms (60 Hz) @ 3 ft
Blenders (kitchen)	20–120 μT rms (60 Hz) @ 1 inch 0.03–0.11 μT rms (60 Hz) @ 3 ft
Airport walk-through metal detector	< 100 μT rms (< 1 MHz)
Television sets	2.5–50 μT rms (60 Hz) @ 1 inch < 0.01–0.15 μT rms (60 Hz) @ 3 ft
Clothes washers	0.7–40 μT rms (60 Hz) @ 1 inch 0.01–0.15 μT rms (60 Hz) @ 3 ft
Irons	8–30 μT rms (60 Hz) @ 1 inch 0.01–0.02 μT rms (60 Hz) @ 3 ft
Fans and blowers	1.7–30 μT rms (60 Hz) @ 1 inch < 0.01–0.32 μT rms (60 Hz) @ 3 ft
Coffee makers	1.6–26 μT rms (60 Hz) @ 1 inch < 0.01 μT rms (60 Hz) @ 3 ft
Dishwashers	3.5–20 μT rms (60 Hz) @ 1 inch 0.08–0.25 μT rms (60 Hz) @ 3 ft
Hairdryers (handheld)	6.0–20 μT rms (60 Hz) @ 1 inch < 0.01 μT rms (60 Hz) @ 3 ft
Toasters	7–15 μT rms (60 Hz) @ 1 inch < 0.01–0.015 μT rms (60 Hz) @ 3 ft
Crock pots	1.5–8 μT rms (60 Hz) @ 1 inch < 0.01 μT rms (60 Hz) @ 3 ft
Clothes dryers	0.3–7.8 μT rms (60 Hz) @ 1 inch 0.02–0.05 μT rms (60 Hz) @ 3 ft
Electric ovens	1–5 μT rms (60 Hz) @ 1 inch < 0.01–0.04 μT rms (60 Hz) @ 3 ft
Electric blankets	0.3–5.0 μT rms (60 Hz) @ 1 inch
Refrigerators	0.5–1.5 μT rms (60 Hz) @ 1 inch < 0.01–0.04 μT rms (60 Hz) @ 3 ft

[a] *f* in MHz

3.25 MQS and Perfect Conductors

When is an electromagnetic system magnetoquasistatic? When can a real conductor or real insulator be considered perfect under magnetoquasistatic conditions?[28] [Haus; Adler; Fano]

A pure magneto*static* field is a magnetic field that does not vary with time. A magnetostatic field, however, can vary in space. It is generated by nontime-varying (i.e., dc) current (and magnets). For example, when dc current is "injected" into a perfectly conducting loop of wire surrounded by free space (and sufficient time is allowed to elapse so that the current in the wire is in steady state), only a magnetic field exists around the wire. No electric field is generated by the constant current in the ideal wire.

When a pure static magnetic field is in the presence of a pure static electric field, there is no coupling or interaction between the fields. The electric and magnetic fields do not influence each other in any way when both fields are static. Thus, if a conventional capacitor with a dc applied voltage across it is physically near to an inductor with a dc current through it, the capacitor's electric field is not expected to influence the inductor's magnetic field and vice versa. Although static magnetic fields are interesting, this book emphasizes practical time-varying electric and magnetic fields. In some cases, the coupling between the electric and magnetic fields is very strong, as with radiation from an antenna. In other cases, the coupling between the fields is weak, as with some resistive heaters and electric blankets. Referring to Table 3.5, for nonstatic conditions, the coupling between the electric and magnetic fields is apparent through $d\vec{B}/dt$, which is sometimes referred to as the magnetic induction component, and through $d\vec{D}/dt$, referred to as the electric displacement current. If the electric and magnetic fields are not time varying or are static, both of these time derivatives are zero. For free space, $\vec{B}=\mu_o\vec{H}$ and $\vec{D}=\varepsilon_o\vec{E}$.

Fortunately, in many cases it is not necessary to use Maxwell's equations in their complete unsimplified form to determine a reasonable approximation for the electric and magnetic field distribution. The emphasis of the previous discussions has been on magnetic fields or, more specifically, magnetoquasistatic (MQS) fields. For MQS conditions, the magnetic field is usually the field variable of interest. The electric field is usually not mentioned or is determined as if it has no effect on the magnetic field. However, MQS conditions do *not* imply that the electric and magnetic fields are not time varying. MQS conditions do imply, however, that the magnetic energy in the region of interest is much greater than the electric energy. In those cases where these energies can be determined, this energy relationship can be used as a check on whether MQS conditions prevail.

TABLE 3.5 Basic Laws in Electromagnetics

Common Name	Differential Form	Integral Form	Eqn. #
Faraday's law	$\nabla\times\vec{E}=-\frac{\partial\vec{B}}{\partial t}$	$\oint_C\vec{E}\cdot d\vec{L}=-\frac{d}{dt}\iint_S\vec{B}\cdot d\vec{s}$	(3.136)
Ampère's law	$\nabla\times\vec{H}=\vec{J}+\frac{\partial\vec{D}}{\partial t}$	$\oint_C\vec{H}\cdot d\vec{L}=\iint_S\vec{J}\cdot d\vec{s}+\frac{d}{dt}\iint_S\vec{D}\cdot d\vec{s}$	(3.137)
Gauss' law	$\nabla\cdot\vec{D}=\rho_V$	$\oiint_S\vec{D}\cdot d\vec{s}=\iiint_V\rho_V dv$	(3.138)
Magnetic flux continuity	$\nabla\cdot\vec{B}=0$	$\oiint_S\vec{B}\cdot d\vec{s}=0$	(3.139)
Charge conservation	$\nabla\cdot\vec{J}+\frac{\partial\rho_V}{\partial t}=0$	$\oiint_S\vec{J}\cdot d\vec{s}+\frac{d}{dt}\iiint_V\rho_V dv=0$	(3.140)

[28]This section closely follows the material contained in the book written by Professor Haus and Professor Melcher.

For MQS conditions, the field solution is generally simpler since (3.137) is approximated as

$$\nabla \times \vec{H} \approx \vec{J} \tag{3.141}$$

In other words, the displacement current is negligible. Since

$$\nabla \cdot \vec{B} = \nabla \cdot (\mu \vec{H}) = 0 \tag{3.142}$$

is true for all conditions, the magnetic field can be determined from these two expressions.[29] This magnetic field solution, which is essentially the magnetostatic solution, is referred to as the zero-order solution. The next approximation is that the time-varying charge is negligible in (3.140):

$$\nabla \cdot \vec{J} \approx 0 \tag{3.143}$$

This approximate expression for MQS conditions is quite "convenient" since it is the basis for KCL in circuit analysis. Equations (3.141) through (3.143) are not a function of d/dt (or the volume charge density). As with EQS conditions, this does not imply that the magnetic field is not time varying. For example, $\vec{J}$ can be a function of time and, according to (3.141), so would $\vec{H}$. Rather, the lack of d/dt in these three equations does imply that the present field distribution is not a function of any past field state (i.e., there is no field memory and no initial time conditions). Also, Equation (3.141) indicates that the effect of the time-varying electric field on the magnetic field is negligible. After the magnetic field is determined without regard for the effect of the electric field, an estimate for the electric field can be determined from Faraday's law, (3.136). If the charge distribution is required, it can be determined from Gauss' law, (3.138), using the electric field estimate. This electric field estimate can also be used to determine a better solution for the magnetic field via (3.137).

Many situations have been discussed in this book that assume MQS conditions. It is enriching to see how the electric field and an improved first-order magnetic field expression can be obtained from a magnetostatic result. Because of the relative simplicity of the analysis, the electric field and the error in the magnetic field will be determined for a single-turn rectangular loop. As shown in Figure 3.60, the single-turn loop, constructed of two closely spaced perfect conductors surrounded by free space, is connected to many identical time-varying surface-current sources.[30] These multiple sources help ensure that the magnetic field is uniform along the full depth of the loop and that the current along the plates in the $\pm x$ directions is not a function of y. End effects or nonuniform fields along the outer edges are neglected. The boundary condition involving the surface current

$$\hat{a}_n \times (\vec{H}_2 - \vec{H}_1) = \vec{K} \tag{3.144}$$

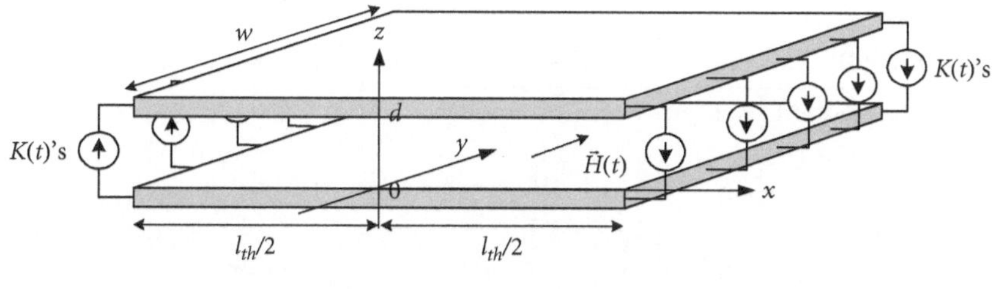

FIGURE 3.60 Many parallel, identical, time-varying current sources applied across two closely spaced parallel plates.

[29]Even when only the curl and the divergence of a vector are specified, both operations which involve differentiation, the vector can still be uniquely specified according to Helmholtz's theorem. (The vector, however, should approach zero at infinity.) This may be surprising since constants are lost in the process of differentiation.

[30]If N identical, ideal current sources $i(t)$ were used on each side to generate the current along the plates, the surface currents would be equal to $Ni(t)/w$.

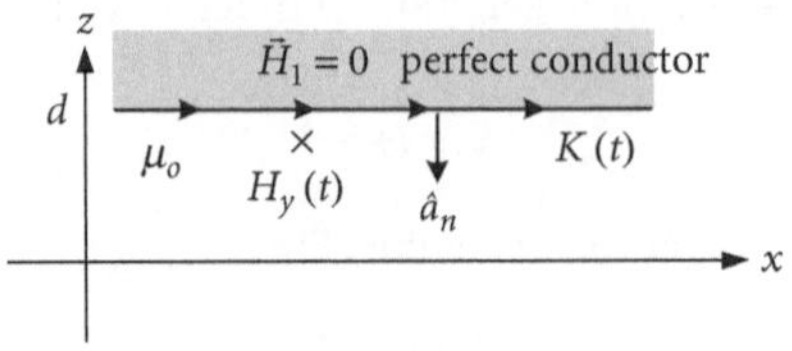

FIGURE 3.61 Magnetic field tangential to the upper (and lower) plate is easily determined using (3.144).

can be used to determine the magnetic field between the plates. The unit vector $\hat{a}_n$ is normal to the interface directed from region 1 to region 2. At the conducting interfaces, the tangential magnetic fields are discontinuous since surface current is present. Referring to Figure 3.61,

$$-\hat{a}_z \times [\vec{H}_2(t) - 0] = -\hat{a}_z \times \vec{H}_2(t) = K(t)\hat{a}_x \quad @z = d$$

Since the surface current is entirely in the x direction, the magnetic field $\vec{H}_2(t)$ must be entirely in the y direction since $-\hat{a}_z \times \hat{a}_y = \hat{a}_x(-\hat{a}_z \times \hat{a}_x = -\hat{a}_y$ and $-\hat{a}_z \times \hat{a}_z = 0)$. Solving for the magnetic field,

$$-\hat{a}_z \times H_y(t)\hat{a}_y = H_y(t)\hat{a}_x = K(t)\hat{a}_x \quad \Rightarrow H_y(t) = K(t) \quad @z = d$$

If this boundary condition is applied to the lower plate at $z = 0$, the same magnetic field is obtained. Outside the plates, the magnetic fields from the lower and upper plate surface currents cancel. Between the plates, the fields add. Since the current density between the plates is zero, $\nabla \times \vec{H} = 0$, ignoring the effect of the electric field. Also, $\nabla \cdot \vec{H} = 0$. Obviously, the magnetic field

$$\vec{H}(t) = K(t)\hat{a}_y \quad \text{if } 0 < z < d \tag{3.145}$$

which is not a function of x, y, or z has zero curl and zero divergence. Equation (3.145) is the standard result for the magnetic field everywhere between the plates. It is considered the zero-order solution for the magnetic field. The electric field can be determined from the integral form in (3.136):

$$\oint_C \vec{E} \cdot d\vec{L} = -\frac{d}{dt} \iint_S \vec{B} \cdot d\vec{s}$$

Referring to the surface and its corresponding contour in Figure 3.62,

$$\underbrace{d[E_z(0)] + (x-0)(0) - d[E_z(x)] - (x-0)(0)}_{\text{line integral of the electric field around the closed path}} = -\frac{d}{dt}\int_0^d \int_0^x \mu_o [K(t)\hat{a}_y] \cdot dx dz \hat{a}_y$$

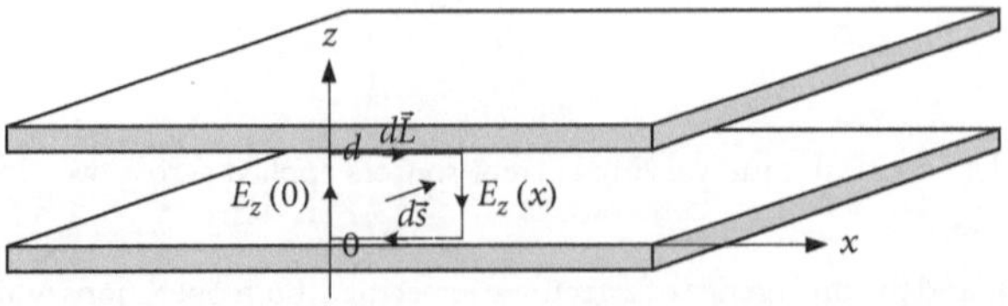

FIGURE 3.62 Electric field between the plates can be determined by using the integral form of Faraday's law. For the given clockwise direction of the line integration contour, the surface vector is in the +y direction. (The current sources are not shown.)

By selecting the surface in the xz plane, $\vec{B}\cdot d\vec{s}$ is easily evaluated since both of these vectors are in the same direction. Furthermore, by the judicious selection of the shape and position of this surface as shown, two sides of the $d\vec{L}$ integration are tangential to the perfectly conducting plates. Since the electric field tangential to the perfectly conducting plates is zero, $\vec{E}\cdot d\vec{L}=0$ for the lower and upper path segments. Solving for the electric field,

$$E_z(x)=E_z(0)+\mu_o x\frac{dK(t)}{dt} \tag{3.146}$$

The term $\mu_o x\, dK(t)/dt$ is an odd function of x. The electric field at the center, $E_z(0)$, is zero. Otherwise, $E_z(x)$ would not be an odd function of x, and the electric field at $x = -l_{th}/2$ would have a different magnitude than the electric field at $x = +l_{th}/2$. Considering the symmetrical nature of the structure and equal surface current sources at both ends of the plates, a symmetrical electric field is reasonable. Now that an estimate for the electric field is known, the integral form of Ampère's law

$$\oint_C \vec{H}\cdot d\vec{L}=\iint_S \vec{J}\cdot d\vec{s}+\frac{d}{dt}\iint_S \vec{D}\cdot d\vec{s}$$

can be used to determine a better approximation for the magnetic field between the plates. The differential surface vector shown in Figure 3.63 is in the z direction. Since the current density in this direction and between the plates is zero, $\vec{J}\cdot d\vec{s}$ is zero. Setting up and then evaluating this integral,

$$\underbrace{wH_y(l_{th}/2)-\left(\frac{l_{th}}{2}-x\right)(0)-w[H_y(x)]+\left(\frac{l_{th}}{2}-x\right)(0)}_{\text{line integral of the magnetic field around the closed path}}=\frac{d}{dt}\left\{\underbrace{\int_0^w\int_x^{l_{th}/2}\varepsilon_o\left[\mu_o x\frac{dK(t)}{dt}\hat{a}_z\right]\cdot dxdy\hat{a}_z}_{\substack{\text{total electric flux over the}\\ \text{given surface}}}\right\}$$

$$H_y(x)=H_y(l_{th}/2)-\frac{\varepsilon_o\mu_o}{2}\left(\frac{l_{th}^2}{4}-x^2\right)\frac{d^2K(t)}{dt^2}$$

The surface currents at $x = l_{th}/2$ forces the magnetic field to be (3.145) at this edge:

$$H_y(x)=K(t)-\frac{\varepsilon_o\mu_o}{2}\left(\frac{l_{th}^2}{4}-x^2\right)\frac{d^2K(t)}{dt^2} \tag{3.147}$$

Equation (3.147) is the improved version of the magnetic field between the plates. It contains the standard magnetostatic zero-order term plus a correction term. The deviation of the magnetic field from the static value increases with the time variation (the second-order time derivative) of the surface current.

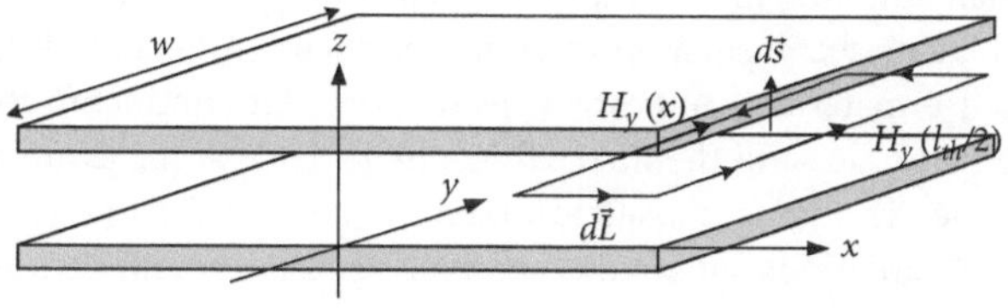

FIGURE 3.63 Correction term for the magnetic field can be obtained using Ampère's law. (The current sources are not shown.)

Probably the most interesting consequence of this work involving several of Maxwell's equations is the error associated with neglecting the correction term for the magnetic field:

$$\left|\frac{H_{MQS}-H_{STATIC}}{H_{STATIC}}\right| = \left|\frac{H_y(x)-H_y\left(\frac{l_{th}}{2}\right)}{H_y\left(\frac{l_{th}}{2}\right)}\right| = \left|\frac{K(t)-\frac{\mu_o\varepsilon_o}{2}\left(\frac{l_{th}^2}{4}-x^2\right)\frac{d^2K(t)}{dt^2}-K(t)}{K(t)}\right|$$

$$= \frac{\frac{\mu_o\varepsilon_o}{2}\left(\frac{l_{th}^2}{4}-x^2\right)\left|\frac{d^2K(t)}{dt^2}\right|}{|K(t)|}$$

If $K(t) = A\cos(\omega t)$, then this ratio is

$$\left|\frac{H_{MQS}-H_{STATIC}}{H_{STATIC}}\right| = \frac{\frac{\mu_o\varepsilon_o}{2}\left(\frac{l_{th}^2}{4}-x^2\right)|-A\omega^2\cos(\omega t)|}{|A\cos(\omega t)|} = \frac{\mu_o\varepsilon_o}{2}\left(\frac{l_{th}^2}{4}-x^2\right)\omega^2$$

The maximum value of this error is at $x = 0$:

$$\left.\left|\frac{H_{MQS}-H_{STATIC}}{H_{STATIC}}\right|\right|_{\max} = \frac{\mu_o\varepsilon_o l_{th}^2\omega^2}{8} = \frac{1}{8}\left(\frac{l_{th}\omega}{c}\right)^2 \quad \text{where } c = \frac{1}{\sqrt{\mu_o\varepsilon_o}} \approx 3\times 10^8 \text{ m/s},$$

$$= \frac{\pi^2}{2}\left(\frac{l_{th}}{\lambda_o}\right)^2 \quad \text{where } \lambda_o = \frac{c}{f} = \frac{2\pi c}{\omega} \text{ m}$$

This reassuring result indicates that when the length of the parallel plates is electrically small, $l_{th}/\lambda_o \ll 1$, the magnetic field expression obtained by neglecting the effects of the electric field is a good approximation to the higher order MQS approximation. For higher frequencies where the structure is not electrically small, the two terms given in (3.147) may be an inadequate description of the magnetic field.

In the previous situation, the two plates were constructed of perfect conductors and the space surrounding the conductors was free space, a perfect insulator. When perfect conductors and perfect insulators are involved, usually the system can be classified as EQS or MQS by allowing the frequency of the applied signal to approach zero or dc. If the voltages disappear leaving only currents, the system is likely MQS. (If the currents disappear leaving only voltage differences, or stationary charges, the system is likely EQS.) This test is reasonable. Many inductors, for example, represented as a multiturn coil or solid single loop, as in the previous example, are (generally) magnetostatic or magnetic energy-storage devices. At dc, ideal inductors are short-circuit zero-voltage devices. Even in crosstalk analysis, when the current is "large" and voltage is "small," the coupling can sometimes be thought of as mainly magnetic or inductive.

Of course, no material is a perfect insulator or perfect conductor. All materials have nonzero and finite conductivities. However, if a material is almost or nearly a perfect conductor, then the current can be considered mainly along its surface and the internal magnetic field can be assumed approximately zero. If a material is almost or nearly a perfect insulator, then the current density in the material is approximately zero. Similar to EQS situations, after studying many cases, it becomes clear that for a material to be considered like a perfect conductor

$$\tau \ll \tau_m \quad \text{good conductor, MQS} \tag{3.148}$$

where τ is the time of interest and τ_m is the magnetic time constant for the material (or system). For a MQS system, the time of interest should be small compared to the magnetic time constant. For EQS situations, the electric time constant is $\tau_e = \varepsilon/\sigma$. Unfortunately, the magnetic time constant is not this simple. The magnetic time constant or magnetic diffusion time is of the form

$$\tau_m = \mu\sigma l_1 l_2$$

where l_1 and l_2 are the characteristic lengths of the system. As the dimensions of the MQS system increase, the system behaves more like a perfect conductor. For example, for a thin, nonmagnetic cylindrical shield of radius R and thickness Δ, the time constant associated with the shielding was given in (3.53) as $\tau_m = \mu_o\sigma R\Delta/2$. In this case, the characteristic lengths could be given as R and $\Delta/2$. Using the circuit concept $\tau = L/R$, the time constant for the single-turn inductor shown in Figure 3.60 is

$$\tau_m = \frac{L}{R} = \frac{\dfrac{\mu_o d l_{th}}{w}}{2\dfrac{l_{th}}{\sigma w t}} = \mu_o \sigma d\left(\frac{t}{2}\right) \qquad (3.149)$$

where t is the thickness of the conductors and σ their conductivity. The variable R is the resistance of both conductors (not the insulation between the conductors). The characteristic lengths for (3.149) could be selected as d and $t/2$.

When the applied signal is sinusoidal with a frequency ω rad/s, a reasonable time of interest, τ, would be the reciprocal of ω since it is proportional to the period of the sinusoidal, $1/\omega = T/(2\pi)$. Therefore, for system to be considered a perfect conductor with a sinusoidal applied signal,

$$\frac{1}{\omega} \ll \mu\sigma l_1 l_2 \quad \Rightarrow \omega\mu\sigma l_1 l_2 \gg 1 \quad \text{good conductor, MQS} \qquad (3.150)$$

Unlike EQS systems where copper satisfies the good-conductor requirement for even very high frequencies, copper is not necessarily a good conductor in MQS systems. For example, a copper cylinder may be entirely inadequate for shielding low-frequency magnetic fields parallel to its surface. According to (3.150), a thick copper shield of large radius is required at low frequencies. At higher frequencies, however, a thin copper shield of smaller radius may provide reasonable shielding. At higher frequencies, the external field will induce larger "bucking" or counter currents in the shield. Of course, the dimensions should still be electrically small to use safely the magnetic time constant and MQS approximations.

Since the concept of skin depth is sometimes incorrectly applied to a MQS situation, the relationship between the skin depth and magnetic time constant will be briefly discussed. The skin depth is defined, for good conductors, as $\delta = 1/\sqrt{\pi f\mu\sigma}$. The skin depth is an extremely helpful parameter in determining the shielding effectiveness of materials to far-field electromagnetic waves. When the shield thickness is large compared to the skin depth, the effectiveness of the ideal shield to far-field radiation is very good. The concept of skin depth also arises when dealing with the diffusion of magnetic fields through materials. Materials that are thick compared to the skin depth can provide near-field magnetic shielding:

$$\delta = \frac{1}{\sqrt{\pi f\mu\sigma}} \ll t \quad \Rightarrow (2\pi f)\mu\sigma t\left(\frac{t}{2}\right) \gg 1 \qquad (3.151)$$

where t is the thickness. When the conductor thickness is large compared to the skin depth, the decay of the magnetic field in the conductor is exponential and a function of the skin depth (e.g., $e^{-x/\delta}$). If "the skin depth is small compared to the dimension of interest," then magnetic shielding will be good and the conductor will appear perfect-like. Notice that this statement is consistent with (3.150). However, a shield does not necessarily need to be thick compared to the skin depth to provide magnetic shielding. A thin shield where $\delta \ll t$ can still provide good magnetic shielding. The thin cylindrical conductor of radius R and thickness Δ previously mentioned provides shielding to magnetic fields when $\omega\tau_m = \omega\mu_o\sigma R\Delta/2 \gg 1$. Written in terms of the skin depth,

$$\omega\mu_o\sigma\frac{R\Delta}{2} = \frac{R\Delta}{\delta^2} \gg 1 \qquad (3.152)$$

If the skin depth is comparable or large compared to the thickness, this inequality can still be satisfied. For example, if the skin depth is an order of magnitude greater than the thickness, then

$$\frac{R\Delta}{\delta^2} = \frac{R}{100\Delta} \gg 1 \quad \text{if } \delta = 10\Delta$$

If the radius of the shield is much greater than 100Δ, then magnetic shielding will occur even though the shield thickness is one-tenth of the skin depth! The skin depth should be used with caution when dealing with magnetic shielding.

As with EQS situations, when analyzing a system, it is not necessary that every part of the system be MQS to utilize the simplifying characteristics associated with MQS laws. This is the key to lumped-circuit analysis. Although no device operating at any nonzero frequency is a true lumped circuit, it is extremely helpful to model circuits using lumped elements. Students (and working engineers) would much rather work with *R*'s, *L*'s, and *C*'s than electric and magnetic field expressions, even with the simplifications associated with EQS and MQS laws.

3.26 Decoupled Time-Varying Electric and Magnetic Fields

When can time-varying electric and magnetic fields be decoupled from each other similar to static fields? [Haus; Paul, '87]

There is an extremely convenient distribution of time-varying electric and magnetic fields that allows the fields to be essentially decoupled from each other as if they were static fields. The principal restriction is that the time-varying electric and magnetic fields should be transverse or perpendicular to the direction of propagation. This mode of operation or field distribution is referred to as TEM (transverse electromagnetic). Under the following conditions,

1. the conductor's cross sections do not vary along their length
2. the conductors are lossless (i.e., infinite conductivity)
3. the material surrounding the conductors is electrically the same everywhere (i.e., constant permittivity and permeability)
4. the spacing between the conductors is electrically small

the electric and magnetic fields between two (or more) conductors of arbitrary cross sections have this interesting decoupling property.

Assume two conductors of any cross section in the *xy* plane have axes parallel to the *z* axis as shown in Figure 3.64 (the cross sections do not have to be the same). The two conductors, for example, can be concentric cylinders (i.e., coax) or parallel plates. Under the four given conditions, even when the electric and magnetic fields are time varying, the fields are decoupled from each other similar to two independent static fields in any *xy* plane. Thus, the time-varying electric field can be determined using electrostatic tools and the time-varying magnetic field can be determined using magnetostatic tools. This permits the traditional EQS voltage to be defined between the conductors in any *xy* plane. Also, the traditional MQS

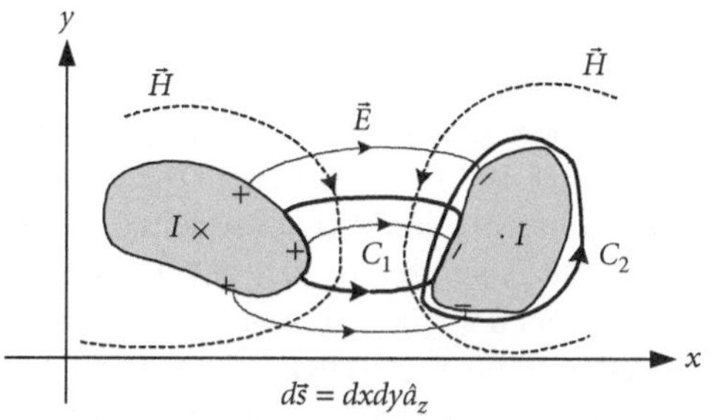

FIGURE 3.64 Rough sketch of the electric and magnetic fields between two arbitrary-shaped conductors that extend in the *z* direction.

current can be defined for each of the conductors. Actually, these conclusions can be readily shown[31] beginning with two of Maxwell's equations, (3.136) and (3.137):

$$\oint_C \vec{E}\cdot d\vec{L} = -\frac{d}{dt}\iint_S \vec{B}\cdot d\vec{s}$$

$$\oint_C \vec{H}\cdot d\vec{L} = \iint_S \vec{J}\cdot d\vec{s} + \frac{d}{dt}\iint_S \vec{D}\cdot d\vec{s}$$

Taking the surface vector $d\vec{s}$ in the *z* direction, these equations reduce to

$$\oint_{C_1} \vec{E}\cdot d\vec{L} = 0 \tag{3.153}$$

$$\oint_{C_2} \vec{H}\cdot d\vec{L} = \iint_{S_2} \vec{J}\cdot d\vec{s} \tag{3.154}$$

since both B_z and D_z are zero (the fields are transverse to the direction of propagation). One closed contour in the *xy* plane for evaluating (3.153) is shown as C_1. Another closed contour in the *xy* plane for evaluating (3.154) is shown as C_2. Since C_2 completely surrounds one of the current-carrying conductors, (3.154) is equal to *I*. Because the conductors are perfect, the tangential electric field along the conductors is zero. The electric field is also zero inside the perfect conductors, but the current density, $\vec{J} = \sigma\vec{E}$, in the *z* direction in the conductors, is still finite because of the infinite conductivity of the conductors.[32] Equations (3.153) and (3.154) are the standard expressions used for electrostatic and magnetostatic conditions, respectively. An important consequence of these results is that when the conductors are electrically short in the *z* direction, they can be exactly modeled using a lumped inductance, *L*, and lumped capacitance, *C*. When the length of the conductors is not electrically short, the transmission line equations can be used, which are based on this same *L* and *C*.

[31]See the references for a more rigorous proof.

[32]The product of infinity and zero is not necessarily zero or infinity!

The previously given four conditions will now be further explained:

1. If the cross sections of the conductors were not constant in the z direction along their length, the field distribution would not be two dimensional and independent of z. In addition, if the cross section varied in the z direction, L and C would vary in the z direction as well.
2. If the loss of the conductors were not zero, then a tangential electric field would exist along the conductors, and the fields would therefore not be TEM. Even at dc, a voltage drop would exist along the length of the conductors implying a nonzero tangential electric field.
3. If the medium surrounding the conductors were not uniform or homogeneous, then the velocity of the signal would be different for each region.[33] This would generate field components that are in the direction of propagation and, thus, the field would not be TEM.[34]
4. If the spacing between the conductors were not electrically small, then other field distributions or modes would (possibly) exist between the conductors. The length of the conductors is not required to be electrically short, which implies that the voltage across and current in the conductors can vary along their length.

The fields determined under this static-like approach should not be used to determine far fields or radiation patterns. The time-varying coupling between the electric and magnetic fields is crucial in determining these far fields and radiation.

It might appear that these four conditions are too restrictive for any practical application. Actually, they can be and have been successfully used to obtain approximate near-field expressions for many real situations. For example, all transmission lines have nonzero conductor loss. If this loss is modeled through R and the inductance modeled as L, then when $R \ll \omega L$ the losses are considered small and the fields can be approximated using the static approach. As another example, for a microstrip line partially embedded in a homogeneous dielectric, if most of the field lines are contained in the dielectric, then the fields will be mostly TEM. Nonhomogeneous medium can usually be adequately modeled using lumped parameters when the spacing between the conductors is electrically small.

---

[33] For a typical coaxial cable, the material between the conductors is homogeneous. Since the fields are entirely between the conductors, and zero outside the ideal cable, the material outside of the cable is irrelevant.

[34] With a nonhomogeneous medium, the normal and tangential electric field boundary conditions along the interface(s) cannot be satisfied unless the electric field has a z component.

4

Additional Shielding Concepts

Before designing or selecting an electromagnetic shield, the basic characteristics of the field must first be determined or estimated. For example, is the field mainly magnetic in nature or plane wave-like? Then, parameters such as the shield material and thickness, number of layers in the shield, and number and size of openings in the shield can be determined for a given shield effectiveness.

4.1 When Is a Shield Flat?

Under what conditions can a curved surface be considered flat? [Kerr; Ramo; Moon, '61]

Many shielding discussions in this book assume that the shielding surface is flat. Of course, most surfaces are not flat, and they have varying curvature and roughness. Analytical modeling of most surfaces that are not flat can be exceedingly difficult. It is therefore important to understand when a real surface can be considered approximately flat.

Imagine that a plane wave is incident upon a metal surface. The surface is not flat but the radii of curvature of any bumps, corrugations, ridges, undulations, or convolutions on the real surface are greater than about one free-space wavelength *and* the metal is either perfectly conducting or its thickness is much greater than the skin depth in the metal.[1] Under these conditions, an approximation for the induced current and voltage along the metal can be obtained by assuming the metal surface is flat.

The radius of curvature at three locations for a two-dimensional profile of a metal surface is shown in Figure 4.1. In this figure, the smallest radius of curvature shown is R_1. Therefore, assuming that R_1 is greater than about one wavelength at the lowest frequency of interest and the material thickness is much greater than the skin depth, δ, this metal can be considered "flat." For some cases, a portion of the surface satisfies this radius of curvature condition over a limited area. In this case, the surface is considered locally "flat" or "gentle." For three-dimensional surfaces, there are two radii of curvature and both must be tested.

Sometimes it is stated that the skin depth should be much less than the radius of curvature at the local point of interest on a surface for the surface to be considered flat. For good conductors, such as copper, the skin depth is several orders of magnitude less than the free-space wavelength for even microwave frequencies. This relationship is shown in Mathcad 4.1. Therefore, a skin depth much less than the smallest radius of curvature

$$\delta \ll R \tag{4.1}$$

[1]All surfaces are not perfectly smooth when viewed microscopically. At very high frequencies, such as in the microwave range, this surface roughness can affect the ac resistance of the metal. Interestingly, a polished high-quality copper surface can have a lower resistive loss than a rough silvered surface. This loss is important in the design of high-Q cavities.

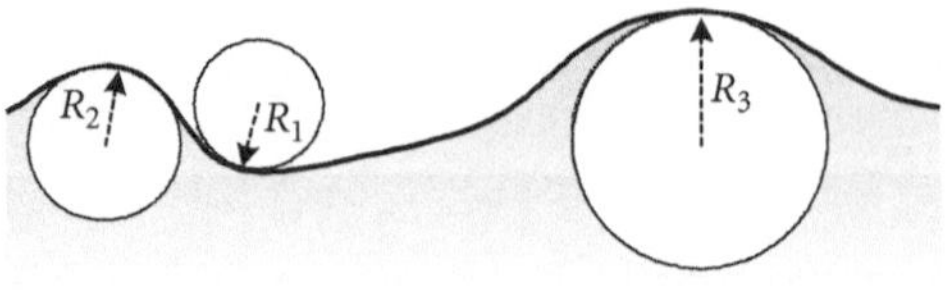

FIGURE 4.1 Surface profile and radius of curvature at select locations. As the curvature or sharpness increases, the radius of curvature decreases.

does not necessarily imply that the radius of curvature is at least one free-space wavelength

$$\lambda_o \le R?$$

since

$$\delta << \lambda_o \tag{4.2}$$

Also shown in Mathcad 4.1 is the wavelength *inside* the metal given by the good-conductor approximation

$$\lambda = \frac{v}{f} \approx \frac{\sqrt{\dfrac{2\omega}{\sigma\mu}}}{\dfrac{\omega}{2\pi}} = 2\pi\sqrt{\frac{2}{\omega\sigma\mu}} \quad \text{if } \sigma >> \omega\varepsilon \tag{4.3}$$

As seen, if $\delta << R$, then $\lambda \le R$ could be satisfied if the wavelength corresponds to inside the metal.

$\mu_o := 4\cdot\pi\cdot10^{-7}$ $\quad \varepsilon_o := 8.854\cdot10^{-12}$ $\quad \sigma_o := 5.8\cdot10^{7}$ $\quad \mu_r := 1$ $\quad \sigma_r := 1$

$x := 30, 31..100$

$$f(x) := \left(x + 1 - 10\cdot\text{floor}\left(\frac{x}{10}\right)\right)\cdot10^{\text{floor}\left(\frac{x}{10}\right)}$$

$$\delta(f) := \frac{1}{\sqrt{\pi\cdot f\cdot\mu_r\cdot\mu_o\cdot\sigma_r\cdot\sigma_o}} \qquad \lambda_o(f) := \frac{3\cdot10^8}{f} \qquad \lambda(f) := 2\cdot\pi\cdot\sqrt{\frac{2}{2\cdot\pi\cdot f\cdot\sigma_r\cdot\sigma_o\cdot\mu_r\cdot\mu_o}}$$

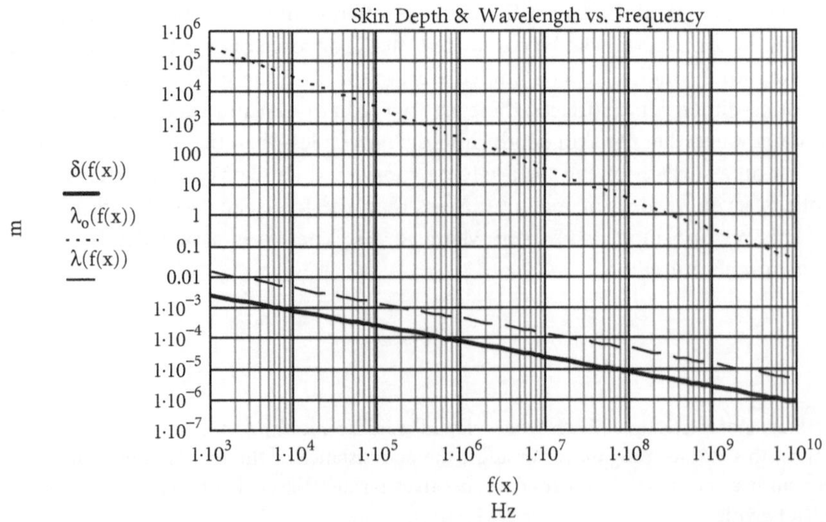

MATHCAD 4.1 Skin depth, free-space wavelength, wavelength in the metal as a function of the frequency for copper.

When using the concept of skin depth to determine the ac resistance of a wire, a good approximation for the ac resistance can be quickly obtained by assuming that most of the current is carried within one skin depth along the perimeter of the wire. However, for this approximation to be good, the skin depth should be small compared to the cross-sectional dimensions of the conductor, and the skin depth should be small compared to the smallest radius of curvature of the cross section of the wire. For a circular wire of radius R, the radius of curvature is R and the cross-section dimension is $2R$. Thus, there is little error in using the skin depth concept, based on a flat thick conducting surface, for a circular wire when the skin depth is much less than the radius of the wire. Clearly, the wire is not a shield to an external incident field but a carrier of current.

When an object is in the far field of a source, the manner that the field is reflected from the object is a function of many factors including the size of the object (relative to the wavelength), shape of the object, and roughness of the surface. Often the complicated details of the object are lumped into a parameter referred to as the effective cross section of the object, which can be experimentally determined. A reflected wave is also referred to as a scattered wave. As a side note of possible interest, even when an incident wave is linearly polarized, the scattered wave from a real nonflat object is often elliptically polarized.

4.2 Performance of a Shielded Room

Collect and tabulate the typical shielding performance of various shielded rooms, including the double-isolated shield. [Hemming, '92; Lindgren; Coombs; ARRL, '96]

Shielded rooms are used for research, development, EMC regulation compliance testing, and product emission and susceptibility reduction. When performing product emission measurements, for example, noise from outside sources can severely interfere with the measurements. This outside noise can be reduced to an acceptable level by working in a shielded room. The noise generated during high-voltage discharge experiments can be very annoying. By working in a shielded enclosure, this noise can be contained so as not to interfere with electrical products outside the room.

The performance of a shield is usually determined by its effectiveness in shielding electric fields (near field), magnetic fields (near field), and plane waves (far field). Often, the magnetic fields are the most difficult to shield, and many older shielded rooms were not designed to shield low-frequency magnetic fields. The shielding performance of a shielded room for magnetic and electric fields is frequently given from a few kHz to 1 MHz while for plane waves from a few MHz to 10 GHz, depending on the application. There are several types of shielded rooms: single shield, cell type, and double electrically isolated. These room types are shown in Figure 4.2.

The single-shielded enclosure frequently consists of a frame of lumber or steel with a solid metal or metal screen attached to one side of the frame. The cell-type enclosure consists of a solid metal or metal screen attached to both sides of a frame with electrical bonding between both sides. The double electrically isolated enclosure consists of a solid metal or metal screen attached to both sides of a frame, but the two sides are

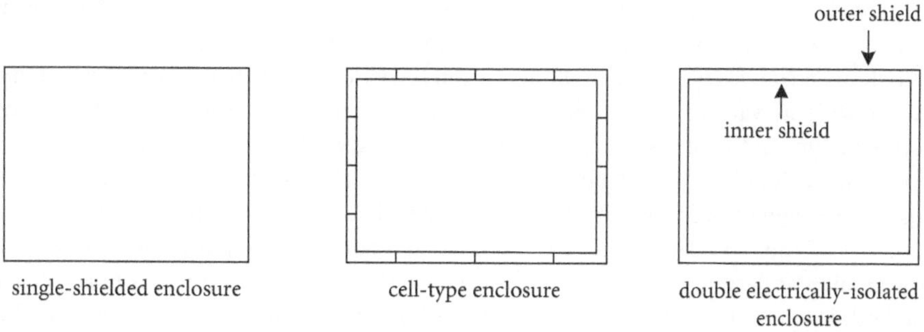

FIGURE 4.2 Three types of shielding enclosures.

TABLE 4.1 Typical Shielding Effectiveness for Magnetic Fields, Electric Fields, and Plane Waves

Enclosure	Magnetic Fields	Electric Fields	Plane Waves
Single shield copper screen	44 dB @ 15 kHz 58 dB @ 100 kHz	65 dB @ 15 kHz	64 dB @ 420 MHz
Cell type copper screen (both sides)	50 dB @ 15 kHz	120 dB @ 15 kHz	120 dB @ 400 MHz
Double electrically isolated bronze screen (both sides)	39 dB @ 15 kHz 69 dB @ 100 kHz	120 dB @ 15 kHz	120 dB @ 450 MHz
Double electrically isolated copper screen (both sides)	68 dB @ 15 kHz 98 dB @ 100 kHz	120 dB @ 15 kHz	120 dB @ 450 MHz
Single shield solid 24 gauge galvanized steel	30 dB @ 10 kHz 85 dB @ 100 kHz	100 dB @ 15 kHz	100 dB @ 100 MHz
Double electrically isolated solid 24 gauge galvanized steel (ext.) and 3 oz copper (int.)	8 dB @ 60 Hz 75–80 dB @ 14 kHz	120 dB @ 14 kHz	120 dB @ 450 MHz
Double electrically isolated solid 24 gauge galvanized steel (ext.) and 24 oz copper (int.)	18–25 dB @ 60 Hz 86–93 dB @ 14 kHz	120 dB @ 14 kHz	120 dB @ 450 MHz
Double electrically isolated solid 24 gauge galvanized steel (both sides)	15 dB @ 60 Hz 80 dB @ 14 kHz	110–120 dB @ 14 kHz	100–110 dB @ 450 MHz
Double electrically isolated laminated solid 24 gauge silicon steel and 24 oz copper (both sides)	36–44 dB @ 60 Hz 110–120 dB @ 14 kHz	120 dB @ 14 kHz	120 dB @ 450 MHz

electrically isolated from each other (including the door). That is, no electrical contact is made between the inner and outer shield. This enclosure is equivalent to a room within a room. By isolating the inner and outer shield, any current induced or present on one shield is not *conductively* transferred to the other shield.

Both metal screen and solid metal are used in shielded rooms. Metals that are used include copper, bronze, brass, aluminum, and galvanized steel. Steel is used in those situations where low-frequency magnetic fields (less than about 200 kHz) require attenuation because of steel's high permeability (at least relative to the other materials listed, which have a relative permeability of nearly one). Steel is not used in those situations where the magnetic field is not to be disturbed (e.g., certain medical imaging equipment). In this case, metals such as copper or aluminum can be used. Screen enclosures have obvious optical and acoustical transparency advantages over solid sheets. Also, screens are of lower cost. However, greater attenuation can be obtained through solid sheets. Sometimes shielded rooms are referred to as screen rooms because of the use of metal screen in some enclosures.

Typical shielding effectiveness values in dB for various types of shield enclosures are given in Table 4.1. It is difficult to compare various designs since vents, door gaskets, field measurement techniques, screen density, workmanship, and many other parameters can strongly influence the results.

For copper films or plating, the thickness of the copper is often given in ounces (oz) or oz/ft^2. This corresponds to the weight of a one square-foot, flat copper sheet of a given thickness. For 1 oz of copper, this thickness is 1.37 mil $\approx$ 35 μm. For 24 oz of copper, the corresponding thickness is 0.835 mm. Also, note that the gauge standard used for iron, steel, nickel, and other magnet (or ferrous) metals is usually the US Standard, listed in Table 4.2, not the American (or B&S) Standard. (The American Standard is used for aluminum, copper, brass, and other nonmagnetic metals.) The relationships given in this table can be used for sheets (thickness), wires (diameter), and rods (diameter).

TABLE 4.2 US Standard Gauge

Gauge	mils	mm
1	281	7.14
2	266	6.75
4	234	5.95
6	203	5.16
8	172	4.37
10	141	3.57
12	109	2.78
14	78.1	1.98
16	62.5	1.59
18	50.0	1.27
20	37.5	0.953
22	31.3	0.794
24	25.0	0.635
26	18.8	0.476
28	15.6	0.397
30	12.5	0.318

4.3 Laminated Shields

Determine and study the expression for the shielding effectiveness of the laminated shield shown in Figure 4.3 consisting of two different conducting materials, in direct contact, each with a different thickness. [Schulz, '88]

The total loss in dB for a flat shield to a normally incident plane wave is the sum of the reflection, absorption, and multiple-reflection losses in dB:

$$TOT_{dB} = R_{dB} + A_{dB} + MR_{dB}$$

The incident waveform is to the left of the material with the impedance η_1. The expressions for the reflection and absorption losses will first be examined since they are very simple to understand. For the two-layer laminated shield shown in Figure 4.3, assuming both materials are good conductors, the reflection loss consists of three terms. Each interface has one interface loss term:

$$R_{dB} = 20\log\left(\frac{1}{2}\left|1+\frac{\eta_1}{\eta_o}\right|\right) + 20\log\left(\frac{1}{2}\left|1+\frac{\eta_2}{\eta_1}\right|\right) + 20\log\left(\frac{1}{2}\left|1+\frac{\eta_o}{\eta_2}\right|\right) \tag{4.4}$$

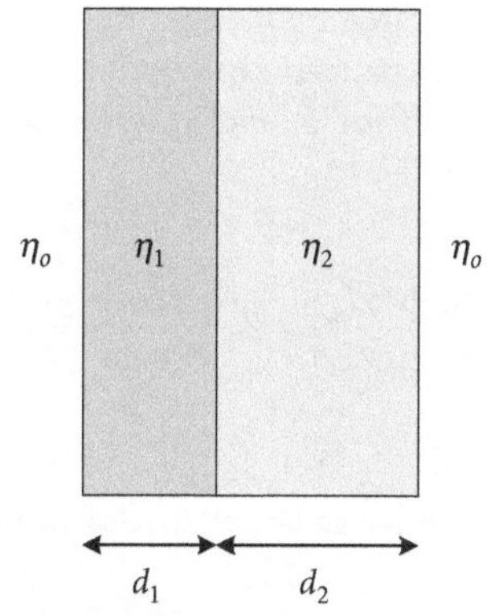

FIGURE 4.3 Two-layer, flat laminated shield surrounded by free space.

The pattern is easily seen. Notice that the reflection loss at the metal-to-metal interface, the middle term in (4.4), is independent of frequency:

$$20\log\left(\frac{1}{2}\left|1+\frac{\eta_2}{\eta_1}\right|\right)\approx 20\log\left[\frac{1}{2}\left|1+\frac{\sqrt{\frac{\omega\mu_2}{2\sigma_2}}(1+j)}{\sqrt{\frac{\omega\mu_1}{2\sigma_1}}(1+j)}\right|\right]=20\log\left(\frac{1}{2}\left|1+\sqrt{\frac{\mu_2\sigma_1}{\mu_1\sigma_2}}\right|\right) \tag{4.5}$$

However, the reflection loss at the metal-to-air interfaces is a function of the frequency since the intrinsic impedance of air, η_o, is constant at 377 Ω.

The absorption loss is the sum of the total loss in dB for each of the two materials:

$$A_{dB}=8.69\left(\frac{d_1}{\delta_1}+\frac{d_2}{\delta_2}\right) \tag{4.6}$$

As expected, as the thickness of each layer increases relative to its skin depth, the losses increase in that layer. Because free space has no loss, there are no absorption loss terms for the free space to the left or right of the laminated shield.

There are two multiple-reflection loss terms. The total multiple-reflection loss is the sum of the losses in the first and second layer:

$$\begin{aligned} MR_{dB}=MR_{1dB}+MR_{2dB}&=20\log\left|1-\left[\frac{(\eta_1-\eta_o)(\eta_1-\eta_{in2})}{(\eta_1+\eta_o)(\eta_1+\eta_{in2})}\right]e^{-2\left(\frac{1}{\delta_1}+j\frac{1}{\delta_1}\right)d_1}\right| \\ &+20\log\left|1-\left[\frac{(\eta_2-\eta_1)(\eta_2-\eta_o)}{(\eta_2+\eta_1)(\eta_2+\eta_o)}\right]e^{-2\left(\frac{1}{\delta_2}+j\frac{1}{\delta_2}\right)d_2}\right| \end{aligned} \tag{4.7}$$

$$\text{where } \eta_{in2}=\eta_2\frac{\eta_o\cosh\left[\left(\frac{1}{\delta_2}+j\frac{1}{\delta_2}\right)d_2\right]+\eta_2\sinh\left[\left(\frac{1}{\delta_2}+j\frac{1}{\delta_2}\right)d_2\right]}{\eta_2\cosh\left[\left(\frac{1}{\delta_2}+j\frac{1}{\delta_2}\right)d_2\right]+\eta_o\sinh\left[\left(\frac{1}{\delta_2}+j\frac{1}{\delta_2}\right)d_2\right]}$$

These expressions are much more complicated than either the reflection or absorption loss expressions. They are easily understood (or at least the pattern of the solution) by analogy to cascaded transmission lines and by comparison to the *MR* term for a single layer shield. The input impedance for a lossy transmission line of length l_{th} is given by

$$Z_{in}=Z_o\frac{Z_L+Z_o\tanh(\gamma l_{th})}{Z_o+Z_L\tanh(\gamma l_{th})}=Z_o\frac{Z_L\cosh(\gamma l_{th})+Z_o\sinh(\gamma l_{th})}{Z_o\cosh(\gamma l_{th})+Z_L\sinh(\gamma l_{th})} \tag{4.8}$$

where Z_o is the impedance of the line and Z_L is its load impedance. By analogy, the input impedance into a slab of material of thickness d with an intrinsic impedance of η_i and a load intrinsic impedance of η_L is

$$\eta_{in}=\eta_i\frac{\eta_L\cosh(\gamma d)+\eta_i\sinh(\gamma d)}{\eta_i\cosh(\gamma d)+\eta_L\sinh(\gamma d)} \tag{4.9}$$

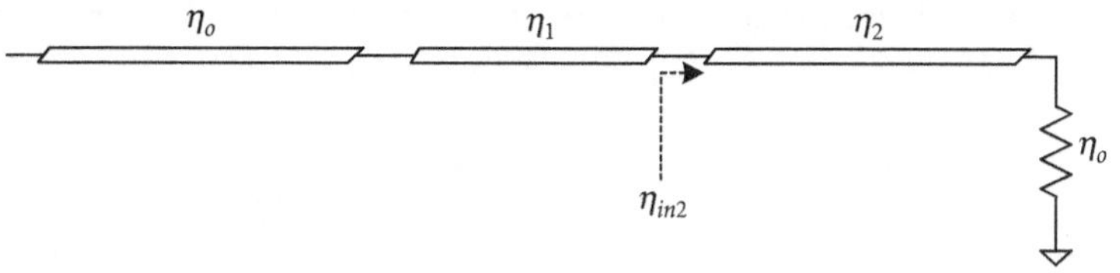

FIGURE 4.4 Modeling a two-layer laminated shield using transmission lines.

$\mu_o := 4\cdot\pi\cdot10^{-7} \quad \varepsilon_o := 8.854\cdot10^{-12} \quad \sigma_o := 5.8\cdot10^{7} \quad \mu_{r1} := 1 \quad \sigma_{r1} := 1 \quad \mu_{r2} := 500 \quad \sigma_{r2} := 0.1 \quad \eta_o := \sqrt{\dfrac{\mu_o}{\varepsilon_o}} \quad j := \sqrt{-1}$

$x := 40, 41 .. 70$

$d_1 := 100\cdot10^{-6} \quad d_2 := 100\cdot10^{-6}$

$$f(x) := \left(x + 1 - 10\cdot\text{floor}\left(\frac{x}{10}\right)\right)\cdot10^{\text{floor}\left(\frac{x}{10}\right)}$$

$$\delta_1(f) := \frac{1}{\sqrt{\pi\cdot f\cdot\mu_{r1}\cdot\mu_o\cdot\sigma_{r1}\cdot\sigma_o}} \qquad \delta_2(f) := \frac{1}{\sqrt{\pi\cdot f\cdot\mu_{r2}\cdot\mu_o\cdot\sigma_{r2}\cdot\sigma_o}} \qquad \eta_1(f) := \sqrt{\frac{2\cdot\pi\cdot f\cdot\mu_{r1}\cdot\mu_o}{2\cdot\sigma_{r1}\cdot\sigma_o}}\cdot(1+j)$$

$$\eta_2(f) := \sqrt{\frac{2\cdot\pi\cdot f\cdot\mu_{r2}\cdot\mu_o}{2\cdot\sigma_{r2}\cdot\sigma_o}}\cdot(1+j) \qquad R_{dB}(f) := 20\cdot\log\left(\frac{1}{2}\cdot\left|1+\frac{\eta_1(f)}{\eta_o}\right|\right) + 20\cdot\log\left(\frac{1}{2}\cdot\left|1+\frac{\eta_2(f)}{\eta_1(f)}\right|\right) + 20\cdot\log\left(\frac{1}{2}\cdot\left|1+\frac{\eta_o}{\eta_2(f)}\right|\right)$$

$$A_{dB}(f) := 8.69\cdot\left(\frac{d_1}{\delta_1(f)} + \frac{d_2}{\delta_2(f)}\right) \qquad \eta_{in2}(f) := \eta_2(f)\cdot\frac{\eta_o\cdot\cosh\left[\left(\frac{1}{\delta_2(f)} + j\cdot\frac{1}{\delta_2(f)}\right)\cdot d_2\right] + \eta_2(f)\cdot\sinh\left[\left(\frac{1}{\delta_2(f)} + j\cdot\frac{1}{\delta_2(f)}\right)\cdot d_2\right]}{\eta_2(f)\cdot\cosh\left[\left(\frac{1}{\delta_2(f)} + j\cdot\frac{1}{\delta_2(f)}\right)\cdot d_2\right] + \eta_o\cdot\sinh\left[\left(\frac{1}{\delta_2(f)} + j\cdot\frac{1}{\delta_2(f)}\right)\cdot d_2\right]}$$

$$MR1_{dB}(f) := 20\cdot\log\left[\left|1-\left(\frac{\eta_1(f)-\eta_o}{\eta_1(f)+\eta_o}\right)\cdot\left(\frac{\eta_1(f)-\eta_{in2}(f)}{\eta_1(f)+\eta_{in2}(f)}\right)\cdot e^{-2\cdot\left(\frac{1}{\delta_1(f)}+j\cdot\frac{1}{\delta_1(f)}\right)\cdot d_1}\right|\right]$$

$$MR2_{dB}(f) := 20\cdot\log\left[\left|1-\left(\frac{\eta_2(f)-\eta_1(f)}{\eta_2(f)+\eta_1(f)}\right)\cdot\left(\frac{\eta_2(f)-\eta_o}{\eta_2(f)+\eta_o}\right)\cdot e^{-2\cdot\left(\frac{1}{\delta_2(f)}+j\cdot\frac{1}{\delta_2(f)}\right)\cdot d_2}\right|\right]$$

$$TOT_{dB}(f) := R_{dB}(f) + A_{dB}(f) + MR1_{dB}(f) + MR2_{dB}(f)$$

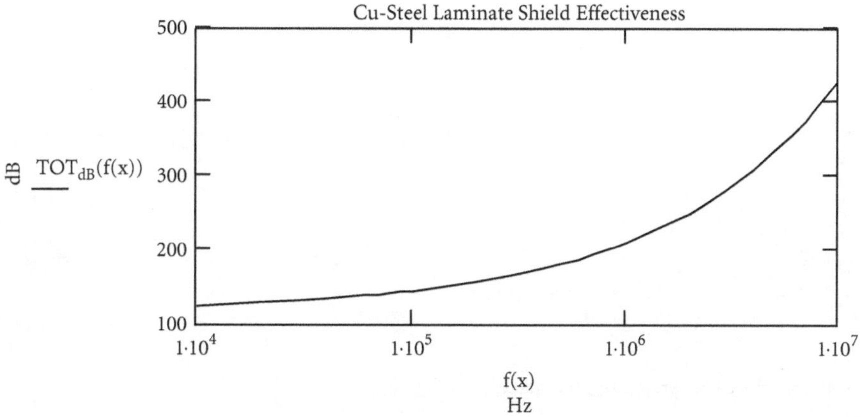

MATHCAD 4.2 SE vs. frequency of a copper-steel laminated shield.

For good conductors, the propagation constant γ for either of the layers can be written in terms of the skin depth:

$$\gamma \approx \sqrt{j\omega\mu\sigma} = \frac{1}{\delta} + j\frac{1}{\delta} \quad \text{where } \delta = \frac{1}{\sqrt{\pi f \mu\sigma}}$$

Each of the multireflection loss terms is a function of a complex exponential of the form $e^{-\gamma d}$. Furthermore, the *MR* term for each layer is a function of the intrinsic impedance of the layer, as well as the intrinsic impedance of the material immediately before the layer (to its left in this example) and the *total load* intrinsic impedance immediately after the layer (to its right in this example). Although this multilayer shield is not a transmission line, Figure 4.4 should assist in understanding the source of the hyperbolic expressions for η_{in2}.[2] For the first layer of impedance η_1, the impedance immediately prior is η_o while the total load impedance is η_{in2}. For the second layer with an impedance of η_2, the impedance immediately prior to this layer is η_1 while the total load impedance immediately after is just η_o.

The results in Mathcad 4.2 indicate that the loss for a copper-steel laminated shield (both metals 100 μm thick and in direct contact) is greater than 100 dB for a wide frequency range. This analysis assumes that the signal to be shielded is a plane wave and normally incident to the flat shield. Although the analysis is not specifically shown, interchanging the positions of the copper and steel does not affect the overall shielding effectiveness (assuming that the permeability of the steel does not significantly change with the applied field). Typically, shielded rooms are not limited in their performance by this ideal flat-shield gain but by nonideal factors such as opening and seam losses. As will be discussed later, when shielding (near-field) magnetic fields, the order of the metal in a laminated shield is important (e.g., Cu-steel vs. steel-Cu).

4.4 Shields with an Air Gap

Determine and study the expression for the shielding effectiveness of the laminated shield shown in Figure 4.5 consisting of two different conducting materials, separated by air, each metal with a different thickness. Is a flat double shield consisting of two metals separated by air always more effective than a single shield consisting of one metal (assuming the thickness of the single shield is the same as the total thickness of the two metals without the air)? [Schulz, '88; Ficchi]

As with the previous laminated shield, the total loss in dB for a flat shield to a normally incident plane wave is the sum of the reflection, absorption, and multiple-reflection losses in dB:

$$TOT_{dB} = R_{dB} + A_{dB} + MR_{dB}$$

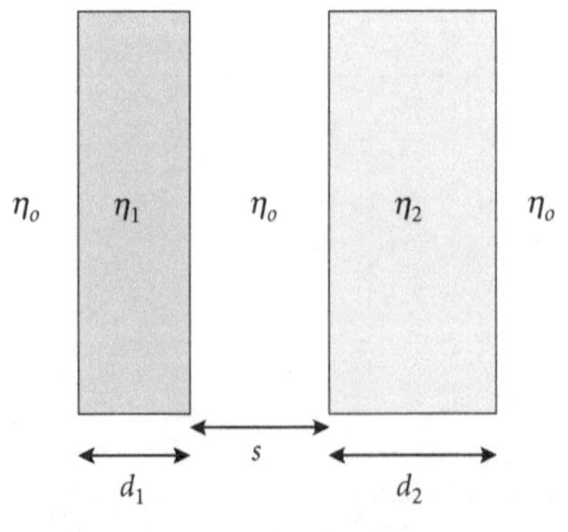

FIGURE 4.5 Flat double shield separated by free space.

[2]The electric and magnetic fields for both plane waves and many lossless transmission lines are TEM, transverse to the direction of propagation and perpendicular to each other.

Referring to Figure 4.5 and assuming that both materials are good conductors, the reflection and absorption losses can be written down by inspection:

$$R_{dB} = 20\log\left(\frac{1}{2}\left|1+\frac{\eta_1}{\eta_o}\right|\right) + 20\log\left(\frac{1}{2}\left|1+\frac{\eta_o}{\eta_1}\right|\right) + 20\log\left(\frac{1}{2}\left|1+\frac{\eta_2}{\eta_o}\right|\right) + 20\log\left(\frac{1}{2}\left|1+\frac{\eta_o}{\eta_2}\right|\right) \quad (4.10)$$

$$A_{dB} = 8.69\left(\frac{d_1}{\delta_1} + \frac{d_2}{\delta_2}\right) \quad (4.11)$$

The absorption loss is still the sum of the absorption losses for each material as with the two-layer laminated shield. (The absorption loss of the free space between the two metals is zero.) There are four terms for the reflection loss, corresponding to the four interfaces. The multiple-reflection loss is more complicated than the shield without the air between the metals. Each of the three regions, including the free-space region between the metals, has a *MR* loss:

$$MR_{dB} = MR_{1dB} + MR_{0dB} + MR_{2dB}$$

$$= 20\log\left|1 - \left[\frac{(\eta_1-\eta_o)(\eta_1-\eta_{in0})}{(\eta_1+\eta_o)(\eta_1+\eta_{in0})}\right] e^{-2\left(\frac{1}{\delta_1}+j\frac{1}{\delta_1}\right)d_1}\right| + 20\log\left|1 - \left[\frac{(\eta_o-\eta_1)(\eta_o-\eta_{in2})}{(\eta_o+\eta_1)(\eta_o+\eta_{in2})}\right] e^{-2j\beta_o s}\right|$$

$$+ 20\log\left|1 - \left(\frac{\eta_2-\eta_o}{\eta_2+\eta_o}\right)\left(\frac{\eta_2-\eta_o}{\eta_2+\eta_o}\right) e^{-2\left(\frac{1}{\delta_2}+j\frac{1}{\delta_2}\right)d_2}\right|$$

$$\text{where } \eta_{in0} = \eta_o \frac{\eta_{in2}\cos(\beta_o s) + j\eta_o \sin(\beta_o s)}{\eta_o \cos(\beta_o s) + j\eta_{in2}\sin(\beta_o s)} \quad (4.12)$$

$$\eta_{in2} = \eta_2 \frac{\eta_o \cosh\left[\left(\frac{1}{\delta_2}+j\frac{1}{\delta_2}\right)d_2\right] + \eta_2 \sinh\left[\left(\frac{1}{\delta_2}+j\frac{1}{\delta_2}\right)d_2\right]}{\eta_2 \cosh\left[\left(\frac{1}{\delta_2}+j\frac{1}{\delta_2}\right)d_2\right] + \eta_o \sinh\left[\left(\frac{1}{\delta_2}+j\frac{1}{\delta_2}\right)d_2\right]}$$

$$\beta_o = \frac{2\pi}{\lambda_o} = \frac{2\pi f}{c}$$

For the two conducting layers,

$$\gamma \approx \sqrt{j\omega\mu\sigma} = \frac{1}{\delta} + j\frac{1}{\delta} \quad \text{where } \delta = \frac{1}{\sqrt{\pi f \mu\sigma}}$$

while for the lossless free-space region between the layers

$$\gamma \approx \alpha + j\beta = 0 + j\beta_o = j\frac{2\pi}{\lambda_o} = j\frac{2\pi f}{c} = j\omega\sqrt{\mu_o\varepsilon_o}$$

The free-space regions before and after the total shield do not have a *MR* loss since the impedance far to the left and right are assumed matched to free space. The source of the terms in (4.12) can be understood by referring to Figure 4.6 and the previous transmission-line analogy discussion.

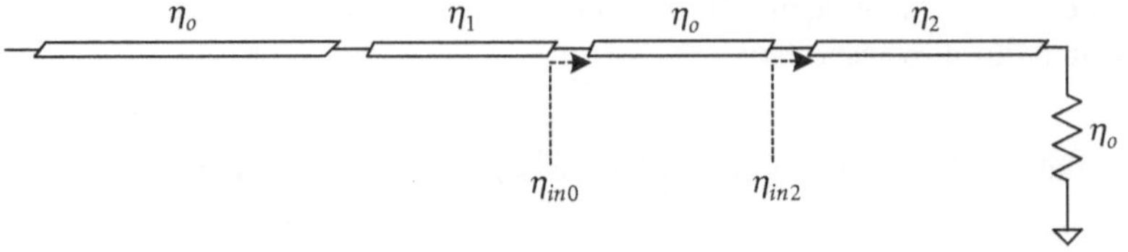

FIGURE 4.6 Modeling a double-layer shield separated by free space using transmission lines.

If both metals of the shield are electrically thick (compared to their respective skin depths), then the multiple-reflection terms *for each of the metals* may be neglected. The strength of the (round-trip) reflected signal is small in comparison to the one-way signal because of the strong attenuation in the thick metal. However, the multiple-reflection term for the free-space region between the two metals is not negligible since the attenuation of free space is zero. The multiple-reflection term for the free-space region between the metals is

$$MR_{0dB} = 20\log\left|1-\left[\frac{(\eta_o-\eta_1)(\eta_o-\eta_{in2})}{(\eta_o+\eta_1)(\eta_o+\eta_{in2})}\right]e^{-2j\beta_o s}\right|$$

$$\text{where } \eta_{in2} = \eta_2\frac{\eta_o\cosh\left[\left(\frac{1}{\delta_2}+j\frac{1}{\delta_2}\right)d_2\right]+\eta_2\sinh\left[\left(\frac{1}{\delta_2}+j\frac{1}{\delta_2}\right)d_2\right]}{\eta_2\cosh\left[\left(\frac{1}{\delta_2}+j\frac{1}{\delta_2}\right)d_2\right]+\eta_o\sinh\left[\left(\frac{1}{\delta_2}+j\frac{1}{\delta_2}\right)d_2\right]} \tag{4.13}$$

When the thickness d_2 is much greater than the skin depth, the impedance looking into the second metal at the left interface, η_{in2}, is approximately η_2 since the reflections from the far interface have negligible effect on the impedance. This can also be shown mathematically by using the standard exponential definitions for the hyperbolic cosine and sine functions:

$$\eta_{in2} = \eta_2\frac{\eta_o\left[\frac{e^{(1+j)\frac{d_2}{\delta_2}}+e^{-(1+j)\frac{d_2}{\delta_2}}}{2}\right]+\eta_2\left[\frac{e^{(1+j)\frac{d_2}{\delta_2}}-e^{-(1+j)\frac{d_2}{\delta_2}}}{2}\right]}{\eta_2\left[\frac{e^{(1+j)\frac{d_2}{\delta_2}}+e^{-(1+j)\frac{d_2}{\delta_2}}}{2}\right]+\eta_o\left[\frac{e^{(1+j)\frac{d_2}{\delta_2}}-e^{-(1+j)\frac{d_2}{\delta_2}}}{2}\right]}$$

For thick metals or large ratios of d_2/δ_2,

$$e^{(1+j)\frac{d_2}{\delta_2}} \gg e^{-(1+j)\frac{d_2}{\delta_2}}$$

Therefore,

$$\eta_{in2} \approx \eta_2\frac{\eta_o\left[\frac{e^{(1+j)\frac{d_2}{\delta_2}}}{2}\right]+\eta_2\left[\frac{e^{(1+j)\frac{d_2}{\delta_2}}}{2}\right]}{\eta_2\left[\frac{e^{(1+j)\frac{d_2}{\delta_2}}}{2}\right]+\eta_o\left[\frac{e^{(1+j)\frac{d_2}{\delta_2}}}{2}\right]} = \eta_2$$

The multiple-reflection term for the free-space region between the metals can then be approximated as

$$MR_{0dB} \approx 20\log\left|1-\left[\frac{(\eta_o-\eta_1)(\eta_o-\eta_2)}{(\eta_o+\eta_1)(\eta_o+\eta_2)}\right]e^{-2j\beta_o s}\right| \tag{4.14}$$

If both metals are of the same composition with an impedance of η, and Euler's formula is used for the complex exponential,

$$MR_{0dB} \approx 20\log\left|1-\frac{(\eta_o-\eta)^2}{(\eta_o+\eta)^2}e^{-2j\beta_o s}\right| = 20\log\left|1-\frac{(\eta_o-\eta)^2}{(\eta_o+\eta)^2}\left[\cos\left(2s\frac{2\pi}{\lambda_o}\right)-j\sin\left(2s\frac{2\pi}{\lambda_o}\right)\right]\right|$$

$$= 20\log\left|1-\frac{\left(1-\dfrac{\eta}{\eta_o}\right)^2}{\left(1+\dfrac{\eta}{\eta_o}\right)^2}\left[\cos\left(2s\frac{2\pi}{\lambda_o}\right)-j\sin\left(2s\frac{2\pi}{\lambda_o}\right)\right]\right|$$

Since the impedance of the metal, η, is much less than $\eta_o = 377\ \Omega$, the binomial expansion can be used to approximate two of the terms in this expression:

$$\begin{aligned} MR_{0dB} &\approx 20\log\left|1-\left(1-2\frac{\eta}{\eta_o}\right)\left(1-2\frac{\eta}{\eta_o}\right)\left[\cos\left(2s\frac{2\pi}{\lambda_o}\right)-j\sin\left(2s\frac{2\pi}{\lambda_o}\right)\right]\right| \\ &\approx 20\log\left|1-\left(1-4\frac{\eta}{\eta_o}\right)\left[\cos\left(4\pi\frac{s}{\lambda_o}\right)-j\sin\left(4\pi\frac{s}{\lambda_o}\right)\right]\right| \end{aligned} \tag{4.15}$$

If the distance between the metals, s, in the free-space region, is much less than the free-space wavelength, λ_o, this expression can be further simplified:

$$MR_{0dB} \approx 20\log\left|1-\left(1-4\frac{\eta}{\eta_o}\right)\left(1-j4\pi\frac{s}{\lambda_o}\right)\right| \approx 20\log\left|4\frac{\eta}{\eta_o}+j4\pi\frac{s}{\lambda_o}\right| \tag{4.16}$$

The argument of the logarithm is much less than one, which implies that the multiple-reflection loss for the free-space region is negative and large (the multiple-reflection loss for the metals is negligible). For thick metals where the distance s is much less than a wavelength, the *MR* reduces the overall shielding effectiveness of the double shield with an air gap. However, when the interspace distance s is not small compared to λ_o, this multiple-reflection term in dB can be positive. For example, at spacings $s=(2k-1)(\lambda_o/4)$ where $k = 1, 2, 3, \ldots$, corresponding to odd multiples of a quarter wavelength,

$$MR_{0dB} \approx 20\log\left|1-\left(1-4\frac{\eta}{\eta_o}\right)\{\cos[\pi(2k-1)]-j\sin[\pi(2k-1)]\}\right| \approx 20\log 2 \approx 6 \tag{4.17}$$

At and near these higher "antiresonant" frequencies, the multiple-reflection term is positive and, thus, the *MR* increases the overall shielding effectiveness of the double shield with an air gap. But, at spacings

$\mu_o := 4 \cdot \pi \cdot 10^{-7}$ $\qquad \varepsilon_o := 8.854 \cdot 10^{-12}$ $\qquad \eta_o := \sqrt{\dfrac{\mu_o}{\varepsilon_o}}$ $\qquad j := \sqrt{-1}$

$\eta := \dfrac{\eta_o}{1000}$

$x := 0.001, 0.0011 .. 1$

$$MR_{dB}(x) := 20 \cdot \log\left[\left|1 - \left(1 - 4 \cdot \frac{\eta}{\eta_o}\right) \cdot \left(\cos(4 \cdot \pi \cdot x) - j \cdot \sin(4 \cdot \pi \cdot x)\right)\right|\right]$$

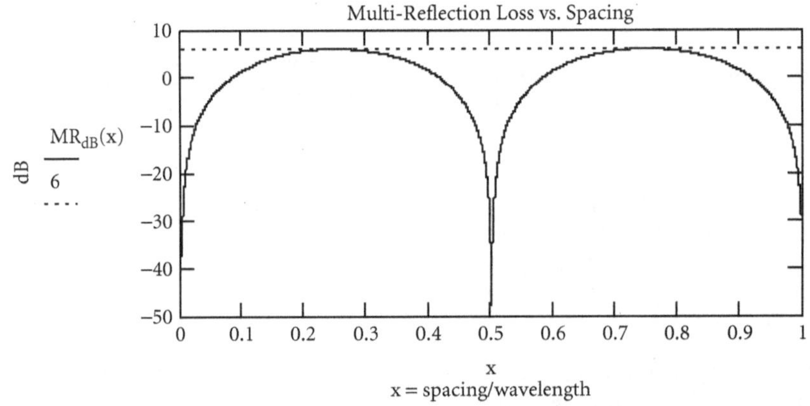

MATHCAD 4.3 Variation of the multiple-reflection loss (of the free-space region between two metals) vs. the ratio of the spacing between the metals to the free-space wavelength. The shielding effectiveness increases with positive values of MR_{dB}.

$s = (2k)(\lambda_o/4)$ where $k = 1, 2, 3, \ldots,$ corresponding to even multiples of a quarter wavelength,

$$\begin{aligned} MR_{0dB} &= 20\log\left|1 - \left(1 - 4\frac{\eta}{\eta_o}\right)[\cos(2\pi k) - j\sin(2\pi k)]\right| \\ &= 20\log\left|\frac{4\eta}{\eta_o}\right| = -20\log\left|\frac{\eta_o}{4\eta}\right| \end{aligned} \tag{4.18}$$

At and near these "resonant" frequencies, the multiple-reflection loss is very large and negative and the *MR* reduces the overall shielding effectiveness of the double shield with an air gap. As will be seen shortly, at these frequencies, the multiple-reflection loss is of equal magnitude but opposite sign to the reflection loss of a *single shield.* The plot in Mathcad 4.3 clearly shows the variation of the multiple-reflection loss with spacing. In this plot, the impedance of the electrically thick metals is one-thousandth of 377 Ω.

Although the absorption loss is only a function of the overall thickness of the metals, the reflection loss is a function of the number of interfaces and size of the impedance discontinuity at these interfaces. When the two metals are constructed of the same type of material, the reflection loss equation, (4.10), reduces to

$$\begin{aligned} R_{dB} &= 2\left[20\log\left(\frac{1}{2}\left|1 + \frac{\eta}{\eta_o}\right|\right) + 20\log\left(\frac{1}{2}\left|1 + \frac{\eta_o}{\eta}\right|\right)\right] = 2\left[20\log\left(\frac{1}{4}\left|1 + \frac{\eta}{\eta_o}\right|\left|1 + \frac{\eta_o}{\eta}\right|\right)\right] \\ &\approx 2\left(20\log\left|\frac{\eta_o}{4\eta}\right|\right) \end{aligned} \tag{4.19}$$

This approximation for R_{dB} is twice the reflection loss of a single shield. This should not be too surprising since most of the reflection loss (for electric fields) occurs at air-to-metal interfaces, and the double shield has twice as many air-to-metal interfaces. Combining the previous absorption, (4.11), reflection, (4.19), and multiple-reflection results, (4.17) and (4.18), it is seen that the shielding effectiveness, TOT_{dB}, for a double shield of electrically-thick metals of the same material varies over the following range:

$$8.69\left(\frac{d_1}{\delta}+\frac{d_2}{\delta}\right)+20\log\left|\frac{\eta_o}{4\eta}\right| < TOT_{dB} < 8.69\left(\frac{d_1}{\delta}+\frac{d_2}{\delta}\right)+2\left(20\log\left|\frac{\eta_o}{4\eta}\right|\right)+6 \tag{4.20}$$

The plots in Mathcad 4.4 shows that a double shield consisting of two 25 mil copper slabs separated by an air gap of 1,000 mil is over 40 dB superior to a single 50 mil copper slab for frequencies between 100 kHz

$\mu_o := 4\cdot\pi\cdot10^{-7}$ $\varepsilon_o := 8.854\cdot10^{-12}$ $\sigma_o := 5.8\cdot10^{7}$ $\mu_r := 1$ $\sigma_r := 1$ $d_1 := 635\cdot10^{-6}$ $d_2 := 635\cdot10^{-6}$ $\eta_o := \sqrt{\frac{\mu_o}{\varepsilon_o}}$

$s := 0.0254$ $j := \sqrt{-1}$

$x := 50, 50.1 .. 70$

$$f(x) := \left(x+1-10\cdot\text{floor}\left(\frac{x}{10}\right)\right)\cdot10^{\text{floor}\left(\frac{x}{10}\right)}$$

$$\delta(f) := \frac{1}{\sqrt{\pi\cdot f\cdot\mu_r\cdot\mu_o\cdot\sigma_r\cdot\sigma_o}} \qquad \eta(f) := \sqrt{\frac{2\cdot\pi\cdot f\cdot\mu_r\cdot\mu_o}{2\cdot\sigma_r\cdot\sigma_o}\cdot(1+j)} \qquad \lambda_o(f) := \frac{3\cdot10^8}{f}$$

$$A_{dB}(f) := 8.69\cdot\left(\frac{d_1}{\delta(f)}+\frac{d_2}{\delta(f)}\right) \qquad R_{dB}(f) := 20\cdot\log\left(\left|\frac{\eta_o}{4\cdot\eta(f)}\right|\right)$$

$$MR_{dB}(f) := 20\cdot\log\left[\left|1-\left(1-4\cdot\frac{\eta(f)}{\eta_o}\right)\cdot\left(\cos\left(4\cdot\pi\cdot\frac{s}{\lambda_o(f)}\right)-j\cdot\sin\left(4\cdot\pi\cdot\frac{s}{\lambda_o(f)}\right)\right)\right|\right]$$

$$TOT_{single}(f) := R_{dB}(f)+A_{dB}(f) \qquad TOT_{double}(f) := 2\cdot R_{dB}(f)+A_{dB}(f)+MR_{dB}(f)$$

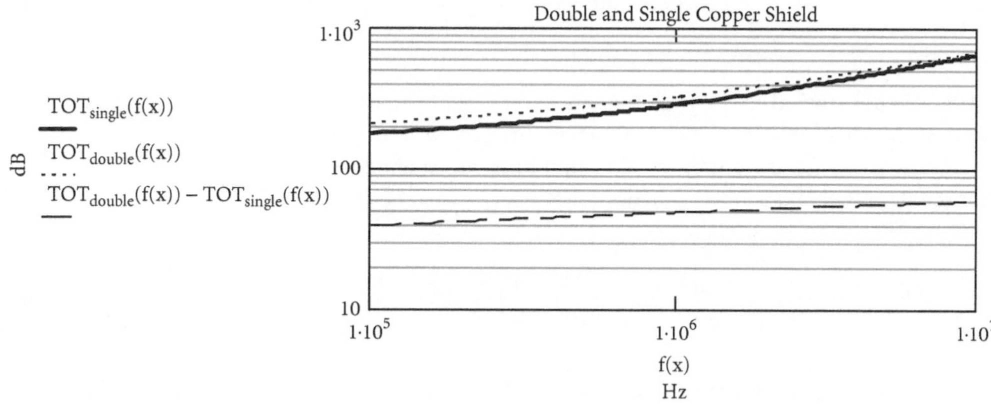

$$\frac{d_1}{\frac{1}{\sqrt{\pi\cdot10^5\cdot\mu_r\cdot\mu_o\cdot\sigma_r\cdot\sigma_o}}} = 3.039$$

MATHCAD 4.4 Total loss of a single shield, 50 mil thick, and a double shield, each conductor 25 mil thick, with a 1,000 mil spacing.

and 10 MHz. For frequencies much less than 100 kHz, 25 mils of copper is not electrically thick, and these expressions should not be used. The first free-space quarter-wavelength, $\lambda_o/4$, frequency corresponding to 1,000 mil is about 3 GHz. Although these equations indicate shielding effectiveness values much greater than 100–200 dB, in practice, the performance of a shield is not much greater than this range because of vents, seams, and other nonideal factors.

An approximate expression for the difference in dB between a double and single shield using good-conducting electrically-thick metals is given by

$$\Delta_{dB} = 20\log\left|1 + j\frac{\pi\frac{s}{\lambda_o}}{\frac{\eta}{\eta_o}}\right| \tag{4.21}$$

where $s \ll \lambda_o$. Therefore, for frequencies such that

$$\pi\frac{s}{\lambda_o} \ll \left|\frac{\eta}{\eta_o}\right| \tag{4.22}$$

the difference in shielding is negligible. In (4.21), the total thickness of metal for the double shield is equal to that for the single shield.

Now, the shielding effectiveness of a double and single shield will be compared for slabs that are not electrically thick. The absorption loss for the double and single shield is the same if the total metal thickness is the same. The reflection loss of the single shield is one-half of the double shield, and this reflection loss is not a function of the metal thickness. The multiple-reflection losses for the double and single shield are not the same. Furthermore, the total multiple-reflection loss for a double shield with slabs that are not electrically thick is not the same as for electrically-thick slabs since the multiple reflections inside the metals should probably not be neglected. The multiple-reflection loss for a single shield of thickness $d_1 + d_2$ is

$$MR_{dB} = 20\log\left|1 - \left(\frac{\eta_o - \eta}{\eta_o + \eta}\right)^2 e^{-2(1+j)\frac{d_1+d_2}{\delta}}\right| \tag{4.23}$$

This MR_{dB} is *not* identical to the multiple-reflection loss of the double shield if the free-space distance, s, is set to zero. When setting the free-space distance equal to zero, the individual reflection and multiple-reflection loss terms for the double shield should not be separately studied. For example, the reflection loss for the double shield is not a function of s:

$$R_{dB} = 20\log\left(\frac{1}{2}\left|1 + \frac{\eta_1}{\eta_o}\right|\right) + 20\log\left(\frac{1}{2}\left|1 + \frac{\eta_o}{\eta_1}\right|\right) + 20\log\left(\frac{1}{2}\left|1 + \frac{\eta_2}{\eta_o}\right|\right) + 20\log\left(\frac{1}{2}\left|1 + \frac{\eta_o}{\eta_2}\right|\right) \tag{4.24}$$

Obviously, if s is equal to zero, the reflections from the "second" interface should be zero, and the second and third terms in the previous expression should cancel or disappear. Notice that they do not cancel. When s is set to zero and the *sum* of the reflection and multiple-reflection losses is determined, the result is identical to the sum of the reflection and multiple-reflection losses for the single shield.

The multiple-reflection loss, reflection loss, absorption loss, and total shielding effectiveness vs. frequency of a double shield consisting of two sheets of 15 mil copper separated by 1,000 mil is given in Mathcad 4.5. The multiple-reflection term associated with the free-space region is much greater (in

$$\mu_o := 4\cdot\pi\cdot10^{-7} \quad \varepsilon_o := 8.854\cdot10^{-12} \quad \sigma_o := 5.8\cdot10^{7} \quad \mu_{r1} := 1 \quad \sigma_{r1} := 1 \quad \mu_{r2} := 1 \quad \sigma_{r2} := 1 \quad \eta_o := \sqrt{\frac{\mu_o}{\varepsilon_o}} \quad j := \sqrt{-1}$$

$$x := 30, 31..\,50 \qquad d_1 := 381\cdot10^{-6} \qquad d_2 := 381\cdot10^{-6}$$

$$f(x) := \left(x + 1 - 10\cdot\text{floor}\left(\frac{x}{10}\right)\right)\cdot10^{\text{floor}\left(\frac{x}{10}\right)}$$

$$\delta_1(f) := \frac{1}{\sqrt{\pi\cdot f\cdot\mu_{r1}\cdot\mu_o\cdot\sigma_{r1}\cdot\sigma_o}} \qquad \delta_2(f) := \frac{1}{\sqrt{\pi\cdot f\cdot\mu_{r2}\cdot\mu_o\cdot\sigma_{r2}\cdot\sigma_o}} \qquad \eta_1(f) := \sqrt{\frac{2\cdot\pi\cdot f\cdot\mu_{r1}\cdot\mu_o}{2\cdot\sigma_{r1}\cdot\sigma_o}}\cdot(1+j) \qquad \beta_o(f) := \frac{2\cdot\pi\cdot f}{3.10^{8}}$$

$$\eta_2(f) := \sqrt{\frac{2\cdot\pi\cdot f\cdot\mu_{r2}\cdot\mu_o}{2\cdot\sigma_{r2}\cdot\sigma_o}}\cdot(1+j) \qquad \eta_{in2}(f) := \eta_2(f)\cdot\frac{\eta_o\cdot\cosh\left[\left(\frac{1}{\delta_2(f)} + j\cdot\frac{1}{\delta_2(f)}\right)\cdot d_2\right] + \eta_2(f)\cdot\sinh\left[\left(\frac{1}{\delta_2(f)} + j\cdot\frac{1}{\delta_2(f)}\right)\cdot d_2\right]}{\eta_2(f)\cdot\cosh\left[\left(\frac{1}{\delta_2(f)} + j\cdot\frac{1}{\delta_2(f)}\right)\cdot d_2\right] + \eta_o\cdot\sinh\left[\left(\frac{1}{\delta_2(f)} + j\cdot\frac{1}{\delta_2(f)}\right)\cdot d_2\right]}$$

$$\eta_{in0}(f, s) := \eta_o\cdot\frac{\eta_{in2}(f)\cdot\cos(\beta_o(f)\cdot s) + j\cdot\eta_o\cdot\sin(\beta_o(f)\cdot s)}{\eta_o\cdot\cos(\beta_o(f)\cdot s) + j\cdot\eta_{in2}(f)\cdot\sin(\beta_o(f)\cdot s)}$$

$$MR1_{dB}(f, s) := 20\cdot\log\left[\left|1 - \left(\frac{\eta_1(f) - \eta_o}{\eta_1(f) + \eta_o}\right)\cdot\left(\frac{\eta_1(f) - \eta_{in0}(f, s)}{\eta_1(f) + \eta_{in0}(f, s)}\right)\cdot e^{-2\cdot\left(\frac{1}{\delta_1(f)} + j\cdot\frac{1}{\delta_1(f)}\right)\cdot d_1}\right|\right]$$

$$MR0_{dB}(f, s) := 20\cdot\log\left[\left|1 - \left(\frac{\eta_o - \eta_1(f)}{\eta_o + \eta_1(f)}\right)\cdot\left(\frac{\eta_o - \eta_{in2}(f)}{\eta_o + \eta_{in2}(f)}\right)\cdot e^{-2\cdot j\cdot\beta_o(f)\cdot s}\right|\right]$$

$$MR2_{dB}(f) := 20\cdot\log\left[\left|1 - \left(\frac{\eta_2(f) - \eta_o}{\eta_2(f) + \eta_o}\right)^2\cdot e^{-2\cdot\left(\frac{1}{\delta_2(f)} + j\cdot\frac{1}{\delta_2(f)}\right)\cdot d_2}\right|\right] \qquad A_{dB}(f) := 8.69\cdot\left(\frac{d_1}{\delta_1(f)} + \frac{d_2}{\delta_2(f)}\right)$$

$$R_{dB}(f) := 20\cdot\log\left(\frac{1}{2}\cdot\left|1 + \frac{\eta_1(f)}{\eta_o}\right|\right) + 20\cdot\log\left(\frac{1}{2}\cdot\left|1 + \frac{\eta_o}{\eta_1(f)}\right|\right) + 20\cdot\log\left(\frac{1}{2}\cdot\left|1 + \frac{\eta_2(f)}{\eta_o}\right|\right) + 20\cdot\log\left(\frac{1}{2}\cdot\left|1 + \frac{\eta_o}{\eta_2(f)}\right|\right)$$

$$TOT_{dB}(f, s) := A_{dB}(f) + R_{dB}(f) + MR1_{dB}(f, s) + MR0_{dB}(f, s) + MR2_{dB}(f)$$

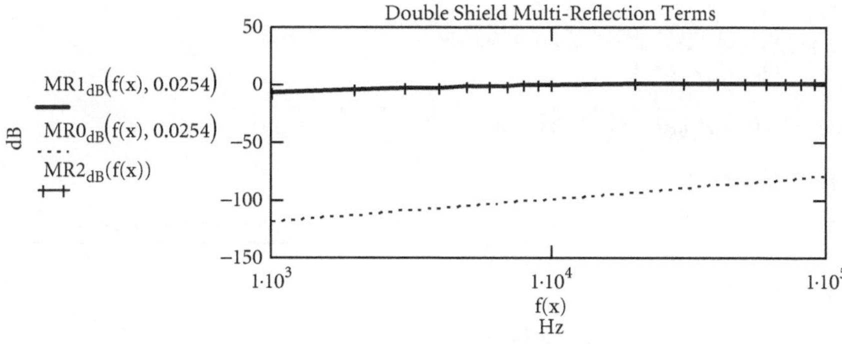

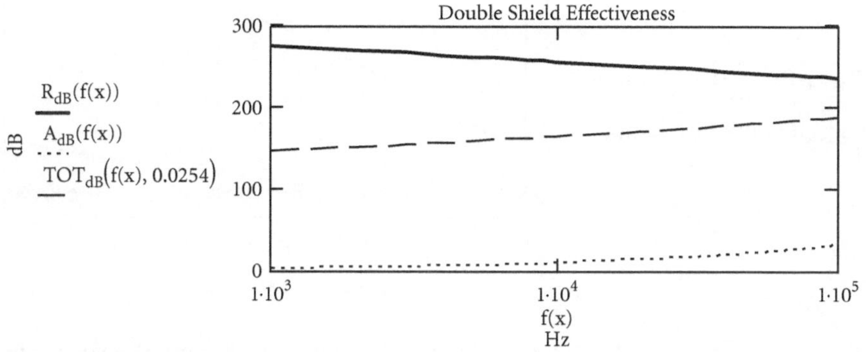

MATHCAD 4.5 Total loss and its components for two 15 mil thick shields separated by a distance of 1,000 mil.

$$\mu_o := 4\cdot\pi\cdot10^{-7} \quad \varepsilon_o := 8.854\cdot10^{-12} \quad \sigma_o := 5.8\cdot10^{7} \quad \mu_{r1} := 1 \quad \sigma_{r1} := 1 \quad \mu_{r2} := 1 \quad \sigma_{r2} := 1 \quad \eta_o := \sqrt{\frac{\mu_o}{\varepsilon_o}} \quad j := \sqrt{-1}$$

$$x := 30, 31..\, 60 \qquad d_1 := 381\cdot10^{-6} \qquad d_2 := 381\cdot10^{-6}$$

$$f(x) := \left(x + 1 - 10\cdot\text{floor}\left(\frac{x}{10}\right)\right)\cdot10^{\text{floor}\left(\frac{x}{10}\right)}$$

$$\delta_1(f) := \frac{1}{\sqrt{\pi\cdot f\cdot\mu_{r1}\cdot\mu_o\cdot\sigma_{r1}\cdot\sigma_o}} \qquad \delta_2(f) := \frac{1}{\sqrt{\pi\cdot f\cdot\mu_{r2}\cdot\mu_o\cdot\sigma_{r2}\cdot\sigma_o}} \qquad \eta_1(f) := \sqrt{\frac{2\cdot\pi\cdot f\cdot\mu_{r1}\cdot\mu_o}{2\cdot\sigma_{r1}\cdot\sigma_o}}\cdot(1+j) \qquad \beta_o(f) := \frac{2\cdot\pi\cdot f}{3.10^{8}}$$

$$\eta_2(f) := \sqrt{\frac{2\cdot\pi\cdot f\cdot\mu_{r2}\cdot\mu_o}{2\cdot\sigma_{r2}\cdot\sigma_o}}\cdot(1+j) \qquad \eta_{in2}(f) := \eta_2(f)\cdot\frac{\eta_o\cdot\cosh\left[\left(\frac{1}{\delta_2(f)} + j\cdot\frac{1}{\delta_2(f)}\right)\cdot d_2\right] + \eta_2(f)\cdot\sinh\left[\left(\frac{1}{\delta_2(f)} + j\cdot\frac{1}{\delta_2(f)}\right)\cdot d_2\right]}{\eta_2(f)\cdot\cosh\left[\left(\frac{1}{\delta_2(f)} + j\cdot\frac{1}{\delta_2(f)}\right)\cdot d_2\right] + \eta_o\cdot\sinh\left[\left(\frac{1}{\delta_2(f)} + j\cdot\frac{1}{\delta_2(f)}\right)\cdot d_2\right]}$$

$$\eta_{ino}(f, s) := \eta_o\cdot\frac{\eta_{in2}(f)\cdot\cos(\beta_o(f)\cdot s) + j\cdot\eta_o\cdot\sin(\beta_o(f)\cdot s)}{\eta_o\cdot\cos(\beta_o(f)\cdot s) + j\cdot\eta_{in2}(f)\cdot\sin(\beta_o(f)\cdot s)}$$

$$MR1_{dB}(f, s) := 20\cdot\log\left[\left|1 - \left(\frac{\eta_1(f) - \eta_o}{\eta_1(f) + \eta_o}\right)\cdot\left(\frac{\eta_1(f) - \eta_{ino}(f, s)}{\eta_1(f) + \eta_{ino}(f, s)}\right)\cdot e^{-2\cdot\left(\frac{1}{\delta_1(f)} + j\cdot\frac{1}{\delta_1(f)}\right)\cdot d_1}\right|\right]$$

$$MR0_{dB}(f, s) := 20\cdot\log\left[\left|1 - \left(\frac{\eta_o - \eta_1(f)}{\eta_o + \eta_1(f)}\right)\cdot\left(\frac{\eta_o - \eta_{in2}(f)}{\eta_o + \eta_{in2}(f)}\right)\cdot e^{-2\cdot j\cdot\beta_o(f)\cdot s}\right|\right] \qquad A_{dB}(f) := 8.69\cdot\left(\frac{d_1}{\delta_1(f)} + \frac{d_2}{\delta_2(f)}\right)$$

$$MR2_{dB}(f) := 20\cdot\log\left[\left|1 - \left(\frac{\eta_2(f) - \eta_o}{\eta_2(f) + \eta_o}\right)^2\cdot e^{-2\cdot\left(\frac{1}{\delta_2(f)} + j\cdot\frac{1}{\delta_2(f)}\right)\cdot d_2}\right|\right]$$

$$R_{dB}(f) := 20\cdot\log\left(\frac{1}{2}\cdot\left|1 + \frac{\eta_1(f)}{\eta_o}\right|\right) + 20\cdot\log\left(\frac{1}{2}\cdot\left|1 + \frac{\eta_o}{\eta_1(f)}\right|\right) + 20\cdot\log\left(\frac{1}{2}\cdot\left|1 + \frac{\eta_2(f)}{\eta_o}\right|\right) + 20\cdot\log\left(\frac{1}{2}\cdot\left|1 + \frac{\eta_o}{\eta_2(f)}\right|\right)$$

$$TOT_{double}(f, s) := A_{dB}(f) + R_{dB}(f) + MR1_{dB}(f, s) + MR0_{dB}(f, s) + MR2_{dB}(f)$$

$$MR_{dB}(f) := 20\cdot\log\left[\left|1 - \left(\frac{\eta_1(f) - \eta_o}{\eta_1(f) + \eta_o}\right)^2\cdot e^{-2\cdot\left(\frac{1}{\delta_1(f)} + j\cdot\frac{1}{\delta_1(f)}\right)\cdot(d_1 + d_2)}\right|\right]$$

$$TOT_{single}(f) := A_{dB}(f) + 20\cdot\log\left(\frac{1}{2}\cdot\left|1 + \frac{\eta_1(f)}{\eta_o}\right|\right) + 20\cdot\log\left(\frac{1}{2}\cdot\left|1 + \frac{\eta_o}{\eta_1(f)}\right|\right) + MR_{dB}(f)$$

$$s := 0.0254$$

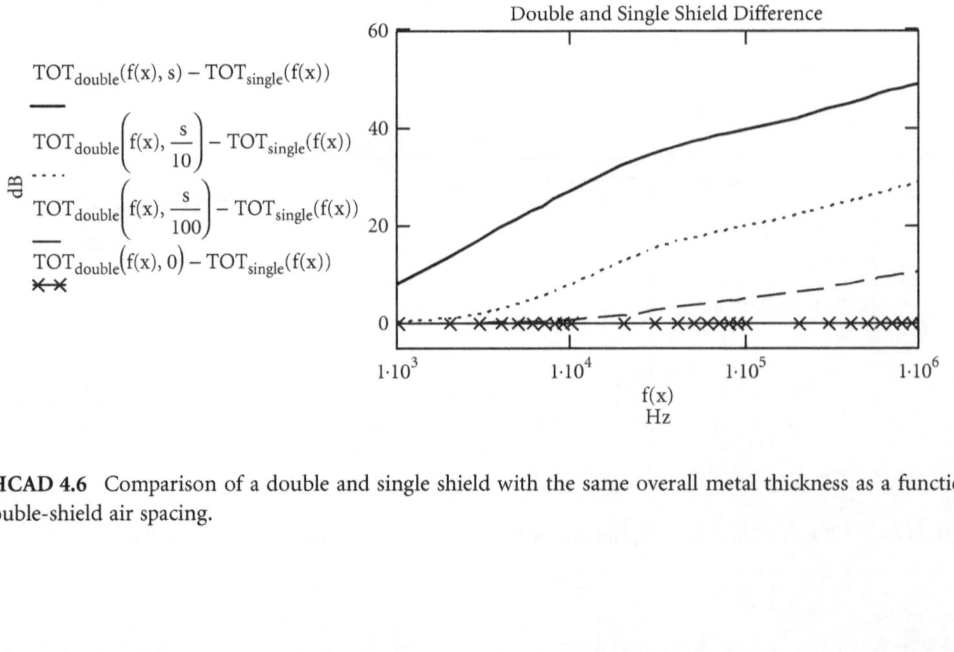

MATHCAD 4.6 Comparison of a double and single shield with the same overall metal thickness as a function of the double-shield air spacing.

magnitude) than the metal multiple-reflection terms over the given frequency range. The metal multiple-reflection terms are much more important at much lower frequencies (but then the incident field would probably not be a plane wave) or for much thinner sheets of copper. For example, for a double shield consisting of two 1.5 mil copper sheets, the *total* multiple-reflection term from the two metals is about one-half of the free-space multiple-reflection term. The second program, Mathcad 4.6, shows that at low frequencies and small spacings, the difference between the double and single shield is negligible.

4.5 Gold Coating on Glass

Discuss the microwave shielding effectiveness of using a very thin gold coating on both sides of glass. [Schulz, '88; Liao; Howard W. Sams; Elliot, '93]

Thin metal coatings are used on glass and plastic substrates to allow visible light through but to attenuate other frequencies. For example, metal films are used for outer-space visors to reduce the strong infrared radiation (IR) while passing the light spectrum (IR is above microwaves but below visible light in the frequency spectrum).

The microwave spectrum is very broad and is generally assumed to range from 1–100 GHz. It is divided into bands, and the IEEE microwave band designations are given in Table 4.3. Other designations are also used. Unfortunately, the labels are sometimes identical but the frequency range may be different (e.g., the New U.S. Military K Band ranges from 20–40 GHz). For this reason, when a microwave band is given, the reference designation (and its age or time of usage) should be determined.

There are several types of glass. One type, silicon dioxide (fused quartz), is a stable material that has a constant dielectric constant of 3.78 from 100 Hz to well into the microwave range.[3] Furthermore, the dissipation factor defined as

$$\tan\delta_{eff} = \frac{\varepsilon''}{\varepsilon'} = \frac{\sigma}{\omega\varepsilon_r\varepsilon_o}$$

TABLE 4.3 IEEE Microwave Band Designations

Microwave Band	Frequency Range	Free-Space Wavelength Range
L	1–2 GHz	30–15 cm
S	2–4 GHz	15–7.5 cm
C	4–8 GHz	7.5–3.75 cm
X	8–12 GHz	3.75–2.5 cm
K_u	12–18 GHz	2.5–1.67 cm
K	18–27 GHz	1.67–1.11 cm
K_a	27–40 GHz	1.11–0.75 cm
Millimeter	40–300 GHz	7.5–1 mm
Submillimeter	> 300 GHz	< 1 mm

[3]Iron-sealing glass has a dielectric constant that varies from 8.40 to 7.84 over this frequency range, and soda-borosilicate has a dielectric constant that varies from about 4.87 to 4.65.

TABLE 4.4 Conductivity and Surface Resistance of Gold

t (Å)[a]	t (m)	σ_f (1/Ω-m)	R_s (Ω)
10	10×10^{-10}	0.24×10^7	420
30	30×10^{-10}	0.54×10^7	62
60	60×10^{-10}	0.87×10^7	19
90	90×10^{-10}	1.11×10^7	10
0.1×10^7	0.1×10^{-3}	4.10×10^7	2.4×10^{-4}
1.0×10^7	1.0×10^{-3}	4.10×10^7	2.4×10^{-5}
10×10^7	10×10^{-3}	4.10×10^7	2.4×10^{-6}

[a] One Ångström, Å, is equal to 10^{-10} m.

varies from 0.00085 at 100 Hz to 0.00025 at 25 GHz for silicon dioxide. Because this ratio of the conduction current to the displacement current is much less than one, glass is a superb insulator over this broad frequency range. Furthermore, these dissipation factors imply an effective conductivity of about 2×10^{-11} 1/Ω-m at 100 Hz to about 1×10^{-3} 1/Ω-m at 25 GHz. These effective conductivities are very small, especially compared to metals, which have conductivities near 5.8×10^7 1/Ω-m. This data clearly indicates that glass (uncontaminated) may be assumed a perfect insulator in this analysis.

Light can pass through very thin films of metal. Also, the conductivity of metal films is less than the standard bulk conductivity typically used for metals. The conductivity of a thin film of gold is equal to

$$\sigma_f = \frac{3t\sigma}{4p}\left[\ln\left(\frac{p}{t}\right) + 0.4228\right] \quad \text{if } p \ll t \tag{4.25}$$

For gold at room temperature, $p = 570$ Å, $\sigma = 4.10 \times 10^7$ 1/Ω-m, and t = thickness of the gold film in Å. For gold, the conductivity varies as shown in Table 4.4. (The surface resistance, $R_s = 1/(\sigma_f t)$, is also given.) As seen, the conductivity increases with thickness. For larger thicknesses, such as 0.1 mm = 100 μm, however, the conductivity is constant and equal to the standard bulk value.

The previous double-shield expressions were for free-space between two metals. Free-space has a dielectric constant of one. Glass has a dielectric constant of about 4. Referring to Figure 4.7, the loss

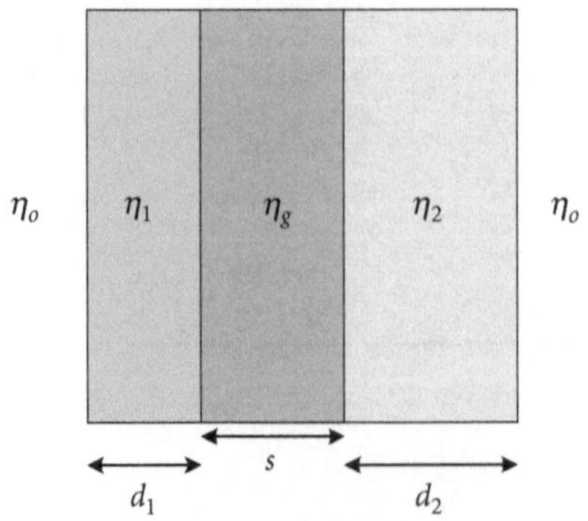

FIGURE 4.7 Shield consisting of glass, of thickness s, sandwiched between two gold layers.

expressions for a double shield consisting of a glass spacer between two metal slabs are

$$R_{dB} = 20\log\left(\frac{1}{2}\left|1+\frac{\eta_1}{\eta_o}\right|\right) + 20\log\left(\frac{1}{2}\left|1+\frac{\eta_g}{\eta_1}\right|\right) + 20\log\left(\frac{1}{2}\left|1+\frac{\eta_2}{\eta_g}\right|\right) + 20\log\left(\frac{1}{2}\left|1+\frac{\eta_o}{\eta_2}\right|\right) \quad (4.26)$$

$$A_{dB} = 8.69\left(\frac{d_1}{\delta_1}+\frac{d_2}{\delta_2}\right) \quad (4.27)$$

$$MR_{dB} = 20\log\left|1-\left[\frac{(\eta_1-\eta_o)(\eta_1-\eta_{ing})}{(\eta_1+\eta_o)(\eta_1+\eta_{ing})}\right]e^{-2\left(\frac{1}{\delta_1}+j\frac{1}{\delta_1}\right)d_1}\right| + 20\log\left|1-\left[\frac{(\eta_g-\eta_1)(\eta_g-\eta_{in2})}{(\eta_g+\eta_1)(\eta_g+\eta_{in2})}\right]e^{-2j\beta s}\right|$$

$$+20\log\left|1-\left(\frac{\eta_2-\eta_g}{\eta_2+\eta_g}\right)\left(\frac{\eta_2-\eta_o}{\eta_2+\eta_o}\right)e^{-2\left(\frac{1}{\delta_2}+j\frac{1}{\delta_2}\right)d_2}\right|$$

$$\text{where } \eta_{ing} = \eta_g\frac{\eta_{in2}\cos(\beta s)+j\eta_g\sin(\beta s)}{\eta_g\cos(\beta s)+j\eta_{in2}\sin(\beta s)} \quad (4.28)$$

$$\eta_{in2} = \eta_2\frac{\eta_o\cosh\left[\left(\frac{1}{\delta_2}+j\frac{1}{\delta_2}\right)d_2\right]+\eta_2\sinh\left[\left(\frac{1}{\delta_2}+j\frac{1}{\delta_2}\right)d_2\right]}{\eta_2\cosh\left[\left(\frac{1}{\delta_2}+j\frac{1}{\delta_2}\right)d_2\right]+\eta_o\sinh\left[\left(\frac{1}{\delta_2}+j\frac{1}{\delta_2}\right)d_2\right]}$$

$$\beta = \frac{2\pi}{\lambda} = \frac{2\pi f}{\frac{c}{\sqrt{\varepsilon_r}}}$$

The parameter η_g is the intrinsic impedance of the glass. The absorption loss has not changed from the free-space double shield since the glass is assumed lossless. The reflective loss at the metal-to-glass and glass-to-metal interfaces has changed since the impedance of glass is not 377 Ω:

$$\eta_g = \sqrt{\frac{j\omega\mu}{\sigma+j\omega\varepsilon}} = \sqrt{\frac{j\omega\mu_o}{0+j\omega\varepsilon_o}} = \sqrt{\frac{\mu_r\mu_o}{\varepsilon_r\varepsilon_o}} = 377\sqrt{\frac{1}{3.78}} \approx 194\,\Omega$$

The multiple-reflection loss has also changed since the impedance seen at the first metal-to-glass interface is affected by the impedance of the glass.

As seen in Mathcad 4.7, extremely small "resonances" are present at these high frequencies. Two glass thicknesses are shown: 250 mil and 2,500 mil. A glass thickness of 250 mil corresponds to one wavelength in the glass at 24 GHz. As seen, for the 250 mil thickness, the first resonance occurs near 6 GHz. A 6 GHz signal has a wavelength of 4×250 mil in glass. The typical first resonance occurs at one-quarter of a wavelength. Although not shown, when the glass is replaced with free space, the minimum and maximum amplitude of the resonances are about 5.5 dB and 7 dB, respectively. Furthermore, when the dielectric constant is increased in value (e.g., 8), the peak-to-peak of the resonances also increases.

$\mu_o := 4\cdot\pi\cdot10^{-7}$ $\varepsilon_o := 8.854\cdot10^{-12}$ $\sigma_o := 5.8\cdot10^{7}$ $\mu_{r1} := 1$ $\sigma_{r1} := 4.1\cdot10^{-2}$ $d_1 := 10\cdot10^{-10}$ $\varepsilon_g := 3.78$ $j := \sqrt{-1}$

$\mu_{r2} := 1$ $\sigma_{r2} := 4.1\cdot10^{-2}$ $d_2 := 10\cdot10^{-10}$

$x := 80, 80.05 .. 110$ $\eta_o := \sqrt{\dfrac{\mu_o}{\varepsilon_o}}$ $\eta_g := \dfrac{\eta_o}{\sqrt{\varepsilon_g}}$

$$f(x) := \left(x + 1 - 10\cdot\text{floor}\left(\frac{x}{10}\right)\right)\cdot10^{\text{floor}\left(\frac{x}{10}\right)}$$

$$\delta_1(f) := \frac{1}{\sqrt{\pi\cdot f\cdot\mu_{r1}\cdot\mu_o\cdot\sigma_{r1}\cdot\sigma_o}} \quad \delta_2(f) := \frac{1}{\sqrt{\pi\cdot f\cdot\mu_{r2}\cdot\mu_o\cdot\sigma_{r2}\cdot\sigma_o}} \quad \eta_1(f) := \sqrt{\frac{2\cdot\pi\cdot f\cdot\mu_{r1}\cdot\mu_o}{2\cdot\sigma_{r1}\cdot\sigma_o}\cdot(1+j)} \quad \beta(f) := \frac{2\cdot\pi\cdot f}{\left(\frac{3\cdot10^8}{\sqrt{\varepsilon_g}}\right)}$$

$$\eta_2(f) := \sqrt{\frac{2\cdot\pi\cdot f\cdot\mu_{r2}\cdot\mu_o}{2\cdot\sigma_{r2}\cdot\sigma_o}\cdot(1+j)} \quad \eta_{in2}(f) := \eta_2(f)\cdot\frac{\eta_o\cdot\cosh\left[\left(\frac{1}{\delta_2(f)} + j\cdot\frac{1}{\delta_2(f)}\right)\cdot d_2\right] + \eta_2(f)\cdot\sinh\left[\left(\frac{1}{\delta_2(f)} + j\cdot\frac{1}{\delta_2(f)}\right)\cdot d_2\right]}{\eta_2(f)\cdot\cosh\left[\left(\frac{1}{\delta_2(f)} + j\cdot\frac{1}{\delta_2(f)}\right)\cdot d_2\right] + \eta_o\cdot\sinh\left[\left(\frac{1}{\delta_2(f)} + j\cdot\frac{1}{\delta_2(f)}\right)\cdot d_2\right]}$$

$$\eta_{ing}(f, s) := \eta_g\cdot\frac{\eta_{in2}(f)\cdot\cos(\beta(f)\cdot s) + j\cdot\eta_g\cdot\sin(\beta(f)\cdot s)}{\eta_g\cdot\cos(\beta(f)\cdot s) + j\cdot\eta_{in2}(f)\cdot\sin(\beta(f)\cdot s)}$$

$$MR1_{dB}(f, s) := 20\cdot\log\left[\left|1 - \left(\frac{\eta_1(f) - \eta_o}{\eta_1(f) + \eta_o}\right)\cdot\left(\frac{\eta_1(f) - \eta_{ing}(f, s)}{\eta_1(f) + \eta_{ing}(f, s)}\right)\cdot e^{-2\cdot\left(\frac{1}{\delta_1(f)} + j\cdot\frac{1}{\delta_1(f)}\right)\cdot d_1}\right|\right]$$

$$MR_{gdB}(f, s) := 20\cdot\log\left[\left|1 - \left(\frac{\eta_g - \eta_1(f)}{\eta_g + \eta_1(f)}\right)\cdot\left(\frac{\eta_g - \eta_{in2}(f)}{\eta_g + \eta_{in2}(f)}\right)\cdot e^{-2\cdot j\cdot\beta(f)\cdot s}\right|\right] \qquad A_{dB}(f) := 8.69\cdot\left(\frac{d_1}{\delta_1(f)} + \frac{d_2}{\delta_2(f)}\right)$$

$$MR2_{dB}(f) := 20\cdot\log\left[\left|1 - \left(\frac{\eta_2(f) - \eta_g}{\eta_2(f) + \eta_g}\right)\cdot\left(\frac{\eta_2(f) - \eta_o}{\eta_2(f) + \eta_o}\right)\cdot e^{-2\cdot\left(\frac{1}{\delta_2(f)} + j\cdot\frac{1}{\delta_2(f)}\right)\cdot d_2}\right|\right]$$

$$R_{dB}(f) := 20\cdot\log\left(\frac{1}{2}\cdot\left|1 + \frac{\eta_1(f)}{\eta_o}\right|\right) + 20\cdot\log\left(\frac{1}{2}\cdot\left|1 + \frac{\eta_g}{\eta_1(f)}\right|\right) + 20\cdot\log\left(\frac{1}{2}\cdot\left|1 + \frac{\eta_2(f)}{\eta_g}\right|\right) + 20\cdot\log\left(\frac{1}{2}\cdot\left|1 + \frac{\eta_o}{\eta_2(f)}\right|\right)$$

$$TOT_{double}(f, s) := A_{dB}(f) + R_{dB}(f) + MR1_{dB}(f, s) + MR_{gdB}(f, s) + MR2_{dB}(f)$$

$$MR_{dB}(f) := 20\cdot\log\left[\left|1 - \left(\frac{\eta_1(f) - \eta_o}{\eta_1(f) + \eta_o}\right)^2\cdot e^{-2\cdot\left(\frac{1}{\delta_1(f)} + j\cdot\frac{1}{\delta_1(f)}\right)\cdot(d_1 + d_2)}\right|\right]$$

$$TOT_{single}(f) := A_{dB}(f) + 20\cdot\log\left(\frac{1}{2}\cdot\left|1 + \frac{\eta_1(f)}{\eta_o}\right|\right) + 20\cdot\log\left(\frac{1}{2}\cdot\left|1 + \frac{\eta_o}{\eta_1(f)}\right|\right) + MR_{dB}(f)$$

$$s := \frac{0.0254}{4}$$

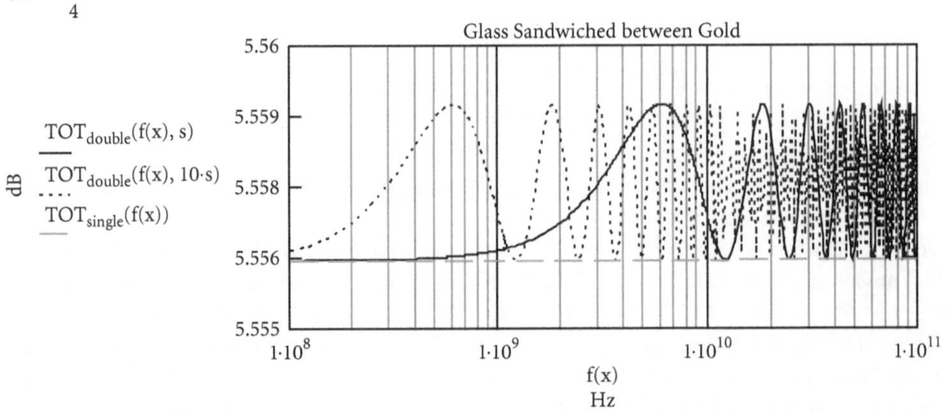

MATHCAD 4.7 SE for a 250 mil and 2,500 mil thick glass with a gold layer on both sides. Also shown is the SE for a single gold layer with the glass removed.

4.6 Laminates for Magnetic Fields

Analyze the various low-frequency and high-frequency relationships available for magnetic shielding using two-shell shields. [Rizk, '77; Thomas; Hoburg, '96]

In most situations, to shield against low-frequency magnetic fields using a single layer is difficult. Even if the shielding material is thick with a high permeability, the shielding effectiveness is not great. To obtain substantial magnetic field shielding at low frequencies (e.g., > 20 dB), multilayer shields are used.

In most of the previous analysis involving shielding, the permeability of the material either was considered equal to that of air (nonmagnetic) or was assumed constant. With all magnetic materials, however, the permeability varies with frequency, magnetic field level, and other factors such as temperature. Generally, the permeability decreases with frequency and maximum saturation level. If the field strength is too strong, the material saturates and the permeability decreases to a very low level.

Explicit (simple) analytical equations for magnetic field shielding effectiveness are only available for a limited number of shapes such as concentric spherical and cylindrical shells. A number of equations for a two-layer (or two-shell) cylindrical and spherical shield will be given in this discussion.

4.6.1 Two Cylindrical Shells in a Parallel Field

The equations for the shielding effectiveness of a two-shell cylindrical shield when the magnetic field outside the shield is tangential or parallel to the axis of the cylinders will first be explored. The cylinders shown in Figure 4.8 are assumed long so that end effects such as leakage into the shield from the ends can be ignored. Assuming that the thickness of each shield is thin relative to their respective radii,

$$d_2 \ll r_2 \quad \text{and} \quad d_1 \ll r_1 \tag{4.29}$$

and the inner radius r_1 and spacing between the shields are much less than the wavelength, the general relationship between the total external (not incident) magnetic field at the outer surface of the double shield, H_{es}, and the internal magnetic field, H_{is}, which is constant from $0 < r < r_1$, is

$$\frac{H_{es}}{H_{is}} = \left[\cosh(\gamma_1 d_1) + \frac{j\omega\mu_o\sigma_1 r_1}{2\gamma_1}\sinh(\gamma_1 d_1)\right] \times \left\{\cosh(\gamma_2 d_2) + \frac{j\omega\mu_o\sigma_2 r_2}{2\gamma_2}\left[1 - \left(\frac{r_1 + d_1}{r_2}\right)^2\right]\sinh(\gamma_2 d_2)\right\}$$
$$+ \frac{(r_1 + d_1)\gamma_1\sigma_2}{r_2\gamma_2\sigma_1}\left[\sinh(\gamma_1 d_1) + \frac{j\omega\mu_o\sigma_1 r_1}{2\gamma_1}\cosh(\gamma_1 d_1)\right]\sinh(\gamma_2 d_2) \tag{4.30}$$

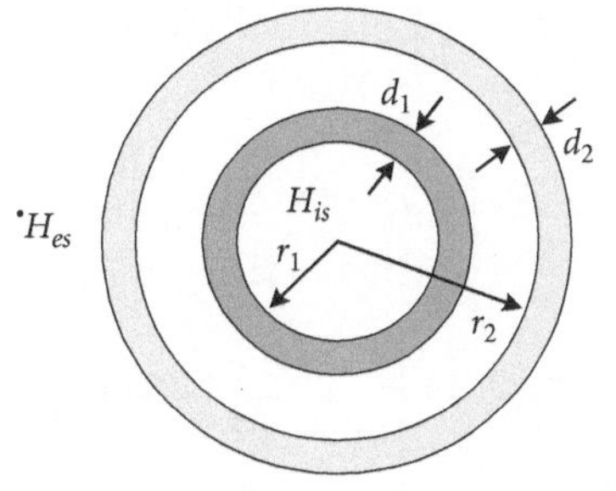

FIGURE 4.8 Two concentric cylindrical shells parallel to an external applied magnetic field.

where the propagation constants for the inner shell, γ_1, and outer shell, γ_2, are (assuming the materials are good conductors)

$$\gamma_1 = \sqrt{j\omega\mu_1\sigma_1} = \frac{1}{\delta_1} + j\frac{1}{\delta_1} \quad \text{where } \delta_1 = \frac{1}{\sqrt{\pi f \mu_1 \sigma_1}}$$

$$\gamma_2 = \sqrt{j\omega\mu_2\sigma_2} = \frac{1}{\delta_2} + j\frac{1}{\delta_2} \quad \text{where } \delta_2 = \frac{1}{\sqrt{\pi f \mu_2 \sigma_2}}$$

The conductivities and permeabilities given in these expressions are the absolute not relative values for these parameters. The expression for the single shield follows directly from this general result by setting $d_2 = 0$ (the hyperbolic sine is equal to zero and hyperbolic cosine is equal to one for those arguments that are a function of d_2).

Frequently, the outer shield is nonmagnetic and inner shield is magnetic, and the level of the magnetic field inside the inner shield is desired since too strong of a field can saturate the magnetic material. Knowledge of the strength of the magnetic field inside the inner shield is therefore of interest. The expression for this field is

$$H_s(r) = H_{is}\left\{\cosh[\gamma_1(r - r_1)] + \frac{j\omega\mu_o\sigma_1 r_1}{2\gamma_1}\sinh[\gamma_1(r - r_1)]\right\} \tag{4.31}$$

This expression is valid for $r_1 \le r \le r_1 + d_1$.

4.6.2 Two Cylindrical Shells in a Parallel Field at Low Frequencies

As previously stated, the magnetic field is parallel to the shells, the cylinders are long so that end effects such as leakage into the shield from the ends can be ignored, the thickness of each shield is thin relative to their respective radii, and the inner radius r_1 and spacing between the shields are much less than the wavelength. At low frequencies where the skin depth is much greater than the thickness of either shield

$$|\gamma_1 d_1| = \frac{\sqrt{2}d_1}{\delta_1} \ll 1 \quad \text{and} \quad |\gamma_2 d_2| = \frac{\sqrt{2}d_2}{\delta_2} \ll 1 \tag{4.32}$$

(4.30) reduces to

$$\begin{aligned}
\frac{H_{es}}{H_{is}} &\approx \left(1 + \frac{j\omega\mu_o\sigma_1 r_1}{2\gamma_1}\gamma_1 d_1\right) \times \left\{1 + \frac{j\omega\mu_o\sigma_2 r_2}{2\gamma_2}\left[1 - \left(\frac{r_1 + d_1}{r_2}\right)^2\right]\gamma_2 d_2\right\} \\
&\quad + \frac{(r_1 + d_1)\gamma_1\sigma_2}{r_2\gamma_2\sigma_1}\left(\gamma_1 d_1 + \frac{j\omega\mu_o\sigma_1 r_1}{2\gamma_1}\right)\gamma_2 d_2 \\
&= \left(1 + \frac{j\omega\mu_o\sigma_1 r_1 d_1}{2}\right) \times \left\{1 + \frac{j\omega\mu_o\sigma_2 r_2 d_2}{2}\left[1 - \left(\frac{r_1 + d_1}{r_2}\right)^2\right]\right\} \\
&\quad + \frac{(r_1 + d_1)}{r_2}\left(j\omega\mu_1\sigma_2 d_2 d_1 + \frac{j\omega\mu_o\sigma_2 d_2 r_1}{2}\right)
\end{aligned} \tag{4.33}$$

since

$$\sinh(\gamma d) \to \gamma d \qquad \left(\sinh x = x + \frac{x^3}{3!} + \frac{x^5}{5!} + \cdots \right)$$

$$\cosh(\gamma d) \to 1 \qquad \left(\cosh x = 1 + \frac{x^2}{2!} + \frac{x^4}{4!} + \cdots \right)$$

when the arguments of the hyperbolic functions are much less than one.

4.6.3 Two Cylindrical Shells with Zero Spacing in a Parallel Field at Low Frequencies

If the spacing between the two shells is set to zero, $r_2 = r_1 + d_1$, then (4.33) further simplifies to

$$\frac{H_{es}}{H_{is}} = 1 + \frac{j\omega\mu_o r_1}{2}(\sigma_1 d_1 + \sigma_2 d_2) + j\omega\mu_1\sigma_2 d_2 d_1 \tag{4.34}$$

This equation states the obvious fact that the shielding effectiveness increases with the thickness and conductivity of either shell. Also, the shielding effectiveness improves with increasing radius of the inner shell. Notice, however, that this expression for a double-laminate shield (with zero spacing between the layers) for low frequencies is not a function of the permeability of the outer shield, μ_2. The shielding effectiveness does increase as the permeability of the inner shield, μ_1, increases. In other words, interchanging the material of the inner and outer shields can change the overall performance of the two-shell shield. Furthermore, as the ratio of the inner radius to the inner shield thickness decreases, the shielding difference associated with interchanging the two materials increases. Finally, according to this low-frequency result, a double (or even single) magnetic or nonmagnetic shell provides no shielding against an external dc magnetic field ($\omega = 0$) that is parallel to its axis. This should not be too surprising since at dc no induced currents are produced in the conducting shells and for a magnetic field tangential to the shell, no magnetic "ducting" or field distortion occurs even when the shield is magnetic. As will be seen later in this discussion, however, for a magnetic field transverse or perpendicular to a magnetic shield, field distortion and dc shielding occurs.

Sometimes the simple expression

$$\frac{H_{eincs}}{H_{is}} = \frac{1}{2}\omega\mu_1\sigma_2 d_2 d_1 \tag{4.35}$$

is seen for the shielding effectiveness of a double shield with zero spacing between the layers, where the variable H_{eincs} corresponds to the external *incident* magnetic field. This field is not the same as the total magnetic field at the outer surface of the outer shield, H_{es}, given in the previous expressions. As will be seen later, the total magnetic field (for a plane wave) along a flat perfect conductor is twice the incident magnetic field. To obtain this simple result from the more general double-layer, zero-spacing, expression, several additional restrictions are required:

$$\omega\mu_1\sigma_2 d_2 d_1 >> \frac{\omega\mu_o r_1}{2}(\sigma_1 d_1 + \sigma_2 d_2) \quad \text{and} \quad \omega\mu_1\sigma_2 d_2 d_1 >> 1 \tag{4.36}$$

The first inequality can be rewritten as

$$2\mu_1\sigma_2 d_2 d_1 >> \mu_o r_1(\sigma_1 d_1 + \sigma_2 d_2)$$

or broken into two inequalities (since both terms on the right side of the expression are positive)

$$2\mu_1\sigma_2 d_2 d_1 >> \mu_o r_1\sigma_1 d_1 \quad \text{and} \quad 2\mu_1\sigma_2 d_2 d_1 >> \mu_o r_1\sigma_2 d_2$$

or

$$\frac{2\mu_1\sigma_2 d_2}{\mu_o r_1\sigma_1} >> 1 \quad \text{and} \quad \frac{2\mu_1 d_1}{\mu_o r_1} >> 1 \tag{4.37}$$

The first inequality is satisfied for many situations involving a high-permeability inner shell and high-conductivity outer shell. However, the second inequality implies that the permeability of the inner shell should be large and its radius small. Furthermore, to use this simple expression for magnetic shielding effectiveness, the other inequality given in (4.36)

$$\omega\mu_1\sigma_2 d_2 d_1 >> 1 \tag{4.38}$$

must also be satisfied. Unless the thickness of each of the shells is quite small and the frequency is very low, this last inequality can be satisfied for an inner shell with a high permeability and an outer shell with a high conductivity.

4.6.4 Two Cylindrical Shells in a Parallel Field at High Frequencies

The complex expression for the total shielding effectiveness of the double shield, (4.30), can also be simplified for high-frequency conditions. At high frequencies where the skin depth is much less than the thickness of either shield

$$|\gamma_1 d_1| = \frac{\sqrt{2}d_1}{\delta_1} >> 1 \quad \text{and} \quad |\gamma_2 d_2| = \frac{\sqrt{2}d_2}{\delta_2} >> 1 \tag{4.39}$$

(but the inner radius r_1 is still much less than a wavelength), the expression reduces to

$$\begin{aligned}\frac{H_{es}}{H_{is}} \approx \frac{e^{\gamma_1 d_1}e^{\gamma_2 d_2}}{4}\left(1+\frac{j\omega\mu_o\sigma_1 r_1}{2\gamma_1}\right)\left\{1+\frac{j\omega\mu_o\sigma_2 r_2}{2\gamma_2}\left[1-\left(\frac{r_1+d_1}{r_2}\right)^2\right]\right\} \\ +\frac{e^{\gamma_1 d_1}e^{\gamma_2 d_2}}{4}\frac{(r_1+d_1)\gamma_1\sigma_2}{r_2\gamma_2\sigma_1}\left(1+\frac{j\omega\mu_o\sigma_1 r_1}{2\gamma_1}\right)\end{aligned} \tag{4.40}$$

since the hyperbolic sine is approximately equal to the hyperbolic cosine when the arguments of the two functions are much greater than one:

$$\sinh(\gamma d) = \frac{e^{\gamma d} - e^{-\gamma d}}{2} \to \frac{e^{\gamma d}}{2}$$

$$\cosh(\gamma d) = \frac{e^{\gamma d} + e^{-\gamma d}}{2} \to \frac{e^{\gamma d}}{2}$$

4.6.5 Two Cylindrical Shells with Zero Spacing in a Parallel Field at High Frequencies

If the spacing between the shells is zero, and

$$\left|\frac{j\omega\mu_o\sigma_1 r_1}{2\gamma_1}\right| = \frac{\omega\mu_o\sigma_1 r_1\delta_1}{2\sqrt{2}} = \frac{\omega\mu_o\sigma_1 r_1}{2\sqrt{2}\sqrt{\pi f\mu_1\sigma_1}} = \frac{\sqrt{\pi\sigma_1\mu_o f}\, r_1}{\sqrt{2\mu_{r1}}} >> 1 \tag{4.41}$$

(4.40) reduces to

$$\frac{H_{es}}{H_{is}} = \frac{e^{\gamma_1 d_1} e^{\gamma_2 d_2}}{4} \frac{j\omega\mu_o\sigma_1 r_1}{2\gamma_1}\left(1+\frac{\gamma_1\sigma_2}{\gamma_2\sigma_1}\right) = \frac{e^{\gamma_1 d_1+\gamma_2 d_2}}{4}\left(\frac{j\omega\mu_o\sigma_1 r_1}{2\gamma_1}+\frac{j\omega\mu_o\sigma_2 r_1}{2\gamma_2}\right) \tag{4.42}$$

Notice from the symmetry of the equation that the two shield materials could be interchanged without affecting the result. This was *not true* for a double shield with zero spacing at low frequencies. Also, the coupling for small spacing is sometimes referred to as additive since the losses are summed: the two close shells are acting as one multilayer thicker shell. As expected, the shielding effectiveness improves with increasing thickness of either shell. Also, it improves as the radius of the inner shell increases.

When the spacing between the two shells is zero, the shells are constructed of the same material, and the total thickness of the shell is d, the expression further reduces to

$$\frac{H_{es}}{H_{is}} = \frac{e^{\gamma d}}{2} \frac{j\omega\mu_o\sigma r}{2\gamma} \tag{4.43}$$

4.6.6 Two Cylindrical Shells with Large Spacing in a Parallel Field at High Frequencies

If the spacing between the two cylindrical shells is large, then

$$(1-x^2) \gg x \quad \text{where } x = \frac{r_1+d_1}{r_2} \tag{4.44}$$

Also, as was assumed before,

$$\left|\frac{j\omega\mu_o\sigma_1 r_1}{2\gamma_1}\right| = \frac{\omega\mu_o\sigma_1 r_1\delta_1}{2\sqrt{2}} = \frac{\omega\mu_o\sigma_1 r_1}{2\sqrt{2}\sqrt{\pi f\mu_1\sigma_1}} = \frac{\sqrt{\pi\sigma_1\mu_o f}\, r_1}{\sqrt{2\mu_{r1}}} \gg 1 \tag{4.45}$$

then it follows that ($r_2 > r_1$)

$$\frac{j\omega\mu_o\sigma_2 r_2}{2\gamma_2} \gg \frac{\gamma_1\sigma_2}{\gamma_2\sigma_1} \quad \text{or} \quad \frac{j\omega\mu_o\sigma_1 r_2}{2\gamma_1} \gg 1 \tag{4.46}$$

The high-frequency, larger spacing approximation for the double shield is from (4.40) (with the inner shell radius and spacing much less than a wavelength)

$$\begin{aligned}\frac{H_{es}}{H_{is}} &\approx \frac{e^{\gamma_1 d_1} e^{\gamma_2 d_2}}{4}\left(\frac{j\omega\mu_o\sigma_1 r_1}{2\gamma_1}\right)\left\{\frac{j\omega\mu_o\sigma_2 r_2}{2\gamma_2}\left[1-\left(\frac{r_1+d_1}{r_2}\right)^2\right]\right\} \\ &+ \frac{e^{\gamma_1 d_1} e^{\gamma_2 d_2}}{4} \frac{(r_1+d_1)\gamma_1\sigma_2}{r_2\gamma_2\sigma_1}\left(\frac{j\omega\mu_o\sigma_1 r_1}{2\gamma_1}\right)\end{aligned} \tag{4.47}$$

$$\frac{H_{es}}{H_{is}} \approx \frac{e^{\gamma_1 d_1} e^{\gamma_2 d_2}}{4}\left(\frac{j\omega\mu_o\sigma_1 r_1}{2\gamma_1}\right)\left\{\frac{j\omega\mu_o\sigma_2 r_2}{2\gamma_2}\left[1-\left(\frac{r_1+d_1}{r_2}\right)^2\right]\right\} \quad \text{if } \frac{r_1+d_1}{r_2} \ll 1 \tag{4.48}$$

The shielding effectiveness of this double shield at high frequencies is always less than the sum of the shields uncoupled from each other.[4] The dB version of this high-frequency large spacing approximation is

$$20\log\left|\frac{H_{es}}{H_{is}}\right| = 20\log\left|\frac{e^{\gamma_1 d_1}}{2}\left(\frac{j\omega\mu_o\sigma_1 r_1}{2\gamma_1}\right)\right| + 20\log\left|\frac{e^{\gamma_2 d_2}}{2}\left(\frac{j\omega\mu_o\sigma_2 r_2}{2\gamma_2}\right)\right| + 20\log\left[1-\left(\frac{r_1+d_1}{r_2}\right)^2\right] \quad (4.49)$$

The first term is the shielding effectiveness of the isolated inner shell, uncoupled from the outer shell (as was shown previously for a zero-spacing shield of total thickness d). The second term is the shielding effectiveness of the isolated outer shell, uncoupled from the inner shell. Since the function

$$\left[1-\left(\frac{r_1+d_1}{r_2}\right)^2\right]$$

is always less than one, the last dB term, which models the coupling between the shells, is always negative. It immediately follows that the nonzero spacing between the two shells at high frequencies reduces the effectiveness of the shield compared to the sum of the shielding effectiveness of two isolated shields.

For these large spacings, the coupling is sometimes referred to as multiplicative since the shielding effectiveness terms from each shell are multiplied (in 4.48). For small spacings, the terms were summed.

In Mathcad 4.8, Mathcad 4.9, and Mathcad 4.10, several of the previous statements are numerically checked. In Mathcad 4.8, the effect of varying the spacing between two steel shells of identical thickness is explored. (In all of these plots, the conductivity and permeability are assumed constant with frequency.) It is seen that for a fixed ratio of inner radii of the two shells the shielding effectiveness increases with increasing radius of the inner shell. In Mathcad 4.9, the shielding effectiveness for a zero-spacing Cu/steel, steel/Cu, and steel/steel shield is shown. The thickness of each of the metals is also identical in this example. As seen in the plots, especially at low frequencies, the combination of copper on the outer shell and steel on the inner shell provides the greatest shielding effectiveness against magnetic fields. This conclusion is in agreement with the previous discussion as well as recommended practice—a high-conductivity metal should be used as the outer layer and a high-permeability metal should be used as the inner layer for low-frequency magnetic shielding. In general, at low frequencies, multilayered shields with alternating layers of high conductivity and high permeability can provide much greater shielding effectiveness than a single layer shield of the same overall thickness composed of only one material. The effect of air-gaps between layers at low frequencies is generally small. For this particular set of parameters, the steel/steel shield outperforms the steel/copper shield. This is not always the case. Also, although not shown, at higher frequencies the shielding effectiveness of the copper/steel shield is similar to the steel/copper shield. In other words, the order of the metals at high frequencies is not important (assuming that the permeability of the metals is not affected by the field strength). Finally, the analysis in Mathcad 4.10 shows that for large spacing and high frequencies, the copper/steel and steel/copper shields have similar shielding effectiveness.

[4]This result should not be applied to high-frequency plane waves where the distance between the shells is not electrically small.

$\mu_o := 4 \cdot \pi \cdot 10^{-7}$ $\quad \sigma_o := 5.8 \cdot 10^7$ $\quad \mu_{r1} := 1000$ $\quad \sigma_{r1} := 0.1$ $\quad \mu_{r2} := 1000$ $\quad \sigma_{r2} := 0.1$ $\quad j := \sqrt{-1}$

$f := 1, 2 .. 300$

$$\delta_1(f) := \frac{1}{\sqrt{\pi \cdot f \cdot \mu_{r1} \cdot \mu_o \cdot \sigma_{r1} \cdot \sigma_o}} \qquad \delta_2(f) := \frac{1}{\sqrt{\pi \cdot f \cdot \mu_{r2} \cdot \mu_o \cdot \sigma_{r2} \cdot \sigma_o}} \qquad \gamma_1(f) := \frac{1+j}{\delta_1(f)} \qquad \gamma_2(f) := \frac{1+j}{\delta_2(f)}$$

$$P_1(f, r_1, d_1, r_2, d_2) := \cosh(\gamma_2(f) \cdot d_2) + \left(\frac{j \cdot 2 \cdot \pi \cdot f \cdot \mu_o \cdot \sigma_{r2} \cdot \sigma_o \cdot r_2}{2 \cdot \gamma_2(f)}\right) \cdot \left[1 - \left(\frac{r_1 + d_1}{r_2}\right)^2\right] \cdot \sinh(\gamma_2(f) \cdot d_2)$$

$$P_2(f, r_1, d_1, r_2, d_2) := \left[\cosh(\gamma_1(f) \cdot d_1) + \left(\frac{j \cdot 2 \cdot \pi \cdot f \cdot \mu_o \cdot \sigma_{r1} \cdot \sigma_o \cdot r_1}{2 \cdot \gamma_1(f)}\right) \cdot \sinh(\gamma_1(f) \cdot d_1)\right] \cdot (P_1(f, r_1, d_1, r_2, d_2))$$

$$P_2(f, r_1, d_1, r_2, d_2) := \left[\frac{(r_1 + d_1) \cdot \gamma_1(f) \cdot \sigma_{r2} \cdot \sigma_o}{r_2 \cdot \gamma_2(f) \cdot \sigma_{r1} \cdot \sigma_o}\right] \cdot \left[\sinh(\gamma_1(f) \cdot d_1) + \left(\frac{j \cdot 2 \cdot \pi \cdot f \cdot \mu_o \cdot \sigma_{r1} \cdot \sigma_o \cdot r_1}{2 \cdot \gamma_1(f)}\right) \cdot \cosh(\gamma_1(f) \cdot d_1)\right] \cdot \sinh(\gamma_2(f) \cdot d_2)$$

$$SE_{dB}(f, r_1, d_1, r_2, d_2) := 20 \cdot \log(|P_2(f, r_1, d_1, r_2, d_2) + P_3(f, r_1, d_1, r_2, d_2)|) \qquad r_1 := 0.1 \qquad d_1 := 1 \cdot 10^{-3}$$

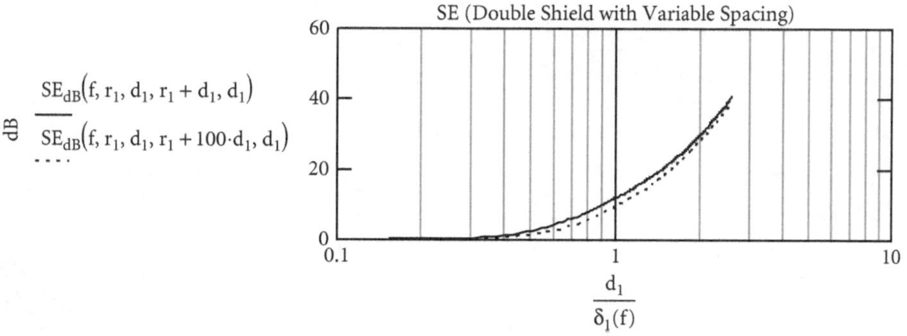

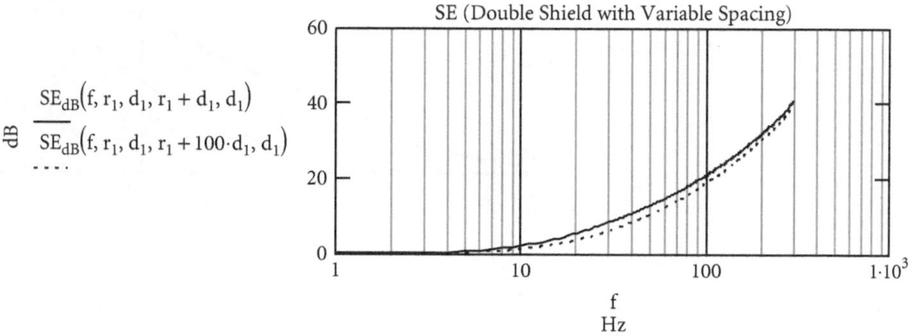

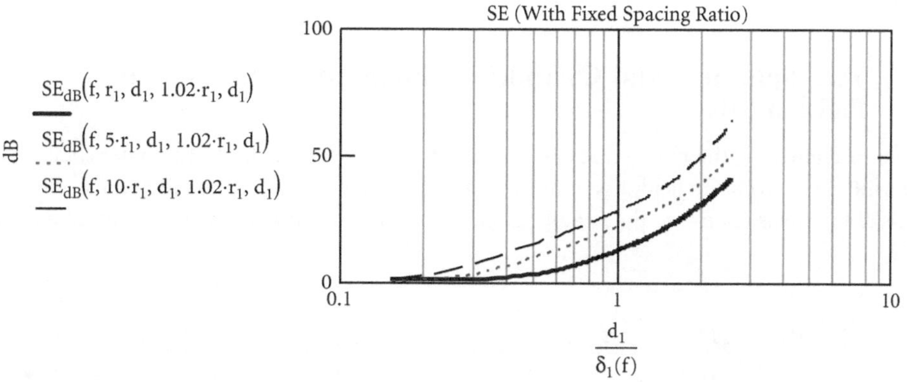

MATHCAD 4.8 SE of two steel shells (of the same thickness) vs. the thickness relative to the skin depth. Also shown is the variation with spacing and frequency.

$\mu_o := 4\cdot\pi\cdot10^{-7} \qquad \sigma_o := 5.8\cdot10^{7} \qquad r_1 := 0.1 \qquad d_1 := 1\cdot10^{-3} \qquad r_2 := r_1 + d_1 \qquad d_2 := d_1 \qquad j := \sqrt{-1}$

$f := 1, 1.5..\, 1000$

$$\delta_1(f,\mu_{r1},\sigma_{r1}) := \frac{1}{\sqrt{\pi\cdot f\cdot\mu_{r1}\cdot\mu_o\cdot\sigma_{r1}\cdot\sigma_o}} \qquad \delta_2(f,\mu_{r2},\sigma_{r2}) := \frac{1}{\sqrt{\pi\cdot f\cdot\mu_{r2}\cdot\mu_o\cdot\sigma_{r2}\cdot\sigma_o}}$$

$$\gamma_1(f,\mu_{r1},\sigma_{r1}) := \frac{1+j}{\delta_1(f,\mu_{r1},\sigma_{r1})} \qquad \gamma_2(f,\mu_{r2},\sigma_{r2}) := \frac{1+j}{\delta_2(f,\mu_{r2},\sigma_{r2})}$$

$$P_1(f,\mu_{r1},\sigma_{r1},\mu_{r2},\sigma_{r2}) := \cosh(\gamma_2(f,\mu_{r2},\sigma_{r2})\cdot d_2) + \left(\frac{j\cdot 2\cdot\pi\cdot f\cdot\mu_o\cdot\sigma_{r2}\cdot\sigma_o\cdot r_2}{2\cdot\gamma_2(f,\mu_{r2},\sigma_{r2})}\right)\cdot\left[1-\left(\frac{r_1+d_1}{r_2}\right)^2\right]\cdot\sinh(\gamma_2(f,\mu_{r2},\sigma_{r2})\cdot d_2)$$

$$P_2(f,\mu_{r1},\sigma_{r1},\mu_{r2},\sigma_{r2}) := \left[\cosh(\gamma_1(f,\mu_{r1},\sigma_{r1})\cdot d_1) + \left(\frac{j\cdot 2\cdot\pi\cdot f\cdot\mu_o\cdot\sigma_{r1}\cdot\sigma_o\cdot r_1}{2\cdot\gamma_1(f,\mu_{r1},\sigma_{r1})}\right)\cdot\sinh(\gamma_1(f,\mu_{r1},\sigma_{r1})\cdot d_1)\right]\cdot P_1(f,\mu_{r1},\sigma_{r1},\mu_{r2},\sigma_{r2})$$

$$P_3(f,\mu_{r1},\sigma_{r1},\mu_{r2},\sigma_{r2}) := \left[\sinh(\gamma_1(f,\mu_{r1},\sigma_{r1})\cdot d_1) + \left(\frac{j\cdot 2\cdot\pi\cdot f\cdot\mu_o\cdot\sigma_{r1}\cdot\sigma_o\cdot r_1}{2\cdot\gamma_1(f,\mu_{r1},\sigma_{r1})}\right)\cdot\cosh(\gamma_1(f,\mu_{r1},\sigma_{r1})\cdot d_1)\right]\cdot\sinh(\gamma_2(f,\mu_{r2},\sigma_{r2})\cdot d_2)$$

$$P_4(f,\mu_{r1},\sigma_{r1},\mu_{r2},\sigma_{r2}) := \frac{(r_1+d_1)\cdot\gamma_1(f,\mu_{r1},\sigma_{r1})\cdot\sigma_{r2}\cdot\sigma_o}{r_2\cdot\gamma_2(f,\mu_{r2},\sigma_{r2})\cdot\sigma_{r1}\cdot\sigma_o}\cdot P_3(f,\mu_{r1},\sigma_{r1},\mu_{r2},\sigma_{r2})$$

$$SE_{dB}(f,\mu_{r1},\sigma_{r1},\mu_{r2},\sigma_{r2}) := 20\cdot\log(|P_2(f,\mu_{r1},\sigma_{r1},\mu_{r2},\sigma_{r2}) + P_4(f,\mu_{r1},\sigma_{r1},\mu_{r2},\sigma_{r2})|)$$

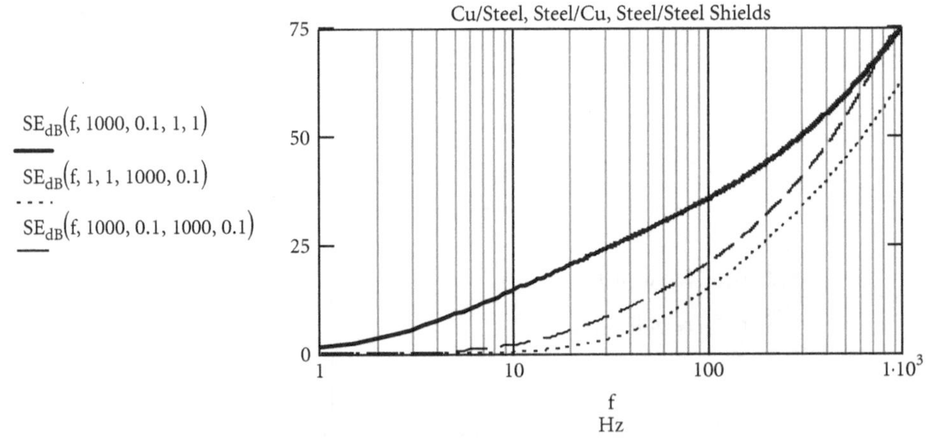

MATHCAD 4.9 SE vs. frequency for zero spacing between Cu/steel, steel/Cu, and steel/steel cylindrical shells.

4.6.7 Two Spherical and Cylindrical Shells in a Perpendicular Field at DC

As shown in Figure 4.9, the magnetic field is transverse or perpendicular to the outer shell.[5] For double spheres with an external dc field, assuming that the relative permeability of the two shells is high and thickness of the shells relative to their radius is small, the ratio of the external to the internal field is given by

$$\frac{H_e}{H_i} \approx 1 + \frac{2}{3}\frac{\mu_{r2}d_2}{r_2} + \frac{2}{3}\frac{\mu_{r1}d_1}{r_1} + \left(\frac{2}{3}\frac{\mu_{r2}d_2}{r_2}\right)\left(\frac{2}{3}\frac{\mu_{r1}d_1}{r_1}\right)\left(1-\frac{r_1^3}{r_2^3}\right) \tag{4.50}$$

[5]Actually, the applied field is perpendicular to a plane passing through the axis of the cylindrical shells.

$\mu_o := 4\cdot\pi\cdot10^{-7}$ $\sigma_o := 5.8\cdot10^{7}$ $r_1 := 0.1$ $d_1 := 1\cdot10^{-3}$ $r_2 := 2\cdot r_1$ $d_2 := d_1$ $j := \sqrt{-1}$

$f := 1000, 2000..\ 10000$

$$\delta_1(f, \mu_{r1}, \sigma_{r1}) := \frac{1}{\sqrt{\pi\cdot f\cdot\mu_{r1}\cdot\mu_o\cdot\sigma_{r1}\cdot\sigma_o}} \qquad \delta_2(f, \mu_{r2}, \sigma_{r2}) := \frac{1}{\sqrt{\pi\cdot f\cdot\mu_{r2}\cdot\mu_o\cdot\sigma_{r2}\cdot\sigma_o}}$$

$$\gamma_1(f, \mu_{r1}, \sigma_{r1}) := \frac{1+j}{\delta_1(f, \mu_{r1}, \sigma_{r1})} \qquad \gamma_2(f, \mu_{r2}, \sigma_{r2}) := \frac{1+j}{\delta_2(f, \mu_{r2}, \sigma_{r2})}$$

$$P_1(f, \mu_{r1}, \sigma_{r1}, \mu_{r2}, \sigma_{r2}) := \cosh(\gamma_2(f, \mu_{r2}, \sigma_{r2})\cdot d_2) + \left(\frac{j\cdot2\cdot\pi\cdot f\cdot\mu_o\cdot\sigma_{r2}\cdot\sigma_o\cdot r_2}{2\cdot\gamma_2(f, \mu_{r2}, \sigma_{r2})}\right)\cdot\left[1-\left(\frac{r_1+d_1}{r_2}\right)^2\right]\cdot\sinh(\gamma_2(f, \mu_{r2}, \sigma_{r2})\cdot d_2)$$

$$P_2(f, \mu_{r1}, \sigma_{r1}, \mu_{r2}, \sigma_{r2}) := \left[\cosh(\gamma_1(f, \mu_{r1}, \sigma_{r1})\cdot d_1) + \left(\frac{j\cdot2\cdot\pi\cdot f\cdot\mu_o\cdot\sigma_{r1}\cdot\sigma_o\cdot r_1}{2\cdot\gamma_1(f, \mu_{r1}, \sigma_{r1})}\right)\cdot\sinh(\gamma_1(f, \mu_{r1}, \sigma_{r1})\cdot d_1)\right]\cdot P_1(f, \mu_{r1}, \sigma_{r1}, \mu_{r2}, \sigma_{r2})$$

$$P_3(f, \mu_{r1}, \sigma_{r1}, \mu_{r2}, \sigma_{r2}) := \left[\sinh(\gamma_1(f, \mu_{r1}, \sigma_{r1})\cdot d_1) + \left(\frac{j\cdot2\cdot\pi\cdot f\cdot\mu_o\cdot\sigma_{r1}\cdot\sigma_o\cdot r_1}{2\cdot\gamma_1(f, \mu_{r1}, \sigma_{r1})}\right)\cdot\cosh(\gamma_1(f, \mu_{r1}, \sigma_{r1})\cdot d_1)\right]\cdot\sinh(\gamma_2(f, \mu_{r2}, \sigma_{r2})\cdot d_2)$$

$$P_4(f, \mu_{r1}, \sigma_{r1}, \mu_{r2}, \sigma_{r2}) := \frac{(r_1+d_1)\cdot\gamma_1(f, \mu_{r1}, \sigma_{r1})\cdot\sigma_{r2}\cdot\sigma_o}{r_2\cdot\gamma_2(f, \mu_{r2}, \sigma_{r2})\cdot\sigma_{r1}\cdot\sigma_o}\cdot P_3(f, \mu_{r1}, \sigma_{r1}, \mu_{r2}, \sigma_{r2})$$

$$SE_{dB}(f, \mu_{r1}, \sigma_{r1}, \mu_{r2}, \sigma_{r2}) := 20\cdot\log(|P_2(f, \mu_{r1}, \sigma_{r1}, \mu_{r2}, \sigma_{r2}) + P_4(f, \mu_{r1}, \sigma_{r1}, \mu_{r2}, \sigma_{r2})|)$$

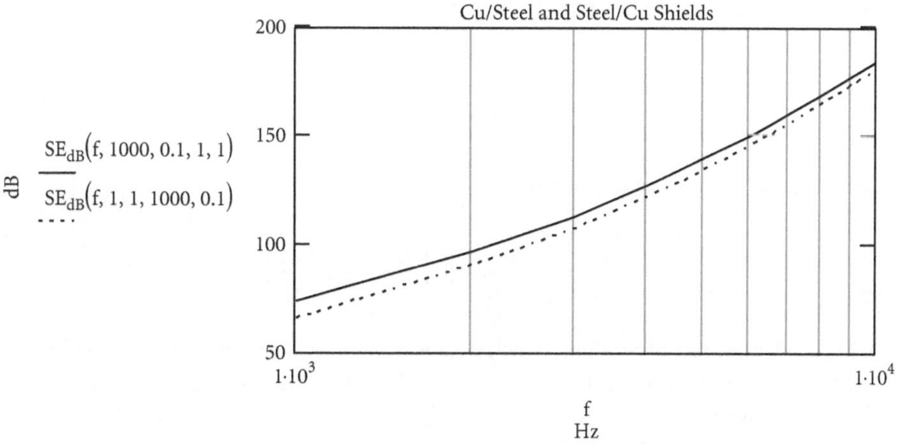

MATHCAD 4.10 SE vs. frequency for Cu/steel and steel/Cu cylindrical shells. The spacing between the two shells is not small.

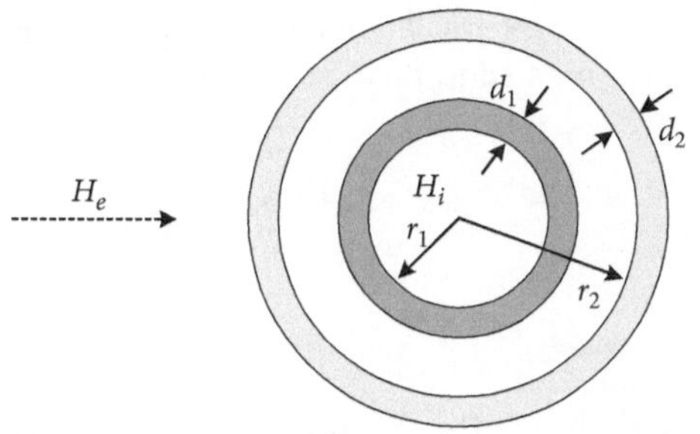

FIGURE 4.9 Two concentric spherical or cylindrical shells perpendicular to an applied external dc magnetic field.

and for double cylinders

$$\frac{H_e}{H_i} \approx 1 + \frac{1}{2}\frac{\mu_{r2}d_2}{r_2} + \frac{1}{2}\frac{\mu_{r1}d_1}{r_1} + \left(\frac{1}{2}\frac{\mu_{r2}d_2}{r_2}\right)\left(\frac{1}{2}\frac{\mu_{r1}d_1}{r_1}\right)\left(1 - \frac{r_1^2}{r_2^2}\right) \tag{4.51}$$

This shielding is due to the magnetic "ducting" effect (i.e., field distortion from the magnetic material). Obviously, because the frequency is dc, there cannot be any shielding due to induced Eddy currents in the shield.

For the spherical two-shell shield, for small spacings $(r_1 \approx r_2)$ the last term in (4.50) is negligible, and the shielding from each shell is additive:

$$\frac{H_e}{H_i} \approx 1 + \frac{2}{3}\frac{\mu_{r2}d_2}{r_2} + \frac{2}{3}\frac{\mu_{r1}d_1}{r_1} \tag{4.52}$$

For double cylinders, it is also additive:

$$\frac{H_e}{H_i} \approx 1 + \frac{1}{2}\frac{\mu_{r2}d_2}{r_2} + \frac{1}{2}\frac{\mu_{r1}d_1}{r_1} \tag{4.53}$$

When the spacing between the spherical shells is large $(r_1 \ll r_2)$ and the individual $\mu_{rx}d_x/r_x$ terms are large, the shielding from each shell is multiplicative:

$$\frac{H_e}{H_i} \approx \left(\frac{2}{3}\frac{\mu_{r2}d_2}{r_2}\right)\left(\frac{2}{3}\frac{\mu_{r1}d_1}{r_1}\right)\left(1 - \frac{r_1^3}{r_2^3}\right) \tag{4.54}$$

For double cylinders, it is also multiplicative:

$$\frac{H_e}{H_i} \approx \left(\frac{1}{2}\frac{\mu_{r2}d_2}{r_2}\right)\left(\frac{1}{2}\frac{\mu_{r1}d_1}{r_1}\right)\left(1 - \frac{r_1^2}{r_2^2}\right) \tag{4.55}$$

The shielding effectiveness of a double-shell, steel/steel spherical and cylindrical shield vs. the ratio of the shell radii is given in Mathcad 4.11. As is seen in the first plot, the effectiveness of the spherical double shell is a few dB greater than the cylindrical shell. Also, there is an optimum ratio for the radii ratio. In the second plot, the increase in the shielding effectiveness with shell thickness is clearly seen.

4.6.8 Two Spherical and Cylindrical Shells in a Perpendicular Field

The more general expressions for the spherical and cylindrical double shell are similar in form to the dc expressions. Assuming the dimensions are small compared to a wavelength, external field is transverse to the surface, and shells are thin compared to their respective radius, the expression for spheres is

$$\frac{H_{es}}{H_{is}} \approx 1 + S_1 + S_2 + S_1 S_2\left(1 - \frac{r_1^3}{r_2^3}\right) \tag{4.56}$$

where for each spherical shell

$$S_x = \frac{(\gamma_x r_x)^2\left\{\left[\frac{1}{\mu_{rx}} + \frac{2\mu_{rx}}{(\gamma_x r_x)^2}\right]\sinh(\gamma_x d_x) + \frac{3}{\gamma_x r_x}\cosh(\gamma_x d_x)\right\}}{3\gamma_x r_x} - 1 \tag{4.57}$$

$$SE_{spdB}(\mu_{r1}, r_1, d_1, \mu_{r2}, r_2, d_2) := 20\cdot\log\left[\left|1+\frac{2}{3}\cdot\frac{\mu_{r1}\cdot d_1}{r_1}+\frac{2}{3}\cdot\frac{\mu_{r2}\cdot d_2}{r_2}+\left(\frac{2}{3}\cdot\frac{\mu_{r1}\cdot d_1}{r_1}\right)\cdot\left(\frac{2}{3}\cdot\frac{\mu_{r2}\cdot d_2}{r_2}\right)\cdot\left(1-\frac{r_1^{\,3}}{r_2^{\,3}}\right)\right|\right]$$

$$SE_{cydB}(\mu_{r1}, r_1, d_1, \mu_{r2}, r_2, d_2) := 20\cdot\log\left[\left|1+\frac{1}{2}\cdot\frac{\mu_{r1}\cdot d_1}{r_1}+\frac{1}{2}\cdot\frac{\mu_{r2}\cdot d_2}{r_2}+\left(\frac{1}{2}\cdot\frac{\mu_{r1}\cdot d_1}{r_1}\right)\cdot\left(\frac{1}{2}\cdot\frac{\mu_{r2}\cdot d_2}{r_2}\right)\cdot\left(1-\frac{r_1^{\,2}}{r_2^{\,2}}\right)\right|\right]$$

$\mu_{r1} := 1000 \qquad \mu_{r2} := 1000 \qquad r_1 := 0.1 \qquad d_1 := 1\cdot 10^{-3} \qquad d_2 := d_1$

$r_2 := r_1 + d_1,\ r_1 + 2\cdot d_1 ..\ 5\cdot r_1$

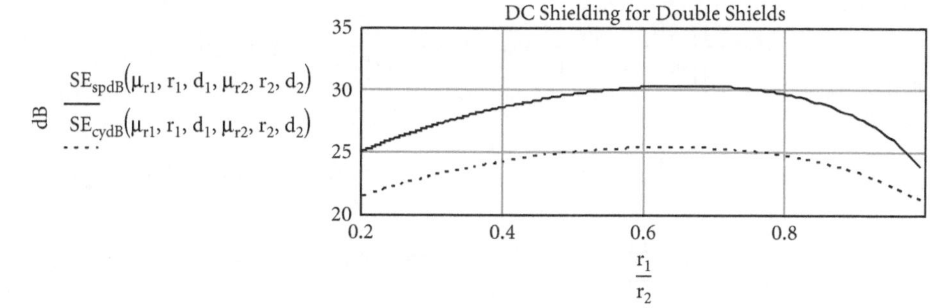

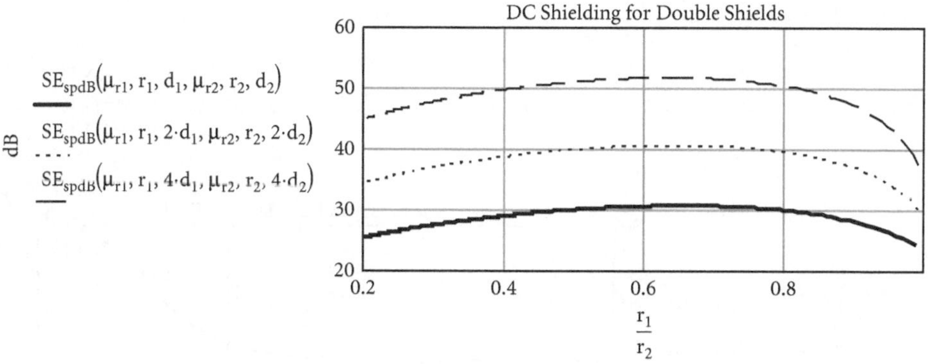

MATHCAD 4.11 SE for steel/steel spherical and cylindrical shells as a function of the ratio of the radii.

For cylinders,

$$\frac{H_{es}}{H_{is}} \approx 1 + S_1 + S_2 + S_1 S_2\left(1-\frac{r_1^2}{r_2^2}\right) \tag{4.58}$$

where for each cylindrical shell

$$S_x = \cosh(\gamma_x d_x) + \frac{1}{2}\left(\frac{\gamma_x r_x}{\mu_{rx}} + \frac{\mu_{rx}}{\gamma_x r_x}\right)\sinh(\gamma_x d_x) - 1 \tag{4.59}$$

In Mathcad 4.12, the shielding effectiveness of a two-shell, steel spherical shield and cylindrical shield (when the field is perpendicular) vs. frequency is compared. The radii of the spherical shells are the same as for the cylindrical shells. As seen, the spherical shield is a few dB greater than the (infinitely long) cylindrical shield. The second program, Mathcad 4.13, compares the shielding effectiveness of a two-shell steel cylindrical shield when the external field is tangential (or parallel) vs. perpendicular (or transverse) to the shield's surface. As seen in the plot, for very low frequencies, the shield is more effective

$$\mu_o := 4\cdot\pi\cdot10^{-7} \quad \sigma_o := 5.8\cdot10^{7} \quad \mu_{r1} := 1000 \quad \sigma_{r1} := 0.1 \quad \mu_{r2} := 1000 \quad \sigma_{r2} := 0.1 \quad j := \sqrt{-1}$$

$$f := 1, 2..\ 300$$

$$\delta_1(f) := \frac{1}{\sqrt{\pi\cdot f\cdot\mu_{r1}\cdot\mu_o\cdot\sigma_{r1}\cdot\sigma_o}} \quad \delta_2(f) := \frac{1}{\sqrt{\pi\cdot f\cdot\mu_{r2}\cdot\mu_o\cdot\sigma_{r2}\cdot\sigma_o}} \quad \gamma_1(f) := \frac{1+j}{\delta_1(f)} \quad \gamma_2(f) := \frac{1+j}{\delta_2(f)}$$

$$S_{s1}(f, r_1, d_1) := \frac{(\gamma_1(f)\cdot r_1)^2\cdot\left[\left[\frac{1}{\mu_{r1}} + \frac{2\cdot\mu_{r1}}{(\gamma_1(f)\cdot r_1)^2}\right]\cdot\sinh(\gamma_1(f)\cdot d_1) + \frac{3}{\gamma_1(f)\cdot r_1}\cdot\cosh(\gamma_1(f)\cdot d_1)\right]}{3\cdot\gamma_1(f)\cdot r_1} - 1$$

$$S_{s2}(f, r_2, d_2) := \frac{(\gamma_2(f)\cdot r_2)^2\cdot\left[\left[\frac{1}{\mu_{r2}} + \frac{2\cdot\mu_{r2}}{(\gamma_2(f)\cdot r_2)^2}\right]\cdot\sinh(\gamma_2(f)\cdot d_2) + \frac{3}{\gamma_2(f)\cdot r_2}\cdot\cosh(\gamma_2(f)\cdot d_2)\right]}{3\cdot\gamma_2(f)\cdot r_2} - 1$$

$$SE_{sdB}(f, r_1, d_1, r_2, d_2) := 20\cdot\log\left[\left|1 + S_{s1}(f, r_1, d_1) + S_{s2}(f, r_2, d_2) + S_{s1}(f, r_1, d_1)\cdot S_{s2}(f, r_2, d_2)\cdot\left(1 - \frac{r_1^{\,3}}{r_2^{\,3}}\right)\right|\right]$$

$$S_{c1}(f, r_1, d_1) := \cosh(\gamma_1(f)\cdot d_1) + \frac{1}{2}\cdot\left(\frac{\gamma_1(f)\cdot r_1}{\mu_{r1}} + \frac{\mu_{r1}}{\gamma_1(f)\cdot r_1}\right)\cdot\sinh(\gamma_1(f)\cdot d_1) - 1$$

$$S_{c2}(f, r_2, d_2) := \cosh(\gamma_2(f)\cdot d_2) + \frac{1}{2}\cdot\left(\frac{\gamma_2(f)\cdot r_2}{\mu_{r2}} + \frac{\mu_{r2}}{\gamma_2(f)\cdot r_2}\right)\cdot\sinh(\gamma_2(f)\cdot d_2) - 1$$

$$SE_{cdB}(f, r_1, d_1, r_2, d_2) := 20\cdot\log\left[\left|1 + S_{c1}(f, r_1, d_1) + S_{c2}(f, r_2, d_2) + S_{c1}(f, r_1, d_1)\cdot S_{c2}(f, r_2, d_2)\cdot\left(1 - \frac{r_1^{\,2}}{r_2^{\,2}}\right)\right|\right]$$

$$r_1 := 0.08 \quad d_1 := 1\cdot10^{-3} \quad r_2 := 0.1 \quad d_2 := d_1$$

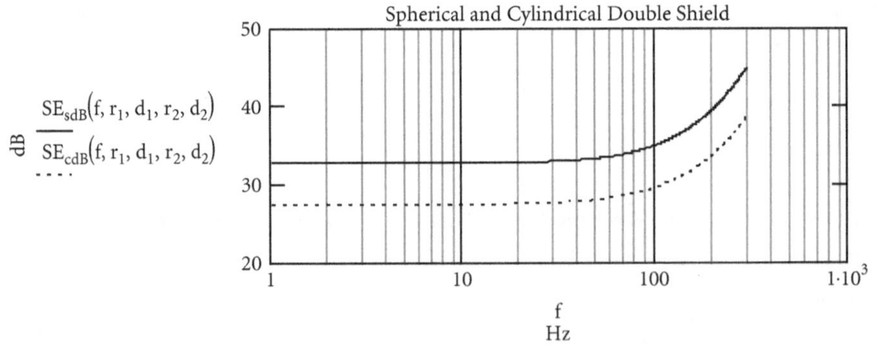

MATHCAD 4.12 SE for a spherical and cylindrical double shield vs. frequency. The field is perpendicular to the cylindrical surface.

when oriented perpendicular to the field: at very low frequencies only flux shunting or magnetic "ducting" can produce any significant shielding. The resultant field distortion occurs when the applied field is perpendicular to the shield's surface.

4.7 Rust Never Sleeps—Corrosion

What is corrosion, and how can its rate be reduced? [Schweitzer]

A topic that is frequently neglected in electrical engineering programs, but is extremely important, is the topic of corrosion. There are many types of corrosion, and corrosion can take place because of chemical or electrochemical reaction between a material (e.g., metal) and its environment. Chemical

$\mu_o := 4\cdot\pi\cdot10^{-7}$ $\quad\sigma_o := 5.8\cdot10^{7}$ $\quad\mu_{r1} := 1000$ $\quad\sigma_{r1} := 0.1$ $\quad\mu_{r2} := 1000$ $\quad\sigma_{r2} := 0.1$ $\quad j := \sqrt{-1}$

$f := 1, 2..\, 300$

$$\delta_1(f) := \frac{1}{\sqrt{\pi\cdot f\cdot\mu_{r1}\cdot\mu_o\cdot\sigma_{r1}\cdot\sigma_o}} \qquad \delta_2(f) := \frac{1}{\sqrt{\pi\cdot f\cdot\mu_{r2}\cdot\mu_o\cdot\sigma_{r2}\cdot\sigma_o}} \qquad \gamma_1(f) := \frac{1+j}{\delta_1(f)} \qquad \gamma_2(f) := \frac{1+j}{\delta_2(f)}$$

$$P_1(f, r_1, d_1, r_2, d_2) := \cosh(\gamma_2(f)\cdot d_2) + \left(\frac{j\cdot2\cdot\pi\cdot f\cdot\mu_o\cdot\sigma_{r2}\cdot\sigma_o\cdot r_2}{2\cdot\gamma_2(f)}\right)\cdot\left[1-\left(\frac{r_1+d_1}{r_2}\right)^2\right]\cdot\sinh(\gamma_2(f)\cdot d_2)$$

$$P_2(f, r_1, d_1, r_2, d_2) := \left[\cosh(\gamma_1(f)\cdot d_1) + \left(\frac{j\cdot2\cdot\pi\cdot f\cdot\mu_o\cdot\sigma_{r1}\cdot\sigma_o\cdot r_1}{2\cdot\gamma_1(f)}\right)\cdot\sinh(\gamma_1(f)\cdot d_1)\right]\cdot(P_1(f, r_1, d_1, r_2, d_2))$$

$$P_3(f, r_1, d_1, r_2, d_2) := \left[\frac{(r_1+d_1)\cdot\gamma_1(f)\cdot\sigma_{r2}\cdot\sigma_o}{r_2\cdot\gamma_2(f)\cdot\sigma_{r1}\cdot\sigma_o}\right]\cdot\left[\sinh(\gamma_1(f)\cdot d_1) + \left(\frac{j\cdot2\cdot\pi\cdot f\cdot\mu_o\cdot\sigma_{r1}\cdot\sigma_o\cdot r_1}{2\cdot\gamma_1(f)}\right)\cdot\cosh(\gamma_1(f)\cdot d_1)\right]\cdot\sinh(\gamma_2(f)\cdot d_2)$$

$$SE_{ctandB}(f, r_1, d_1, r_2, d_2) := 20\cdot\log(|P_2(f, r_1, d_1, r_2, d_2) + P_3(f, r_1, d_1, r_2, d_2)|)$$

$$S_{c1}(f, r_1, d_1) := \cosh(\gamma_1(f)\cdot d_1) + \frac{1}{2}\cdot\left(\frac{\gamma_1(f)\cdot r_1}{\mu_{r1}} + \frac{\mu_{r1}}{\gamma_1(f)\cdot r_1}\right)\cdot\sinh(\gamma_1(f)\cdot d_1) - 1$$

$$S_{c2}(f, r_2, d_2) := \cosh(\gamma_2(f)\cdot d_2) + \frac{1}{2}\cdot\left(\frac{\gamma_2(f)\cdot r_2}{\mu_{r2}} + \frac{\mu_{r2}}{\gamma_2(f)\cdot r_2}\right)\cdot\sinh(\gamma_2(f)\cdot d_2) - 1$$

$$SE_{cperpdB}(f, r_1, d_1, r_2, d_2) := 20\cdot\log\left[\left|1 + S_{c1}(f, r_1, d_1) + S_{c2}(f, r_2, d_2) + S_{c1}(f, r_1, d_1)\cdot S_{c2}(f, r_2, d_2)\cdot\left(1-\frac{r_1^{\,2}}{r_2^{\,2}}\right)\right|\right]$$

$r_1 := 0.08$ $\quad d_1 := 1\cdot10^{-3}$ $\quad r_2 := 0.1$ $\quad d_2 := d_1$

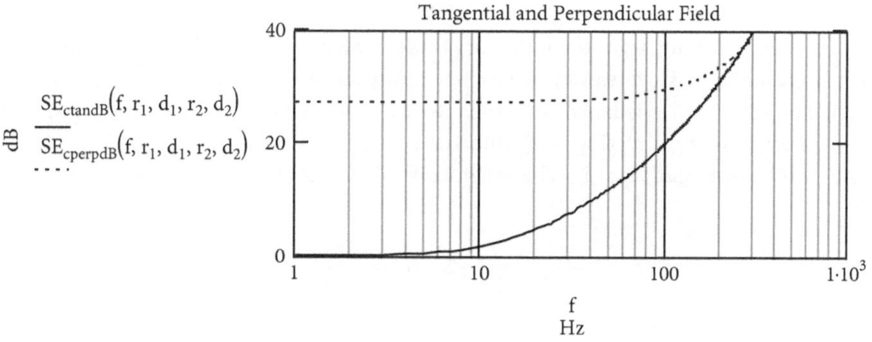

MATHCAD 4.13 SE for a double cylindrical shield vs. frequency for both tangential and perpendicular fields.

corrosion usually results between a metal surface and gas at high temperatures. Electrochemical corrosion is more common and is a consequence of the reaction between a metal and an electrolytic environment (such as water, humid atmosphere, or wet soil). In colloquial language, corrosion of a metal is the undesirable attack on the metal by its surroundings.

The corrosion or rusting of unprotected steel and iron is familiar to most individuals. Although corrosion cannot be stopped, the rate of corrosion can be controlled. For example, for steel the rate of corrosion can be reduced by painting the metal or by providing a "sacrificial" coating such as zinc (i.e., galvanizing the metal). It is important to state that all metals corrode, but some metals such as bronze, brass, and stainless steel corrode slowly.[6]

[6]For some metals such as gold, it takes a "wild" environment for corrosion to occur.

Galvanic corrosion is the corrosion that occurs when two dissimilar metals are in contact and in the presence of moisture. The greater the separation of two metals on the Galvanic series (or electrochemical series), the greater the probability of corrosion. For example, the distance between silver and aluminum in this series is large; therefore, corrosion is expected if the environment contains oxygen and water. To reduce the effects of corrosion, the anode metal (the metal corroding) area should not be small in comparison to the cathode metal area. Or, the dissimilar metals should be insulated from each other. In most situations, corrosion is undesirable. For example, the corrosion of conductors and contacts in an antenna system will reduce the efficiency of the antenna over time.

4.8 Surface Impedance

What is the surface impedance, and under what circumstances can the commonly seen expressions for it be used? [Ramo; Miner; Liao; Weston; Schelkunoff]

Surface impedance (also sometimes known as shield, internal, or metal impedance) is the impedance of a good-conducting material to a plane wave. The surface impedance or resistance is an excellent measure of the shielding quality of conductive materials. The definition for the surface impedance can be derived in several different ways. One approach is to begin with the definition for the intrinsic impedance:

$$\eta = \sqrt{\frac{j\omega\mu}{\sigma + j\omega\varepsilon}}$$

By assuming that the material is a good conductor, $\sigma/(\omega\varepsilon) >> 1$, this expression reduces to

$$\eta \approx \sqrt{\frac{\pi f \mu}{\sigma}}(1+j) = \frac{1}{\delta\sigma}(1+j) = Z_s \tag{4.60}$$

where δ is the skin depth for a good conductor. This expression defines the surface impedance, Z_s, for a metal. It has the units of Ω; however, in practice, the units are frequently given as $\Omega/\square$ or ohms per square (sq). The intrinsic impedance is defined as the ratio of the electric field to magnetic field for a plane wave traveling in one direction. This impedance has the units of $(\text{V/m})/(\text{A/m}) = \Omega$. The $\square$ portion of the units will be discussed shortly. The magnitude of the surface impedance is

$$|Z_s| = \frac{1}{\delta\sigma}\sqrt{2} \approx 0.369\sqrt{\frac{\mu_r f}{\sigma_r}}\ \mu\Omega \tag{4.61}$$

where μ_r is the permeability relative to copper and σ_r is the conductivity relative to copper. The surface resistance is the real part of the surface impedance:

$$R_s = \text{Re}(Z_s) = \frac{1}{\delta\sigma} = \sqrt{\frac{\pi f \mu}{\sigma}}\ \Omega \tag{4.62}$$

The *internal* inductance of the metal can be obtained from the imaginary part of the surface impedance, which is positive, corresponding to inductive reactance:

$$\omega L_{iAC} = \text{Im}(Z_s) = \frac{1}{\delta\sigma} = \sqrt{\frac{\pi f \mu}{\sigma}} \quad \Rightarrow L_{iAC} = \sqrt{\frac{\mu}{4\pi f \sigma}}\ \text{H} \tag{4.63}$$

The expression for the surface impedance can also be obtained by determining the ratio of the voltage to the current at the surface. When a plane wave is normally incident to a flat surface and traveling in the positive *z* direction, the electric and magnetic fields inside the material can be described as (assuming that the material is sufficiently thick so that any reflections from the material's other side can be neglected)

$$\vec{E}_{2s} = C_s e^{-\gamma z}\hat{a}_x, \ \vec{H}_{2s} = \frac{C_s}{\eta} e^{-\gamma z}\hat{a}_y$$

where the propagation constant, γ, for a good conductor can be approximated as

$$\gamma = \alpha + j\beta \approx \frac{1}{\delta} + j\frac{1}{\delta}$$

For a good conductor, both the electric and magnetic fields quickly decay into the conductor. After several skin depths, the fields in the conductor are usually considered negligible. The current density in the conductor is in the same direction as the electric field and is defined as

$$\vec{J}_{2s} = \sigma \vec{E}_{2s} = \sigma C_s e^{-\gamma z}\hat{a}_x$$

This current density does not vary with *x* since the wave incident to the material is a plane wave. It does, however, vary in the *z* direction, into the material. Referring to Figure 4.10, the total current in the material over a width *w* in the *y* direction is (assuming the material is very thick in the *z* direction)

$$I_s = \iint \vec{J}_{2s} \cdot d\vec{s} = \int_0^\infty \int_0^w \sigma C_s e^{-\gamma z}\hat{a}_x \cdot dydz\hat{a}_x = \sigma C_s w \int_0^\infty e^{-\gamma z} dz = \frac{\sigma C_s w}{\gamma} \tag{4.64}$$

The change in the voltage along the surface can be obtained from the standard *electrostatics* definition (assuming that the length of the path associated with *dL* is electrically short):

$$V = -\int \vec{E} \cdot d\vec{L}$$

(For time-varying fields, this definition should be used carefully, especially when the distance between the two measuring points is not electrically short.) The voltage difference, at the surface ($z = 0$), over the

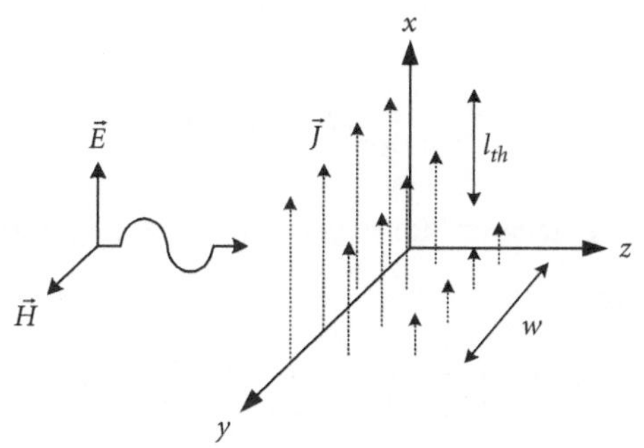

FIGURE 4.10 Sketch of the direction and magnitude of the current density inside a metal when exposed to a plane wave.

length l_{th} in the x direction is then

$$V_s = -\int_{l_{th}}^{0} C_s e^{-\gamma z}\hat{a}_x \cdot dx\hat{a}_x\Big|_{z=0} = -C_s \int_{l_{th}}^{0} dx = C_s l_{th} \tag{4.65}$$

assuming that $l_{th} < \lambda/10$. For conductors, this voltage should be small since the electric field tangential to a conductor is small. For a perfect conductor, the tangential electric field is zero. The impedance seen at the surface of the material is this voltage divided by the total current:

$$Z_s = \frac{V_s}{I_s} = \frac{C_s l_{th}}{\frac{\sigma C_s w}{\gamma}} = \frac{l_{th}}{w}\frac{\gamma}{\sigma} \tag{4.66}$$

For a good conductor,

$$Z_s = \frac{l_{th}}{w}\frac{\frac{1}{\delta} + j\frac{1}{\delta}}{\sigma} = \frac{l_{th}}{w}\frac{1}{\delta\sigma}(1+j) \tag{4.67}$$

Now the meaning of per square can be discussed. Imagine that two parallel test electrodes are in contact with the same side of a surface. The electrodes are separated by a distance of l_{th} and have a width of w. A sinusoidal voltage is applied across the electrodes, and the resultant sinusoidal current between them is measured. The ratio of this voltage to current, in the frequency domain, is the surface impedance. Furthermore, if $l_{th} = w$, then the surface impedance is not a function of any dimension:

$$Z_s = \frac{1}{\delta\sigma}(1+j)\,\Omega \quad \text{if } l_{th} = w \tag{4.68}$$

This should not be too surprising since according to the standard expression for the resistance for a uniform current distribution (where A is the cross-sectional area and t is the thickness of the body)

$$R = \frac{l_{th}}{\sigma A} = \frac{l_{th}}{\sigma w t} \tag{4.69}$$

If l_{th} is doubled, the resistance is doubled, but if w is doubled, the resistance is halved. Doubling both l_{th} and w thus produces no net change in resistance. Increasing the size of the *square* does not affect the net resistance. Also, the expression for the ac surface resistance is

$$R_s = \frac{1}{\delta\sigma}\,\Omega \quad \text{if } l_{th} = w \tag{4.70}$$

Therefore, the ac resistance of the material (for a square region) is equivalent to its dc resistance when $l_{th} = w$ and $t = \delta$:

$$R = \frac{l_{th}}{\sigma w t} = \frac{l_{th}}{\sigma l_{th}\delta} = \frac{1}{\delta\sigma}\,\Omega \quad \text{if } l_{th} = w \tag{4.71}$$

The ac resistance, where the current exponentially decays into the material, is essentially equivalent to the dc resistance when the uniformly distributed dc current is confined to one skin depth into the material.

To use properly the given expressions for the surface impedance, current, and voltage, all of the major assumptions in the derivation should be carefully noted. First, the body should be several skin depths thick (e.g., 5δ) so that any reflections from the other "outgoing" interface are negligible. Second, the surface of the body to the incoming wave should be relatively flat so that current distribution is independent of x and y. Although discussed previously, for a surface to be considered relatively flat, the smallest radius of curvature of the body should be much larger than the skin depth. (The radius of curvature of a round wire of radius r_w is equal to r_w.) As will be discussed, if the material is not several skin depths thick, then the thickness of the conductor should be specified when the surface resistance is given. Finally, the body should be electrically short so that the current and voltage distribution along it is essentially constant. If the body is not electrically short, then the current and voltage distribution will be approximately sinusoidal with a distance between maximums or minimums about equal to λ_o, the free-space wavelength of the incoming sinusoidal wave.

The ac resistance and internal inductance expressions given for round solid conductors at *high frequencies* can be derived from the surface resistance expression (4.62) and internal inductance expression (4.63). For a round wire of radius r_w and length l_{th}, the ratio of the length (l_{th}) to the width ("w" $= 2\pi r_w$) is not usually one. Therefore, the resistance and internal inductance per square relationships should be multiplied by l_{th}/w to obtain the nonsquare high-frequency resistance and internal inductance:

$$r_{AC} = \frac{l_{th}}{w}\sqrt{\frac{\pi f\mu}{\sigma}} = \frac{l_{th}}{2\pi r_w}\sqrt{\frac{\pi f\mu}{\sigma}} = \frac{l_{th}}{2r_w}\sqrt{\frac{f\mu}{\pi\sigma}} = r_{DC}\frac{r_w}{2\delta} \tag{4.72}$$

$$l_{iAC} = \frac{l_{th}}{w}\sqrt{\frac{\mu}{4\pi f\sigma}} = \frac{l_{th}}{2\pi r_w}\sqrt{\frac{\mu}{4\pi f\sigma}} = \frac{l_{th}}{4\pi r_w}\sqrt{\frac{\mu}{\pi\sigma f}} \tag{4.73}$$

These results are identical to the standard high-frequency expressions for a round wire.

The surface resistance is plotted for copper for a wide frequency range in Mathcad 4.14. Also, the thickness corresponding to 5δ is plotted over this same frequency range so that the range where these results may be used can be determined. For thicknesses greater than 5δ, the thickness of the material is not a significant factor in determining the surface resistance since the current density is quite small beyond this depth. For copper, the expression for the surface resistance is

$$R_s = \sqrt{\frac{\pi f\mu}{\sigma}} = \sqrt{\frac{\pi f(4\pi\times10^{-7})}{5.8\times10^{7}}} \approx 2.6\times10^{-7}\sqrt{f}\ \Omega \quad \text{if } l_{th} = w \tag{4.74}$$

Again, note that this expression is not a function of the thickness of the material. If the flat conducting body is used as a reference plane, increasing the thickness of the material will not significantly affect its quality if it is already much thicker than the skin depth.

The concept of surface resistance is also used when the thickness of the material is not several skin depths. If the material thickness is less than the skin depth, the surface resistance is a measure of the resistance of the material of a given thickness, t, assuming that $l_{th} = w$:

$$R_s = \frac{l_{th}}{\sigma wt} = \frac{l_{th}}{\sigma l_{th}t} = \frac{1}{\sigma t}\ \Omega \quad \text{if } l_{th} = w \tag{4.75}$$

This surface resistance is equal to the standard bulk resistance of the material assuming a uniform current distribution. To use this expression, the frequency of the signal should be sufficiently low so that the current is approximately uniformly distributed over the cross section of the material. (Note that it is not equal to the previous high-frequency surface-resistance expression $R_s = 1/(\delta\sigma)$). The low-frequency resistance obtained from (4.75) is identical to the dc resistance of the material.

$$\mu := 4\cdot\pi\cdot 10^{-7} \qquad \sigma := 5.8\cdot 10^{7}$$

$$x := 40, 40.1 .. 80$$

$$f(x) := \left(x + 1 - 10\cdot \text{floor}\left(\frac{x}{10}\right)\right)\cdot 10^{\text{floor}\left(\frac{x}{10}\right)}$$

$$\delta(f, \mu, \sigma) := \frac{1}{\sqrt{\pi\cdot f\cdot \mu\cdot \sigma}} \qquad R_s(f) := \frac{1}{\delta(f, \mu, \sigma)\cdot \sigma}$$

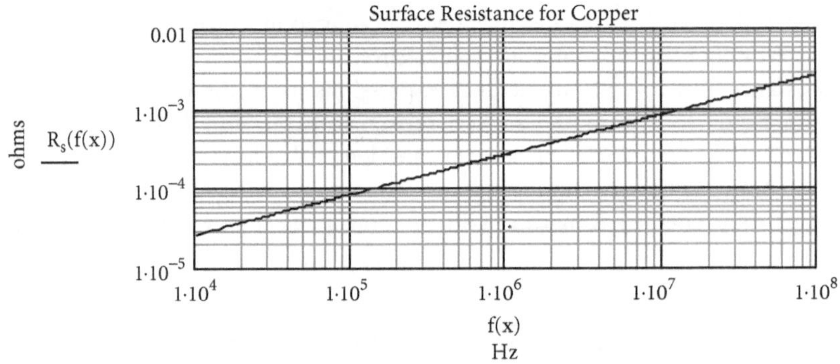

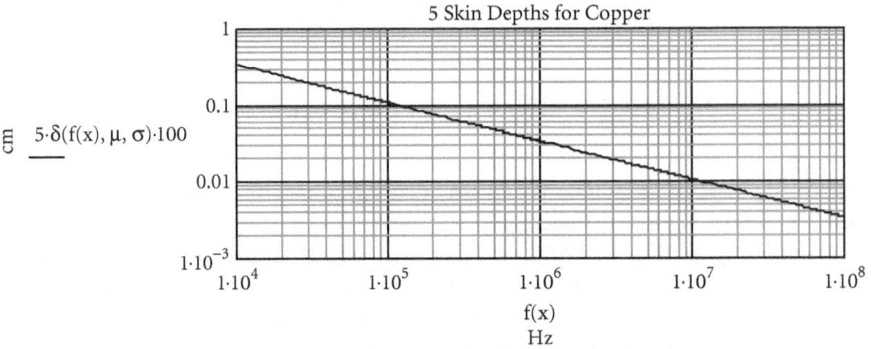

MATHCAD 4.14 Surface resistance and 5δ for copper vs. frequency.

For very thin metallic films, the conductivity, σ, given in (4.75) is not equal to the standard bulk conductivity for the material. When the film thickness is much greater than the electron mean-free-path (the average distance an electron travels before colliding with another molecule), the bulk and film conductivities are about the same. When the film thickness is about the same as the electron mean-free-path or less, electron scattering is important and the film conductivity is less than the bulk conductivity. The expression for the film conductivity is

$$\sigma_f = \frac{3t\sigma}{4p}\left[\ln\left(\frac{p}{t}\right) + 0.4228\right] \quad \text{if } p \ll t \tag{4.76}$$

For copper at room temperature, $p = 420$ Å, $\sigma = 5.8\times 10^7$ 1/Ω-m, and t = thickness of the copper film in Å. The surface resistance and conductivity at room temperature are given in Table 4.5 for copper for various thicknesses. Frequently for metallic films, gold will be used because of its noncorrosive properties. The additional cost of using gold is usually negligible if a small quantity is used. For reference purposes,

TABLE 4.5 Conductivity and Surface Resistance of Copper

t (Å)	t (m)	σ_f (1/Ω-m)	R_s (Ω)
10	10×10^{-10}	0.43×10^{7}	230
30[a]	30×10^{-10}	0.95×10^{7}	35
60	60×10^{-10}	1.47×10^{7}	11
90	90×10^{-10}	1.83×10^{7}	6.1
0.1×10^{7}	0.1×10^{-3}	5.80×10^{7}	170×10^{-6}
1.0×10^{7}	1.0×10^{-3}	5.80×10^{7}	17×10^{-6}
10×10^{7}	10×10^{-3}	5.80×10^{7}	1.7×10^{-6}

[a]About 90% of light is transmitted through 30 Å of copper!

p = 570 Å and $\sigma = 6.17 \times 10^{7}$ 1/Ω-m for silver at room temperature. The parameters for gold were previously given.

A general expression for the surface resistance and internal inductance for a flat conductor (compared to the radius of curvature) of thickness t is (if $w = l_{th}$)

$$R_s = \frac{1}{\delta\sigma}\left[\frac{\sinh\left(\frac{2t}{\delta}\right)+\sin\left(\frac{2t}{\delta}\right)}{\cosh\left(\frac{2t}{\delta}\right)-\cos\left(\frac{2t}{\delta}\right)}\right]\Omega,\ L_{iAC} = \frac{1}{2\pi f\delta\sigma}\left[\frac{\sinh\left(\frac{2t}{\delta}\right)+\sin\left(\frac{2t}{\delta}\right)}{\cosh\left(\frac{2t}{\delta}\right)-\cos\left(\frac{2t}{\delta}\right)}\right]\text{H} \tag{4.77}$$

where sinh(x) and cosh(x) are the hyperbolic sine and hyberbolic cosine functions, respectively. In these expressions, the thickness of the metal is not assumed thick and the current distribution is not assumed uniform. However, the metal is assumed a good conductor. A much simpler approximate expression for the surface resistance is

$$R_s \approx \frac{1}{\delta\sigma}\frac{1}{1-e^{-\frac{t}{\delta}}}\,\Omega \quad \text{if } w = l_{th} \tag{4.78}$$

There is less than 1% difference between these two expressions for $t/\delta > 4.63$ and $t/\delta < 0.02$. The sum $R_s + j\omega L_{iAC}$ is sometimes referred to as the barrier impedance of the metal. As a simple check on (4.78), for low frequencies where the skin depth is large compared to the thickness,

$$R_s \approx \frac{1}{\delta\sigma}\frac{1}{1-e^{-\frac{t}{\delta}}} \approx \frac{1}{\delta\sigma}\frac{1}{1-\left(1-\frac{t}{\delta}\right)} = \frac{1}{\sigma t}\,\Omega \quad \text{if } \delta \gg t, w = l_{th} \tag{4.79}$$

since $e^{x} \approx 1+x$ when $x \ll 1$. This last approximation is the standard expression for the dc surface resistance.

In Mathcad 4.15, the general expression for the surface resistance is plotted for various thicknesses that are a function of the skin depth. Also, for comparison, the surface resistance for a thick shield, $R_s = 1/(\delta\sigma)$, is plotted. As expected, the surface resistance increases with decreasing shield thickness. Also plotted is the surface resistance of 1 mm and 0.1 mm thick copper.

$$\mu := 4\cdot\pi\cdot 10^{-7} \qquad \sigma := 5.8\cdot 10^{7}$$

$$x := 40, 40.1 .. 80$$

$$f(x) := \left(x + 1 - 10\cdot\text{floor}\left(\frac{x}{10}\right)\right)\cdot 10^{\text{floor}\left(\frac{x}{10}\right)}$$

$$\delta(f,\mu,\sigma) := \frac{1}{\sqrt{\pi\cdot f\cdot\mu\cdot\sigma}} \qquad R_s(f) := \frac{1}{\delta(f,\mu,\sigma)\cdot\sigma} \qquad R(f,t) := R_s(f)\cdot\frac{\sinh\left(\frac{2\cdot t}{\delta(f,\mu,\sigma)}\right)+\sin\left(\frac{2\cdot t}{\delta(f,\mu,\sigma)}\right)}{\cosh\left(\frac{2\cdot t}{\delta(f,\mu,\sigma)}\right)-\cos\left(\frac{2\cdot t}{\delta(f,\mu,\sigma)}\right)}$$

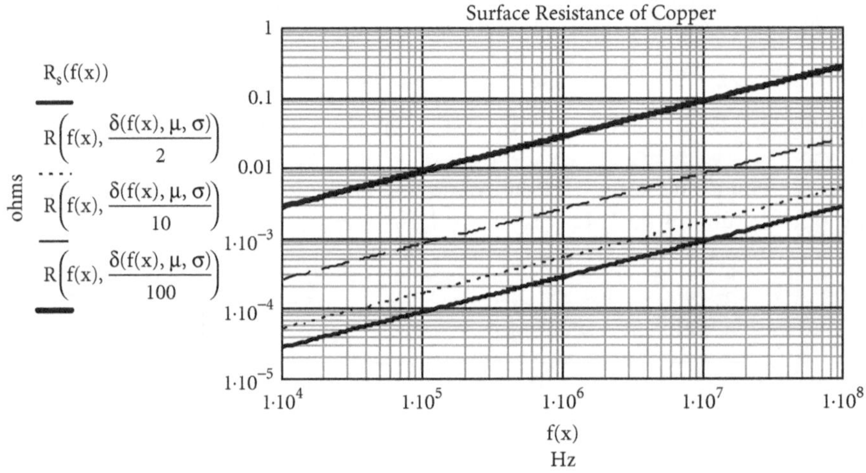

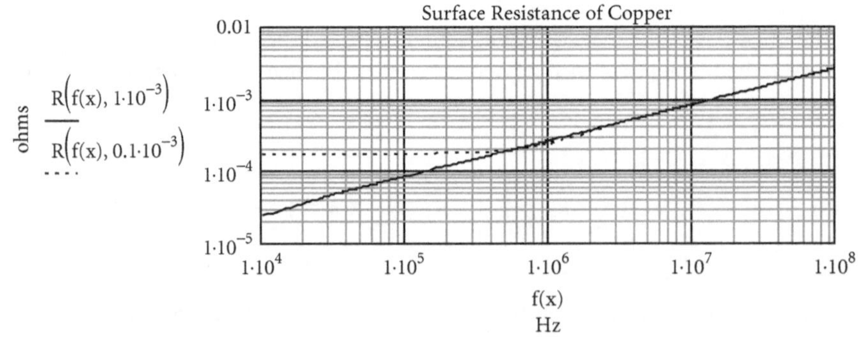

MATHCAD 4.15 Surface resistance of copper vs. frequency as a function of thickness.

4.9 Voltage and Current along a Chassis

A 144 MHz plane wave with a field strength of 300 V/m is normally incident upon a 0.1 mm thick, flat chassis constructed of aluminum. What are the expected amplitudes of the current and voltage along this chassis due to this plane wave?

This plane wave will induce a current and voltage along the chassis. Before determining this current and voltage, the skin depth at this frequency in aluminum will be determined:

$$\delta = \frac{1}{\sqrt{\pi f \mu \sigma}} = \frac{1}{\sqrt{\pi f \mu \sigma_r \sigma_o}} = \frac{1}{\sqrt{\pi (144\times 10^{6})(4\pi \times 10^{-7})(0.61\times 5.8\times 10^{7})}} \approx 7.1\,\mu\text{m}$$

For the chassis to be considered flat, the smallest radius of curvature of the chassis where the current and voltage are to be estimated should be much greater than 7.1 μm. Also, notice that the thickness of the chassis is much greater than the skin depth so the standard expression for the surface resistance may be used:

$$R_s = \frac{1}{\delta\sigma} \approx 4\ \mathrm{m}\Omega$$

Furthermore, the reflections from the side of the chassis not directly exposed to the field can be ignored since they will have a very small magnitude. The shield is about 14 skin depths thick so the absorption through the chassis will be tremendous. The electric field tangential to the chassis can be determined by using either the reflection or transmission coefficient. Using the transmission coefficient, T,

$$E_{ts} = E_{is}\mathrm{T} = E_i\left(\frac{2\eta_L}{\eta_L - \eta_o}\right) = 300\left[\frac{2\sqrt{\frac{\omega\mu}{2\sigma}}(1+j)}{\sqrt{\frac{\omega\mu}{2\sigma}}(1+j) - 377}\right] \Rightarrow |E_{ts}| \approx 9.0\ \mathrm{mV/m}$$

As expected for a conducting surface, the tangential electric field is small. The magnitude of the surface impedance, which is equal to the surface resistance multiplied by $\sqrt{2}$,

$$|Z_s| = \left|\frac{1}{\delta\sigma}(1+j)\right| = \sqrt{2}\frac{1}{\delta\sigma} = \sqrt{2}R_s\ \Omega \tag{4.80}$$

can be used to determine the magnitude of the current along the surface:

$$|I_s| = \left|\frac{V_s}{Z_s}\right| = \frac{|E_{ts}|l_{th}}{\sqrt{2}R_s} = \frac{(9\times10^{-3})l_{th}}{\sqrt{2}(4\times10^{-3})} \approx 1.6 l_{th}$$

where l_{th} is the length (or distance) along the chassis. Because the surface resistance per square was used, this result assumes that the width w of the probes used to measure the current is equal to the distance between the probes, l_{th}. The magnitude of the voltage is

$$|V_s| = |E_{ts}|l_{th} = (9\times10^{-3})l_{th}$$

Before determining the voltage and current for a specific length, l_{th}, the free-space wavelength of the 144 MHz signal will be determined:

$$\lambda = \frac{c}{f} = \frac{3\times10^8}{144\times10^6} \approx 2.1\ \mathrm{m}$$

Therefore, objects that are smaller than about 2.1/10 = 21 cm are electrically short. The difference in voltage across 10 cm of the chassis is small as expected at 0.9 mV, but the current is not small at 160 mA. Incident waves typically produce little voltage variation along conducting surfaces, but they can generate important currents and counter or induced fields. When the object is not electrically short, the current and voltage can vary considerably over its dimensions. The characteristic impedance of the chassis (if it can be reasonably defined) could possibly be used to help determine the voltage and current distribution when the object is not electrically short.

If the voltage is measured across two points along a chassis or other piece of metal, care should be taken. The measurement leads may actually be acting as part of a pickup loop (with the chassis acting as one segment of the loop). The loop area generated by the leads and chassis should be minimized to reduce this probe or antenna effect.

4.10 Impedance of Coated Conductors

When do the electrical properties of the substrate or base metal influence the impedance of the metal coating? [Ramo]

Metals are quite often coated with a metal of another type. For example, some metals are coated with another metal to increase their hardness, their conductivity, their shielding effectiveness, or their lifetime. For the large flat conductor shown in Figure 4.11 of permeability μ_1, conductivity σ_1, and thickness t that is placed on an infinitely thick conductor of permeability μ_2 and conductivity σ_2, the expression for the surface impedance of the composite conductor is given by

$$Z_s = R_s + j\omega L_{iAC} = R_{s1}(1+j)\left[\frac{\sinh(\gamma_1 t)+\left(\frac{R_{s2}}{R_{s1}}\right)\cosh(\gamma_1 t)}{\cosh(\gamma_1 t)+\left(\frac{R_{s2}}{R_{s1}}\right)\sinh(\gamma_1 t)}\right]\Omega \quad \text{if } l_{th} = w \tag{4.81}$$

$$\text{where } \gamma_1 = \frac{1+j}{\delta_1} = (1+j)\sqrt{\pi f \mu_1 \sigma_1},\ R_{s1} = \frac{1}{\sigma_1 \delta_1} = \sqrt{\frac{\pi f \mu_1}{\sigma_1}},\ R_{s2} = \frac{1}{\sigma_2 \delta_2} = \sqrt{\frac{\pi f \mu_2}{\sigma_2}}$$

where $\sinh x$ and $\cosh x$ are the hyperbolic sine and hyberbolic cosine functions, respectively. The length, l_{th}, and width, w, are as previously defined for a single layer slab. The real part of Z_s is the surface resistance.

The real and imaginary parts of the surface impedance normalized by the surface resistance of the upper layer of thickness t are plotted (in Mathcad 4.16) as a function of this thickness divided by the skin depth of the upper layer. The upper layer is silver and lower substrate layer is copper. As seen, both the surface resistance and internal inductive reactance stabilize or remain relatively constant after t is greater than about $2\delta_1$. As expected, when the higher conductivity silver layer is very thin, relative to its skin depth, the normalized impedance is higher. In this case, the impedance is mainly determined by the lower conductivity copper substrate material. However, when the silver layer is about a skin depth or so thick, the normalized impedance stabilizes to a smaller value. If the substrate layer were silver and upper layer were gold, a material with a conductivity lower than silver, then the normalized impedance would be initially smaller for a thin gold layer and rise to a larger impedance for a thick gold layer. The second plot clearly shows that the surface resistance increases with frequency.

When the coating material is very thick compared to the skin depth (of the coating), the surface impedance is mainly determined by the electrical properties of the coating. The substrate properties are

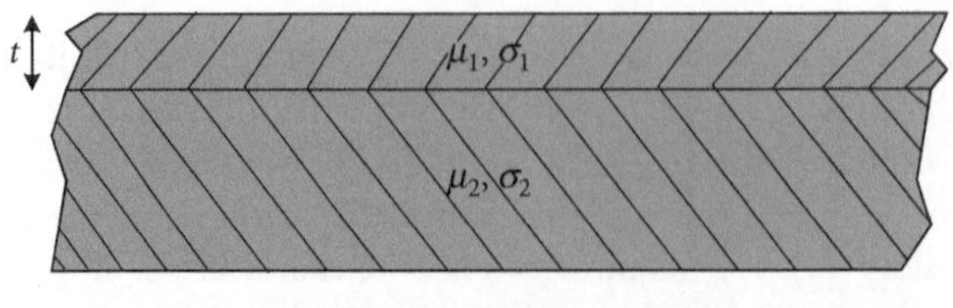

FIGURE 4.11 Two-layer composite metal.

$\mu_o := 4\cdot\pi\cdot 10^{-7} \qquad \sigma_o := 5.8\cdot 10^{7} \qquad t := 200\cdot 10^{-6} \qquad j := \sqrt{-1}$

$x := 10, 10.5 .. 70$

$$f(x) := \left(x + 1 - 10\cdot \mathrm{floor}\left(\frac{x}{10}\right)\right)\cdot 10^{\mathrm{floor}\left(\frac{x}{10}\right)}$$

$$\delta_1(f,\mu_1,\sigma_1) := \frac{1}{\sqrt{\pi\cdot f\cdot \mu_1\cdot \sigma_1}} \qquad \delta_2(f,\mu_2,\sigma_2) := \frac{1}{\sqrt{\pi\cdot f\cdot \mu_2\cdot \sigma_2}}$$

$$R_{s1}(f,\mu_1,\sigma_1) := \frac{1}{\sigma_1\cdot \delta_1(f,\mu_1,\sigma_1)} \qquad R_{s2}(f,\mu_2,\sigma_2) := \frac{1}{\sigma_2\cdot \delta_2(f,\mu_2,\sigma_2)} \qquad \gamma_1(f,\mu_1,\sigma_1) := \frac{1+j}{\delta_1(f,\mu_1,\sigma_1)}$$

$$Z_s(f,\mu_1,\sigma_1,\mu_2,\sigma_2,t) := R_{s1}(f,\mu_1,\sigma_1)\cdot(1+j)\cdot\left(\frac{\sinh(\gamma_1(f,\mu_1,\sigma_1)\cdot t)\ ... + \dfrac{R_{s2}(f,\mu_2,\sigma_2)}{R_{s1}(f,\mu_1,\sigma_1)}\cdot\cosh(\gamma_1(f,\mu_1,\sigma_1)\cdot t)}{\cosh(\gamma_1(f,\mu_1,\sigma_1)\cdot t)\ ... + \dfrac{R_{s2}(f,\mu_2,\sigma_2)}{R_{s1}(f,\mu_1,\sigma_1)}\cdot\sinh(\gamma_1(f,\mu_1,\sigma_1)\cdot t)}\right)$$

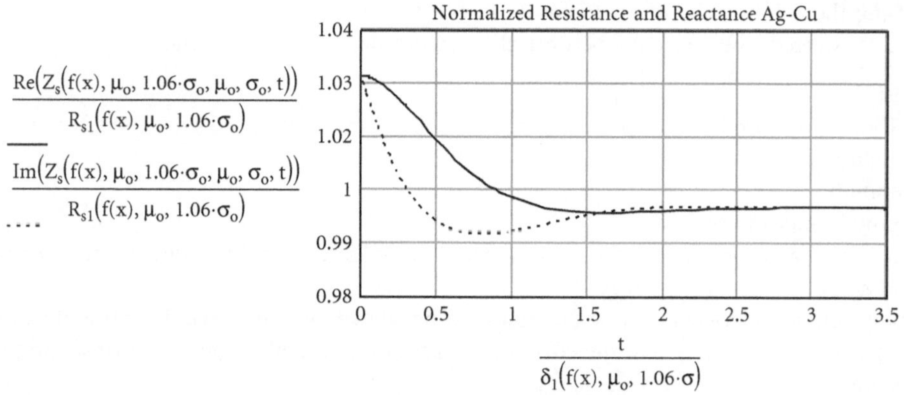

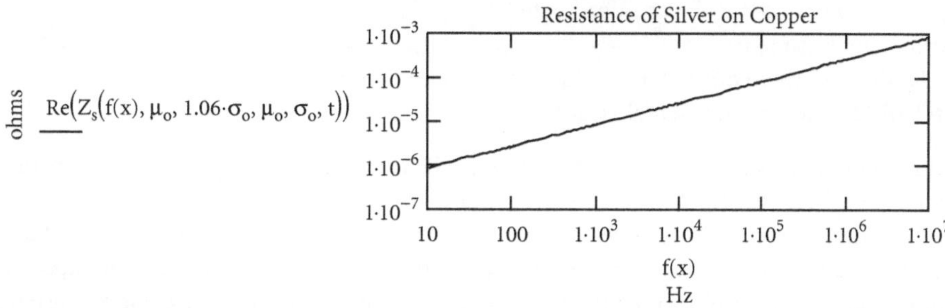

MATHCAD 4.16 Normalized surface resistance of a silver-coated copper substrate vs. the ratio of the silver thickness to its skin depth. Also plotted is the unnormalized surface resistance vs. frequency.

not important (from a surface impedance standpoint). In the silver coating example, the conductivity of the silver is nearly equal to the conductivity of the copper base. Hence, if the resistance of the composite is plotted vs. frequency for different silver layer thicknesses, very little difference in the surface resistance is noted. This small difference is to be expected since the normalized impedance values for the silver coating on copper varies from only about 0.99 to 1.03. In some situations, such as with high-*Q* resonant cavities, a pure silver coating may be worthwhile.

4.11 Nontraditional Shielding Materials

Compare the various nontraditional shielding materials such as copper paint and metallized cloth. [Ott; Weston; White, '88; Chatterton; Enthone; product data sheets]

Many electronic devices, such as cellular phones and notebook computers, are partially or completely enclosed in a metallic enclosure for shielding and charge dissipation. However, this metallic enclosure is often not a solid piece of, for example, steel or aluminum. Frequently, conductive coatings, conductive plastics, and other nontraditional materials are used to provide an enclosure with some shielding ability and conductive properties. These nontraditional materials and methods are used to reduce cost, to increase aesthetic appeal, and to increase the ease of application. In some cases, as with conductive windows, only thin conductive coatings and fine meshes can provide shielding while allowing light through.

Most of these conductive shields are effective against far-field plane waves and near-field electric fields but not against near-field magnetic fields (unless the shield material is magnetic in nature). When the shielding effectiveness is provided for these materials vs. frequency in data sheets, it is normally measured for near-field electric fields. Unless the same method is used to determine the shielding effectiveness of a material or method of application, it is difficult to compare values for dB shielding effectiveness. As discussed in this book, generally the apertures in the shield (e.g., shield defects) are the major source of loss in shielding effectiveness, not the material itself.

Increasing the conductivity of an enclosure has certain advantages and disadvantages. Factors that should be considered when selecting between the various methods and materials include:

1) cost
2) thickness, uniformity, and surface resistivity—all of which affect the shielding effectiveness
3) stability
4) durability
5) aesthetic appearance
6) shape, size, and complexity of the object to be shielded or coated (including the number, position, and size of any "hidden" areas)
7) composition and cleanliness of the object to be coated (which affects how the object is to be prepared before coating and the effect of the preparation method on the intrinsic properties of the object)
8) physical strength of the object to be coated
9) expected environment
10) bondability to other conductors, including contact resistance
11) recyclabililty and environmental issues.

Different methods are used to measure the shielding effectiveness. A common method, probably because of its simplicity, is to measure the low-frequency resistance between two specified points on the surface of the object. This measurement, which is referred to as the point-to-point test, can be repeated at multiple locations. The resistance between two points will be a function of the thickness, conductivity, and uniformity of the coating or material, as well as the distance between the two points (and the contact resistance of the probes used to perform the test). Although it will probably not provide a strong indication of the uniformity of the coating over the entire surface, the point-to-point method is a quick and simple method of comparison. Another method of comparing the effectiveness of a material or method is through the surface resistivity, given as ρ_s in this discussion. The surface resistivity is given in Ω in this section, but it is commonly given as Ω/sq or $\Omega/\square$ in the literature. Unless this test is performed over many (flat) locations of the actual object, enclosure, or chassis, there can be significant variation over the part based on its shape and features (e.g., vents, bosses, and joints). For comparison purposes, the surface resistivity of stamped steel ranges from 0.1–0.3 Ω for thicknesses that range from 375–1,500 μm.[7] The surface resistivity values provided in this discussion are only estimates and should only be used for initial calculations.

[7]To convert meters to mils (thousandths of an inch), divide by 2.54×10^{-5}.

Although mostly the high-conductivity, low-resistivity, shielding materials and coatings are discussed, higher resistivity materials are available that are used in electrostatic discharge control. For example, germanium, silicon oxide, and indium tin oxide are used when higher resistivities are desired.

4.11.1 Vacuum Metallization

A thin film of a boiling metal, such as aluminum or gold, is deposited or sputtered in a vacuum chamber directly on various materials such as plastics and glass without a base coat. The resultant metallic film thickness is usually in the micron range; thus, the mass of the coating is usually quite small compared to the mass of the base object. No solvents are required for vacuum metallization. Mixed layers of stainless and copper, or aluminum and stainless, can be generated. For very thin deposits, diffusion can occur between different metal layers at higher temperatures. A thin layer of aluminum is susceptible to abrasion and scratching. The metal film can be sandwiched between two materials to help protect it from corrosion and physical damage. Also, a clear top coat can also be applied to help reduce corrosion and increase abrasion resistance. Partially transparent optical windows can be obtained through vacuum metallization. Thicknesses up to 12 μm are used in shielding applications. A thin layer of indium tin oxide, with a surface resistivity from 60–250 Ω, can be applied on clear polyester, 127–177 μm thick.

$$\rho_s = 0.1 - 0.2\ \Omega \text{ for } 5 - 7.5\ \mu\text{m thick aluminum}$$

4.11.2 Metal Mesh

Metal meshes (and thin films of metal) are used to provide shielding where some degree of optical transparency is desired. Meshes are frequently considered visually inferior. The appearance of an embedded metal mesh between two layers of glass or acrylic, for example, is usually considered inferior in windows although the transparency is high (up to 98%). For high mesh density, the mesh generally provides a greater shielding effectiveness than a *thin* metallic film. For corrosion and abrasion protection, the metal mesh is often coated, embedded between other sheets of plastic or glass, or placed on the side not directly accessible (e.g., back side of a monitor screen). The effectiveness is a function of the number of openings per unit area and wire diameter. Stainless steel based materials and copper are used. For copper coating over a black resin, ρ_s can be large and about 20 kΩ (for electrostatic discharge applications). For other meshes, the surface resistivity can range from 5–20 Ω. The bonding of the wires at the crossover points must be carefully performed if nonlinear junctions are to be avoided and good electrical contact made.

4.11.3 Embedded Plastic

A conductive material can be included with the resin before a final plastic object is produced. As the concentration of conductive filler increases, the conductivity of the resin or base material generally increases (but eventually the conductivity saturates). Carbon, metallized glass, and metal fibers are used in the resin. However, the conductive particles can significantly affect the mechanical and aesthetic properties of the plastic, especially at higher filler levels. Also, stress on the plastic can break some of the conductive paths. In some cases, low-resistance continuity via the surface of the object may not be possible. Mating contact with a chassis constructed of this material might be a problem. The shielding effectiveness is good and the resistance to abrasion and scratching is excellent since the conductive particles are embedded in the resin and are not merely a coating. The surface resistivity can range from $10–10^6$ Ω.

4.11.4 Conductive Paints

Conductive metal flakes (and powders) are added to a binding agent to produce a liquid conductive paint. These metal-loaded paints can be sprayed (inexpensively) with conventional equipment. Copper, nickel, silver, graphite, and hybrid metals are available in this form. This is a very popular and common method of increasing the conductivity of an object, and this paint is seen on the inner surface of enclosures or chassis. One disadvantage of conductive paints is that the uniformity of the coating is a strong function of the experience of the applicator (if hand coated). Another disadvantage of spraying conductive paints is the overspray loss. This can be reduced, however, with robotic sprayers. The shielding effectiveness of conductive paints is very good, but the surface can be easily scratched unless a protective coating is applied. Graphite coatings are usually adequate for bleeding off charge (e.g., electrostatic discharge applications). Conductive inks can also be applied to plastic, formable plastic sheets. Solvents are usually required when spraying, but water-based paints are expected to reduce some of the environmental and safety costs associated with solvents. Copper paint has a high conductivity and lower film thickness compared to nickel, but nickel is harder and more durable with an excellent resistance to corrosion. The conductivity of silver is higher than copper and less can be applied, but it is more expensive and softer. Silvered copper or silver-plated copper, a hybrid metal, is becoming more popular because of its good conductivity, compared to copper, and hardness, compared to silver.

$$\rho_s = 0.03\text{–}0.05\ \Omega \text{ for } 13\text{–}37\ \mu\text{m thick silver}$$

$$\rho_s = 0.2\text{–}0.5\ \Omega \text{ for } 50\text{–}75\ \mu\text{m thick copper}$$

$$\rho_s = 0.03\text{–}0.05\ \Omega \text{ for } 37\text{–}75\ \mu\text{m thick silvered copper}$$

$$\rho_s > 0.5\ \Omega \text{ for } 37\text{–}75\ \mu\text{m thick nickel}$$

$$\rho_s \approx 150\ \Omega \text{ graphite}$$

4.11.5 Electroless Plating

The chemical deposition of metal in this fashion can produce a very uniform, very thin (0.5 μm) precise coating with little waste. It can be applied on many metal, plastic, and ceramic substrates. The shielding effectiveness of the coating can be excellent with the appropriate thickness. The abrasion and scratch resistance is excellent for an electroless nickel coating over copper. Hidden locations such as recesses and blind holes can be coated since this process is not dependent on line-of-sight coating as with spraying and vacuum metallization. With double-sided electroless coating, both sides of the substrate are coated. Thicknesses up to 5 μm can be applied using copper. If greater thicknesses are required, the electroless process can be supplemented with normal electroplating.

$$\rho_s = 0.02\text{–}0.03\ \Omega \text{ for } 2\text{–}2.5\ \mu\text{m thick copper, selective plating}$$

4.11.6 Conductive Fabrics/Foams

Copper plated fabric consists of nylon and polyester filaments coated with metal. Obviously, as the percentage of the metal in the fabric increases, the surface resistance decreases. Conductive fabrics are available woven and nonwoven. Conductive tapes, gaskets, and cable wraps are also available. If nickel is used, because of its lower conductivity, more metal is required compared to copper. Conductive foams have knitted fabric or nonwoven material on both sides of polyurethane foam. The surface resistivities range from 0.03–0.70 Ω.

$$\rho_s = 0.09\ \Omega \text{ for } 10\%\text{ Cu, low-metal Cu/nylon nonwoven}$$

$$\rho_s = 0.047\ \Omega \text{ for } 30\%\text{ Cu, high-metal Cu/nylon nonwoven}$$

4.11.7 Metallic Foils/Tapes

Various metal foils of aluminum and copper are available with an adhesive backing. For production work, the process of applying the tape is too slow, but for objects of simple shapes, when prototyping or experimenting, these foils and tapes are useful. Large rolls are available that can be applied like wallpaper to produce a low-cost shielded room!

$$\rho_s = 0.05\text{–}0.15\ \Omega \text{ for } 125\text{–}1{,}250\ \mu\text{m thick copper}$$

$$\rho_s < 5\ \Omega \text{ for } 160\ \mu\text{m thick carbon-based tape}$$

4.11.8 Thermal Flame/Arc Spray Coating

Metals such as zinc and copper are melted by an electric arc and then sprayed with special equipment. The coating is dense with excellent conductivity, but the surface is rough and textured (unless polished). This process is primarily used for coating metal on metal. Many metals when sprayed in this manner will oxidize, which will affect the contact resistance to the coating.

4.12 Shielding Effectiveness vs. Surface Resistance

Plot the shielding effectiveness (SE) in dB vs. frequency for a wide range of surface resistances in Ω. Assume that the shield is thin and that the incident field is a plane wave normally incident to the shield.

The total loss in dB for a flat shield to a normally incident plane wave is the sum of the reflection, absorption, and multiple-reflection losses in dB:

$$TOT_{dB} = R_{dB} + A_{dB} + MR_{dB}$$

For a good-conducting shield of thickness *t*, the approximate expressions for these losses in dB are

$$R_{dB} = 20\log\left|\frac{2.6\times10^{8}}{\sqrt{\dfrac{\mu_r f}{\sigma_r}}}\right|$$

$$A_{dB} = 131t\sqrt{\mu_r\sigma_r f}$$

$$MR_{dB} = 10\log\left|1 - 2e^{-30.3t\sqrt{\mu_r\sigma_r f}}\cos\left(30.3t\sqrt{\mu_r\sigma_r f}\right) + e^{-60.6t\sqrt{\mu_r\sigma_r f}}\right|$$

The surface resistance for a shield that is thin compared to the skin depth ($t \ll \delta$) is

$$R_s = \frac{l_{th}}{\sigma wt} = \frac{l_{th}}{\sigma l_{th} t} = \frac{1}{\sigma t}\ \Omega$$

(This expression is also referred to as the low-frequency, uniform-current, surface resistance.) As seen in Mathcad 4.17 for copper shields, the absorption loss is negligible for thin shields but the multiple-reflection loss is *not* negligible. Since the highest frequency plotted on these graphs is 1 GHz, the shield should be much less than 2 μm, which is the skin depth of copper at 1 GHz. Notice that the total loss, TOT_{dB}, is constant with frequency for these large surface resistances. Because the surface resistance is so large, the thickness of the shield is extremely small. For example, for a surface resistance of 5 Ω,

$$t = \frac{1}{R_s\sigma} \approx 3.4\ \text{nm}$$

(Of course, when the thickness of the metal is this thin, the conductivity is not equal to the bulk conductivity used in this expression.) This thickness is much less than the skin depth at 1 GHz. The total

$\mu_o := 4\cdot\pi\cdot10^{-7}$ $\quad \varepsilon_o := 8.854\cdot10^{-12}$ $\quad \sigma_o := 5.8\cdot10^{7}$ $\quad \mu_r := 1$ $\quad \sigma_r := 1$ $\quad j := \sqrt{-1}$

$x := 40, 40.1 .. 90$

$$f(x) := \left(x + 1 - 10\cdot\text{floor}\left(\frac{x}{10}\right)\right)\cdot 10^{\text{floor}\left(\frac{x}{10}\right)}$$

$$t(R_s) := \frac{1}{\sigma_o\cdot R_s} \qquad \delta(f) := \frac{1}{\sqrt{\pi\cdot f\cdot\mu_r\cdot\mu_o\cdot\sigma_r\cdot\sigma_o}} \qquad R_{dB}(f) := 20\cdot\log\left(\left|\frac{2.6\cdot10^{8}}{\sqrt{\frac{\mu_r\cdot f}{\sigma_r}}}\right|\right) \qquad A_{dB}(f, R_s) := 131\cdot t(R_s)\cdot\sqrt{\mu_r\cdot\sigma_r\cdot f}$$

$$MR_{dB}(f, R_s) := 10\cdot\log\left(1 - 2\cdot e^{-30.3\cdot t(R_s)\cdot\sqrt{\mu_r\cdot\sigma_r\cdot f}}\cdot\cos\left(30.3\cdot t(Rs)\cdot\sqrt{\mu_r\cdot\sigma_r\cdot f}\right) + e^{-60.6\cdot t(R_s)\cdot\sqrt{\mu_r\cdot\sigma_r\cdot f}}\right)$$

$$TOT_{dB}(f, R_s) := R_{dB}(f) + A_{dB}(f, R_s) + MR_{dB}(f, R_s)$$

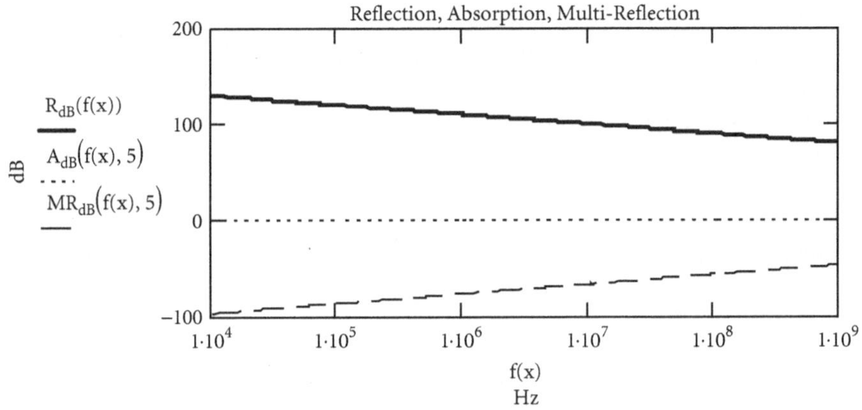

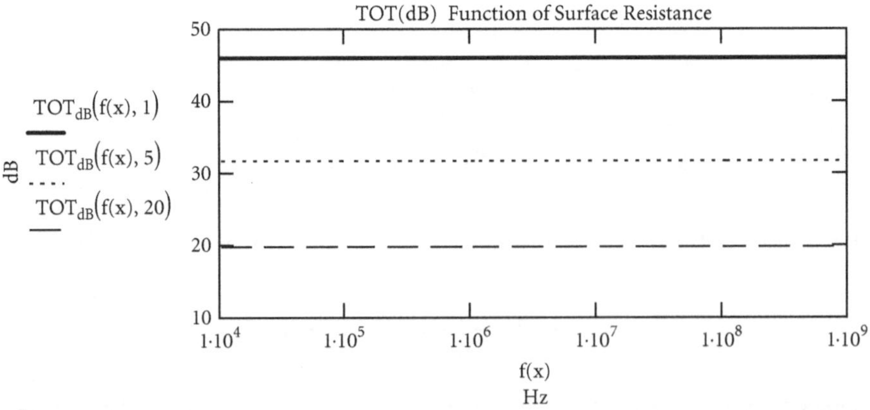

$$\frac{1}{\sqrt{\pi\cdot10^{9}\cdot\mu_r\cdot\mu_o\cdot\sigma_r\cdot\sigma_o}} = 2.09\times10^{-6}$$

MATHCAD 4.17 Total loss and its components for a copper shield vs. frequency as a function of the surface resistance.

loss can be simply written as a function of the surface resistance if the approximation for the total loss for thin shields is used:

$$TOT_{dB} = 20\log\left(\frac{377\sigma t}{2\sqrt{2}}\right) = 20\log\left(\frac{377}{2\sqrt{2}R_s}\right) \approx 42.5 - 20\log(R_s) \tag{4.82}$$

As expected, as the shielding resistance increases, the shielding effectiveness decreases. For example, doubling R_s results in about a 6 dB decrease in the shielding effectiveness.

4.13 Near-Field Electric Shielding Effectiveness

Plot the shielding effectiveness (SE) in dB vs. frequency for a wide range of surface resistances in Ω. Assume that the shield is thin and in the presence of a *near-field electric field.* [Lasitter]

The total loss in dB (or shielding effectiveness) is the sum of the reflection, absorption, and multiple-reflection losses in dB:

$$TOT_{dB} = R_{dB} + A_{dB} + MR_{dB} \tag{4.83}$$

Similar to the plane-wave situation, the thin-shield absorption loss is negligible for near-field electric fields while the reflection and multiple-reflection losses are both important. The equations given in the previous discussion assume that the wave impedance of the incident wave is 377 Ω, independent of frequency. However, for near-field electric fields, the wave impedance is a function of the frequency and distance to the source of the field. For a Hertzian dipole located along the z axis, the wave impedance in the near field is

$$|Z_w| = \frac{\lambda_o \eta_o}{2\pi r} = \frac{3\times 10^8 (377)}{2\pi f r} \approx \frac{1.8\times 10^{10}}{fr}\ \Omega \quad \text{if } r \ll \frac{\lambda_o}{2\pi} = \frac{3\times 10^8}{2\pi f} \tag{4.84}$$

(The wave impedance was defined as E_θ/H_ϕ in the near field. Obviously, there are other components in the near field, but only these two components produce real radiated power.) Since the impedance mismatch produces reflective losses at the two interfaces of the shield, the shielding effectiveness should be a function of the frequency and position of the shield relative to the source. For plane waves and thin shields, the shielding effectiveness is not a strong function of frequency.

To determine the expression for the shielding effectiveness for near-field electric fields, a more general expression for the ratio of the incident to the transmitted field must be used. This expression is a function of the near-field wave impedance. For a good-conducting shield of thickness t, assuming that the impedance of the incoming signal is η_o,

$$\frac{E_{is}}{E_{ts}} = \frac{(\eta_o + \eta)^2}{4\eta_o \eta}\left[1 - \left(\frac{\eta_o - \eta}{\eta_o + \eta}\right)^2 e^{-2\frac{t}{\delta}} e^{-j2\frac{t}{\delta}}\right] e^{\frac{t}{\delta}} e^{j\frac{t}{\delta}} e^{-j\beta_o t}$$

Replacing the free-space impedance, η_o, with the near-field impedance results in the expression

$$\frac{E_{is}}{E_{ts}} = \frac{\left(\frac{\lambda_o \eta_o}{2\pi r} + \eta\right)^2}{4\frac{\lambda_o \eta_o}{2\pi r}\eta}\left[1 - \left(\frac{\frac{\lambda_o \eta_o}{2\pi r} - \eta}{\frac{\lambda_o \eta_o}{2\pi r} + \eta}\right)^2 e^{-2\frac{t}{\delta}} e^{-j2\frac{t}{\delta}}\right] e^{\frac{t}{\delta}} e^{j\frac{t}{\delta}} e^{-j\beta_o t} \tag{4.85}$$

When a shield is thin relative to the skin depth,

$$\frac{t}{\delta} \ll 1$$

Therefore, the two real exponentials that are raised to a small power, t/δ, are about one:

$$e^{-2\frac{t}{\delta}} \approx 1 \quad \text{and} \quad e^{\frac{t}{\delta}} \approx 1$$

Also, when the shield is a good conductor, its impedance is much less than that of an electric field in the near field:

$$|\eta| \approx \left|\sqrt{\frac{\omega\mu}{2\sigma}}(1+j)\right| << \frac{\lambda_o \eta_o}{2\pi r}$$

These approximations dramatically reduce the complexity of (4.85):

$$\frac{E_{is}}{E_{ts}} \approx \frac{\left(\frac{\lambda_o\eta_o}{2\pi r}\right)^2}{4\frac{\lambda_o\eta_o}{2\pi r}\eta}\left[1-\left(\frac{\frac{\lambda_o\eta_o}{2\pi r}}{\frac{\lambda_o\eta_o}{2\pi r}}\right)^2 e^{-j2\frac{t}{\delta}}\right] e^{j\frac{t}{\delta}} e^{-j\beta_o t} = \frac{\frac{\lambda_o\eta_o}{2\pi r}}{4\eta}\left(1-e^{-j2\frac{t}{\delta}}\right) e^{j\frac{t}{\delta}} e^{-j\beta_o t} \tag{4.86}$$

The complex exponential can be reduced to a simpler form by using Euler's formula:

$$e^{-jx} = \cos x - j\sin x \approx 1 - jx \quad \text{when } x \text{ is small}$$

($\cos x \approx 1$ and $\sin x \approx x$ if x is small). Furthermore, the magnitude of a complex exponential is always one:

$$|e^{-jx}| = \sqrt{\cos^2 x + \sin^2 x} = 1$$

Using these relationships, the magnitude of the previous expression reduces to a simple form:

$$\left|\frac{E_{is}}{E_{ts}}\right| \approx \left|\frac{\frac{\lambda_o\eta_o}{2\pi r}}{4\eta}\left[1-\left(1-j2\frac{t}{\delta}\right)\right]\right| \approx \left|\frac{\frac{\lambda_o\eta_o}{2\pi r}}{2\eta}\right|\frac{t}{\delta} = \left|\frac{\frac{\lambda_o\eta_o}{2\pi r}}{2\sqrt{\frac{\omega\mu}{2\sigma}}(1+j)}\right| t\sqrt{\pi f\mu\sigma}$$
$$= \frac{\frac{3\times 10^8\eta_o}{4f\pi r}}{\sqrt{\frac{2\pi f\mu}{\sigma}}} t\sqrt{\pi f\mu\sigma} \approx \frac{(6.36\times 10^9)t\sigma}{fr} = \frac{6.36\times 10^9}{R_s fr} \tag{4.87}$$

The low-frequency definition for the surface resistance was substituted in this expression. The shielding effectiveness in dB is then

$$TOT_{dB} = 20\log\left|\frac{E_{is}}{E_{ts}}\right| = 20\log\left(\frac{6.36\times 10^9}{R_s fr}\right) \approx 196 - 20\log(R_s fr) \tag{4.88}$$

$r := 0.0254$

$x := 40, 40.1 .. 90$

$$f(x) := \left(x + 1 - 10 \cdot \text{floor}\left(\frac{x}{10}\right)\right) \cdot 10^{\text{floor}\left(\frac{x}{10}\right)}$$

$$TOT_{dB}(f, R_s) := 196 - 20 \cdot \log(R_s \cdot f \cdot r)$$

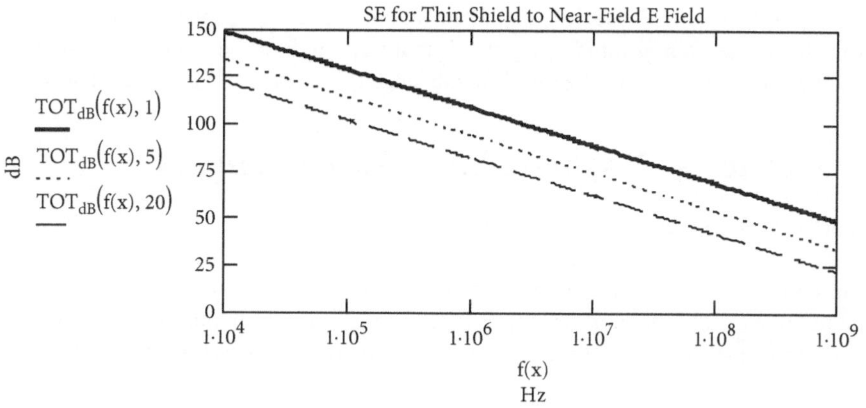

MATHCAD 4.18 SE for a thin shield to near-field electric fields vs. frequency as the surface resistance is varied.

The shielding effectiveness for a thin shield located 2.54 cm from the source and oriented so that the shield surface is perpendicular to the direction of propagation[8] is plotted in Mathcad 4.18. As previously noted, the shielding effectiveness is a function of the frequency for thin shields since the wave impedance is a function of the frequency. It should also be repeated that there are other nonradiating, energy-storage, fields that are present in the near field of electric field sources. The shielding of these field components is not included in the derived shielding effectiveness equations, (4.85) through (4.88).

4.14 An Equipotential Surface

The electrical properties of an object, frequency of the applied signal, and current magnitude through the object all determine whether its surface can be considered at a constant potential. Under what conditions may a surface be considered a constant potential?

For an object to have a reasonably constant potential (or equipotential) surface, one condition that must be satisfied is that the ratio of the conduction current to the displacement current must be much greater than one:

$$\frac{\sigma}{\omega\varepsilon} \gg 1$$

In other words, the material should be a good conductor. In most shielding applications, the shield is a metal and good conductor, and it easily satisfies this inequality.

A second condition that must be satisfied (in many cases) is that the object should be electrically small at the highest frequency of interest. When the shield is not electrically small, the voltage along the shield can vary even if the shield is a good conductor. Standing waves can exist along a metallic

[8]In the near field of most magnetic field sources, it is not always easy to know whether the direction of propagation is perpendicular to the shield's surface.

object (unless the object is somehow impedance "matched" at its "load" similar to setting $Z_L = Z_o$ for a transmission line). If standing waves exist, the object would not have an equipotential surface. The distributed inductance and capacitance, implied by the existence of the standing waves, allow the voltage to vary along a good-conducting surface in the same manner as along a mismatched transmission line. Of course, the voltage along a metal is also a function of the current along or through it, even at dc.

Sometimes several locations along a shield are connected to a ground or reference (preferably, at the same reference point) to "force" the potential along an electrically-long shield to be more uniform. The problem with this single-point grounding scheme is that the length of the grounding wires to the various locations on the shield may not be electrically small since the shield itself is not electrically small.

4.15 Electric vs. Magnetic Field Measurements

When a plane wave is normally incident to a good conductor, is it easier to measure the magnetic field or electric field near the conductor? [Smith, '77]

To determine the strength of a plane wave traveling in one direction, either the electric field (e.g., through the use of a dipole antenna) or magnetic field (e.g., through the use of a loop antenna) can be measured. Once one quantity is measured, the other can be determined via the impedance of the medium:

$$\eta = \frac{E}{H}$$

In practice, accurate measurement of these fields can be challenging. For example, nearby objects, including the ground and instrument, can produce reflections or scattering fields.

An appreciation for the effect of reflections and the difference between the electric field and magnetic field variation near a metallic surface can be obtained by analyzing a simple situation. Imagine that a plane wave, traveling in the $+z$ direction, is normally incident to a perfect ground plane located at $z = 0$. Assuming the field is polarized in the x direction, the electric and magnetic fields traveling in the $+z$ direction can be described as (for $z < 0$)

$$\vec{E}_{is} = A_s e^{-j\beta_o z}\hat{a}_x, \quad \vec{H}_{is} = \frac{A_s}{\eta_o} e^{-j\beta_o z}\hat{a}_y$$

where $\eta_o = 377\,\Omega$ and $\beta_o = 2\pi/\lambda_o$. The reflected signals, traveling in the $-z$ direction can be described as (for $z < 0$)

$$\vec{E}_{rs} = B_s e^{j\beta_o z}\hat{a}_x, \quad \vec{H}_{rs} = -\frac{B_s}{\eta_o} e^{j\beta_o z}\hat{a}_y$$

The total electric and magnetic fields in the frequency domain are then

$$\vec{E}_{is} + \vec{E}_{rs} = A_s e^{-j\beta_o z}\hat{a}_x + B_s e^{j\beta_o z}\hat{a}_x, \quad \vec{H}_{is} + \vec{H}_{rs} = \frac{A_s}{\eta_o} e^{-j\beta_o z}\hat{a}_y - \frac{B_s}{\eta_o} e^{j\beta_o z}\hat{a}_y \quad \text{for } z < 0$$

To determine the relationship between A_s and B_s, a boundary condition is required. One of the boundary conditions is that the total tangential electric field along a perfect conductor is zero:

$$E_{xis} + E_{xrs}\Big|_{z=0} = 0 = A_s e^{-j\beta_o(0)} + B_s e^{j\beta_o(0)} \quad \Rightarrow B_s = -A_s$$

Therefore, the total electric and magnetic fields are

$$\vec{E}_{is} + \vec{E}_{rs} = A_s(e^{-j\beta_o z} - e^{j\beta_o z})\hat{a}_x,\ \vec{H}_{is} + \vec{H}_{rs} = \frac{A_s}{\eta_o}(e^{-j\beta_o z} + e^{j\beta_o z})\hat{a}_y \quad \text{for } z < 0$$

Through the use of Euler's formula, $e^{jx} = \cos x + j\sin x$, these equations can be rewritten as

$$\vec{E}_{is} + \vec{E}_{rs} = -j2A_s \sin(\beta_o z)\hat{a}_x,\ \vec{H}_{is} + \vec{H}_{rs} = \frac{2A_s}{\eta_o}\cos(\beta_o z)\hat{a}_y \quad \text{for } z < 0 \tag{4.89}$$

The amplitude of the incident electric field is $|A_s|$, and the amplitude of the incident magnetic field is $|A_s/\eta_o|$. Near the perfect conductor,

$$\vec{E}_{is} + \vec{E}_{rs} \approx -j2A_s\beta_o z\hat{a}_x,\ \vec{H}_{is} + \vec{H}_{rs} \approx \frac{2A_s}{\eta_o}\hat{a}_y \quad \text{near } z = 0 \tag{4.90}$$

(The standard series approximations for the cosine and sine functions were used.) In the time domain, assuming $A_s = A$ is entirely real,

$$\vec{E}_i(t) + \vec{E}_r(t) = \text{Re}(-j2A\beta_o z\hat{a}_x e^{j\omega t}) = 2A\beta_o z\sin(\omega t)\hat{a}_x \quad \text{near } z = 0 \tag{4.91}$$

$$\vec{H}_i(t) + \vec{H}_r(t) = \text{Re}\left(\frac{2A}{\eta_o}\hat{a}_y e^{j\omega t}\right) = \frac{2A}{\eta_o}\cos(\omega t)\hat{a}_y \quad \text{near } z = 0 \tag{4.92}$$

These results are valid for a normally incident plane wave onto a perfect conductor. Of course, the signals are in sinusoidal steady state and, therefore, cannot be directly used for transient signals. Because of the approximations used in (4.90), these equations are valid if the distance between the test instrument and conductor is electrically small:

$$\beta_o z = \frac{2\pi}{\lambda_o} z \ll 1$$

Near the conductor, the ratio of the electric field amplitude to the magnetic field amplitude is much less than η_o:

$$\frac{2A\beta_o z}{\frac{2A}{\eta_o}} = \beta_o z\eta_o = \frac{2\pi}{\lambda_o} z\eta_o \ll \eta_o \tag{4.93}$$

The ratio of the electric field to magnetic field for a single traveling wave in free space is η_o. Therefore, compared to a single traveling wave, the magnetic field is much greater than the electric field near the conducting surface (or the electric field is much less than the magnetic field).

The electric field near the conductor varies with distance from the conductor while the magnetic field does not. If a short dipole probe were used to measure the incident electric field, its output voltage would vary with distance. If a small loop probe were used to measure the incident magnetic field, its output voltage would not vary (or, at least, not as much). It would seem that the measurement of the magnetic field is easier than the measurement of the electric field. However, there are many other factors that should be considered, including the sensitivity and impedance of the electric field probe or antenna vs. the magnetic field probe or antenna.

Ideally, if a lossy material that would not affect the incoming wave were available, then the temperature of the lossy material could be an indirect measure of the power flow and, hence, electric and magnetic field strength. The material should be lossy (e.g., containing graphite) so that the electric field could impart sufficient energy to the medium and increase its temperature.

The measurement of fields near a conducting surface is difficult. There are many sources of error including polarization uncertainty, coupling of the probe to nearby objects and other instrument "loading," and reflections and emissions from any connecting cables.

4.16 Single-Conductor Transmission Line

Can a single conductor act as a transmission line?

Yes! If the single conductor consists of a hollow conducting structure, then a single conductor can act like a transmission line. A hollow "single-conductor" transmission line designed to guide the field to its destination is referred to as a waveguide. Unlike the typical application of a two-conductor transmission line, waveguides pass certain field distributions above a specific frequency— the cutoff frequency. Below this cutoff frequency, the field is attenuated. This attenuation is useful in shielding.

A simple example involving a single-conductor transmission line is visible light passing through a pipe. Light, an electromagnetic wave, can travel through a hollow straight pipe as long as the pipe is sufficiently large in diameter. The pipe is acting as a waveguide.

4.17 TEM, TE, and TM Waves

What is the difference between a TEM, TE, and TM wave? Which of these wave types normally travel on a two-conductor transmission line? Which of these wave types cannot travel in a single-conductor transmission line?

The acronyms TEM, TE, and TM refer to Transverse Electromagnetic, Transverse Electric, and Transverse Magnetic, respectively. A simple example of a TEM wave is a plane wave. With a plane wave, both the electric and magnetic fields are transverse, or perpendicular, to the direction of propagation. (The air surrounding the wave, however, is not normally thought of as a transmission line.) With a lossless two-conductor transmission line with homogeneous medium surrounding the conductors, the fields are also usually TEM. The electric field is directed from one conductor to the other while the magnetic fields encircle each conductor. The direction of propagation is down the line, parallel to the conductors. Again, the electric and magnetic fields are transverse to the direction of propagation.

When a transmission line consists of two very large, good-conducting parallel plates, the fields between the two conductors can be TEM, TE, and TM. For the TEM wave, the electric field is directed from one plate to the other plate. This electric field is the normal or expected field for a transmission line. For the TE wave, the electric field is parallel to the two plates and transverse to the direction of propagation. However, this electric field parallel to the plates must vanish at the plates since the plates are good conductors and the tangential electric field must vanish along the good conductors. The magnetic field for this TE wave has both a transverse component, directed from one plate to the other plate, and a component in the direction of propagation, parallel to the plates. Thus, for this TE wave, the electric field is entirely transverse to the direction of propagation but the magnetic field is not. For TM waves, the magnetic field has only one component transverse to the direction of propagation, while the electric field has a component both transverse to and in the direction of propagation.

For TEM waves to be carried on a transmission system, *two or more* lossless (or low-loss) conductors must be involved with a homogeneous medium between them. The field distribution for a TEM wave is identical to the superposition of the electrostatic and magnetostatic fields. A waveguide cannot support a TEM wave since the static fields inside a hollow conducting structure must be zero: the total charge or current enclosed by the waveguide is zero. Therefore, a waveguide, a single-conductor structure, cannot carry a TEM wave.

For TE or TM waves to be carried on a transmission system, *one or more* conductors are required. A waveguide can carry both TE and TM waves. In addition to TEM waves, a coaxial cable, for example, at sufficiently high frequencies can also carry TE and TM waves.

4.18 Cutoff Frequency of a Waveguide

Beginning from Maxwell's equations, derive the cutoff frequency equation for a rectangular waveguide operating in the TE_{m0} mode. Generalize this result to TE_{mn} and TM_{mn} modes. [Kraus, '92]

The cutoff frequency can be determined from Maxwell's equations written in the frequency domain:

$$\nabla \times \vec{E}_s = -j\omega \vec{B}_s = -j\omega\mu \vec{H}_s$$

$$\nabla \times \vec{H}_s = \vec{J}_s + j\omega \vec{D}_s = \sigma \vec{E}_s + j\omega\varepsilon \vec{E}_s$$

$$\nabla \cdot \vec{D}_s = \nabla \cdot \varepsilon \vec{E}_s = \rho_V$$

$$\nabla \cdot \vec{B}_s = \nabla \cdot \mu \vec{H}_s = 0$$

If the medium inside the waveguide is lossless free space, then the current density, $\vec{J}_s$, and net volume charge density, ρ_V, are both zero and the permittivity and permeability are equal to their free-space values. In this case, Maxwell's equations reduce to

$$\nabla \times \vec{E}_s = -j\omega\mu_o \vec{H}_s \tag{4.94}$$

$$\nabla \times \vec{H}_s = j\omega\varepsilon_o \vec{E}_s \tag{4.95}$$

$$\nabla \cdot \varepsilon_o \vec{E}_s = 0 \tag{4.96}$$

$$\nabla \cdot \mu_o \vec{H}_s = 0 \tag{4.97}$$

It is actually quite easy to obtain one equation that is only a function of the electric field or magnetic field. If both sides of (4.94) are "multiplied" by the same vector mathematical operator, the curl, then

$$\nabla \times (\nabla \times \vec{E}_s) = \nabla \times (-j\omega\mu_o \vec{H}_s) = -j\omega\mu_o \nabla \times \vec{H}_s$$

and a vector identity used for the curl of the curl of a vector,

$$\nabla \times (\nabla \times \vec{E}_s) = \nabla \cdot (\nabla \cdot \vec{E}_s) - \nabla^2 \vec{E}_s = -j\omega\mu_o \nabla \times \vec{H}_s$$

Utilizing (4.95) and (4.96), this equation reduces to

$$\nabla \cdot (0) - \nabla^2 \vec{E}_s = -\nabla^2 \vec{E}_s = -j\omega\mu_o (j\omega\varepsilon_o \vec{E}_s) = \omega^2 \mu_o \varepsilon_o \vec{E}_s$$

$$\nabla^2 \vec{E}_s + \omega^2 \mu_o \varepsilon_o \vec{E}_s = 0 \tag{4.98}$$

This equation along with its companion equation

$$\nabla^2 \vec{H}_s + \omega^2 \mu_o \varepsilon_o \vec{H}_s = 0 \tag{4.99}$$

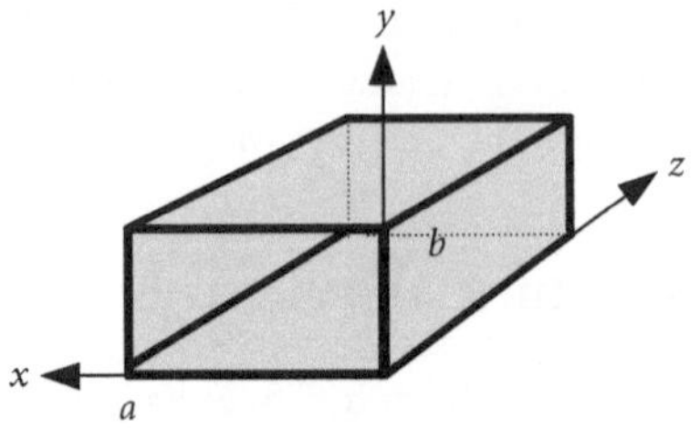

FIGURE 4.12 Waveguide of cross-sectional dimensions a by b.

are referred to as the wave equations. All possible sinusoidal steady-state fields in a free-space, current-free, charge density-free region must satisfy these equations.

In order to determine the cutoff frequency inside a rectangular waveguide of dimensions a by b, it is necessary to solve for the electric field (or magnetic field) inside of a waveguide. The rectangular waveguide shown in Figure 4.12 will be analyzed since it is the simplest case to study, and yet it still illustrates the important concepts that are also present with other waveguides, such as circular waveguides. To determine the electric field distribution inside the waveguide, the walls of the waveguide must be incorporated into the solution of (4.98). It is important to realize that many possible electric field distributions can exist inside the waveguide. One possible configuration, or mode, in the waveguide assumes that the electric field travels in the +z direction ($e^{-j\beta z}$) but is oriented in the y direction ($\hat{a}_y$):

$$\vec{E}_s = Ae^{-j\beta z}\hat{a}_y \tag{4.100}$$

The variable β is the phase constant. The tangential electric field along any of the walls must be zero (since the walls are assumed perfect conductors). The form of the electric field, (4.100), automatically satisfies the solution for zero tangential electric field along the bottom and top walls,

$$\begin{aligned} E_{xs}(y=0) &= 0 \\ E_{xs}(y=b) &= 0 \end{aligned} \tag{4.101}$$

since the field is entirely in the y direction. However, (4.100) is not automatically zero along the left and right walls since the field is y directed. Because the electric field must vanish at the walls, the amplitude, A, must be at least a function of x:

$$\vec{E}_s = E_y(x)e^{-j\beta z}\hat{a}_y \tag{4.102}$$

Again, this field is only one possible distribution that can exist inside the waveguide. One interesting characteristic of waveguides is that many field distributions can exist inside a waveguide simultaneously.

By substituting (4.102) into the wave equation for the electric field, the solution for the electric field, specifically $E_y(x)$, can be determined:

$$\nabla^2\vec{E}_s + \omega^2\mu_o\varepsilon_o\vec{E}_s = \left(\frac{\partial^2}{\partial x^2} + \frac{\partial^2}{\partial y^2} + \frac{\partial^2}{\partial z^2}\right)[E_y(x)e^{-j\beta z}\hat{a}_y] + \omega^2\mu_o\varepsilon_o E_y(x)e^{-j\beta z}\hat{a}_y = 0$$

$$\frac{d^2E_y(x)}{dx^2}e^{-j\beta z}\hat{a}_y + E_y(x)(-j\beta)^2e^{-j\beta z}\hat{a}_y + \omega^2\mu_o\varepsilon_o E_y(x)e^{-j\beta z}\hat{a}_y = 0$$

$$\frac{d^2E_y(x)}{dx^2} + \left[\left(\frac{\omega}{c}\right)^2 - \beta^2\right]E_y(x) = 0 \quad \text{where } c = \frac{1}{\sqrt{\mu_o\varepsilon_o}} = 3\times10^8 \text{ m/s} \tag{4.103}$$

This classical differential equation is of second order. The solution, not surprisingly, contains a cosine and sine function:

$$E_y(x) = F\cos(kx) + G\sin(kx) \quad \text{where } k = \pm\sqrt{\left(\frac{\omega}{c}\right)^2 - \beta^2} \tag{4.104}$$

To determine the variables F and G, it is necessary to apply the zero tangential field boundary conditions along the right and left walls:

$$E_y(0) = F\cos(k0) + G\sin(k0) = F = 0$$

$$E_y(a) = F\cos(ka) + G\sin(ka) = 0 + G\sin(ka) = 0$$

In order for $G\sin(ka)$ to be always zero, either G is zero (which is the trivial solution) or

$$\sin(ka) = 0 \quad \Rightarrow ka = m\pi \quad \Rightarrow k = \frac{m\pi}{a} = \pm\sqrt{\left(\frac{\omega}{c}\right)^2 - \beta^2} \tag{4.105}$$

where $m = 1, 2, 3, \ldots$ Solving for the phase constant,

$$\beta = \pm\sqrt{\left(\frac{\omega}{c}\right)^2 - \left(\frac{m\pi}{a}\right)^2} \tag{4.106}$$

Since the speed of a signal is a function of β, this result clearly shows that the speed of a signal in a waveguide is a function of both ω and m.

The electric field solution for this particular field distribution is

$$\vec{E}_s = G\sin(kx)e^{-j\beta z}\hat{a}_y \tag{4.107}$$

It satisfies the boundary conditions for all four walls of the waveguide. In the time domain, the electric field is

$$\begin{aligned}\vec{E}(t) &= \mathrm{Re}[G\sin(kx)e^{-j\beta z}\hat{a}_y e^{j\omega t}] = \mathrm{Re}[G\sin(kx)e^{j(\omega t - \beta z)}\hat{a}_y] \\ &= G\sin(kx)\cos(\omega t - \beta z)\hat{a}_y\end{aligned} \tag{4.108}$$

This wave is directed (polarized) in the y direction, traveling in the $+z$ direction, and varying in the x direction. However, in the frequency-domain to time-domain conversion, it was assumed that β was real. In other words,

$$\left(\frac{\omega}{c}\right)^2 \geq \left(\frac{m\pi}{a}\right)^2$$

$$\omega \geq \frac{m\pi}{a}c$$

The frequency, ω_c, is the cutoff frequency of this waveguide for this particular field distribution:

$$\omega_c = \frac{m\pi}{a}c \tag{4.109}$$

When the frequency of the signal in the waveguide is less than this critical cutoff frequency, then β is imaginary. To understand the implications of an imaginary β, rewrite this imaginary β as

$$\beta=-\sqrt{\left(\frac{\omega}{c}\right)^2-\left(\frac{m\pi}{a}\right)^2}=-\sqrt{-\left[\left(\frac{m\pi}{a}\right)^2-\left(\frac{\omega}{c}\right)^2\right]}=-j\sqrt{\left(\frac{m\pi}{a}\right)^2-\left(\frac{\omega}{c}\right)^2}=-j\beta_e \tag{4.110}$$

$$\text{where } \beta_e=\sqrt{\left(\frac{m\pi}{a}\right)^2-\left(\frac{\omega}{c}\right)^2}>0 \text{ and real}$$

and substitute it back into the electric field expression before converting to the time domain:

$$\vec{E}_s=G\sin(kx)e^{-j(-j\beta_e)z}\hat{a}_y=G\sin(kx)\,e^{-\beta_e z}\hat{a}_y \tag{4.111}$$

$$\vec{E}(t)=\mathrm{Re}[G\sin(kx)e^{-\beta_e z}\hat{a}_y e^{j\omega t}]=Ge^{-\beta_e z}\sin(kx)\cos(\omega t)\hat{a}_y \tag{4.112}$$

If the frequency of the signal is less than the cutoff frequency, it is attenuated as it passes down the waveguide. It has the form of a decaying oscillatory signal.

The waveguide acts like a high-pass filter. Frequencies greater than the cutoff frequency are passed while frequencies below the cutoff frequency are attenuated. The cutoff frequency for this particular configuration, (4.109), is a function of the dimension of the waveguide in the x direction and a function of the mode number m. As the size of the waveguide increases in the x direction, the cutoff frequency decreases; that is, the wavelength of the signals that can pass down the waveguide increases as the size of the waveguide increases. In layman's terms, the signal must be small enough to "fit" in the waveguide to travel down the waveguide. Above the cutoff frequency, the shape of the magnitude of the electric field along the x axis is a function of

$$|G\sin(kx)|=\left|G\sin\left(\frac{m\pi}{a}x\right)\right|$$

When the mode number $m = 1$, then

$$|G\sin(kx)|=\left|G\sin\left(\frac{\pi}{a}x\right)\right|$$

When the mode number $m = 2$, then

$$|G\sin(kx)|=\left|G\sin\left(\frac{2\pi}{a}x\right)\right|$$

The meaning of the mode number is apparent from these results. The mode number m is equal to the number of maximums in the amplitude of the field along the x direction. These maximums are illustrated in Figure 4.13 for $m = 1$, 2, and 3. As the mode number increases, the variation in the field increases. Also, it is interesting to compare the dimension a of the waveguide with the wavelength corresponding to the cutoff frequency:

$$\lambda_c=\frac{c}{f_c}=\frac{c}{\frac{\omega_c}{2\pi}}=\frac{c}{\frac{\frac{m\pi}{a}c}{2\pi}}=\frac{2a}{m} \tag{4.113}$$

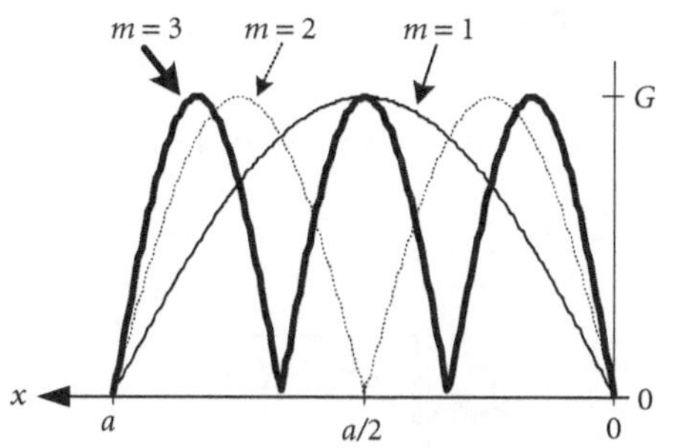

FIGURE 4.13 Variation in the magnitude of the electric field along the x axis of a waveguide as a function of the mode number.

For $m = 1$, the cutoff wavelength is $2a$. One-half of a wavelength of the signal at the cutoff frequency "fits" into the waveguide for this mode. At $m = 2$, the cutoff wavelength is a. A "full" one wavelength "fits" into the waveguide for this mode.

In the beginning of the derivation, it was stated that one possible field distribution inside the waveguide is of the form

$$\vec{E}_s = E_y(x)e^{-j\beta z}\hat{a}_y$$

Although it will not be shown, another possible solution to the wave equation inside the waveguide is

$$\vec{E}_s = E_x(x,y)e^{-j\beta z}\hat{a}_x + E_y(x,y)e^{-j\beta z}\hat{a}_y \quad (4.114)$$

This allows for field variation of the x and y directed electric field to be a function of both x and y. The cutoff frequency corresponding to this TE_{mn} mode is

$$\omega_c = \pi c\sqrt{\left(\frac{m}{a}\right)^2 + \left(\frac{n}{b}\right)^2} \quad (4.115)$$

where m corresponds to the number of maximums in the amplitude of the electric field in the x direction and n corresponds to the number of maximums in the amplitude of the electric field in the y direction. The combination mn is referred to as the mode number. The electric field is still entirely transverse to the direction of propagation, but it now varies sinusoidally in both the x and y directions. Again, all the necessary boundary conditions must be satisfied along all four walls of the waveguide. It is also important to note that the cutoff frequency corresponding to the TM_{mn} modes, when the magnetic field is transverse to the direction of propagation, is identical to (4.115).

4.19 Attenuation Beyond Cutoff

A waveguide with a length five times its diameter (or equivalent width) can be quite effective in attenuating certain signals. Determine the minimum attenuation of this structure in dB. What frequencies are attenuated by this structure? [Kraus, '92; Ramo]

If the frequency of the input signal to a rectangular waveguide is less than the cutoff frequency,

$$\omega_c = \pi c\sqrt{\left(\frac{m}{a}\right)^2 + \left(\frac{n}{b}\right)^2}$$

then the signal is attenuated. For a TE_{m0} signal operating below cutoff, the electric field is described by

$$\vec{E}(t) = Ge^{-\beta_e z}\sin(kx)\cos(\omega t)\hat{a}_y$$

where

$$\beta_e = \sqrt{\left(\frac{m\pi}{a}\right)^2 - \left(\frac{\omega}{c}\right)^2}$$

When below the cutoff frequency of a waveguide, a signal is referred to as beyond cutoff or evanescent (not effervescent).[9] The attenuation provided by a waveguide when the frequency of operation is below the cutoff frequency is the ratio of the signal at the input of the waveguide to the signal at the output of the waveguide. If the length of the waveguide is l_{th}, this attenuation ratio is (neglecting any reflections from the open ends)

$$attenuation = \frac{Ge^{-\beta_e(0)}\sin(kx)\cos(\omega t)}{Ge^{-\beta_e l_{th}}\sin(kx)\cos(\omega t)}$$

As long as the signals are measured at the same x position at the input and output of the guide and the same method is used to measure the signals (e.g., time-average or peak), this ratio reduces to

$$attenuation = \frac{1}{e^{-\beta_e l_{th}}} \tag{4.116}$$

or in dB

$$\begin{aligned} attenuation_{dB} &= 20\log\left(\frac{1}{e^{-\beta_e l_{th}}}\right) = 20\log 1 - 20\log\left(e^{-\beta_e l_{th}}\right) = 20\beta_e l_{th}\log e^1 \approx 8.69\beta_e l_{th} \\ &= 8.69 l_{th}\sqrt{\left(\frac{m\pi}{a}\right)^2 - \left(\frac{\omega}{c}\right)^2} = 8.69 l_{th}\sqrt{\left(\frac{\omega_c}{c}\right)^2 - \left(\frac{\omega}{c}\right)^2} \\ &= \frac{8.69 l_{th}\omega}{c}\sqrt{\left(\frac{\omega_c}{\omega}\right)^2 - 1} \end{aligned} \tag{4.117}$$

where

$$\omega_c = \frac{m\pi}{a}c \tag{4.118}$$

As the length of the guide increases, the attenuation increases. Also, as the ratio of the cutoff frequency to the signal frequency increases, the attenuation increases. If the square of this ratio is much greater than one, then

$$attenuation_{dB} \approx \frac{8.69 l_{th}\omega_c}{c} = \frac{8.69 l_{th} m\pi}{a} \approx 27m\frac{l_{th}}{a} \quad \text{if} \left(\frac{\omega_c}{\omega}\right)^2 \gg 1 \tag{4.119}$$

[9]The terminology "beyond cutoff" is not clear, but it usually implies a signal *wavelength* beyond the cutoff wavelength. In other words, operating "beyond cutoff" usually implies operating *below* the cutoff frequency.

When the length of the guide is five times the width of the guide,

$$attenuation_{dB} = \frac{8.69(5a)m\pi}{a} \approx 136m \tag{4.120}$$

When $m = 1$, the attenuation is about 136 dB.

For a general TE_{mn} or TM_{mn} mode, the same steps can be repeated. It is easy to show that the attenuation equation is identical to (4.117) with

$$\omega_c = \pi c\sqrt{\left(\frac{m}{a}\right)^2 + \left(\frac{n}{b}\right)^2} \tag{4.121}$$

The lowest cutoff frequency is when $m = 1$ and $n = 0$ if $a > b$, which is referred to as the dominant mode. The attenuation for the TE_{01} mode, assuming that the signal frequency is significantly less than the cutoff frequency, is from (4.119),

$$attenuation_{dB} \approx 27\frac{l_{th}}{a} \tag{4.122}$$

This result is not a function of the frequency. For other modes, the cutoff frequency and attenuation are greater.

For circular waveguides, the cutoff frequency is a more complicated expression involving the Bessel function. The mode with the lowest cutoff frequency is TE_{11}. Its cutoff frequency is

$$\omega_c = \frac{1.17\pi c}{d} \approx \frac{3.7c}{d} \tag{4.123}$$

where d is the diameter of the circular waveguide. Therefore, the minimum attenuation for signal frequencies much less than the dominant mode cutoff frequency is

$$attenuation_{dB} = \frac{8.69 l_{th}\omega}{c}\sqrt{\left(\frac{\frac{1.17\pi c}{d}}{\omega}\right)^2 - 1} \approx \frac{32 l_{th}}{d} \quad \text{if} \left(\frac{\omega_c}{\omega}\right)^2 >> 1 \tag{4.124}$$

This result is independent of the frequency.

It is common when using the previous equations to assume that the input and output are decoupled; that is, the amplitude of the reflection from the "open-circuited" impedance mismatch at the output is negligible after arriving at the input of the waveguide. If the thickness, l_{th}, is greater than the width, a or d, for the rectangular and circular waveguide, respectively, the one-way attenuation is at least 27 dB and 32 dB, respectively. If l_{th} is much less than a or d, then the opening is not acting as a waveguide but instead a slot antenna or an aperture.

4.20 Seepage through a Seam

A nonideal overlap joint is modeled as a waveguide operating beyond cutoff (below the cutoff frequency). How effective is this nonideal joint in attenuating electric fields perpendicular to the opening and parallel to the opening? Assume the overlap is long compared to its height and short circuited at both ends. [Boxleitner; product data sheets]

The cross-sectional view of the opening is shown in Figure 4.14 and Figure 4.15. The upper and lower plates touch at the side walls, idealized as flat end caps in this model. In Figure 4.14, the electric field is in the y direction and labeled E_y. In Figure 4.15, the electric field is in the x direction and labeled E_x. As shown in both of these figures, the electric field tangential to any of the good-conducting walls of the

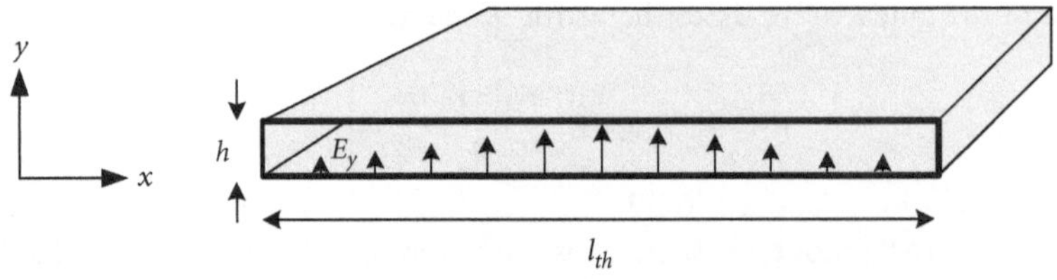

FIGURE 4.14 Electric field perpendicular to the longer cross-sectional dimension.

structure is small. Referring to these two figures, the *largest* wavelength signal that can pass for each of the field polarizations is

$$\frac{\lambda}{2} = l_{th} \quad \Rightarrow \lambda = 2l_{th} \text{ for } E_y \quad \text{and} \quad \frac{\lambda}{2} = h \quad \Rightarrow \lambda = 2h \text{ for } E_x$$

Therefore, the *smallest* frequency that can pass is

$$f = \frac{c}{\lambda} = \frac{c}{2l_{th}} \text{ for } E_y \quad \text{and} \quad f = \frac{c}{\lambda} = \frac{c}{2h} \text{ for } E_x \tag{4.125}$$

where $c = 3\times10^8$ m/s. Assuming the length of the opening or seam, l_{th}, is much greater than its height, h, the smallest frequency corresponding to the perpendicular field, E_y, is much less than the smallest frequency corresponding to the parallel field, E_x. A greater "number" of electric fields perpendicular to the opening will pass through it than electric fields parallel to the opening; in other words, the opening is more effective in attenuating electric fields parallel to it than perpendicular to it. This opening is acting like a waveguide with a cutoff frequency

$$\omega_c = \pi c\sqrt{\left(\frac{m}{l_{th}}\right)^2 + \left(\frac{n}{h}\right)^2} \tag{4.126}$$

Since $l_{th} >> h$, the lowest cutoff frequency is TE_{10}, corresponding to an electric field entirely in the y direction that is varying in the x direction.

The leakage into (and out of) an opening is a function of the area of the opening. However, the longest dimension of the opening determines the lowest frequency that can pass through it. One reference recommends that the longest dimension should be less than $\lambda/20$ or even $\lambda/50$. Except for very demanding shielding applications, the $\lambda/50$ guideline is probably excessive.

Doors, cracks, seams, joints, and other openings require special shielding techniques. For openings such as doors that frequently open and close, an enormous variety of products is available to provide various degrees of electrical closure. A summary of these products are given in Table 4.6. Generally, the mating

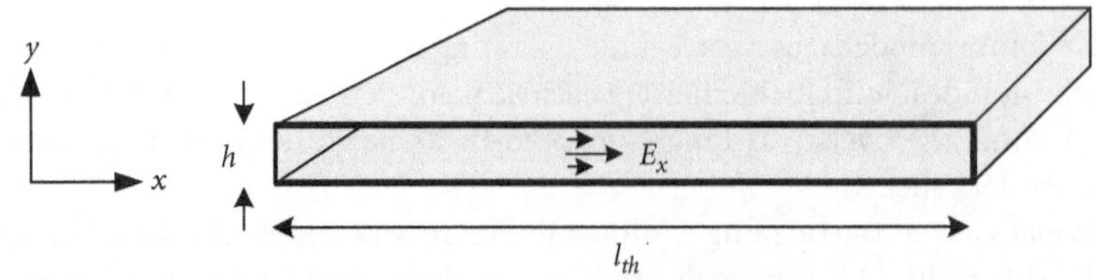

FIGURE 4.15 Electric field parallel to the longer cross-sectional dimension.

TABLE 4.6 Summary of Products for Electrically Sealing an Opening

Product	Description	Comments
Adhesive	Nickel-coated graphite, silver, silver-plated aluminum, silver-plated copper, and silver-plated nickel conductive particle-loaded sealant	Excellent adhesion with some flexibility, used to repair gaskets and adhere conductive tape and wire mesh to objects, cure time specified
Caulk	Copper and silver-plated glass particles in a resin	Easy to apply using conventional equipment including putty knife, excellent adhesion to metals, used to repair gaskets and fill-in cabinet openings, environmental seal, cure time specified
Epoxy	Two materials mixed to generate a silver-filled conductive paste	Used to join and bond (including dissimilar) metals and for circuit board trace repair, sometimes used as a "cold solder," working and cure times specified
Fingerstock	Beryllium-copper, phosphor bronze, and stainless steel flexible spring-like set of mating gaskets	Huge selection of sizes, shapes, profiles, and mounting methods are available, used for sealing doors, can be used in wiping action or in compression, compressibility of the gasket along its length should be considered
Gasket	Aluminum, carbon, monel,[a] nickel-coated graphite, silver, silver-plated aluminum, silver-plated copper, and silver-plated nickel embedded in an elastomer[b] or plastic	Die-cut, molded, and formed in-place gaskets available, sponge and solid densities, provide electrical and environmental seal, metal fillers mechanically weaken the material
Grease	Silver-filled silicone conductive viscous paste	Used on switches to reduce arcing and "hot" spots, usually thick, unlimited pot life, relatively inert and stable over broad environmental conditions
Knitted-wire fabric	Aluminum, beryllium copper, monel, phosphor bronze, stainless steel, and tin plated copper-clad steel knitted-wire mesh	Suitable for uneven surfaces with infrequent access, available in flat, round, and rectangular shapes, can also be knitted over an elastomer core, environmental seal not provided

(*Continued*)

TABLE 4.6 (Continued)

Product	Description	Comments
Tape	Aluminum, copper, copper-nickel, silver-plated, and tin-plated foil (and fabric) tape with conductive adhesive backing	Convenient to use, easy to wrap around cables and between transformer windings or to tape seams of cabinets and shielded rooms, environmental seal not generally provided

[a]Monel is an alloy containing copper, nickel, and other materials that are very resistant to corrosion.

[b]An elastomer is a polymer with rubber-like elastic characteristics.

surfaces should be clean, dry, oil-free, conductive, and, in some cases, flat and smooth.[10] Field leakage of these products can increase at very high frequencies if any openings in the product are not electrically small. For example, some loosely knitted-mesh gaskets can "leak" at a few GHz. Of course, because of the high conductivity of the materials used in these products, they can also reduce electrostatic buildup when used to bridge conductors that may be electrically isolated or floating. Depending on the frequency to be shielded, it is not always necessary to seal the entire opening. Whenever two conductors are in contact, the electrochemical compatibility of the materials should be considered; otherwise, the resultant corrosion can lead to unexpected noise (and possibly deterioration in the shielding over a long period of time). For those products containing solvents, such as those with some conducting adhesives, their potential reaction with any contacting materials, especially plastics, should be determined beforehand.

4.21 One Large Hole vs. Several Smaller Holes

Show that plane-wave leakage through a circular aperture on a metal shield is reduced by a factor of $10\log N$ when the aperture is divided into N smaller circular apertures having an equal total area. [Chatterton; Ramo]

When shielding, apertures such as windows, cracks, and ports can significantly reduce the effectiveness of the shield. The size of the aperture determines the degree of this shielding degradation. When a uniform plane wave is incident upon a metal shield with an electrically-small circular opening of radius a, as shown in Figure 4.16, the power transmitted through the opening is equal to

$$P_T = P_I G \tag{4.127}$$

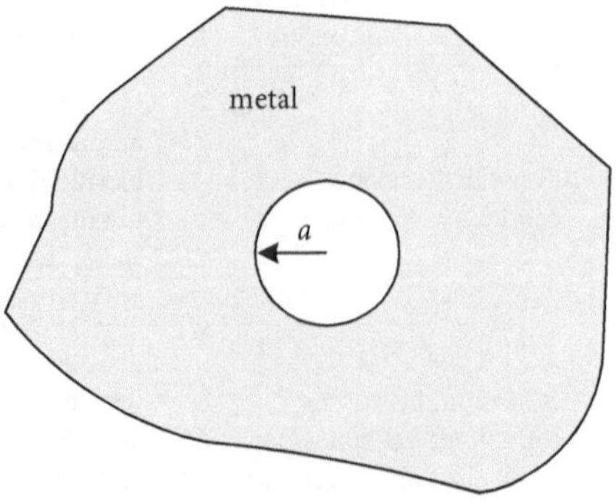

FIGURE 4.16 Large metal plate with an electrically-small circular opening.

[10]In some applications involving epoxies and adhesives, the surfaces should be slightly roughened up.

where P_I is the incident power to the opening and G is the "gain" of the opening. The incident power is the incident power density, P_D, multiplied by the area of the opening:

$$P_I = P_D(\pi a^2) \tag{4.128}$$

The gain of a circular aperture when uniformly illuminated by a plane wave is

$$G = \left(\frac{2\pi a}{\lambda}\right)^2 \tag{4.129}$$

The power transmitted through the opening is therefore

$$P_T = P_D(\pi a^2)\left(\frac{2\pi a}{\lambda}\right)^2 \tag{4.130}$$

If the aperture of radius a is broken into N smaller circular openings, the radius of these smaller openings should be

$$a_N = \frac{a}{\sqrt{N}} \tag{4.131}$$

if the sum of each of the N areas is equal to the area of the one larger opening:

$$\pi a^2 = N(\pi a_N^2)$$

In Figure 4.17, one large circular opening is broken into 16 smaller openings. The power transmitted through *one* of these smaller apertures is

$$P_{TN} = P_D\left[\pi\left(\frac{a}{\sqrt{N}}\right)^2\right]\left(\frac{2\pi\frac{a}{\sqrt{N}}}{\lambda}\right)^2 = \frac{1}{N^2}P_D(\pi a^2)\left(\frac{2\pi a}{\lambda}\right)^2 \tag{4.132}$$

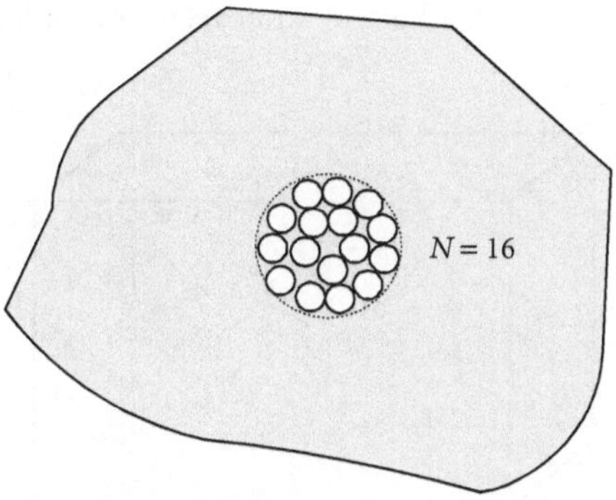

FIGURE 4.17 Large metal plate with many, closely spaced electrically-small circular openings.

Since the wavelength of the incident signal is large compared to *a*, the total power transmitted through *all* of these smaller apertures is then

$$NP_{TN} = \frac{1}{N} P_D(\pi a^2)\left(\frac{2\pi a}{\lambda}\right)^2 = \frac{1}{N} P_T \tag{4.133}$$

This assumes that the fields transmitted through the openings are coherent (or in phase). Equation (4.133) clearly indicates the advantage of subdividing a large opening into several smaller openings. The power leakage, or shielding effectiveness, is improved by

$$10\log\left(\frac{P_T}{NP_{TN}}\right) = 10\log N \tag{4.134}$$

For example, there is a 10 dB improvement for ten holes. This improvement should be expected since the gain of the circular aperture decreases as the radius of the opening decreases.

4.22 Honeycomb Ventilation Openings

A honeycomb metallic structure allows for ventilation in a shielded room while attenuating electromagnetic fields with frequencies below the cutoff frequency of each of the "waveguides" in the structure. Determine the shielding effectiveness of this structure for N waveguide openings. [Weston; Chatterton; Goedbloed]

A metallic enclosure with no openings is an excellent shield against plane waves (and near-field electric fields). An opening in the shield, unfortunately, often significantly reduces its shielding effectiveness. When the thickness of the shield is small compared to the opening width, then the signal power transmitted through the opening is a function of factors such as the gain of the aperture. When the thickness of the shield is not small, as shown in Figure 4.18, the opening also functions as a waveguide. If the wavelength of the incident uniform plane wave is comparable to (or larger than) the width of the opening, as previously discussed, then the waveguide is operating below its cutoff frequency. The signal is then attenuated as it travels down the opening.

When a larger opening is subdivided into N smaller openings, the shielding power loss is improved by a factor of $10\log N$. When the frequency is sufficiently low and shield is sufficiently thick, the holes act like waveguides operating below the cutoff frequency. The attenuation of the electric field for circular waveguides operating below the cutoff frequency was given in (4.124) as

$$attenuation_{dB} = \frac{32 l_{th}}{d}$$

where l_{th} is the thickness of the shield and d is the diameter of each hole. If this equation is used for the honeycomb structure, the total electric field attenuation relative to a hole with the same approximate

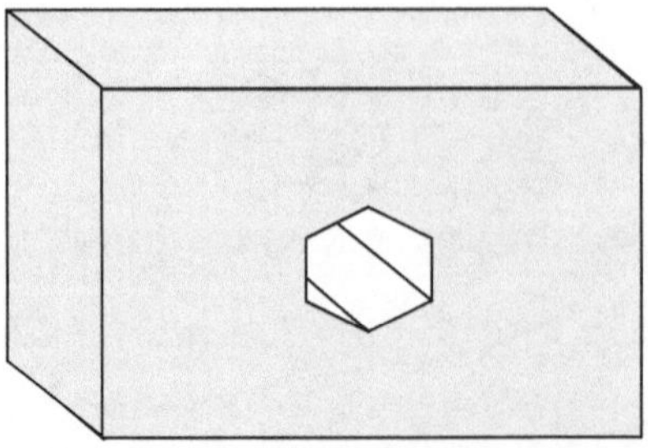

FIGURE 4.18 One aperture in a thick metallic plate.

size as the entire honeycomb structure is

$$total_{dB} \approx \frac{32 l_{th}}{d} + 20 \log N \qquad (4.135)$$

Again, this assumes that the signal output from each of the openings is in phase so that the powers can be directly added.

When a wave hits a honeycomb structure, like that shown in Figure 4.19, a percentage of the wave is reflected off the structure. The second term in (4.135) involving N is this loss. The waves that enter the structure that are below the cutoff frequency are attenuated. The first factor involving l_{th} is this loss for waves far below the cutoff frequency. Once the waves reach the output of the structure, a fraction of them are reflected. This loss is not included in this equation. For honeycomb filters where $l_{th} > 4d$, the *absorption* loss is greater than 100 dB. At these theoretical shielding levels, the shielding is frequently limited by nonidealities such as cracks, seams, and other openings. The reflection loss is often secondary in importance to the absorption loss.

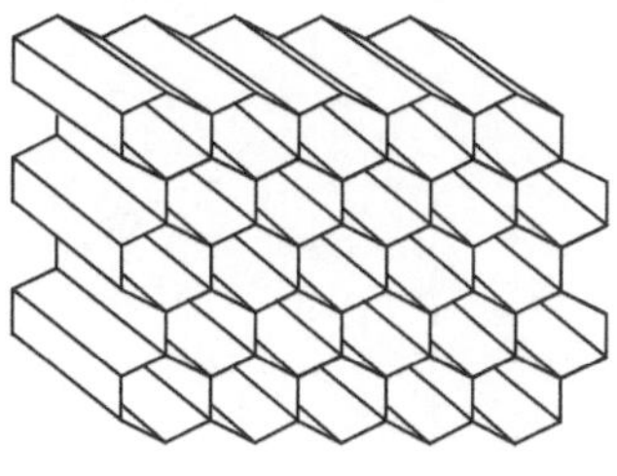

FIGURE 4.19 Honeycomb structure.

4.23 Coupling through an Aperture

First, referring to the Jackson reference, examine the classical simple expressions for the electric field and magnetic field through an aperture. Then, locate and discuss the reasonableness of the models used in the coupling of these electric and magnetic fields to a long conductor, parallel to the aperture. [Jackson; Lee, '77; Butler, '78; Lee, '86; Meulenaere; Cohn, '52; Taylor, '86; Weber]

Even very thin metallic shields of aluminum are extremely effective for reducing near-field electric fields (and far-field plane waves). Openings or apertures such as air vents, plastic viewing windows, cracks, or conductors entering the shield, however, dramatically weaken the shield. Simple relationships, which estimate the coupling between the external fields and any internal conductors, are helpful. These relationships are useful when dealing with electromagnetic pulses, or EMPs, from nuclear explosions.

Simple expressions exist for two of the components of the electric field for a circular aperture in a flat plane. Assume that far below a perfectly conducting plane, located at $z = 0$, the electric field is uniform with a value of $E_o \hat{a}_z$. (The sketch in Figure 4.20 does not show the local distortion of the electric field

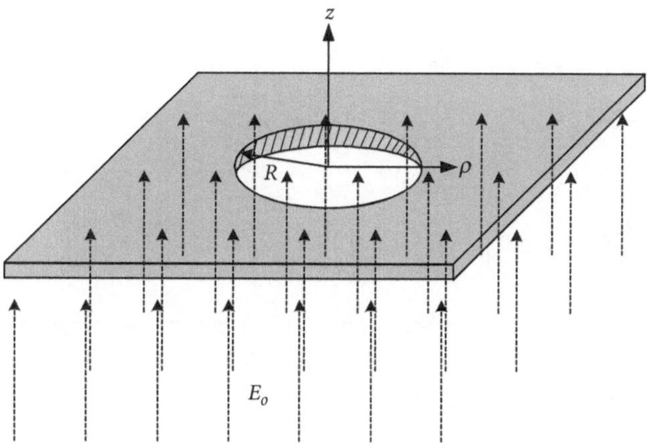

FIGURE 4.20 Perfectly conducting large plate with a circular aperture placed in a uniform electric field originating below the plate. The local distortion of the field near the aperture is not shown.

near the aperture.) The plane has a circular aperture or hole with a radius of R centered on the z axis. (The hole diameter is electrically short.) Free space exists everywhere around the plane including in the aperture opening. An integral expression is available for the potential distribution on both sides of the sheet. It reduces to a simple form along the z axis and along the $z = 0$ plane:

$$\Phi(0,z) = \frac{E_o}{\pi} R\left[1 - \frac{z}{R}\tan^{-1}\left(\frac{R}{z}\right)\right] \quad \text{if } 0 \le z \tag{4.136}$$

$$\Phi(\rho,0) = \frac{E_o}{\pi}\sqrt{R^2 - \rho^2} \quad \text{if } 0 \le \rho < R \tag{4.137}$$

$$\Phi(\rho,0) = 0 \quad \text{if } R \le \rho \tag{4.138}$$

The electric field along the z axis and the tangential electric field at $z = 0$ can be obtained by differentiation since $\vec{E} = -\nabla\Phi$:

$$E_z(0,z) = -\frac{d\Phi(0,z)}{dz} = \frac{RE_o}{\pi}\left[\frac{1}{R}\tan^{-1}\left(\frac{R}{z}\right) - \frac{1}{z\left(1 + \frac{R^2}{z^2}\right)}\right] \quad \text{if } 0 \le z \tag{4.139}$$

$$E_\rho(\rho,0) = -\frac{d\Phi(\rho,0)}{d\rho} = \frac{E_o\rho}{\pi\sqrt{R^2 - \rho^2}} \quad \text{if } 0 \le \rho < R \tag{4.140}$$

$$E_\rho(\rho,0) = 0 \quad \text{if } R \le \rho \tag{4.141}$$

Note that the tangential electric field at $z = 0$ is very large at $\rho = R^-$. Because of the sudden sharp edge at this location, this field is expected. (If the field is sufficiently strong, arcing is also most likely to occur at this sharp edge.) The tangential component at $z = 0$ for $\rho > R$ is zero since the electric field along a perfect conductor must be entirely normal to the surface. Two checks on the z component of the electric field along the z axis can be performed. For great distances from the hole, the electric field is zero:

$$E_z(0,z) \approx \frac{RE_o}{\pi}\left(\frac{1}{R}\frac{R}{z} - \frac{1}{z}\right) = 0 \quad \text{if } z \gg R$$

At the hole, the electric field is the average of the field values for $z \gg R$ and $z \ll -R$:

$$E_z(0,z) = \frac{RE_o}{\pi}\left[\frac{1}{R}\frac{\pi}{2} - \lim_{z\to 0}\left(\frac{z^2}{z^3 + R^2 z}\right)\right] = \frac{RE_o}{\pi}\left[\frac{1}{R}\frac{\pi}{2} - \lim_{z\to 0}\left(\frac{2z}{3z^2 + R^2}\right)\right] \quad \text{if } z = 0$$

$$= \frac{RE_o}{\pi}\left(\frac{1}{R}\frac{\pi}{2} - 0\right) = \frac{E_o}{2} \quad \text{if } z = 0$$

(L'Hospital's rule was used to determine the limit.) As is apparent in Mathcad 4.19, the electric field rapidly decreases along the z axis away from the hole. At a distance of one R from the center of the hole, the field is about one-tenth of the applied field:

$$E_z(0,R) = \frac{E_o}{\pi}\left(\frac{\pi}{4} - \frac{1}{2}\right) \approx 0.1E_o \quad \text{if } z = R$$

$R := 1 \qquad E_o := 1$

$z := 0, 0.01..\ 10$

$$E_z(z) := \frac{R \cdot E_o}{\pi} \cdot \left[\frac{1}{R} \cdot \operatorname{atan}\left(\frac{R}{z}\right) - \frac{1}{z \cdot \left(1 + \frac{R^2}{z^2}\right)} \right]$$

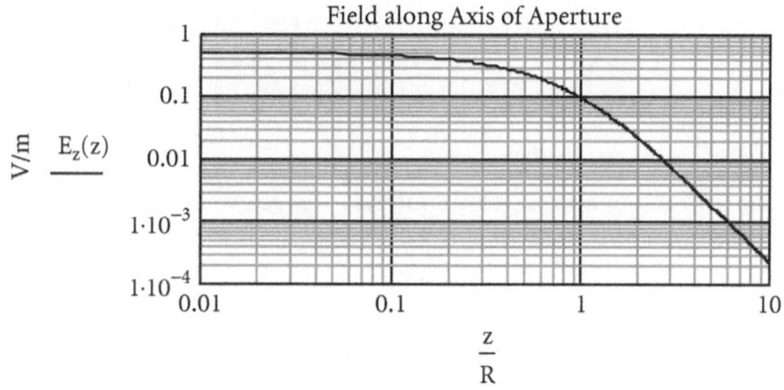

MATHCAD 4.19 Electric field along the axis of the circular aperture in Figure 4.20 vs. distance relative to the aperture radius.

Simple expressions exist for the components of the magnetic field in a circular aperture located in a flat plane. Assume that far below a perfectly conducting plane, located at $z = 0$, the magnetic field is uniform with a value of $H_o\hat{a}_y$. The plane has a circular aperture or hole with a radius of R centered on the z axis. (The hole diameter is electrically short.) Free space exists everywhere around the plane including in the aperture opening. Referring to Figure 4.21,

$$H_z(\rho, \phi, 0) = \frac{2H_o}{\pi} \frac{\rho}{\sqrt{R^2 - \rho^2}} \sin\phi \quad \text{if } 0 \le \rho < R \tag{4.142}$$

$$H_\rho(\rho, \phi, 0) = \frac{H_o}{2} \quad \text{if } 0 \le \rho < R \tag{4.143}$$

The plane must be perfectly conducting since dc magnetic fields pass through nonmagnetic conductors. Therefore, at low frequencies these expressions have limited use. (Note that these expressions are not a

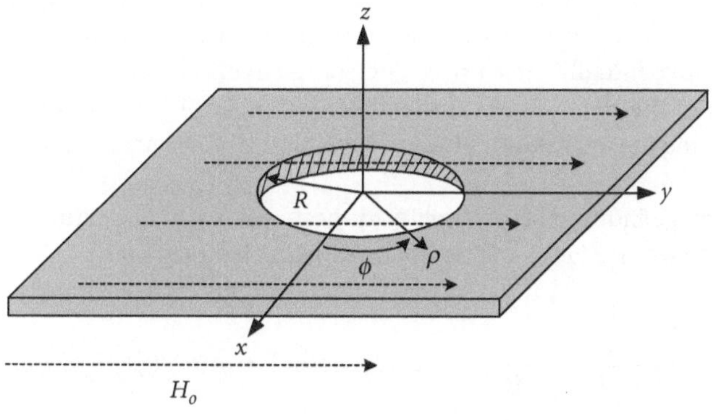

FIGURE 4.21 Perfectly conducting large plate with a circular aperture placed in a uniform magnetic field originating below the plate. The local distortion of the field near the aperture is not shown.

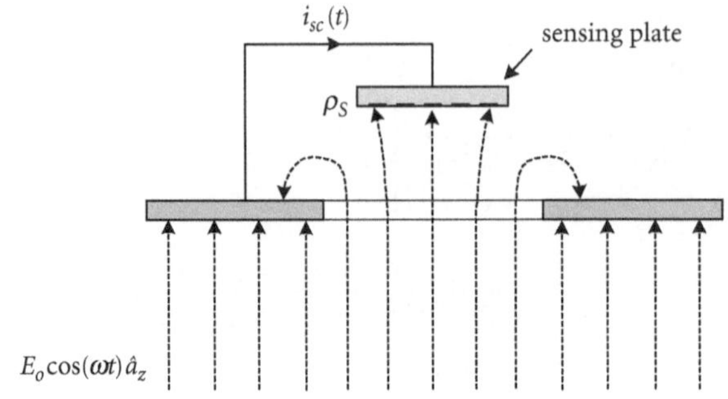

FIGURE 4.22 Electric field through the aperture induces charge on the sensing plate. The induced charge shown assumes the field is in the given direction.

function of the frequency, so they are not considering the effect of induced or "bucking" currents in the plate.)

Before the expressions for the induced short-circuit current and open-circuit voltage for a conductor near an aperture as shown in Figure 4.24 are given, very crude estimates for the current and voltage will be determined. Referring to Figure 4.22, imagine that a sensing plate of area *A* is located behind the aperture. The plate is symmetrically placed above the aperture opening, and its area is less than or about equal to the area of the aperture opening, πR^2. The applied electric field is slowly time varying so that electroquasistatic conditions can be assumed. Ignoring the variation of the electric field in both the *z* and ρ directions, the surface charge per unit area on the plate is given by, using the boundary condition $\rho_S = D_{n2} - D_{n1}$,

$$\rho_S = 0 - \varepsilon_o E_o \cos(\omega t) = -\varepsilon_o E_o \cos(\omega t) \tag{4.144}$$

Since the electric field in the conducting sensing plate is zero, $D_{n2} = 0$. The total charge on the plate is $q = A\rho_S$, where *A* is the area of the bottom side of the plate. The short-circuit current is

$$i_{sc}(t) = A\frac{d\rho_S}{dt} = A\omega\varepsilon_o E_o \sin(\omega t) \tag{4.145}$$

To check the sign of this current, (4.144) indicates that initially after $t = 0$, the charge on the plate is negative and decreasing in magnitude. Since charge is conserved, negative charge must be leaving the plate, which corresponds to positive current toward the plate. According to (4.145), the current toward the plate is indeed positive initially after $t = 0$. The short-circuit current increases with frequency and, of course, the strength of the electric field at the plate. Although (4.139) does not consider the effect of a sensing plate outside the opening, it does indicate the field will decrease with increasing distance from the aperture.

The open-circuit voltage induced across a wire connected across the aperture and oriented normal to the magnetic field as shown in Figure 4.23 is obtained using Faraday's law:

$$v_{oc}(t) = -\frac{d\Phi}{dt} = -\frac{d}{dt}[(h2R)\mu_o H_o \cos(\omega t)] = \omega(h2R)\mu_o H_o \sin(\omega t) \tag{4.146}$$

As with the electric field sensing plate, all variation in the magnetic field over the pickup area of the loop is neglected. The magnetic field is assumed uniform and normal to the surface of the loop. Actually,

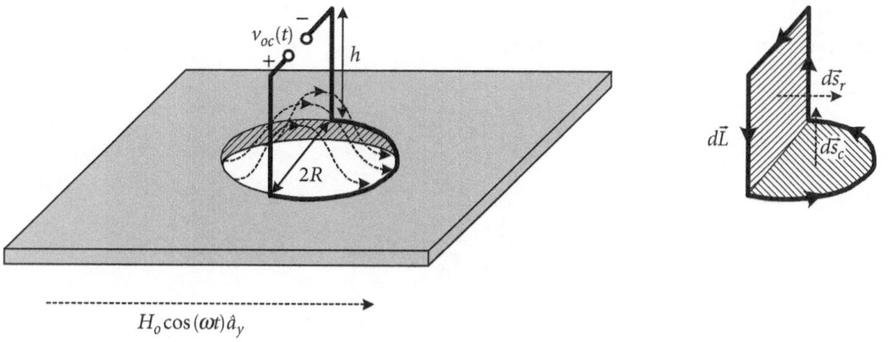

FIGURE 4.23 Open-circuit voltage induced across the loop is a function of the orientation (and shape and size) of the loop.

(4.146) also assumes that the loop is rectangular with dimensions h by $2R$, which it is not as seen in the figure. Assuming the uniform magnetic field is entirely normal to both the rectangular portion and half-circle portion of the closed loop, the voltage is

$$v_{oc}(t) = \omega\left(h2R - \frac{\pi R^2}{2}\right)\mu_o H_o \sin(\omega t) \tag{4.147}$$

In Figure 4.23, the direction of the two surface vectors is also shown for the given integration path along the loop. Since the half-circle portion $d\vec{s}_c$ is in the opposite direction to the magnetic field passing through it, its flux subtracts from the flux collected by the rectangular portion. When h is small compared to R,

$$v_{oc}(t) \approx -\omega\frac{\pi R^2}{2}\mu_o H_o \sin(\omega t) \quad \text{if } h \ll R \tag{4.148}$$

For the circular conductor shown in Figure 4.24, which is parallel to the large conducting plane, equations are available for the induced voltage and current. Far from the electrically-small aperture, the electric field is similar to that from an electric dipole, and the magnetic field is similar to that from a magnetic dipole. The expressions derived by Lee and Yang for the induced voltage and induced current

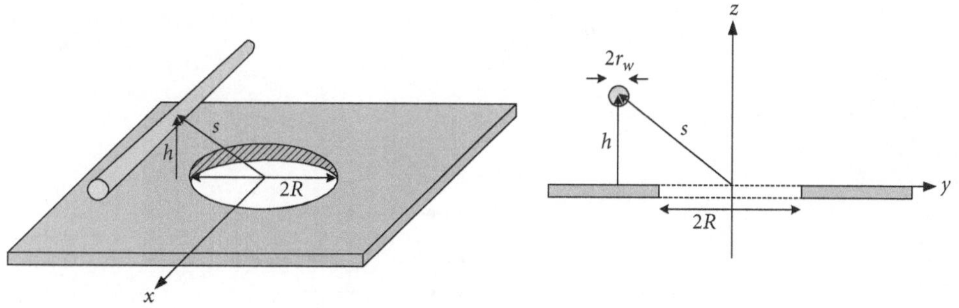

FIGURE 4.24 Long parallel conductor located a distance s from the center of the aperture.

on the conductor far from the hole, and the effective-loading inductance and capacitance of the hole on the conductor, are (in the frequency domain)

$$V_{ocs} = j\omega\mu_o\left(\frac{h}{\pi s^2}\right)\hat{a}_y \cdot \left(\overline{\overline{\alpha}}_m \cdot \vec{H}_{scs}\right) \quad \text{if } s >> R \tag{4.149}$$

$$I_{scs} = j\omega\varepsilon_o\left(\frac{h}{\pi s^2}\right)\frac{1}{Z_o}\sqrt{\frac{\mu_o}{\varepsilon_o}}\hat{a}_z \cdot (\alpha_{e,zz}\vec{E}_{scs}) \quad \text{if } s >> R \tag{4.150}$$

$$L_a = \mu_o\alpha_{m,yy}\left(\frac{h}{\pi s^2}\right)^2 \tag{4.151}$$

$$C_a = -\mu_o\frac{\alpha_{e,zz}}{Z_o^2}\left(\frac{h}{\pi s^2}\right)^2 \tag{4.152}$$

where Z_o is the characteristic impedance of the circular conductor above an infinite ground plane:

$$Z_o = 60\cosh^{-1}\left(\frac{h}{r_w}\right) \approx 60\ln\left(\frac{2h}{r_w}\right)$$

The equivalent circuit is shown in Figure 4.25. The vector $\vec{H}_{scs}$ is the magnetic field, in the frequency domain, tangential to the lower surface of the flat conductor when the aperture is not present. This field is also referred to as the short-circuited magnetic field since it corresponds to the field below the flat conductor when the aperture is short circuited. In Figure 4.21, $\vec{H}_{scs} = H_o\hat{a}_y$. Along a perfect conductor, the magnetic field is entirely tangential and can be in both the x and y directions for the conducting surface shown. The variable $\vec{E}_{scs}$ is the electric field, in the frequency domain, when the aperture is not present. It is referred to as the short-circuited electric field. Along a perfect conductor, the electric field must be entirely normal to the surface. In Figure 4.20, $\vec{E}_{scs} = E_o\hat{a}_z$.

The variable $\alpha_{e,zz}$ is the electric polarizability. Although the aperture appears like an electric dipole far from the opening, not surprisingly, the actual strength of the field at the conductor is a function of the size and shape of the aperture. The relationship between the magnetic field at the conductor and the aperture size and shape is more complicated since there are two components of the magnetic field tangential to the flat conductor. For this reason, the magnetic polarizability variable, $\overline{\overline{\alpha}}_m$, is a matrix:

$$\overline{\overline{\alpha}}_m = \begin{bmatrix} \alpha_{m,xx} & \alpha_{m,xy} \\ \alpha_{m,yx} & \alpha_{m,yy} \end{bmatrix} \tag{4.153}$$

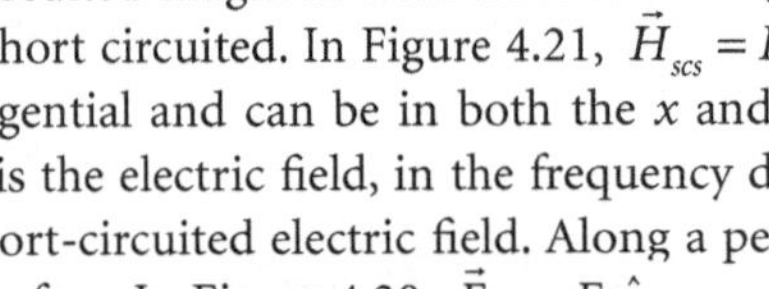

FIGURE 4.25 Model of the electric and magnetic coupling to the long conductor in Figure 4.24.

TABLE 4.7 Electric and Magnetic Polarizabilities for a Few Simple Symmetrically Shaped Apertures Located in the *xy* Plane ($K(e)$ and $E(e)$ are Complete Elliptic Integrals of the First and Second Kind, respectively)

	Shape	$\alpha_{e,zz}$	$\alpha_{m,xx}$	$\alpha_{m,yy}$
Circle	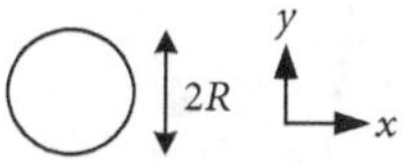	$\frac{2}{3}R^3$	$\frac{4}{3}R^3$	$\frac{4}{3}R^3$
Ellipse $e=\sqrt{1-\left(\frac{w}{l}\right)^2}$	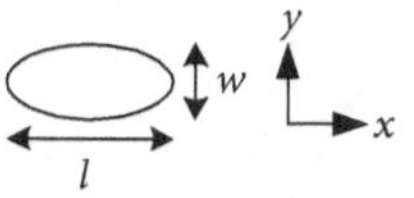	$\frac{\pi}{24}\frac{w^2 l}{E(e)}$	$\frac{\pi}{24}\frac{l^3 e^2}{K(e)-E(e)}$	$\frac{\pi}{24}\frac{l^3 e^2}{\left(\frac{l}{w}\right)^2 E(e)-K(e)}$
Narrow ellipse ($w \ll l$)	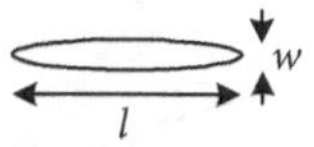	$\frac{\pi}{24}w^2 l$	$\frac{\pi}{24}\frac{l^3}{\ln\left(\frac{4l}{w}\right)-1}$	$\frac{\pi}{24}w^2 l$
Narrow slit ($w \ll l$)	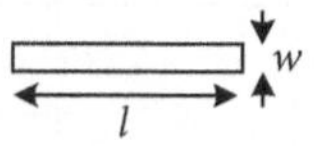	$\frac{\pi}{16}w^2 l$	$\frac{\pi}{24}\frac{l^3}{\ln\left(\frac{4l}{w}\right)-1}$	$\frac{\pi}{16}w^2 l$

For the symmetrical shapes provided in Table 4.7, $\overline{\overline{\alpha}}_m$ is a diagonal or $\alpha_{m,xy}=\alpha_{m,yx}=0$. For apertures of this type, where $\vec{H}_{scs}=H_{scxs}\hat{a}_x+H_{scys}\hat{a}_y$,

$$\overline{\overline{\alpha}}_m\cdot\vec{H}_{scs}=\begin{bmatrix}\alpha_{m,xx} & 0\\ 0 & \alpha_{m,yy}\end{bmatrix}\begin{bmatrix}H_{scxs}\\ H_{scys}\end{bmatrix}=\begin{bmatrix}\alpha_{m,xx}H_{scxs}\\ \alpha_{m,yy}H_{scys}\end{bmatrix}=\alpha_{m,xx}H_{scxs}\hat{a}_x+\alpha_{m,yy}H_{scys}\hat{a}_y$$

To develop a sense of the reasonableness of these equations, assume that the aperture is circular, $\vec{H}_{scs}=H_o\hat{a}_y$, $\vec{E}_{scs}=E_o\hat{a}_z$, and the conductor is directly above the aperture so $s=h$. Thus, (4.149)–(4.152) reduce to

$$V_{ocs}\approx j\omega\mu_o\frac{H_o}{2.4}\left(\frac{R^3}{h}\right),\quad I_{scs}\approx j\omega\varepsilon_o 1.3E_o\left[\frac{R^3}{h\ln\left(\frac{2h}{r_w}\right)}\right]$$

$$L_a\approx\mu_o\frac{1}{7.4}\left(\frac{R^3}{h^2}\right),\quad C_a\approx-\mu_o\frac{1}{53{,}000}\left\{\frac{R^3}{\left[h\ln\left(\frac{2h}{r_w}\right)\right]^2}\right\}$$

As expected, as R increases or h decreases, both of the induced sources increase. As the frequency increases, the coupling to the conductor also increases. The inductance and capacitance loading effect of the aperture can sometimes be neglected for large values of s relative to R. The negative shunt capacitance implies

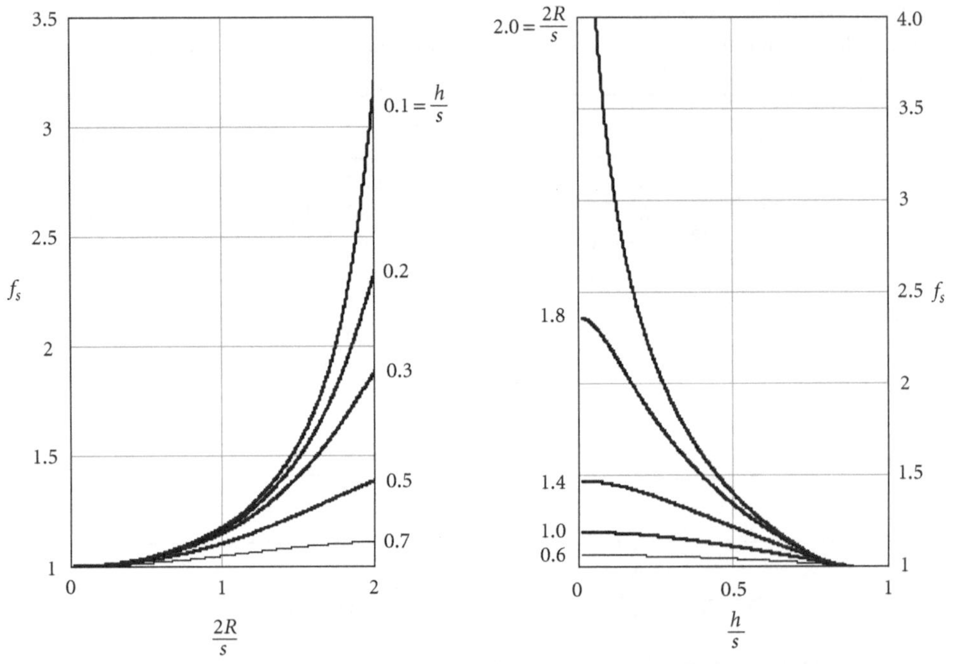

FIGURE 4.26 Correction factor as a function of the aperture radius, conductor height, and aperture-to-conductor spacing.

positive shunt inductance since

$$\frac{1}{j\omega(-|C_a|)} = \frac{j}{\omega|C_a|} = j\omega L_{eq,shunt} \quad \Rightarrow L_{eq,shunt} = \frac{1}{\omega^2|C_a|} \tag{4.154}$$

When the conductor is not far from the hole, a correction factor, f_s, has been derived for both the induced voltage and current sources. The new induced voltage is $f_s V_{ocs}$ and the new induced current is $f_s I_{scs}$. This correction factor is plotted in Figure 4.26. As seen in the left-most graph, for the range of h/s shown, if the distance from the center of the aperture to the conductor is greater than about the diameter of the aperture, $s > 2R$, the correction factor is close to one.

4.24 Radio in a Metal Box

A battery-powered radio is turned on, tuned to a strong AM station, and then placed inside a six-sided metal box with no openings (e.g., a steel tool box). Will the AM signal still be heard?

A common misconception is "because of Faraday shielding" that no electromagnetic signal can pass through the box. It is true that *stationary* charges outside of a conducting enclosure with no openings will not produce electric fields inside the enclosure, whether the enclosure is floating or grounded. This is referred to as Faraday shielding. However, the electromagnetic fields from an AM station are time varying. The charges on the broadcast antenna that produce these time-varying fields are not static but time varying.

Assuming that the antenna of the portable radio is not in contact with the inner surface of the metal box, the metal box will provide some shielding and reduction in the signal level received by the radio. For higher frequencies, such as FM stations, this attenuation will be greater. The quality of the reception depends on the dynamic range of the receiver and signal strength of the station. If the antenna is in contact with the inner surface of the metal enclosure or the enclosure has holes and other openings, the effectiveness of the shielding is generally reduced.

4.25 Lightning Protection Inside an Automobile

Discuss the validity of the statement, "When lightning hits a car, its passengers are safe since the tires insulate the car from the ground." Is the automobile electrically small? [Greenwald; Cote; Marshall, '73; Uman]

The inside of the automobile is a relatively safe haven during a lightning storm, if certain precautions are taken, but not because of its tires! Assuming that the roof and most of the body of the automobile is constructed of metal, the car body acts like a Faraday cage and shield. Recall that a metal enclosure or shell (with no openings) essentially "keeps out" external electrostatic fields. For time-varying fields, some of the field passes through the shield.

Lightning is a high-current, small rise time, transient energy signal. The fields from lightning are not electrostatic. If the lightning strikes nearby but not directly on the car, the electric and magnetic fields from the lightning can pass through the windows and other openings in the automobile. Obviously, this nearby strike is not as severe as a direct hit. There can be consequences of this nearby strike, depending on its proximity, including tingling sensations and eye damage. A nearby strike can also cause a potential difference or voltage gradient along the ground near the car. If an individual steps outside the vehicle, they can feel a shock due to a step voltage. If the lightning strikes the car directly, then the discharge current can pass along the car body, along glass surfaces (especially if they are wet or dirty), and even along tires. Although it tends to travel directly to the ground, it can take other unexpected paths depending on the impedance of these paths.

Because lighting is a transient pulse, it does not have a single frequency, but it does have an energy spectrum. Most of the energy is contained in the several MHz range. The higher frequency energy will travel mostly along the outer surface of the car body, and its depth of penetration is a function of the skin depth of the material of the car body.[11] The lower frequency portion of the lightning's energy, because of its large skin depth, can travel or diffuse through the metal body, and appear along the inner surface of the car body. An external surface current induces a tangential electric field (via the surface transfer impedance) in the body's metal interior. This can generate a voltage drop along certain parts of the interior of the car.

A metal body and roof does provide some protection for its passengers. However, it is prudent to follow the "hands in the lap" safety guideline. Because the lightning discharge will mostly follow along paths of low impedance to the earth, hands and other parts of the body should be within the automobile's protective Faraday cage. Safety groups also recommend that the windows be up. The windows provide another path to the lower portions of the car body. Although clean glass is a good insulator, the surface impedance of the windows is reduced by dirt, grease, and water. Also, by closing the windows, the conductive path that may result from rainwater wetting the interior of the car is reduced or eliminated. Touching different metallic elements in the car can cause a shock, especially if these elements have a direct connection to the outer body of the automobile. Therefore, to be safe, car radios, door handles, door locks, gearshifts, and other metallic or thinly coated metallic parts should not be touched.

If the rise time of the lightning discharge is about 0.1 μs, then the highest frequency of interest, assuming a double-exponential model of the discharge, is $1/(2\pi\tau_r) \approx 2$ MHz. This frequency has a free-space wavelength of about 150 m. One-tenth of this wavelength is 15 m, which is much larger than any

[11]Currently, metal roof and doors are about 1/32" (≈ 0.79 mm) thick. For sheet metal, assuming that $\mu_r \sigma_r = 100$, the skin depth at 100 kHz is about 0.02 mm and at 1 MHz is about 0.007 mm.

dimension of an automobile.[12] Thus, the automobile will appear electrically small to most of the energy contained in the lightning discharge.

A common myth is that the rubber tires on a car, since they are "insulating," prevent lightning from striking a car. Lightning, however, does hit cars. It is true that when the tires are dry, the tires may temporarily insulate the vehicle from the earth. As a result, the voltage of the vehicle will rise for a short time. Because of its intensity, however, the lightning will often arc or jump across the tires (or low parts of the vehicle), damaging one or more of the tires. When the tires are wet or dirty, the impedance of the path to ground is lower. Chains have been used to help discharge lightning (and electrostatic charges) and to provide a lower impedance to ground. Although a chain might help in providing a lower impedance path to ground, the path taken by lightning is probably a function of many parameters including the highest point on the vehicle.

Vehicles that have a nonmetallic (e.g., plastic) roof or no roof (e.g., convertible) offer no or very little protection from lightning. The lightning can puncture through plastic and thin materials (including sheet metal if it is sufficiently thin). The current generally follows conductive bodies in the car or the car frame to reach the earth. Although this discussion has concentrated on automobiles, the same principles apply to heavy equipment such as tractors and smaller equipment such as golf cars and riding lawnmowers. These vehicles are not immune to lightning strikes just because they are elevated from the earth via tires. The vehicles can provide some protection if they have metal cabins or enclosures.

[12]In Hollywood, very long extended or stretched limousines may not be electrically small.

References

Abramowitz, Milton, and Irene A. Stegun, *Handbook of Mathematical Functions with Formulas, Graphs, and Mathematical Tables*, National Bureau of Standards, 1964.

Abrie, Pieter L.D., *The Design of Impedance-Matching Networks for Radio-Frequency and Microwave Amplifiers*, Artech House, 1985.

Adamczewski, Ignacy, *Ionization, Conductivity, and Breakdown in Dielectric Liquids*, Barnes & Noble, 1969.

Adams, John W. and Dennis S. Friday, *Measurement Procedures for Electromagnetic Compatibility Assessment of Electroexplosive Devices*, IEEE Transactions on Electromagnetic Compatibility, Vol. 30, No. 4, November 1988.

Adler, Richard B., Lan Jen Chu, and Robert M. Fano, *Electromagnetic Energy Transmission and Radiation*, John Wiley & Sons, 1960.

Agarwal, A., C.J. Reddy, and M.Y. Joshi, *Coupled Bars in Rectangular Coaxial*, Electronics Letters, Vol. 25, No. 1, January 1989.

AIEE Committee Report, *Voltage Gradients Through the Ground Under Fault Conditions*, Transactions of AIEE, Vol. 77, Part III, October 1958.

Amin, M.B. and I.A. Benson, *Nonlinear Effects in Coaxial Cables at Microwave Frequencies*, Electronics Letters, Vol. 13, No. 25, December 1977.

Analog Devices, *Instrumentation Amplifier Data Sheets*, Analog Devices, 1999.

Anderson, Dan, *Electrostatic Discharge*, Richmond Technology, Videocassette, 1994.

Anderson, Edwin M., *Electric Transmission Line Fundamentals*, Reston, 1985.

Anderson, John M., *Wide Frequency Range Current Transformers*, The Review of Scientific Instruments, Vol. 42, No. 7, July 1971.

Anderson, Leonard R. and Jack Macneill, *Electric Machines & Transformers*, Prentice-Hall, 1988.

Andrew Corporation, *Andrew Catalog 37*, 1995.

ANSI/IEEE Grounding of Industrial and Commercial Power Systems, Standard 142, *Green Book*, 1982.

Arnold, A.H.M., *The Alternating-Current Resistance of Parallel Conductors of Circular Cross-Section*, Journal of IEE, Vol. 77, No. 463, July 1935.

ARRL Antenna Compendium: Vol. 4, American Radio Relay League, 1995.

ARRL Antenna Handbook, American Radio Relay League, 1997.

ARRL Handbook for Radio Amateurs, American Radio Relay League, 1996.

Asano, Kazutoshi, *On the Electrostatic Problem Consisting of a Charged Nonconducting Disk and a Metal Sphere*, Journal of Applied Physics, Vol. 48, No. 3, March 1977.

ASM International, *Electronic Materials Handbook, Volume 1: Packaging*, 1989.

ASM International, *Metals Handbook Vol. 2 Properties and Selection: Nonferrous Alloys and Special-Purpose Materials*, 1990.

ASM International, *Metals Handbook*, 1998.

Attwood, Stephen S., *Electric and Magnetic Fields*, Dover, 1967.

Auger, Raymond W., *The Relay Guide*, Reinhold, 1960.

Baboian, Robert, ed., *Corrosion Tests and Standards: Application and Interpretation*, ASTM, 1995.

Ballou, Glen M., ed., *Handbook for Sound Engineers*, Howard W. Sams, 1991.

Banks, Robert S. and Tieu Vinh, *An Assessment of the 5-mA 60 Hz Contact Current Safety Level*, IEEE Transactions on Power Apparatus and Systems, Vol. PAS-103, No. 12, December 1984.

Barbano, Normand, *The Aerodiscone Antenna*, The Microwave Journal, November 1966.

Bardell, P.R., *Magnetic Materials in the Electrical Industry*, Purnell and Sons, 1955.

Barke, Erich, *Line-to-Ground Capacitance Calculation for VLSI: A Comparison*, IEEE Transactions on Computer-Aided Design, Vol. 7, No. 2, February 1988.

Barnes, John R., *Electronic System Design: Interference and Noise Control Techniques*, Prentice-Hall, 1987.

Bazelyan, E.M. and Yu P. Raizer, *Spark Discharge*, CRC Press, 1998.

Beeman, Donald, ed., *Industrial Power Systems Handbook*, McGraw-Hill, 1955.

Belden Wire and Cable Company, *Master Catalog*, 1995.

Bell Telephone Laboratories, *Integrated Device and Connection Technology*, Vol. 3, Prentice-Hall, 1971.

Bell Telephone Laboratories, *Physical Design of Electronic Systems*, Vol. 1, Prentice-Hall, 1970.

Bendjamin, J., R. Thottappillil, and V. Scuka, *Time Varying Magnetic Fields Generated by Human Metal (ESD) Electrostatic Discharges*, Journal of Electrostatics, Vol. 46, 1999.

Benner, Linda S., T. Suzuki, K. Meguro, and S. Tanaka, ed., *Precious Metals: Science and Technology*, International Precious Metals Institute, 1991.

Berghe, S. Van den, F. Olyslager, D. De Zutter, J. De Moerloose, and W. Temmerman, *Power Plane Resonances as a Source of Delta-I Noise and the Influence of Decoupling Capacitors*, IEEE International Symposium on Electromagnetic Compatibility, 1997.

Bernstein, Theodore, *Electrical Shock Hazards and Safety Standards*, IEEE Transactions on Education, Vol. 34, No. 3, August 1991.

Bhartia, P., K.V.S. Rao, and R.S. Tomar, *Millimeter-Wave Microstrip and Printed Circuit Antennas*, Artech House, 1991.

Biddle Instruments, *"Getting Down to Earth…" A Manual on Earth-Resistance Testing for the Practical Man*, Biddle Instruments, 1981.

Bird, Dick, *The G4ZU "Bird Cage" Aerial*, CQ, April 1960.

Blackburn, John F., ed., *Components Handbook*, McGraw-Hill, 1949.

Blinchikoff, Herman J. and Anatol I. Zverev, *Filtering in the Time & Frequency Domains*, John Wiley & Sons, 1976.

Blood, William R., Jr., *MECL System Design Handbook*, Motorola Semiconductor Products, Mesa, Arizona, 1972.

Blume, L.F., A. Boyajian, G. Camilli, T.C. Lennox, S. Minneci, and V.M. Montsinger, *Transformer Engineering*, John Wiley & Sons, 1951.

Boast, Warren B., *Vector Fields*, Harper & Row, 1964.

Bodson, Dennis, *Electromagnetic Pulse and the Radio Amateur* — Part 4, QST, November 1986.

Bogle, A.G., *The Effective Inductance and Resistance of Screened Coils*, Journal of IEE, Vol. 87, No. 525, September 1940.

Bonsen, Robert, *Building a Semi-anechoic Chamber: An Overview*, ITEM 1999.

Bosch, *Interference Suppression* (Technical Instruction), Delta Press, Great Britain, 1979.

Bosich, Joseph F., *Corrosion Prevention for Practicing Engineers*, Barnes & Noble, 1970.

Bouwers, A. and P.G. Cath, *The Maximum Electric Field Strength For Several Simple Electrode Configurations*, Philips Technical Review, Vol. 6, No. 9, September 1941.

Boxleitner, Warren, *Electrostatic discharge and electronic equipment: a practical guide for designing to prevent ESD problems*, IEEE Press, 1989.

Bozorth, Richard M., *Ferromagnetism*, D. Van Nostrand, 1951.

Bradley, Fennimore N., *Materials for Magnetic Functions*, Hayden, 1971.

Brailsford, F., *Physical Principles of Magnetism*, D. Van Nostrand, 1966.

Brandes, E.A. and G.B. Brook, ed., *Smithells Metals Reference Book*, Butterworths-Heinemann, 1988.

Bridges, Jack E., *An Update on the Circuit Approach to Calculate Shielding Effectiveness*, IEEE Transactions on Electromagnetic Compatibility, Vol. 30, No. 3, August 1988.

Bridges, Jack E., *New Development in Electrical Shock Safety*, IEEE International Symposium on Electromagnetic Compatibility, 1994.

Britt, Scott D., David M. Hockanson, Fei Sha, James L. Drewniak, Todd H. Hubing, and Thomas P. Van Doren, *Effects of Gapped Groundplanes and Guard Traces on Radiated EMI*, IEEE International Symposium on Electromagnetic Compatibility, 1997.

Bronaugh, Edwin L. and William S. Lambdin, *Electromagnetic Interference Test Methodology and Procedures*, Interference Control Technologies, Vol. 6, 1988.

Brotherton, M., *Capacitors: Their Use in Electronic Circuits*, D. Van Nostrand, 1946.

Brown, Robert Grover, Robert A. Sharpe, and William Lewis Hughes, *Lines, Waves, and Antennas: The Transmission of Electric Energy*, Ronald Press, 1961.

Browne, Thomas E., Jr., ed., *Circuit Interruption: Theory and Techniques*, Marcel Dekker, 1984.

Burberry, R.A., *VHF and UHF Antennas*, The Redwood Press, 1992.

Burgett, Richard R., Richard E. Massoll, and Donald R. Van Uum, *Relationship Between Spark Plugs and Engine-Radiated Electromagnetic Interference*, IEEE Transactions on Electromagnetic Compatibility, Vol. EMC-16, No. 3, August 1974.

Butler, Chalmers M., *Capacitance of a Finite-Length Conducting Cylindrical Tube*, Journal of Applied Physics, Vol. 51, No. 11, November 1980.

Butler, Chalmers M., Yahya Rahmat-Samii, and Raj Mittra, *Electromagnetic Penetration Through Apertures in Conducting Surfaces*, IEEE Transactions on Antennas and Propagation, Vol. AP-26, No. 1, January 1978.

Cadick, John, Mary Capelli-Schellpfeffer, and Dennis Neitzel, *Electrical Safety Handbook*, McGraw-Hill, 2000.

Cahill, L.W., *Approximate Formulae for Microstrip Transmission Lines*, Proceedings of the I.R.E., Australia 35, No. 10, October 1974.

Campbell, George A. and Ronald M. Foster, *Fourier Integrals for Practical Applications*, D. Van Nostrand, 1948.

Campbell, Peter, *Comments on Energy Stored in Permanent Magnets*, IEEE Transactions on Magnetics, Vol. 36, No. 1, January 2000.

Caola, R.J., D.W. Deno, and V.S.W. Dymek, *Measurements of Electric and Magnetic Fields In and Around Homes Near a 500 kV Transmission Line*, IEEE Transactions on Power Apparatus and Systems, Vol. PAS-102, No. 10, October 1983.

Carlisle, Ben H., *A New Way to Prevent Electrocution*, Machine Design, Vol. 60, No. 7, April 7, 1988.

Carr, Clifford C., *Electric Machinery*, John Wiley & Sons, 1958.

Carr, Joseph J., *Automobile Electronics Servicing Guide*, Howard W. Sams, 1973.

Carr, Joseph J., *Practical Antenna Handbook*, McGraw-Hill, 1998.

Carson, John R., *A Generalization of the Reciprocal Theorem*, The Bell System Technical Journal, Vol. 3, July 1924.

Carson, John R., *Reciprocal Theorems in Radio Communication*, Proceedings of the IRE, Vol. 17, No. 6, June 1929.

Cebik, L.B., W4RNL, www.cebik.com, 2001.

Ceramic Source '86, The American Ceramic Society, 1985.

Chapman, Stephen J., *Electric Machinery Fundamentals*, McGraw-Hill, 1991.

Chatterjee, R., *Elements of Microwave Engineering*, Ellis Horwood Limited, 1986.

Chatterton, P.A. and M.A. Houlden, *EMC: Electromagnetic Theory to Practical Design*, John Wiley & Sons, 1992.

Chen, Wai-Kai and Run-Sheng Liang, *A General n-Port Network Reciprocity Theorem*, IEEE Transactions on Education, Vol. 33, No. 4, November 1980.

Cheng, David K., *Field and Wave Electromagnetics*, Addison-Wesley, 1989.

Chikazumi, Soshin and Stanley H. Charap, *Physics of Magnetism*, John Wiley & Sons, 1964.

Chipman, Robert A., *Theory and Problems of Transmission Lines*, Schaum's Outline Series, McGraw-Hill, 1968.

Chou, Chung-Kwang, Arthur W. Guy, and John A. McDougall, *Shielding Effectiveness of Improved Microwave-Protective Suits*, IEEE Transactions on Microwave Theory and Techniques, Vol. MTT-35, No. 11, November 1987.

Chow, Y.L. and M.M. Yovanovich, *The Shape Factor of the Capacitance of a Conductor*, Journal of Applied Physics, Vol. 53, No. 12, December 1982.

Christiansen, Donald, ed., *Electronic Engineers' Handbook*, McGraw-Hill, 1996.

Chubb, J.N. and G.J. Butterworth, *Charge Transfer and Current Flow Measurements in Electrostatic Discharges*, Journal of Electrostatics, Vol. 13, 1982.

Cleveland, Robert F., Jr. and Jerry L. Ulcek, *Questions and Answers about Biological Effects and Potential Hazards of Radiofrequency Electromagnetic Fields*, Federal Communications Commission, OET Bulletin 56, August 1999.

Cleveland, Robert F., Jr., David M. Sylvar, and Jerry L. Ulcek, *Evaluating Compliance with FCC Guidelines for Human Exposure to Radiofrequency Electromagnetic Fields*, Federal Communications Commission, OET Bulletin 65, August 1997.

Cogger, Steve and Hugh Hemphill, *Filter Selections Simplified with 0.1/100-Ohm Data*, ITEM 1997.

Cohn, Seymour B., *Characteristic Impedances of Broadside-Coupled Strip Transmission Lines*, IRE Transactions on Microwave Theory and Techniques, Vol. MTT-8, No. 6, November 1960.

Cohn, Seymour B., *The Electric Polarizability of Apertures of Arbitrary Shape*, Proceedings of the IRE, Vol. 40, No. 9, September 1952.

Collin, Robert E., *Antennas and Radiowave Propagation*, McGraw-Hill, 1985.

Collin, Robert E. and Francis J. Zucker, *Antenna Theory Part 1 and 2*, McGraw-Hill, 1969.

Collis, Mike, *Omni-Gain Vertical Collinear for VHF and UHF*, 73, August 1990.

COMAR, *Possible Health Hazards From Exposure to Power-Frequency Electric and Magnetic Fields — A COMAR Technical Information Statement*, IEEE Engineering in Medicine and Biology Magazine, Vol. 19, No. 1, January/February 2000.

Cooley, William W., *Low-Frequency Shielding Effectiveness of Nonuniform Enclosures*, IEEE Transactions on Electromagnetic Compatibility, Vol. EMC-10, No. 1, March 1968.

Coombs, Clyde F., ed., *Printed Circuits Handbook*, McGraw-Hill, 1988.

Cooper, W. Fordham and D.A. Dolbey Jones, *Electrical Safety Engineering*, 1993.

Cooperman, Gene, *A New Current-Voltage Relation for Duct Precipitators Valid for Low and High Current Densities*, IEEE Transactions on Industry Applications, Vol. 1A-17, No. 2, March/April 1981.

Cooperman, P., *A Theory for Space-Charge-Limited Currents with Application to Electrical Precipitation*, Transactions of the AIEE, Vol. 79, Part 1, 1960.

Copson, David A., *Microwave Heating*, AVI, 1975.

Corona, Paolo, Gaetano Latmiral, and Enrico Paolini, *Performance and Analysis of a Reverberating Enclosure with Variable Geometry*, IEEE Transactions on Electromagnetic Compatibility, Vol. EMC-22, No. 1, February 1980.

Corona, Paolo, John Ladbury, and Gaetano Latmiral, *Reverberation-Chamber Research — Then and Now: A Review of Early Work and Comparison With Current Understanding*, IEEE Transactions on Electromagnetic Compatibility, Vol. 44, No. 1, February 2002.

Corp, M. Bruce, *ZZAAP Taming ESD, RFI, and EMI*, Academic Press, 1990.

Corson, Dale R. and Paul Lorrain, *Introduction to Electromagnetic Fields and Waves*, W.H. Freeman, 1962.

Cote, Authur E., ed., *Fire Protection Handbook*, 1991.

Crawford, Myron L., *Generation of Standard EM Fields Using TEM Transmission Cells*, IEEE Transactions on Electromagnetic Compatibility, Vol. EMC-16, No. 4, November 1974.

Crawford, Myron L., John L. Workman, and Curtis L. Thomas, *Expanding the Bandwidth of TEM Cells for EMC Measurements*, IEEE Transactions on Electromagnetic Compatibility, Vol. EMC-20, No. 3, August 1978.

Creed, Frank C., *The Generation and Measurement of High Voltage Impulses*, Center Book Publishers, 1989.

Cross, Jean, *Electrostatics: Principles, Problems and Applications*, Adam Hilger, 1987.

Crowley, Joseph M., *Fundamentals of Applied Electrostatics*, John Wiley & Sons, 1986.

Crowley, Joseph M., *The Electrostatics of Static-Dissipative Worksurfaces*, Journal of Electrostatics, Vol. 24, 1990.

Cruft Laboratory Electronics Training Staff, *Electronic Circuits and Tubes*, McGraw-Hill, 1947.

Cullity, B.D., *Introduction to Magnetic Materials*, Addison-Wesley, 1972.

Cuming Microwave Corporation, *High Performance Broadbanded Pyramidal RF Absorber Data Sheet,* Technical Bulletin 390–1.

Cumming, W.A., *A Nonresonant Endfire Array for VHF and UHF,* IRE Transactions-Antennas and Propagation, Vol. AP-3, No. 2, April 1955.

Dalziel, C.F. and F.P. Massoglia, *Let-Go Currents and Voltages,* Transactions of AIEE, Vol. 75, Part II, 1956.

Dalziel, Charles F., *A Study of Hazards of Impulse Currents,* Transactions of AIEE, Vol. 72, Part III, 1953.

Dalziel, Charles F., *Electric shock hazard,* IEEE Spectrum, Vol. 9, No. 2, February 1972.

Dalziel, Charles F. and W.R. Lee, *Lethal Electric Currents,* IEEE Spectrum, Vol. 6, No. 2, February 1969.

Das, B. N., S.B. Chakrabarty, and K. Siva Ramo Rao, *Capacitance of Transmission Line of Parallel Cylinders in the Presence of Dielectric Coating,* IEEE Transactions on Electromagnetic Compatibility, Vol. 37, No. 1, 1995.

Das, Sisir K. and B.K. Sinha, *Numerical Solution of Higher Order Mode Cut-Off Frequencies in Symmetric TEM Cells Using Finite Element Method,* IEEE Transactions on Electromagnetic Compatibility, Vol. 32, No. 4, November 1990.

Dascalescu, Lucian, Patrick Ribardière, Claude Duvanaud, and Jean-Marie Paillot, *Electrostatic Discharges from Charged Spheres Approaching a Grounded Surface,* Journal of Electrostatics, Vol. 47, 1999.

Daugherty, Kevin M., *Analog-to-Digital Conversion: A Practical Approach,* McGraw-Hill, 1995.

Davidson, J.L., T.J. Williams, A.G. Bailey, and G.L. Hearn, *Characterisation of Electrostatic Discharges from insulating surfaces,* Journal of Electrostatics, Vol. 51–52, 2001.

Davis, Fred E., *Speaker Testing for Audibility Cables,* Audio, July 1993.

DeGauque, Pierre and Joel Hamelin, ed., *Electromagnetic Compatibility,* Oxford University Press, 1993.

Demarest, Kenneth R., *Engineering Electromagnetics,* Prentice-Hall, 1998.

DeMaw, M.F., *Ferromagnetic Core Design & Applications Handbook,* Prentice-Hall, 1981.

Deno, Don W., *Currents Induced in the Human Body by High Voltage Transmission Line Electric Field — Measurement and Calculation of Distribution and Dose,* IEEE Transactions on Power Apparatus and Systems, Vol. PAS-96, No. 5, September/October 1977.

Deno, Don W., *Electrostatic Effect Induction Formulae,* IEEE Transactions on Power Apparatus and Systems, Vol. PAS-94, No. 5, September/October 1975.

Destefan, Dennis E., *Calibration and Testing Facility for Resistance Welding Current Monitors,* IEEE Transactions on Instrumentation and Measurement, Vol. 45, No. 2, April 1996.

Devoldere, John, *Antennas and Techniques for Low-Band DXing,* ARRL, 1994.

Doetsch, Gustav, *Guide to the Applications of the Laplace and Z-Transforms,* Van Nostrand Reinhold, 1971.

Domininghaus, Hans, *Plastics for Engineers: Materials, Properties, Applications,* Hanser, 1988.

Donovan, John E., *Triboelectric Noise Generation in Some Cables Commonly Used with Underwater Electroacoustic Transducers,* The Journal of the Acoustical Society of America, Vol. 48, No. 3 (part 2), 1970.

Dorf, Richard D., ed., *The Electrical Engineering Handbook,* CRC Press, 1993.

Dorsett, Henry G., Jr. and John Nagy, *Dust Explosibility of Chemicals, Drugs, Dyes, and Pesticides,* Bureau of Mines, Report of Investigations 7132, 1968.

Drabowitch, S., A. Papiernik, H. Griffiths, and J. Encinas, *Modern Antennas,* Chapman & Hall, 1998.

Dummer, G.W.A., *Materials for Conductive and Resistive Functions,* Hayden, 1970.

Dunn, John M., *Local, High-Frequency Analysis of the Fields in a Mode-Stirred Chamber,* IEEE Transactions on Electromagnetic Compatibility, Vol. 32, No. 1, February 1990.

Durham, Marcus O., Karen D. Durham, and Robert A. Durham, *TVSS Design,* IEEE Industry Applications Magazine, Vol. 8, No. 5, September/October 2002.

Dwight, H.B., *Calculation of Resistances to Ground,* AIEE Transactions, Vol. 55, December 1936.

Dwight, Herbert B., *Effective Resistance of Isolated NonMagnetic Rectangular Conductors,* AIEE Transactions, Vol. 66, 1947.

Dworsky, Lawrence N., *Modern Transmission Line Theory and Applications,* John Wiley & Sons, 1979.

Eaton, Robert J., *Electric Power Transmission Systems,* Prentice-Hall, 1972.

Eggers, Bruce A., *An Analysis of the Balun,* QST, April 1980.

Eilers, Keith W., Mark Wingate, and Eric Pham, *Application and Safety Issues for Transient Voltage Surge Suppressors*, IEEE Transactions on Industry Applications, Vol. 36, No. 6, November/December 2000.

Electronic Materials Handbook: Volume 1 Packaging, ASM International, 1989.

Elliot, Robert S., *An Introduction to Guided Waves and Microwave Circuits*, Prentice-Hall, 1993.

Elliot, Robert S., *Antenna Theory and Design*, Prentice-Hall, 1981.

Elliott, S.R., *The Physics and Chemistry of Solids*, John Wiley & Sons, 1998.

Elmore, W.C., *The Transient Response of Damped Linear Networks with Particular Regard to Wideband Amplifiers*, Journal of Applied Physics, Vol. 19, January 1948.

Emerson, William H., *Electromagnetic Wave Absorbers and Anechoic Chambers Through the Years*, IEEE Transactions on Antennas and Propagation, Vol. AP-21, No. 4, July 1973.

Engdahl, Tomi, *Telephone Line Audio Interface Circuits*, www.hut.fi/Misc/Electronics/circuits/teleinterface.html, 1999.

Enthone Electronic Materials, *A Comparison of Conductive Coatings for EMI Shielding Applications*, 2001.

Everitt, W.L. and G.E. Anner, *Communication Engineering*, McGraw-Hill, 1956.

Fair-Rite Products Corp., *Fair-Rite Soft Ferrites*, 1994.

Fano, Robert M., Lan Jen Chu, and Richard B. Adler, *Electromagnetic Fields, Energy, and Forces*, John Wiley & Sons, 1960.

Felici, N.J., *Electrostatics and Electrostatic Engineering*, Proceedings of the Static Electrification Conference, Institute of Physics and the Physical Society, 1967.

Feng, James Q., *An analysis of corona currents between two concentric cylindrical electrodes*, Journal of Electrostatics, Vol. 46, 1999.

Fenn, John B., *Lean Flammability Limit and Minimum Spark Ignition Energy*, Industrial and Engineering Chemistry, Vol. 43, No. 12, December 1951.

Ferronics Incorporated, *Ferrite Cores Design Guide and Product Catalog*, 1991.

Ficchi, Rocco F., ed., *Practical Design for Electromagnetic Compatability*, Hayden, 1971.

Fink, Donald G. and H. Wayne Beaty, *Standard Handbook for Electrical Engineers*, McGraw-Hill, 1978.

Fischer Custom Communications, Inc., *Current Probes Catalog*, Torrance, CA.

Flanagan, William M., *Handbook of Transformer Applications*, McGraw-Hill, 1986.

Flügge, S., ed., *Electric Fields and Waves, Encyclopedia of Physics*, Vol. 16, Springer-Verlag, 1958.

Fowler, E.P., *Microphony of Coaxial Cables*, Proceedings of the IEE, Vol. 123, No. 10, October 1976.

Frankel, Sidney, *Characteristic Impedance of Parallel Wires in Rectangular Troughs*, Proceedings of the I.R.E., Vol. 30, No. 4, April 1942.

Freeman, Roger L., *Practical Data Communications*, John Wiley & Sons, 1995.

Freeman, Roger L., *Reference Manual for Telecommunication Engineering*, John Wiley & Sons, 1994.

Frerking, Marvin E., *Crystal Oscillator Design and Temperature Compensation*, Van Nostrand Reinhold, 1978.

Gaertner, Reinhold, Karl-Heinz Helling, Gerhard Biermann, Erich Brazda, Roland Haberhauer, Wilfried Koehl, Richard Mueller, Werner Niggemeier, and Bernhard Soder, *Grounding Personnel via the Floor/Footwear System*, Electrostatic Overstress/Electrostatic Discharge Symposium Proceedings, 1997.

Gandhi, O.P. and I. Chatterjee, *Radio-Frequency Hazards in the VLF and MF Band*, Proceedings of the IEEE, Vol. 70, No. 12, December 1982.

Gandhi, Om P. and Abbas Riazi, *Absorption of Millimeter Waves by Human Beings and Its Biological Implications*, IEEE Transactions on Microwave Theory and Techniques, Vol. MTT-34, No. 2, February 1986.

Gandhi, Om P., ed., *Biological Effects and Medical Applications of Electromagnetic Energy*, Prentice-Hall, 1990.

Gardner, Murray F. and John L. Barnes, *Transients in Linear Systems*, Vol. 1, John Wiley & Sons, 1942.

Gauger, J.R., *Household Appliance Magnetic Field Survey*, IEEE Transactions on Power Apparatus and Systems, Vol. PAS-104, No. 9, September 1985.

Geddes, Leslie A., ed., *Handbook of Electrical Hazards and Accidents*, CRC Press, 1995.

General Motors Corporation, *Automotive Component EMC Specification Overview of Test Requirements: GM9100P*, 1993.

Gerber, Eduard A. and Arthur Ballato, *Precision Frequency Control, Volume 2: Oscillators and Standards*, Academic Press, 1985.

German, Robert F., Henry W. Ott, and Clayton R. Paul, *Effect of an Image Plane on Printed Circuit Board Radiation*, IEEE International Symposium on Electromagnetic Compatibility, 1990.

Gibson, Norbert, *Static Electricity — An Industrial Hazard Under Control?*, Journal of Electrostatics, Vol. 40–41, 1997.

Glasband, Martin, *"Lifting" the Grounding Enigma*, MIX Magazine, November 1994.

Glasford, Glenn M., *Fundamentals of Television Engineering*, McGraw-Hill, 1955.

Glor, Martin, *Electrostatic Hazards in Powder Handling*, Research Studies Press, 1988.

Goedbloed, Jasper, *Electromagnetic Compatibility*, Prentice Hall, 1992.

Golde, R.H., *Lightning: Lightning Protection*, Vol. 2, Academic Press, 1977.

Goldman, Alex, *Modern Ferrite Technology*, Van Nostrand Reinhold, 1990.

Gray, Dwight E., ed., *American Institute of Physics Handbook*, McGraw-Hill, 1972.

Greason, William D., *Electrostatic Damage in Electronics: Devices and Systems*, Research Studies Press, 1987.

Greason, William D., *Generalized Model of Electrostatic Discharge (ESD) for Bodies in Approach: Analyses of Multiple Discharges and Speed of Approach*, Journal of Electrostatics, Vol. 54, 2002.

Greason, William D., *Quasi-static Analysis of Electrostatic Discharge (ESD) and the Human Body Using a Capacitance Model*, Journal of Electrostatics, Vol. 35, 1995.

Green, Harry E., *Determination of the Cutoff of the First Higher Order Mode in a Coaxial Line by the Transverse Resonance Technique*, IEEE Transactions on Microwave Theory and Techniques, Vol. 37, No. 10, October 1989.

Greene, Frank M., *NBS Field-Strength Standards and Measurements (30 Hz to 1000 MHz)*, Proceedings of the IEEE, Vol. 55, No. 6, June 1967.

Greenwald, E.K., ed., *Electrical Hazards and Accidents: Their Cause and Prevention*, Van Nostrand Reinhold, 1991.

Grigsby, L.L., ed., *The Electric Power Engineering Handbook*, CRC Press LLC, 2001.

Grossner, Nathan R., *Transformers for Electronic Circuits*, McGraw-Hill, 1983.

Grover, Frederick W., *Inductance Calculations*, Dover Publications, 1973.

Gruber, Mike, *Learn how the ARRL Laboratory Evaluates New Products — and What Those Numbers Mean to You!*, QST, October 1994.

Gunn, Ross, *The Electrification of Precipitation and Thunderstorms*, Proceedings of the IRE, Vol. 45, No. 10, 1957.

Guofu, Zhou and Gong Lian, *An Improved Analytical Model for Braided Cable Shields*, IEEE Transactions on Electromagnetic Compatibility, Vol. 32, No. 2, May 1990.

Gupta, K.C., Ramesh Garg, and I.J. Bahl, *Microstrip Lines and Slotlines*, Artech House, 1979.

Gupta, K.C., Ramesh Garg, and Rakesh Chadha, *Computer-Aided Design of Microwave Circuits*, Artech House, 1981.

Gupta, Someshwar C., *Transform and State Variable Methods in Linear Systems*, John Wiley & Sons, 1966.

Guru, Bhag Singh and Hüseyin R. Hiziroğlu, *Electromagnetic Field Theory Fundamentals*, PWS, 1998.

Gustafson, W.G., *Magnetic Shielding of Transformers at Audio Frequencies*, The Bell System Technical Journal, Vol. 17, No. 3, July 1938.

Guy, Arthur W., Chung-Kwang Chou, John A. McDougall, and Carrol Sorensen, *Measurement of Shielding Effectiveness of Microwave-Protective Suits*, IEEE Transactions on Microwave Theory and Techniques, Vol. MTT-35, No. 11, November 1987.

Haase, Heinz, *Electrostatic Hazards: Their Evaluation and Control*, Veriag Chemie-Weinheim, 1977.

Hagen, Jon B., *Radio-Frequency Electronics: Circuits and Applications*, Cambridge University Press, 1996.

Hall, Stephen H., Garret W. Hall, and James A. McCall, *High-Speed Digital System Design: A Handbook of Interconnect Theory and Design Practices*, John Wiley & Sons, 2000.

Hansen, R.C., *Shielding Formulas for Near Fields*, Microwave Journal, Vol. 43, No. 11, November 2000.

Hansson, Kjell, Monica Sandström, and Anders Johnsson, *Measured 50 Hz Electric and Magnetic Fields in Swedish and Norwegian Residential Buildings*, IEEE Transactions on Instrumentation and Measurement, Vol. 45, No. 3, June 1996.

Hara, Masanori and Masanori Akazaki, *A Method for Prediction of Gaseous Discharge Threshold Voltage in the Presence of a Conducting Particle*, Journal of Electrostatics, Vol. 2, 1976/1977.

Hare, Ed and Robert Schetgen, ed., *Radio Frequency Interference: How to Find It and Fix It*, Publication No. 149, American Radio Relay League, 1991.

Harper, Charles A., ed., *Handbook of Wiring, Cabling, and Interconnecting for Electronics*, McGraw-Hill, 1972.

Harper, Charles A., ed., *High Performance Printed Circuit Boards*, McGraw-Hill, 2000.

Harper, Charles A., ed., *Passive Electronic Component Handbook*, McGraw-Hill, 1997.

Harper, W.R., *Contact and Frictional Electrification*, Oxford University Press, 1967.

Harrington, Roger F., *Time-Harmonic Electromagnetic Fields*, McGraw-Hill, 1961.

Harris Corporation, *Applications Notes 9304, 9307, 9310, 9768, and 9771*, 1998–1999.

Harris, John W., and Horst Stocker, *Handbook of Mathematics and Computational Science*, Springer, 1998.

Hartmann, Irving, John Nagy, and Murray Jacobson, *Explosive Characteristics of Titanium, Zirconium, Thorium, Uranium and Their Hyrides*, Bureau of Mines, Report of Investigations 4835, 1951.

Hartmann, Irving, *Recent Research on the Explosibility of Dust Dispersion*, Industrial and Engineering Chemistry, Vol. 40, 1948.

Hasselgren, Lennart and Jorma Luomi, *Geometrical Aspects of Magnetic Shielding at Extremely Low Frequencies*, IEEE Transactions on Electromagnetic Compatibility, Vol. 37, No. 3, August 1995.

Hatsuda, Takeshi and Tadashi Matsumoto, *Computation of Impedance of Partially Filled and Slotted Coaxial Line*, IEEE Transactions on Microwave Theory and Techniques, Vol. MTT-15, No. 11, November 1967.

Haus, Herman A. and James R. Melcher, *Electromagnetic Fields and Energy*, Prentice-Hall, 1989.

Haussmann, Gary, Marty Matthews, and Franz Gisin, *Impact on Radiated Emissions on Printed Circuit Board Stitching*, IEEE International Symposium on Electromagnetic Compatibility, 1999.

Haykin, S.S., *Active Network Theory*, Addison-Wesley, 1970.

Hayt, William H., Jr. and John A. Buck, *Engineering Electromagnetics*, McGraw-Hill, 2001.

Hayward, W.H., *Introduction to Radio Frequency Design*, Prentice-Hall, 1982.

Heck, Carl, *Magnetic Materials and Their Applications*, Butterworth, 1974.

Heidelberg, E., *Generation of Igniting Brush Discharges by Charged Layers on Earthed Conductors*, Proceedings of the Static Electrification Conference, Institute of Physics and the Physical Society, 1967.

Heidler, F., J.M. Cvetić, and B.V. Stanić, *Calculation of Lightning Current Parameters*, IEEE Transactions on Power Delivery, Vol. 14, No. 2, April 1999.

Heiland, C.A., *Geophysical Exploration*, Prentice-Hall, 1946.

Helfrick, A.O., *Modern Aviation Electronics*, Prentice-Hall, 1994.

Heller, Peter, *Analog Demonstrations of Ampere's Law and Magnetic Flux*, American Journal of Physics, Vol. 60, No. 1, January 1992 (figure corrections in March 1992).

Helszajn, J., *Microwave Planar Passive Circuits and Filters*, John Wiley & Sons, 1994.

Hemming, LeLand H., *Architectural Electromagnetic Shielding Handbook: A Design and Specifications Guide*, IEEE Press, 1992.

Hemming, LeLand H., *Electromagnetic Anechoic Chambers: A Fundamental Design and Specifications Guide*, IEEE Press, 2002.

Heumann, Klemens, *Magnetic Potentiometer of High Precision*, IEEE Transactions on Instrumentation and Measurement, Vol. IM-15, No. 4, December 1966.

Heydt, G.T., *Electric Power Quality*, Stars in a Circle Publications, 1991.

Hilberg, Wolfgang, *Electrical Characteristics of Transmission Lines*, Artech House, 1979.

Hill, David A., *Electromagnetic Theory of Reverberation Chambers*, National Institute of Standards and Technology, Technical Note 1506, 1998.

Hill, David A., *Electronic Mode Stirring for Reverberation Chambers*, IEEE Transactions on Electromagnetic Compatibility, Vol. 36, No. 4, November 1994.

Hingorani, Narain G., *Introducing Custom Power*, IEEE Spectrum, Vol. 32, No. 6, June 1995.

Hippel, Arthur von, ed., *Dielectric Materials and Applications*, Artech House, 1995.

Hiziroğlu, Hüseyin R., Unpublished Results, January 1986.

Hoburg, J.F., *A Computational Methodology and Results for Quasistatic Multilayered Magnetic Shielding*, IEEE Transactions on Electromagnetic Compatibility, Vol. 38, No. 1, February 1996.

Hoburg, J.F., *Principles of Quasistatic Magnetic Shielding with Cylindrical and Spherical Shields*, IEEE Transactions on Electromagnetic Compatibility, Vol. 37, No. 4, November 1995.

Hoeft, Lothar O. and Joseph S. Hofstra, *Measured Electromagnetic Shielding Performance of Commonly Used Cables and Connectors*, IEEE Transactions on Electromagnetic Compatibility, Vol. 30, No. 3, August 1988.

Hoer, Cletus and Carl Love, *Exact Inductance Equations for Rectangular Conductors With Applications to More Complicated Geometries*, Journal of Research of the National Bureau of Standards–C. Engineering and Instrumentation, Vol. 69C, No. 2, April–June 1965.

Hoffmann, Reinmut K., *Handbook of Microwave Integrated Circuits*, Artech House, 1987.

Holloway, Christopher L. and Edward F. Kuester, *Net and Partial Inductance of a Microstrip Ground Plane*, IEEE Transactions on Electromagnetic Compatibility, Vol. 40, No. 1, February 1998.

Holloway, Christopher L., Paul M. McKenna, Roger A. Dalke, Rodney A. Perala, and Charles L. Devor, Jr., *Time-Domain Modeling, Characterization, and Measurements of Anechoic and Semi-Anechoic Electromagnetic Test Chambers*, IEEE Transactions on Electromagnetic Compatibility, Vol. 44, No. 1, February 2002.

Holloway, Christopher L., Ronald R. DeLyser, Robert F. German, Paul McKenna, and Motohisa Kanda, *Comparison of Electromagnetic Absorbers Used in Anechoic and Semi-Anechoic Chambers for Emissions and Immunity Testing of Digital Devices*, IEEE Transactions on Electromagnetic Compatibility, Vol. 39, No. 1, February 1997.

Holm, Ragnar, *Electric Contacts: Theory and Application*, Springer-Verlag, 1967.

Horenstein, Mark N., *Microelectronic Circuits and Devices*, Prentice-Hall, 1996.

Horowitz, Paul and Winfield Hill, *The Art of Electronics*, Cambridge University Press, 1989.

Horváth, T. and I. Berta, *Static Elimination*, Research Studies Press, 1982.

Hotte, P.W., G. Gela, J.D. Mitchell, Jr., and P.F. Lyons, *Electrical Performance of Conductive Suits*, IEEE Transactions on Power Delivery, Vol. 12, No. 3, July 1997.

Howard W. Sams, *Reference Data for Radio Engineers*, 1975.

Huang, Yi, *Conducting Triangular Chambers for EMC Measurements*, Measurement Science Technology, Vol. 10, 1999.

Hubing, Todd H., James L. Drewniak, Thomas P. Van Doren, and David M. Hockanson, *Power Bus Decoupling on Multilayer Printed Circuit Boards*, IEEE Transactions on Electromagnetic Compatibility, Vol. 37, No. 2, 1995.

Hubing, Todd H., Thomas P. Van Doren, and James L. Drewniak, *Identifying and Quantifying Printed Circuit Board Inductance*, IEEE International Symposium on Electromagnetic Compatibility, 1994.

Hughes, J.F., *Electrostatic Powder Coating*, Research Studies Press, 1984.

Hunt, William T., Jr. and Robert Stein, *Static Electromagnetic Devices*, Allyn and Bacon, 1963.

Hymers, W., *Star-delta Method of Earth Electrode Measurements*, Electrical Review, Vol. 196, No. 1, January 1975.

IEC Report Publication, *Effects of Current Passing Through the Human Body, Part 1: General Aspects*, Publication 479-1, Geneva: International Electrotechnical Commission, 1984.

Inan, Umran S. and Aziz S. Inan, *Engineering Electromagnetics*, Addison-Wesley, 1999.

IPCS, Environmental Health Criteria 137-Electromagnetic Fields (300 Hz to 300 GHz), International Programme on Chemical Safety-World Health Association, 1992.

IPCS, Environmental Health Criteria 69, International Programme on Chemical Safety-World Health Association, 1987.

Jackson, J.D., *Classical Electrodynamics*, John Wiley & Sons, 1975.

Jacobson, Murray, Austin R. Cooper, and John Nagy, *Explosibility of Metal Powders*, Bureau of Mines, Report of Investigations 6516, 1964.

Jacobson, Murray, John Nagy, and Austin R. Cooper, *Explosibililty of Dusts Used in the Plastics Industry*, Bureau of Mines, Report of Investigations 5971, 1962.

Jacobson, Murray, John Nagy, Austin R. Cooper, and Frank J. Ball, *Explosibililty of Agricultural Dusts*, Bureau of Mines, Report of Investigations 5753, 1961.

Janssen W. and F. Nilber, *High-Frequency Circuit Engineering*, IEE, 1996.

Jefimenko, Oleg D., *Electricity and Magnetism*, Electret Scientific, 1966.

Johnson, Howard and Martin Graham, *High-Speed Signal Propagation: Advanced Black Magic*, Prentice-Hall, 2003.

Johnson, Howard W. and Martin Graham, *High-Speed Digital Design: A Handbook of Black Magic*, Prentice-Hall, 1993.

Johnson, Richard C. and Henry Jasik, ed., *Antenna Engineering Handbook*, McGraw-Hill, 1961.

Johnson, Walter C., *Transmission Lines and Networks*, McGraw-Hill, 1950.

Jonassen, Niels, *Electrostatics*, Chapman & Hall, 1998.

Jones, T.B. and S. Chan, *Charge Relaxation in Partially Filled Vessels*, Journal of Electrostatics, Vol. 22, 1989.

Jones, Thomas B. and Jack L. King, *Powder Handling and Electrostatics: Understanding and Preventing Hazards*, Lewis, 1991.

Jones, Thomas B. and Kit-Ming Tang, *Charge Relaxation in Powder Beds*, Journal of Electrostatics, Vol. 19, 1987.

Jordon, Edward C. and Keith G. Balmain, *Electromagnetic Waves and Radiating Systems*, Prentice-Hall, 1968.

Joyner, Kenneth H., Paul R. Copeland, and Ian P. MacFarlane, *An Evaluation of a Radiofrequency Protective Suit and Electrically Conductive Fabrics*, IEEE Transactions on Electromagnetic Compatibility, Vol. 31, No. 2, May 1989.

Kaiser, Kenneth L., *A Study of Free-Surface Electrohydrodynamics*, Ph.D. Thesis, Purdue University, 1989.

Kallman, Raymond, *Realities of Wrist Strap Monitoring Systems*, Electrostatic Overstress/Electrostatic Discharge Symposium Proceedings, 1994.

Kam, Dong Gun, Heeseok Lee, Seungyong Baek, Bongcheol Park, and Joungho Kim, *Enhanced Immunity against Crosstalk and EMI Using GHz Twisted Differential Line Structure on PCB*, IEEE International Symposium on Electromagnetic Compatibility, 2002.

Kanda, Motohisa, *Electromagnetic-Field Distortion Due to a Conducting Rectangular Cylinder in a Transverse Electromagnetic Cell*, IEEE Transactions on Electromagnetic Compatibility, Vol. EMC-24, No. 3, August 1982.

Kaplan, Wilfred, *Operational Methods for Linear Systems*, Addision-Wesley, 1962.

Karrer, Nicolas and Patrick Hofer-Noser, *A New Current Measuring Principle for Power Electronic Applications*, Proceedings of ISPSD, 1999.

Keiser, Bernhard, *Principles of Electromagnetic Compatibility*, Artech House, 1987.

Keithley, *Low Level Measurements*, Keithley Test Instruments, 1993.

Kerr, Donald E., ed., *Propagation of Short Radio Waves*, McGraw-Hill, 1951.

Kessler, LeAnn and W. Keith Fisher, *A Study of the Electrostatic Behavior of Carpets Containing Conductive Yarns*, Journal of Electrostatics, Vol. 39, 1997.

Khalifa, M., ed., *High-Voltage Engineering: Theory and Practice*, Marcel Dekker, 1990.

Kibble, B.P. and G.H. Raymer, *Coaxial AC Bridges*, Adam Hilger, 1984.

Kimbark, Edward Wilson, *Electrical Transmission of Power and Signals*, John Wiley & Sons, 1949.

Kimmel, William D. and Daryl D. Gerke, *Electromagnetic Compatibility in Medical Equipment: A Guide for Designers and Installers*, IEEE, 1995.

Kimmel, William D., *Wide Frequency Impedance Modeling of EMI Ferrites*, IEEE International Symposium on Electromagnetic Compatibility, 1994.

King, Ronald W. P., *Transmission Line Theory*, McGraw-Hill, 1955.

Kinsman, Robert G., *Crystal Filters: Design, Manufacture, and Application*, John Wiley & Sons, 1987.

Klein, A.G., *Demonstration of Ampere's Circuital Law Using a Rogowski Coil*, American Journal of Physics, Vol. 43, No. 4, April 1975.

Kobb, Bennett Z., *Spectrum Guide: Radio Frequency Allocations in the United States 30 MHz-300 GHz*, New Signals Press, 1995.

Korn, Granino A., and Theresa M. Korn, *Mathematical Handbook for Scientists and Engineers*, McGraw-Hill, 1968.

Koucherng, Roger Lee, John E. Bennett, William H. Pinkston, and J.E. Bryant, *New Method for Assessing EED Susceptibility to Electromagnetic Radiation*, IEEE Transactions on Electromagnetic Compatibility, Vol. 33, No. 4, November 1991.

Koyler, John M. and Donald E. Watson, *ESD from A to Z: Electrostatic Discharge Control for Electronics*, Van Nostrand Reinhold, 1990.

Kraus, John D., *Antennas*, McGraw-Hill, 1988.

Kraus, John D., *Electromagnetics*, McGraw-Hill, 1992.

Krauss, Herbert L. and Charles W. Allen, *Designing Toroidal Transformers to Optimize Wideband Performance*, Electronics, August 16, 1973.

Krevelen, D.W. Van, *Properties of Polymers*, Elsevier, 1990.

Ku, Y.H., *Transient Circuit Analysis*, D. Van Nostrand, 1961.

Kumar, A., *Fixed and Mobile Terminal Antennas*, Artech House, 1991.

Kussy, Frank W. and Jack L. Warren, *Design Fundamentals for Low-Voltage Distribution and Control*, Marcel Dekker, 1987.

Lacanette, Kerry, *A Basic Introduction to Filters-Active, Passive, and Switched Capacitor*, National Semiconductor Application Note 779, April 1991.

Ladbury, John M. and Galen H. Koepke, *Reverberation Chamber Relationships: Corrections and Improvements or Three Wrongs Can (almost) Make a Right*, IEEE International Symposium on Electromagnetic Compatibility, 1999.

Ladbury, John M. and Kevin Goldsmith, *Reverberation Chamber Verification Procedures, or, How to Check if Your Chamber Ain't Broke and Suggestions on How to Fix It if It Is*, IEEE International Symposium on Electromagnetic Compatibility, 2000.

Ladbury, John M., Galen H. Koepke, and Dennis G. Camell, *Improvements in the CW Evaluation of Mode-Stirred Chambers*, IEEE International Symposium on Electromagnetic Compatibility, 1997.

Lago, Gladwyn V. and Donald L. Waidelich, *Transients in Electrical Circuits*, Ronald Press, 1958.

Lai, Fang-Shi and J. Kenneth Watson, *A Sectional Spheroid Model for Cylindrical Permanent Magnets*, IEEE Transactions on Magnetics, Vol. MAG-16, No. 2, 1980.

Laport, Edmund A., *Radio Antenna Engineering*, McGraw-Hill, 1952.

LaRocca, Robert L., *Personnel Protection Devices for Use on Appliances, IEEE Transactions on Industry Applications*, Vol. 28, No. 1, January/February 1992.

Larsen, Øystein, Janicke H. Hagen, and Kees van Wingerden, *Ignition of Dust Clouds by Brush Discharges in Oxygen Enriched Atmospheres*, Journal of Loss Prevention in the process industries, Vol. 14, 2001.

Lasitter, Homer A., *Low-Frequency Shielding Effectiveness of Conductive Glass*, IEEE Transactions on Electromagnetic Compatibility, Vol. EMC-6, No. 2, July 1964.

Lawrence, Ralph R., *Principles of Alternating-Current Machinery*, McGraw-Hill, 1953.

Lee, Capt. Paul H., *The Amateur Radio Vertical Antenna Handbook*, CQ Communications, 1984(a).

Lee, K.S.H and F.C. Yang, *A Wire Passing by a Circular Aperture in an Infinite Ground Plane*, EMP Interaction Notes, IN 317, Air Force Weapons Laboratory, Kirtland AFB, NM, February 1977.

Lee, K.S.H., *EMP Interaction: Principles, Techniques, and Reference Data, A Handbook of Technology from the EMP Interaction Notes*, Hemisphere, 1986.

Lee, Kai Fong, *Principles of Antenna Theory*, John Wiley & Sons, 1984(b).

Lee, Thomas H., *The Design of CMOS Radio-Frequency Integrated Circuits*, Cambridge University Press, 1998.

Leferink, Frank B.J. and Marcel J.C.M. van Doorn, *Inductance of Printed Circuit Board Planes*, IEEE International Symposium on Electromagnetic Compatibility, 1993.

Leferink, Frank B.J., *High Field Strength in a Large Volume: The Balanced Stripline TEM Antenna*, IEEE International Symposium on Electromagnetic Compatibility, 1998.

Leferink, Frank B.J., *Inductance Calculations; Methods and Equations*, IEEE International Symposium on Electromagnetic Compatibility, 1995.

Lefferson, Peter, *Twisted Magnetic Wire Transmission Line*, IEEE Transactions on Parts, Hybrids, and Packaging, Vol. PHP-7, No. 4, December 1971.

Lenk, John D., *Handbook of Electronic Circuit Designs*, Prentice-Hall, 1976.

Lerner, C.M., *Problems and Solutions in Electromagnetic Theory*, John-Wiley & Sons, 1985.

LeVasseur, Dave, *Midcom's Tips for Transformer Modeling*, Midcom Technical Note #82, 1998.

Liao, Samuel Y., *Microwave Devices and Circuits*, Prentice-Hall, 1990.

Licari, James, J., *Plastic Coatings for Electronics*, McGraw-Hill, 1970.

Lindgren, Erik A., *Contemporary RF Enclosures*, 1967.

Lo, Y.T. and S.W. Lee, ed., *Antenna Handbook*, Van Nostrand Reinhold, 1988.

Looms, J.S.T., *Insulators for High Voltages*, Peter Peregrinus, 1988.

Lopez, Michael, John L. Prince, and Andreas C. Cangellaris, *Influence of a Floating Plane on Effective Ground Plane Inductance in Multilayer and Coplanar Packages*, IEEE Transactions on Advanced Packaging, Vol. 22, No. 2, May 1999.

Lovatt, Howard C. and Peter A. Watterson, *Energy Stored in Permanent Magnets*, IEEE Transactions on Magnetics, Vol. 35, No. 1, January 1999.

Lubkin, Yale Jay, *Filter Systems and Design: Electrical, Microwave, and Digital*, Addison-Wesley, 1970.

Lüttgens, Günter and Norman Wilson, *Electrostatic Hazards*, Butterworth-Heinemann, 1977.

MIT Staff, *Magnetic Circuits and Transformers*, The MIT Press, 1943.

Ma, Mark T., Motohisa Kanda, Myron L. Crawford, and Ezra B. Larsen, *A Review of Electromagnetic Compatibility/Interference Measurement Methodologies*, Proceedings of the IEEE, Vol. 73, No. 3, March 1985.

Macalpine, W.W. and R.O. Schildknecht, *Coaxial Resonators with Helical Inner Conductor*, Proceedings of the IRE, Vol. 47, No. 12, December 1959.

Macatee, Stephen R., *Considerations in Grounding and Shielding Audio Devices*, Journal of Audio Engineering Society, Vol. 43, No. 6, June 1995.

Mager, A., *Magnetic Shielding Efficiencies of Cylindrical Shells with Axis Parallel to the Field*, Journal of Applied Physics, Vol. 39, No. 3, February 1968.

Mager, Albrecht J., *Magnetic Shields*, IEEE Transactions on Magnetics, Vol. MAG-6, No. 1, March 1970.

Magnusson, Philip C., *Transmission Lines and Wave Propagation*, Allyn and Bacon, 1970.

Magrab, Edward B. and Donald S. Blomquist, *The Measurement of Time-Varying Phenomena*, John Wiley & Sons, 1971.

Maissel, Leon I. and Reinhard Glang, *Handbook of Thin Film Technology*, McGraw-Hill, 1970.

Mardiguian, Michel, *Controlling Radiated Emissions by Design*, Van Nostrand Reinhold, 1992.

Mardiguian, Michel, *Electromagnetic Control in Components and Devices, Vol. 5*, Interference Control Technologies, 1988(a).

Mardiguian, Michel, *EMI Troubleshooting Techniques*, McGraw-Hill, 2000.

Mardiguian, Michel, *Grounding and Bonding, Vol. 2*, Interference Control Technologies, 1988(b).

Mardiguian, Michel, *Interference Control in Computers and Microprocessor-Based Equipment*, Don White Consultants, 1984.

Marshall, J.L., *Lightning Protection*, John Wiley & Sons, 1973.

Marshall, S.V., Richard E. DuBroff, and G.G. Skitek, *Electromagnetic Concepts and Applications*, Prentice-Hall, 1996.

Martin, R.E., *Electrical Interference in Electronic Systems: Its Avoidance within High-Voltage Substations and Elsewhere*, Research Studies Press, 1979.

Maruvada, P. Sarma, A. Turgeon, D.L. Goulet, and C. Cardinal, *An Experimental Study of Residential Magnetic Fields in the Vicinity of Transmission Lines*, IEEE Transactions on Power Delivery, Vol. 13, No. 4, October 1998.

Matisoff, Bernard S., *Wiring and Cable Designer's Handbook*, TAB, 1987.

Maurer, B., M. Glor, G. Lüttgens, and L. Post, *Hazards Associated with Propagating Brush Discharges on Flexible Intermediate Bulk Containers, Compounds and Coated Materials*, Electrostatics 1987, Institute of Physics Conference Series, No. 85, 1987.

McCaig, Malcolm, *Permanent Magnets in Theory and Practice*, John Wiley & Sons, 1977.

McCleery, D.K., *Introduction to Transients*, John Wiley & Sons, 1961.

McDonald, Jack R., Wallace B. Smith, and Herbert W. Spencer, III, *A Mathematical Model for Calculating Electrical Conditions in Wire-Duct Electrostatic Precipitation Devices*, Journal of Applied Physics, Vol. 48, No. 6, June 1977.

McElroy, P.K., *Designing Resistive Attenuating Networks*, Proceedings of the IRE, Vol. 23, No. 3, March 1935.

McGranaghan, Mark F., David R. Mueller, and Marek J. Samotyj, *Voltage Sags in Industrial Systems*, IEEE Transactions on Industrial Applications, Vol. 29, No. 2, March/April 1993.

McLachlan, N.W., *Bessel Functions for Engineers*, Oxford University Press, 1934.

Measures, Richard L., *A Balanced Antenna Tuner*, QST, February 1990.

Meeldijk, Victor, *Electronic Components*, John Wiley & Sons, 1995.

Melcher, James R., *Continuum Electromechanics*, MIT Press, 1981.

Meliopoulos, A.P. Sakis, *Power System Grounding and Transients: An Introduction*, Marcel Dekker, 1988.

Metaxas, A.C. and R.J. Meredith, *Industrial Microwave Heating*, Peter Peregrinus, 1983.

Meulenaere, Filip De and Jean Van Bladel, *Polarizability of Some Small Apertures*, IEEE Transactions on Antennas and Propagation, Vol. AP-25, No. 2, March 1977.

Mild, Kjell Hansson, *Occupational Exposure to Radio Frequency Electromagnetic Fields*, Proceedings of the IEEE, Vol. 68, No. 1, January 1980.

Mills, Jeffrey P., *Electro-Magnetic Interference Reduction in Electronic Systems*, Prentice-Hall, 1993.

Miner, Gayle F., *Lines and Electromagnetic Fields for Engineers*, Oxford University Press, 1996.

Mitchell, M.J., *Analyzing the Double Exponential Pulse*, EMC Test & Design, Vol. 5, No. 3, March 1994.

Mizuno, A., *Electrostatic Precipitation*, IEEE Transactions on Dielectrics and Electrical Insulation, Vol. 7, No. 5, October 2000.

Montrose, Mark I., *EMC and the Printed Circuit Board*, IEEE Press, 1999.

Montrose, Mark I., *Printed Circuit Board Techniques For EMC Compliance*, IEEE Press, 2000.

Moon, Parry and Domina Eberle Spencer, *Field Theory for Engineers*, D. Van Nostrand, 1961.

Moon, Parry and Domina Eberle Spencer, *Foundations of Electrodynamics*, D. Van Nostrand, 1960.

Moore, A.D., ed., *Electrostatics and Its Applications*, John Wiley & Sons, 1973.

Moore, Richard K., *Traveling-Wave Engineering*, McGraw-Hill, 1960.

Morrish, Allan H., *The Physical Principles of Magnetism*, John Wiley & Sons, 1965.

Morrison, Ralph, *Grounding and Shielding Techniques*, John Wiley & Sons, 1998.

Morrison, Ralph, *Solving Interference Problems in Electronics*, John Wiley & Sons, 1995.

Morse, Philip M. and Herman Feshbach, *Methods of Theoretical Physics*, McGraw-Hill, 1953.

Moser, J. Ronald and Ralph F. Spencer, Jr., *Predicting the Magnetic Fields from a Twisted-Pair Cable*, IEEE Transactions on Electromagnetic Compatibility, Vol. EMC-10, No. 3, September 1968(a).

Moser, J. Ronald, *An Empirical Study of ELF and VLF Shield Cans*, IEEE Transactions on Electromagnetic Compatibility, Vol. EMC-10, No. 1, March 1968(b).

Moser, J. Ronald, *Low-Frequency Shielding of a Circular Loop Electromagnetic Field Source*, IEEE Transactions on Electromagnetic Compatibility, Vol. EMC-9, No. 1, March 1967.

Mullin, William F., *abc's of Capacitors*, Howard W. Sams, 1967.

MuRata, *EMI Suppression Filter*, Cat. No. C30E, Murata.

Nagy, John and Harry C. Verakis, *Development and Control of Dust Explosions*, Marcel Dekker, 1983.

Nagy, John, Austin R. Cooper, and Henry G. Dorsett, Jr., *Explosibility of Miscellaneous Dusts*, Bureau of Mines, Report of Investigation 7208, 1968.

Nagy, John, Henry G. Dorsett, Jr., and Austin R. Cooper, *Explosibility of Carbonaceous Dusts*, Bureau of Mines, Report of Investigations 6597, 1965.

Nahman, Norris S. and Donald R. Holt, *Transient Analysis of Coaxial Cables Using the Skin Effect Approximation* $A+B\sqrt{s}$, IEEE Transactions on Circuit Theory, Vol. 19, No. 5, September 1972.

Nanevicz, J.E., *Some Techniques for the Elimination of Corona Discharge Noise in Aircraft Antennas*, Proceedings of the IEEE, Vol. 52, No. 1, January 1964.

Nanevicz, Joseph E., *Static Charging and its Effects on Avionic Systems*, IEEE Transactions on Electromagnetic Compatibility, Vol. EMC-24, No. 2, May 1982.

National Association of Relay Manufacturers, *Engineer's Relay Handbook*, Hayden, 1960.

Nave, Mark J., *Power Line Filter Design for Switched-Mode Power Supplies*, Van Nostrand Reinhold, 1991.

Nelson, W.R., *Interference Handbook*, Radio Publications, 1981.

Nicholson, J.R. and J.A. Malack, *RF Impedance of Power Lines and Line Impedance Stabilization Networks in Conducted Interference Measurements*, IEEE Transactions on Electromagnetic Compatibility, EMC-15, No. 2, May 1973.

Nilsson, James W. and Susan A. Riedel, *Electric Circuits*, Prentice-Hall, 2001.

Nishiyama, Hitoshi and Mitsunobu Nakamura, *Capacitance of Disk Capacitors*, IEEE Transactions on Components, Hybrids, and Manufacturing Technology, Vol. 16, No. 3, May 1993.

Nixon, Floyd E., *Handbook of Laplace Transformation: Fundamentals, Applications, Tables, and Examples*, Prentice-Hall, 1965.

Novak, Istvan, *Reducing Simultaneous Switching Noise and EMI on Ground/Power Planes by Dissipative Edge Termination*, IEEE Transactions on Advanced Packaging, Vol. 22, No. 3, August 1999.

Nussbaum, Allen, *Electromagnetic Theory for Engineers and Scientists*, Prentice-Hall, 1965.

O'Riley, Ronald P., *Electrical Grounding: Bringing Grounding Back to Earth*, Delmar, 1999.

Oliver, Frank J., *Practical Instrumentation Transducers*, Hayden, 1971.

Olsen, E., *Applied Magnetism: A Study in Quantities*, Springer-Verlag, 1966.

Olsen, R.G. and T.A. Pankaskie, *On the Exact, Carson and Image Theories for Wires at or Above the Earth's Interface*, IEEE Transactions on Power Apparatus and Systems, Vol. PAS-102, No. 4, April 1983.

Olsen, Robert G., *Some Observations About Shielding Extremely Low-Frequency Magnetic Shields by Finite Width Shields*, IEEE Transactions on Electromagnetic Compatibility, Vol. 38, No. 3, August 1996.

Olsher, Dick, *Linday-Geyer Highly Magnetic Cable*, Stereophile, February 1991.

Oppenheim, Alan V., Alan S. Willsky, and S. Hamid Nawab, *Signals & Systems*, Prentice-Hall, 1997.

Ott, Henry W., *Noise Reduction Techniques in Electronic Systems*, John Wiley & Sons, 1988.

Ozenbaugh, Richard Lee, *EMI Filter Design*, Dekker, 1996.

Page, H., *Principles of Aerial Design*, D. Van Nostrand, 1966.

Papakanellos, Panagiotis J., Dimitra I. Kaklamani, and Christos N. Capsalis, *Analysis of an Infinite Current Source Above a Semi-Infinite Lossy Ground Using Fictitious Current Auxiliary Sources in Conjunction With Complex Image Theory Techniques*, IEEE Transactions on Antennas and Propagation, No. 49, No. 10, October 2001.

Parker, Carole U., *Choosing a Ferrite for the Suppression of EMI*, Fair-Rite Technical Paper, Fair-rite Products Corp., 1991.

Parker, Rollin J. and Robert J. Studders, *Permanent Magnets and Their Application*, John Wiley & Sons, 1962.

Paul, Clayton R. and Jack W. McKnight, *Prediction of Crosstalk Involving Twisted Pairs of Wires-Part I: A Transmission-Line Model for Twisted-Wire Pairs*, IEEE Transactions on Electromagnetic Compatibility, Vol. EMC-21, No. 2, May 1979.

Paul, Clayton R. and Jack W. McKnight, *Prediction of Crosstalk Involving Twisted Pairs of Wires-Part II: A Simplified Low-Frequency Prediction Model*, IEEE Transactions on Electromagnetic Compatibility, Vol. EMC-21, No. 2, May 1979.

Paul, Clayton R. and Keith B. Hardin, *Diagnosis and Reduction of Conducted Noise Emissions*, IEEE Transactions on Electromagnetic Compatibility, Vol. 30, No. 4, November 1988.

Paul, Clayton R. and Syed A. Nasar, *Introduction to Electromagnetic Fields*, McGraw-Hill, 1987.

Paul, Clayton R., *Analysis of Multiconductor Transmission Lines*, John Wiley & Sons, 1994.

Paul, Clayton R., *Effectiveness of Multiple Decoupling Capacitors*, IEEE Transactions on Electromagnetic Compatibility, Vol. EMC-34, No. 2, May 1992(a).

Paul, Clayton R., *Frequency Response of Multiconductor Transmission Lines Illuminated by an Electromagnetic Field*, IEEE Transactions on Electromagnetic Compatibility, Vol. EMC-18, No. 4, November 1976.

Paul, Clayton R., *Introduction to Electromagnetic Compatibility*, John Wiley & Sons, 1992(b).

Paul, Clayton R., *Prediction of Crosstalk in Ribbon Cables*, IEEE International Symposium on Electromagnetic Compatibility, 1978.

Peek, F.W., *Dielectric Phenomena in High Voltage Engineering*, McGraw-Hill, 1929.

Perez, Reinaldo, ed., *Handbook of Electromagnetic Compatibility*, Academic Press, 1995.

Perls, Thomas A., *Electrical Noise from Instrument Cables Subjected to Shock and Vibration*, Journal of Applied Physics, Vol. 23, No. 6, June 1952.

Perry, Michael P., *Low Frequency Electromagnetic Design*, Marcel Dekker, 1985.

Perry, Tekla S., *Today's View of Magnetic Fields*, IEEE Spectrum, Vol. 31, No. 12, December 1994.

Perz, M.C. and M.R. Raghuveer, *Transmission Lines — Effects of Earth Resistivity on Magnetic Field, Images and Equivalent Circuit*, IEEE Transactions on Power Apparatus and Systems, Vol. PAS-98, No. 6, November/December 1979.

Pettersson, Per, *Image Representation of Wave Propagation on Wires Above, On, and Under Ground*, IEEE Transactions on Power Delivery, Vol. 9, No. 2, April 1994.

Pidoll, Ulrich von, Helmut Krämer, and Heino Bothe, *Avoidance of Electrostatic Hazards during Refueling of Motorcars*, Journal of Electrostatics, Vol. 40–41, 1997.

Plastics for Electronics: Desk-top Data Bank, The International Plastics Selector, 1979.

Plonsey, Robert and Robert E. Collin, *Principles and Applications of Electromagnetic Fields*, McGraw-Hill, 1961.

Pratt, Thomas H., *Electrostatic Ignition Hazards*, Burgoynes, 1995.

Prentiss, Stan, *Modern Television: Service and Repair*, Prentice-Hall, 1989.

Pucel, Robert A., Daniel J. Massé, and Curtis P. Hartwig, *Losses in Microstrip*, IEEE Transactions on Microwave Theory and Techniques, Vol. MTT-16, No. 6, June 1968.

Pugh, Emerson M. and Emerson W. Pugh, *Principles of Electricity and Magnetism*, Addison-Wesley, 1970.

Püschner, H., *Heating With Microwaves: Fundamentals, Components and Circuit Technique*, Springer-Verlag, 1966.

Queen, Ralph H., *Common or Differential-Mode Noise? It Makes a Difference in Your Filter*, EMC Technology, Vol. 4, No. 3, July/September 1985.

Räde, Lennart, and Bertil Westergren, *Beta Mathematics Handbook*, Studentlitteratur, 1992.

Radun, Arthur, *An Alternative Low-Cost Current-Sensing Scheme for High-Current Power Electronics Circuits*, IEEE Transactions on Industrial Electronics, Vol. 42, No. 1, February 1995.

Rakov, Vladimir A. and Martin A. Uman, *Review and Evaluation of Lightning Return Stroke Models Including Some Aspects of Their Application*, IEEE Transactions on Electromagnetic Compatibility, Vol. EMC-40, No. 4, November 1998.

Raloff, Janet, *EMFs Run Aground*, Science News, Vol. 144, August 1993.

Ramboz, John D., *Machinable Rogowski Coil, Design, and Calibration*, IEEE Transactions on Instrumentation and Measurement, Vol. 45, No. 2, April 1996.

Ramo, Simon, John R. Whinnery, and Theodore Van Duzer, *Fields and Waves in Communication Electronics*, John Wiley & Sons, 1965.

Rao, Nannapaneni Narayana, *Elements of Engineering Electromagnetics*, Prentice-Hall, 1994.

Ratz, Alfred G., *Triboelectric Noise*, Instrument Society of America Transactions, Vol. 9, No. 2, 1970.

Rauch, Gregory B., Gary Johnson, Paul Johnson, Andreas Stamm, Seietsu Tomita, and John Swanson, *A Comparison of International Residential Grounding Practices and Associated Magnetic Fields*, IEEE Transactions on Power Delivery, Vol. 7, No. 2, April 1992.

Raychem Wire and Cable, *Electrical Screening*, Tyco Electronics Corporation.

Redl, Richard, Paolo Tenti, and J. Daan Van Wyk, *Power Electronic's Polluting Effects*, IEEE Spectrum, Vol. 34, No. 5, May 1997.

Reed, Michael and Ron Rohrer, *Applied Introductory Circuit Analysis for Electrical and Computer Engineers*, Prentice-Hall, 1999.

Reilly, J. Patrick, *Electrical Stimulation and Electropathology*, Cambridge University Press, 1992.

Riggs, Olen L., Jr. and Carl E. Locke, *Anodic Protection*, Plenum Press, 1981.

Rikitake, T., *Magnetic and Electromagnetic Shielding*, D. Reidel, 1987.

Riley, Karl, *Tracing EMFs in Building Wiring and Grounding: A Practical Guide for Reducing AC Magnetic Fields Produced by Building Wiring and Grounding Practices*, Magnetic Sciences International, 1995.

Rizk, F.A.M., *Low-Frequency Shielding Effectiveness of a Double Cylinder Enclosure*, IEEE Transactions on Electromagnetic Compatibility, Vol. EMC-19, No. 1, February 1977.

Rizk, Farouk A.M., *Effect of Floating Conducting Objects on Critical Switching Impulse Breakdown of Air Insulation*, IEEE Transactions on Power Delivery, Vol. 10, No. 3, July 1995.

Roberts, Willmar K., *A New Wide-Band Balun*, Proceedings of the IRE, Vol. 45, No. 12, December 1957.

Rochester, Louis, *Coaxial Cables Application Data*, Boston Technical, 1969.

Rogers, Walter E., *Introduction to Electric Fields*, McGraw-Hill, 1954.

Roseberry, B.E. and R.B. Schulz, *A Parallel-Strip Line for Testing RF Susceptibility*, IEEE Transactions on Electromagnetic Compatibility, EMC-7, No. 2, June 1965.

Rosenstark, Sol, *Transmission Lines in Computer Engineering*, McGraw-Hill, 1994.

Roters, Herbert C., *Electromagnetic Devices*, John Wiley & Sons, 1941.

Rotman, Walter and Nicholas Karas, *Printed Circuit Radiators: The Sandwich Wire Antenna*, The Microwave Journal, August 1959.

Rowland, Phillip W., *Industrial System Grounding for Power, Static, Lightning, and Instrumentation, Practical Applications*, IEEE Transactions on Industry Applications, Vol. 31, No. 6, November/December 1995.

Rudge, A.W., K. Milne, A.D. Olver, and P. Knight, *The Handbook of Antenna Design, Volume 1 and 2*, Peter Peregrinus Ltd, 1982–83.

Ruehli, A.E., *Inductance Calculations in a Complex Integrated Circuit Environment*, IBM J. Res. Develop., September 1972.

Ruehli, Albert, Clayton Paul, and Jan Garrett, *Inductance Calculations using Partial Inductances and Macromodels*, IEEE International Symposium on Electromagnetic Compatibility, 1995.

Rulf, Benjamin and Gregory A. Robertshaw, *Understand Antennas for Radar, Communications, and Avionics*, Van Nostrand Reinhold, 1987.

Ruston, Henry and Joseph Bordogna, *Electric Networks: Functions, Filters, and Analysis*, McGraw-Hill, 1966.

Ruthroff, C.L., *Some Broad-Band Transformers*, Proceedings of the IRE, Vol. 47, No. 8, August 1959.

Rutledge, David B., *The Electronics of Radio*, Cambridge University Press, 1999.

Saums, Harry L. and Wesley W. Pendleton, *Materials for Electrical Insulating and Dielectric Functions*, Hayden, 1973.

Schelkunoff, S.A., *The Electromagnetic Theory of Coaxial Transmission Lines and Cylindrical Shields*, The Bell System Technical Journal, Vol. 13, No. 4, October 1934.

Schlabbach, J., D. Blume, and T. Stephanblome, *Voltage Quality in Electrical Power Systems*, The Institution of Electrical Engineers, 2001.

Schneider, M.V., Bernard Glance, and W.F. Bodtmann, *Microwave and Millimeter Wave Hybrid Integrated Circuits for Radio Systems*, The Bell System Technical Journal, Vol. 48, No. 6, July–August 1969.

Schultz, John J., *The G4ZU X Beam for 20*, CQ, June 1965.

Schulz, Richard B., George C. Huang, and Walter L. Williams, *RF Shielding Design*, IEEE Transactions on Electromagnetic Compatibility, Vol. EMC-10, No. 1, March 1968.

Schulz, Richard B., *Shielding Theory and Practice*, IEEE Transactions on Electromagnetic Compatibility, Vol. EMC-30, No. 3, August 1988.

Schwab, Adolf J., *High-Voltage Measurement Techniques*, The MIT. Press, 1972.

Schwarz, S.J., *Analytical Expressions for the Resistance of Grounding Systems*, Transactions of AIEE, Part III-B, August 1954.

Schweitzer, Philip A., ed., *Corrosion and Corrosion Protection Handbook*, Marcel Dekker, 1988.

Sclater, Neil, *Electrostatic Discharge Protection for Electronics*, TAB, 1990.

Secker, P.E. and J.N. Chubb, *Instrumentation for Electrostatic Measurements*, Journal of Electrostatics, Vol. 16, 1984.

Sedra, Adel S. and Kenneth C. Smith, *Microelectronic Circuits*, Saunders, 1991.

Seshadri, S.R., *Fundamentals of Transmission Lines and Electromagnetic Fields*, Addison-Wesley, 1971.

Setian, Leo, *Practical Communication Antennas with Wireless Applications*, Prentice-Hall, 1998.

Sevick, Jerry, *Building Baluns and Ununs: Practical Designs for the Experimenter*, CQ Communications, 1994.

Sevick, Jerry, *Transmission Line Transformers*, The American Radio Relay League, 1987.

Shen, L.C., S.A. Long, M.R. Allerding, and M.D. Walton, *Resonant Frequency of a Circular Disc, Printed-Circuit Antenna*, IEEE Transactions on Antennas and Propagation, Vol. AP-25, No. 4, July 1977.

Shenfeld, Saul, *Magnetic Fields of Twisted-Wire Pairs*, IEEE Transactions on Electromagnetic Compatibility, Vol. EMC-11, No. 4, November 1969.

Shenfeld, Saul, *Shielding of Cylindrical Tubes*, IEEE Transactions on Electromagnetic Compatibility, Vol. EMC-10, No. 1, March 1968.

Siebert, William McC., *Circuits, Signals, and Systems*, McGraw-Hill, 1986.

Silver, Samuel, ed., *Microwave Antenna Theory and Design*, Peter Peregrinus, 1984.

Silvester, P., *Modern Electromagnetic Fields*, Prentice-Hill, 1968.

Sim, A.C., *New H.F. Proximity-Effect Formula*, Wireless Engineer, Vol. 30, No. 7, August 1953.

Simpson, Ted L., *Effect of a Conducting Shield on the Inductance of an Air-Core Solenoid*, IEEE Transactions on Magnetics, Vol. 35, No. 1, January 1999.

Single Side Band, Collins Radio Company, 1959.

Sjögren, C.T.H., *Application of the Wave Impedance Concept when Calculating Shielding Effectiveness*, Ninth International Conference on Electromagnetic Compatibility, No. 396, September 1994.

Skilling, Hugh Hildreth, *Electric Transmission Lines*, McGraw-Hill, 1951.

Slade, Paul G., ed., *Electrical Contacts: Principles and Applications*, Marcel Dekker, 1999.

Smith, Albert A., Jr., *Coupling of External Electromagnetic Fields to Transmission Lines*, John Wiley & Sons, 1977.

Smith, Douglas C., *High Frequency Measurements and Noise in Electronic Circuits*, Van Nostrand Reinhold, 1993.

Smith, Douglas C., *Unusual Forms of ESD and Their Effects*, Electrostatic Overstress/Electrostatic Discharge Symposium Proceedings, 1999.

Smith, Glenn S., *Radiation Efficiency of Electrically Small Multiturn Loop Antennas*, IEEE Transactions on Antennas and Propagation, Vol. AP-20, No. 5, September 1972.

Smits, F.M., *Measurement of Sheet Resistivities with the Four-Point Probe*, The Bell System Technical Journal, No. 3, May 1958.

Smythe, William R., *Static and Dynamic Electricity*, McGraw-Hill, 1968.

Snelling, E.C., *Soft Ferrites: Properties and Applications*, CRC Press, 1969.

Soliman, Samir S. and Mandyam D. Srinath, *Continuous and Discrete Signals and Systems*, Prentice-Hall, 1998.

Somerville, J.M., *The Electric Arc*, John Wiley & Sons, 1959.

Southwick, Roger, *EMI Signal Measurements at Open Area Test Sites*, EMC Test & Design, Vol. 3, No. 4, July/August 1992.

Spangenberg, Karl R., *Vacuum Tubes*, McGraw-Hill, 1948.

Spencer, Ned A., *An Antenna for 30-50 MC Service Having Substantial Freedom from Noise Caused by Precipitation Static and Corona*, IRE Transactions on Vehicular Communications, VC-9, No. 2, August 1960.

Spiegel, Murray R., and John Liu, *Mathematical Handbook of Formulas and Tables*, Schaum's Outline Series, 1999.

Standler, Ronald B., *Protection of Electronic Circuits from Overvoltages*, John Wiley & Sons, 1989.

Static Control Components, Inc., *Choosing the Right Static Bag*, Technical Bulletin P1, 1996.

Stevenson, William D., Jr., *Elements of Power System Analysis*, McGraw-Hill, 1975.

Streetman, Ben G. and Sanjay Banerjee, *Solid State Electronic Devices*, Prentice-Hall, 2000.

Stremler, Ferrel G., *Introduction to Communication Systems*, Addison-Wesley, 1982.

Strojny, Jan A., *Some Factors Influencing Electrostatic Discharge from a Human Body*, Journal of Electrostatics, Vol. 40–41, 1997.

Stuchly, M.A. and S.S. Stuchly, *Dielectric Properties of Biological Substances — Tabulated*, Journal of Microwave Power, Vol. 15, No. 1, 1980.

Stutzman, Warren L. and Gary A. Thiele, *Antenna Theory and Design*, John Wiley & Sons, 1998.

Su, Kendall L., *Time-Domain Synthesis of Linear Networks*, Prentice-Hall, 1971.

Sze, S.M., *Physics of Semiconductor Devices*, John Wiley & Sons, 1981.

Tagg, G.F., *Earth Resistances*, Pitman, 1964.

Takuma, T., M. Yashima, and T. Kawamoto, *Principle of Surface Charge Measurement for Thick Insulating Specimens*, IEEE Transactions on Dielectrics and Electrical Insulation, Vol. 5, No. 4, August 1998.

Taylor, Clayborne D., Earl Harper, Charles W. Harrison, Jr., and William A. Davis, *On Bounding the Excitation of a Terminated Wire Behind an Aperture in a Shield*, IEEE Transactions on Antennas and Propagation, Vol. AP-34, No. 2, February 1986.

Taylor, D.M. and P.E. Secker, *Industrial Electrostatics: Fundamentals and Measurements*, Research Studies Press, 1994.

Taylor, D.M., *Measuring Techniques for Electrostatics*, Journal of Electrostatics, Vol. 51–52, 2001.

Tektronix, *ABCs of Probes*, Tektronix, 1997.

Tell, Richard A. and Edwin D. Mantiply, *Population Exposure to VHF and UHF Broadcast Radiation in the United States*, Proceedings of the IEEE, Vol. 68, No. 1, January 1980.

Thomas, Alan K., *Magnetic Shielded Enclosure Design in the DC and VLF Region*, IEEE Transactions on Electromagnetic Compatibility, Vol. EMC-10, No. 1, March 1968.

Thomson, J.J. and G.P. Thomson, *Conduction of Electricity Through Gases*, Vol. II, Dover, 1969.

Tippet, John C. and David C. Chang, *Characteristic Impedance of a Rectangular Coaxial Line with Offset Inner Conductor*, IEEE Transactions on Microwave Theory and Techniques, Vol. MTT-26, No. 11, November 1978.

Tolson, P., *Assessing the Safety of Electrically Powered Static Eliminators for Use in Flammable Atmospheres*, Journal of Electrostatics, Vol. 11, 1981.

Tomboulian, D.H., *Electric and Magnetic Fields*, Harcourt, Brace, & World, 1965.

Tremaine, Howard M., *Audio Cyclopedia*, Howard W. Sams, 1969.

Tsaliovich, Anatoly, *Cable Shielding for Electromagnetic Compatibility*, Van Nostrand Reinhold, 1995.

Uitert, LeGrand G. Van, *Dielectric Properties of and Conductivity in Ferrites*, Proceedings of the IRE, Vol. 44, No. 10, October 1956.

Ulaby, Fawwaz T., *Applied Electromagnetics*, Prentice-Hall, 1999.

Uman, Martin A., *Understanding Lightning*, Bek Technical Publications, 1971.

Upton, Miles E.G., and Andrew C. Marvin, *The Fields Due to a Small Loaded Loop in Free Space*, IEEE Transactions on Electromagnetic Compatibility, Vol. EMC-36, No. 1, February 1994.

Valdes, L.B., *Resistivity Measurements on Germanium for Transistors*, Proceedings of the IRE, February 1954.

Valley, George E., Jr. and Henry Wallman, *Vacuum Tube Amplifiers*, McGraw-Hill, 1948.

Vance, Edward F., *Coupling to Shielded Cables*, John Wiley & Sons, 1978.

Vendelin, George D., Anthony M. Pavio, and Ulrich L. Rohde, *Microwave Circuit Design Using Linear and Nonlinear Techniques*, John Wiley & Sons, 1990.

Vines, Roger M., H. Joel Trussell, Kenneth C. Shuey, and J.B. O'Neal, Jr., *Impedance of the Residential Power-Distribution Circuit*, IEEE Transactions on Electromagnetic Compatibility, Vol. EMC-27, No. 1, February 1985.

Vines, Roger M., H. Joel Trussell, Louis J. Gale, and J. Ben O'Neal, Jr., *Noise on Residential Power Distribution Circuits*, IEEE Transactions on Electromagnetic Compatibility, Vol. EMC-26, No. 4, November 1984.

Violette, J.L. Norman, Donald R.J. White, and Michael F. Violette, *Electromagnetic Compatibility Handbook*, Van Nostrand Reinhold, 1987.

Vo, Joe, *A Comparison of Differential Termination Techniques*, National Semiconductor Application Note 903, August 1993.

Vosteen, William E., *A Review of Current Electrostatic Measurement Techniques and Their Limitations*, Monroe Electronics, 1984.

Wadell, Brian C., *Transmission Line Design Handbook*, Artech House, 1991.

Wait, James R. and David A. Hill, *Electromagnetic Shielding of Sources Within a Metal-Cased Bore Hole*, IEEE Transactions on Geoscience Electronics, Vol. GE-15, No. 2, April 1977.

Wait, James R., *Electromagnetic Induction Technique for Locating a Buried Source*, IEEE Transactions on Geoscience Electronics, Vol. GE-9, No. 2, April 1971.

Wait, James R., *The False Image of a Line Current Within a Conducting Half-Space*, IEEE Transactions on Electromagnetic Compatibility, Vol. 39, No. 3, August 1997.

Wakerly, John F., *Digital Design: Principles and Practices*, Prentice-Hall, 2001.

Walker, Charles S., *Capacitance, Inductance, and Crosstalk Analysis*, Artech House, 1990.

Warne, L.K. and K.S.H. Lee, *Some Remarks on Antenna Response in a Reverberation Chamber*, IEEE Transactions on Electromagnetic Compatibility, Vol. 43, No. 2, May 2001.

Watkins, J., *Circular Resonant Structures in Microstrip*, Electronics Letters, Vol. 5, No. 21, October 1969.

Watson, J.K., *Applications of Magnetism*, self published, 1985.

Watt, Arthur D., *VLF Radio Engineering*, Pergamon Press, 1967.

Watterson, Peter A., *Energy Calculation of a Permanent Magnetic System by Surface and Flux Integrals (the Flux-mmf Method)*, IEEE Transactions on Magnets, Vol. 36, No. 2, March 2000.

Weber, Ernst, *Electromagnetic Theory: Static Fields and Their Mapping*, Dover, 1965.

Weeks, W.L., *Antenna Engineering*, McGraw-Hill, 1968.

Weeks, W.L., *Electromagnetic Theory for Engineering Applications*, John Wiley & Sons, 1964.

Welsby, V.G., *The Theory and Design of Inductance Coils*, MacDonald, 1960.

Wenner, Frank, *A Method of Measuring Earth Resistivity*, Bulletin of the Bureau of Standards, Vol. 12, No. 4, October 11, 1915.

Weston, David A., *Electromagnetic Compatibility: Principles and Applications*, Marcel Dekker, 1991.

Wheeler, Harold A., *Formulas for the Skin Effect*, Proceedings of the IRE, Vol. 30, No. 9, September 1942.

Wheeler, Harold A., *Inductance Formula for Circular and Square Coils*, Proceedings of the IEEE, Vol. 70, No. 12, December 1982.

Wheeler, Harold A., *The Spherical Coil as an Inductor, Shield, or Antenna*, Proceedings of the IRE, Vol. 46, No. 9, September 1958.

Wheeler, Harold A., *Transmission-Line Properties of a Round Wire in a Polygon Shield*, IEEE Transactions on Microwave Theory and Technique, Vol. MTT-27, No. 8, August 1979.

Whitaker, Jerry C., *AC Power Systems Handbook*, CRC Press LLC, 1999.

White, Donald R.J., *EMI Test Instrumentation and Systems, Vol. 4*, Don White Consultants, 1971.

White, Joseph F., *Microwave Semiconductor Engineering*, Van Nostrand Reinhold, 1982.

White, R.J. and Michel Mardiguian, *Electromagnetic Shielding, Vol. 3*, Interference Control Technologies, 1988.

Whitehouse, A.C.D., *Screening: New Wave Impedance for the Transmission-Line Analogy*, Proceedings of the IEE, Vol. 116, No. 7, July 1969.

Whiteside, H. and R. King, *The Loop Antenna as a Probe*, IEEE Transactions on Antennas and Propagation, Vol. AP-12, No. 3, May 1964.

Whitlock, Bill, *Answers to Common Questions About Audio Transformers*, Jensen Application Note AN-002.

Whitlock, Bill, *Interconnection of Balanced and Unbalanced Equipment*, Jensen Application Note AN-003.

Whitlock, Bill, *Subtleties Count in Wide-Dynamic-Range Analog Interfaces*, EDN, June 4, 1998.

Williams, George M. and Thomas H. Pratt, *Characteristics and hazards of electrostatic discharges in air*, Journal of Loss Prevention in the Process Industries, Vol. 3, October 1990.

Wilson, Perry F. and Mark T. Ma, *Simple Approximate Expressions for Higher Order Mode Cutoff and Resonant Frequencies in TEM Cells*, IEEE Transactions on Electromagnetic Compatibility, Vol. EMC-28, No. 3, August 1986.

Wilson, Robert, *The Offset Multiband Trapless Antenna (OMTA)*, QST, October 1995.

Winburn, D.C., *Practical Electrical Safety*, Marcel Dekker, 1988.

Winder, Steve, *Filter Design*, Newnes, 1997.

Wolff, Edward A., *Antenna Analysis*, John Wiley & Sons, 1967.

Wolff, Ingo and Norbert Knoppik, *Rectangular and Circular Microstrip Disk Capacitors and Resonators*, IEEE Transactions on Microwave Theory and Techniques, Vol. MTT-22, No. 10, October 1974.

Wood, Jody W., Personal Communications, New England Wire Technologies, 2003.

Xu, Xiao-Bang and Xiao-Mei Yang, *A Simple Computational Method for Predicting Magnetic Field in the Vicinity of a Three-Phase Underground Cable with a Fluid-Filled Steel-Pipe Enclosure*, IEEE Transactions on Power Delivery, Vol. 10, No. 1, January 1995.

Xu, Xiao-Bang and Xiao-Mei Yang, *Computation of the Magnetic Field Generated by an Underground Pipe-Type Cable*, IEEE Transactions on Power Delivery, Vol. 11, No. 2, April 1996.

Young, F.J., D.J. Boomgaard, and D.A. Colling, *Transfer Impedance of Soft Magnetic Materials*, Journal of Applied Physics, Vol. 42, No. 4, March 1971.

Yu, Luke, Mark Chow, and Jay Bowen, *Safety and Ground Fault Protection in Electrical Systems*, IEEE Industry Applications Magazine, March/April 1998.

Zaborszky, John and Joseph W. Rittenhouse, *Electric Power Transmission*, Ronald Press, 1954.

Zaffanella, L.E., *Pilot Study of Residential Power Frequency Magnetic Fields*, Electric Power Research Institute, EL-6509, Research Project 2942, September 1989.

Zahn, Markus, *Electromagnetic Field Theory: a problem solving approach*, Robert E. Krieger, 1979.

Zemanian, A.H., *Distribution Theory and Transform Analysis: An Introduction to Generalized Functions, with Applications*, Dover, 1965.

Zverev, Anatol I., *Handbook of Filter Synthesis*, John Wiley & Sons, 1967.

Index

D

E

K

L

M

N

O

P

Q

R

S

T

U

V

W

Z